The Experimental Animal in Biomedical Research

Volume I
A Survey of Scientific and Ethical Issues for Investigators

Editor

Bernard E. Rollin

Professor of Philosophy
Professor of Physiology and Biophysics
Director of Bioethical Planning
Colorado State University
Fort Collins, Colorado

Assistant to the Editor

M. Lynne Kesel

Assistant Professor
Department of Clinical Sciences
College of Veterinary Medicine
Colorado State University
Fort Collins, Colorado

CRC Press
Boca Raton Ann Arbor Boston

Library of Congress Cataloging-in-Publication Data

The experimental animal in biomedical research / editor, Bernard E. Rollin;
assistant to the editor, M. Lynne Kesel.
p. cm.
Includes bibliographical references.
Contents: Vol. 1. A survey of scientific and ethical issues for investigators.
ISBN 0-8493-4981-8 (v. 1)
1. Animal experimentation. I. Rollin, Bernard, E. II. Kesel, M. Lynne.
[DNLM: 1. Animals, Laboratory. 2. Research—United States. QY 50 E96]
HV4915.E97 1990
619—dc20
DNLM/DLC
for Library of Congress 90-1698
CIP

This book represents information obtained from authentic and highly regarded sources. Reprinted material is quoted with permission, and sources are indicated. A wide variety of references are listed. Every reasonable effort has been made to give reliable data and information, but the author and the publisher cannot assume responsibility for the validity of all materials or for the consequences of their use.

Direct all inquiries to CRC Press, Inc., 2000 Corporate Blvd., N. W., Boca Raton, Florida, 33431.

International Standard Book Number 0-8493-4981-8 (vol. 1)

Library of Congress Card Number 90-1698
Printed in the United States 3 4 5 6 7 8 9 0
Printed on acid-free paper

PREFACE

The historical development of biomedical science has been closely, indeed inextricably, linked with the use of animals as a principal tool for both basic research and the articulation of new therapeutic modalities. At the same time, as in any other goal-oriented activity, little attention was focused on the tool, save as a means to an end. As long as live animals were plentiful, readily available, inexpensive, and in constant supply, there seemed little need for researchers to concern themselves with their extra-experimental care and treatment, or indeed with their experimental use except as a means to answering a question.

Traditionally, a number of vectors converged to make the animals used in research essentially invisible to researchers. First, society as a whole devoted little attention to the moral status of animals, and the treatment of animals was thus of little social concern, except insofar as it had ramifications for human benefit. As a result, the lowest common denominator ethic for how animals should be treated obtaining in society was indeed minimalistic and was reflected in the fact that virtually the only legal constraints on what could be done to animals were the proscriptions against cruelty codified in anticruelty statutes. Furthermore, cruelty was defined so as to emphasize malicious intent, sadism, or deviance which might lead to harm inflicted on humans if unchecked, society having long been aware that psychopathic behavior often reveals itself first in animal abuse. By definition, activities such as research, agriculture, trapping, hunting, and rodeo, statistically the source of immeasurably more animal suffering than was intentional cruelty, were excluded from the purview of these statutes.

This social failure to reckon with the major sources of animal abuse was further buttressed by traditional humane movement emphasis on "kindness" and "cruelty" as the primary moral categories for evaluating animal treatment, and by the related, sentimentally based tendency of animal advocates to focus their attention on favored animals such as dogs, cats, and horses, with little concern for animals which did not enjoy the same appeal, such as rodents and farm animals. Since the bulk of animals used in biomedicine were not favored animals, the majority of animal research was not a source of concern to society. In short, aside from occasional flurries about dogs and cats, society did not exert a force on researchers which countered research animal invisibility, especially in the light of the wholesale public confidence in science and medicine dominant through most of the burgeoning of research in the 20th century.

Second, biomedical researchers tended to have little awareness of the extent to which animal care, treatment, and husbandry could affect the physiological and metabolic variables in which the researchers were interested. The irrelevance of a whole host of environmental and stress variables affecting the animals that one used to the validity of one's results was simply assumed and unwittingly created a vicious cycle. Researchers were typically not educated concerning the animals they used, except to the fact that they modeled a particular disease or syndrome. This in turn led to absence of regard for their care and treatment in even the best research institutions, which in turn produced generations of researchers insensitive to these issues, who in turn never felt the need to think or teach much about them, etc. Thus, attention to one's animals was relegated to animal caretakers, who were basically seen as fulfilling a function more analogous to a janitorial one than to a scientific one.

Of late, however, new vectors have emerged which militate in favor of greater researcher knowledge of and concern for the animals they use in biomedical investigations. In the first place, the rise of laboratory animal science as a credible branch of knowledge has demonstrated that, not surprisingly, animals are extremely sensitive to a dazzling variety of environmental, psychological, and "stress" variables, the effects of which may be felt in an extraordinary range of physiological and metabolic parameters, as many of the papers in this volume will delineate. Pain, noise, heat, crowding, light, bedding, diet, fear, anxiety, solitude (for social animals), pheromones, subclinical infection, simple and complex experimental manipulations, drugs, personality of handlers and technicians, circadian cycles, environmental chemicals, etc. can have significant effects on virtually all biological parameters. Thus, all sorts of considerations which have been ignored during much of the history of animal research are now known to be highly relevant to research results in areas ranging from reproduction to cancer, from wound healing to learning.

Second, as other papers in this volume delineate, society's attitude toward the moral status of animals in general, and towards research animals in particular, has undergone dramatic evolution in the last two decades. Along with the general extension of enfranchisement to traditionally down-trodden and neglected factions of society has come also a new and generalized examination of animal use and treatment. With the pervasive social demand for accountability in all aspects of life has come a demand for scientific accountability as well, especially since public money supports the majority of biomedical research. These two factors, coupled with the rise of societal skepticism about science as a curer of all ills, have engendered a decline of naive and uncritical public support of science. In the same vein, public confidence in scientific medicine has also been eroded by escalating health care costs; increased social awareness of iatrogenic illness; the burgeoning of alternative health programs and therapies; general skepticism about the expensive, heroic, and artificial prolongation of life; and by widespread social rejection of the "medicalization of evil", as exemplified by the widespread rumblings against the insanity defense.

In any event, all of these factors, and doubtless many others — including the rise of articulate moral-philosophical arguments pressing for the augmentation of the moral status of animals, and questioning the moral defensibility of any invasive animal use, regardless of benefit accruing therefrom (see Sapontzis' contribution to this volume) — have converged to focus far greater attention on the care and treatment of animals in biomedical research, as well as on the rationale for their use. This attention and concern has led to the passage of new and significant legislation governing animal research (as discussed in Newcomer's chapter in this volume), and there is no sign that the demand for further accountability and legislation will abate.

For many years (as detailed in Rollin's contribution to this volume), the research community denied any relevance of value judgments to science, or even the existence of such judgments therein. In actual fact, of course, science did rest on value judgments, including those relating to the morality of animal use, but such judgments were implicit, invisible, unarticulated, and undefended, since virtually all researchers shared them and no one challenged them. Again, the advent of the factors discussed above has forced researchers, like it or not, to abandon the myth that science has no truck with values.

The net result of all this for researchers, then, is that they must be knowledgeable in numerous areas hitherto badly neglected in biomedical research. They must now engage ethical questions as well as scientific questions related to animal use, questions they are very unlikely to have been trained to deal with. They must be conversant with the arguments against the legitimacy of invasive biomedical research on animals; they must understand the moral basis of demands for significant constraints on biomedical research; they must articulate a rationally defensible position of their own on their moral obligations to animals. They must assimilate a significant volume of information regarding regulation and law. They must be prepared to justify their work to the public, and to public representatives on local committees and in government, and delineate the extent of animal pain and suffering engendered by their research and how they are minimizing, mitigating, and controlling it. They must demonstrate that they have seriously examined non-animal alternatives and cannot employ these instead of live animals. They must demonstrate the appropriateness of animal species, number, and experimental design. They must understand the vast number of possible environmental and "stress" variables which can skew research results, and control for them. They must be knowledgeable in proper anesthetic, surgical, euthanasia, and analgesic procedures and concern themselves not only with overt physical pain in animals, but with other modalities of animal suffering: psychological and emotional pain, distress, anxiety, boredom, and so on, many of which notions have never been discussed in scientific literature relating to animals. They must begin to think about housing and husbandry procedures not only in terms of expense and researcher convenience, but also in terms of their effects on the animals' comfort. And they must demonstrate and certify that they have in fact genuinely engaged and thought through these issues. In short, and dramatically put, biomedical (and other) animal research must become what the 18th century called a "moral science", one where empirical and ethical questions are closely intertwined and engaged.

Unfortunately, as indicated above, most of these questions are new and indeed revolutionary for established biomedical researchers. Little if any of their training has been directed towards either the moral or the scientific questions relevant to animal care detailed above. And while those individuals focusing directly on the fields of laboratory animal science and medicine certainly have much of the relevant empirical knowledge, they constitute only a tiny fraction of the biomedical research community (there are, for example, only some 500 board-certified laboratory animal veterinarians in North America), and they too are basically untrained in the moral aspects of biomedical science. Furthermore, there have hitherto been no books focusing on these issues which were explicitly intended for researchers attempting to provide themselves with the tools necessary for feeling comfortable with the questions we have listed above.

This volume is specifically intended to fill that lacuna and provide researchers with at least an overview of areas they are suddenly expected to deal with, but regarding which they are often untrained. All of the authors have been charged with discussing their subjects from both a scientific and social and ethical perspective, in a way which speaks directly to the requirements and concerns of the researchers suddenly in need of this information, and which also speaks directly to the welfare and benefit of the animals used in research. As a result, this volume contains what I believe to be a significant number of bold and innovative attempts to balance ethics and science, and to at least generally chart waters hitherto neglected and unexplored. These papers are not intended to be the last word on the issues they examine, but rather, in many cases, to be the first. At the same time, they are intended to be of immediate and practical value to the researchers who read them, helping them to better understand the issues and, in their daily activities, to practice better animal care and, thereby, better science, for the two indeed go hand in hand.

I am grateful first and foremost to the authors who courageously and willingly undertook a task not "in their job description," and showed themselves to be scientists in the full sense of the term — knowers and thinkers. Despite the risks inherent in entering unexplored country, none of the authors whom I invited to participate refused to do so. And they have responded nobly, bringing their state-of-the-art knowledge to bear on moral and conceptual questions too long neglected.

I am grateful, too, to CRC Press for their vision in chartering these explorations. And I would be remiss in not thanking those scientists and veterinarians who over the past 12 years have patiently taught me enough — and listened to me enough — to give me the confidence to undertake such a project. Most especially, I owe an invaluable debt to the researchers and clinicians at the Colorado State University (CSU) College of Veterinary Medicine who adopted me as a colleague, and to the laboratory animal veterinarians there and all over the world who patiently exposed me to their theory and practice. At the same time, I must thank my colleagues in the CSU philosophy department, who not only tolerated, but supported my ever-increasing involvement in areas traditionally off-limits to philosophers.

Finally, I owe a special debt to my co-editor, Dr. M. L. Kesel, at once my student and my teacher, whose own career and education stand as a shining example of the ultimate unity of science and humanities, and of medicine and morality.

Bernard E. Rollin

THE EDITOR

Bernard E. Rollin, Ph.D., is Professor of Philosophy in the College of Arts, Humanities, and Social Sciences, Professor of Physiology and Biophysics in the College of Veterinary Medicine and Biomedical Sciences, and Director of Bioethical Planning at Colorado State University, Fort Collins.

Dr. Rollin received his B.A. degree from the City College of New York in 1964. He was a Fulbright Fellow at the University of Edinburgh, and earned his doctoral degree at Columbia University, where he was a Woodrow Wilson Fellow. He was appointed Assistant Professor of Philosophy at Colorado State University in 1969, Associate Professor in 1974, and Professor in 1978, when he was also jointly appointed to a professorship in Physiology and Biophysics.

Rollin received the Harris T. Guard Teaching Award in 1981 from CSU, the Waco F. Childres Award in 1982 from the American Humane Association, the Veterinary Service Award from the Colorado Veterinary Medical Association in 1983, and was named Honors Professor in 1983. He has received grants from the National Science Foundation and the National Endowment for the Humanities. Rollin has been an invited and endowed lecturer at hundreds of universities all over the world, and is on the editorial boards of numerous journals in science and philosophy.

Dr. Rollin is the author of more than 100 publications including 3 books he authored, 3 books he edited, and 1 book he co-authored. His current research interests include ethics and animals, animal pain, animal consciousness, and veterinary medical ethics, a field he is generally credited with founding.

THE ASSISTANT TO THE EDITOR

M. Lynne Kesel, D.V.M., is the former Staff Veterinarian for Laboratory Animal Resources and is currently Assistant Professor in Clinical Sciences of the College of Veterinary Medicine and Biomedical Sciences at Colorado State University, Fort Collins.

Dr. Kesel earned her B.A. in Fine Arts/English at Colorado State College in 1969 and her M.A. in Fine Arts from the University of Northern Colorado in 1970. After completion of 2 years of science requirements, she entered the professional veterinary program at Colorado State University, where she obtained her D.V.M. in 1981. After graduation, she joined the staff of Laboratory Animal Resources at CSU in a training position in laboratory animal medicine, which later became a permanent staff position. In 1987 she was appointed Assistant Professor of Clinical Sciences at CSU.

Dr. Kesel is a member of the American Veterinary Medical Association, the American Society of Laboratory Animal Practitioners, and the American Association for Laboratory Animal Science.

She has published several articles and given several lectures or continuing education workshops to professional groups in the area of basic laboratory animal medicine, handling, and techniques. She is board eligible for the American College of Laboratory Animal Medicine.

CONTRIBUTORS

G. J. Benson, D.V.M., M.S., Dip.A.C.V.A.
Professor
Department of Veterinary Clinical Medicine, Anesthesia Section
College of Veterinary Medicine
University of Illinois
Urbana, Illinois

Emerson L. Besch, Ph.D.
Professor
Department of Physiological Sciences
College of Veterinary Medicine
University of Florida
Gainesville, Florida

Lloyd E. Davis, Ph.D.
Professor
Clinical Pharmacology Studies Unit
College of Veterinary Medicine
University of Illinois
Urbana, Illinois

Rebecca Dresser, J.D.
Associate Professor
Schools of Law and Medicine
Case Western Reserve University
Cleveland, Ohio

Pamela H. Eisele, D.V.M.
Senior Veterinarian
Animal Resources Service
School of Veterinary Medicine
University of California
Davis, California

Andrew F. Fraser, M.R.C.V.S., M.V.Sc.
Professor
Department of Surgery
Memorial University of Newfoundland
St. John's, Newfoundland, Canada

Warren W. Frost, D.V.M., M.S.
Director
Department of Animal Resources
Montana State University
Bozeman, Montana

Michael J. Guinan, M.S.
Staff Research Associate
Department of Veterinary Medicine Anatomy
School of Veterinary Medicine
University of California
Davis, California

Thomas E. Hamm, D.V.M., Ph.D.
Director
Division of Laboratory Animal Medicine
Stanford Medical School
Stanford, California

Martha Lynne Kesel, D.V.M.
Assistant Professor
Department of Clinical Sciences
Colorado State University
Fort Collins, Colorado

Ralph L. Kitchell, D.V.M., Ph.D., V.M.D.h.c.
Professor
Department of Veterinary Medicine Anatomy
School of Veterinary Medicine
University of California
Davis, California

Charles W. Leathers, D.V.M., Ph.D.
Associate Professor
Department of Veterinary Microbiology and Pathology
Washington State University
Pullman, Washington

Seymour Levine, Ph.D.
Department of Psychiatry and Behavioral Sciences
Primate Facility
Stanford University School of Medicine
Stanford, California

Scott Line, D.V.M.
Research Scientist
California Regional Primate Research Center
University of California
Davis, California

Hal Markowitz, Ph.D.
Professor
Department of Biology
San Francisco State University
San Francisco, California

Christian E. Newcomer, V.M.D.
Clinical Associate Professor
Department of Comparative Medicine
Tufts University School of Veterinary Medicine
Boston, Massachusetts

Bernard E. Rollin, Ph.D.
Professor
Department of Philosophy
Colorado State University
Fort Collins, Colorado

Andrew Rowan, Ph.D.
Associate Professor
Department of Environmental Studies
School of Veterinary Medicine
Tufts University
North Grafton, Massachusetts

Harry C. Rowsell, Ph.D.
Executive Director
Canadian Council on Animal Care
Ottawa, Ontario, Canada

Steve F. Sapontzis, Ph.D.
Professor
Department of Philosophy
California State University
Hayward, California

Anthony Schwartz, D.V.M., Ph.D.
Diplomate, American College of Veterinary Surgeons
Associate Dean and Chairman
Department of Surgery
Tufts University School of Veterinary Medicine
North Grafton, Massachusetts

George E. Seidel, Jr., Ph.D.
Professor
Department of Physiology
Colorado State University
Fort Collins, Colorado

Joseph S. Spinelli, D.V.M.
Director
Animal Care Facility
University of California
San Francisco, California

John Thurmon, D.V.M., M.S., Dip.A.C.V.A.
Professor and Head
Department of Veterinary Clinical Medicine, Anesthesia Section
College of Veterinary Medicine
University of Illinois
Urbana, Illinois

Françoise Wemelsfelder, Drs.
Research Associate
Department of Theoretical Biology
Leiden University
Leiden, The Netherlands

Judith F. Woodson
Research Specialist
Department of Internal Medicine
University of Virginia
Charlottesville, Virginia

Eugene M. Wright, Jr., D.V.M.
Former Vice Chairman
Department of Comparative Medicine
University of Virginia
Charlottesville, Virginia

To
Michael David Hume Rollin,
for appreciating the inseparability
of science and ethics.

TABLE OF CONTENTS

Part I
Ethical Issues

Chapter 1

THE CASE AGAINST INVASIVE RESEARCH WITH ANIMALS

S. F. Sapontzis

EDITOR'S PROEM

Much of the recent emphasis on improving care and husbandry of laboratory animals has its roots in rising social unease about any use of animals in invasive research. Biomedical researchers, clear in their own minds about the benefits which animal use has brought to society, have had a tendency to dismiss questions about the right to use animals to advance human health and welfare as emotional, sentimental misanthropy and Luddism, or else to respond simply by cataloging the benefits resulting from animal research. In this paper, Professor Sapontzis lays out a careful, nonemotional argument articulating the moral reasoning which stands at the core of an abolitionist position. It is crucial that biomedical researchers understand and engage this sort of argument, not merely dismiss it out of hand because it challenges presumptions they have taken for granted throughout their careers.

TABLE OF CONTENTS

I. INTRODUCTION

Norman led Jennie into the laboratory and had her sit on a metal table near the windows. She sat there quietly while Norman and Peter fitted her with a helmet containing electrical monitors and couplings for attaching the helmet to other devices. She was watching the people walking across the lawn outside the windows. When they had finished, she had to lie down while the helmet was secured to a large machine and her arms and legs were secured to the table. All she could see now was the ceiling. Peter and Norman hooked up the monitoring devices in the helmet to a large console, checked out their equipment, and then turned it on. Jennie's head was given a tremendous blow by a piston which crashed into her helmet. She was knocked unconscious and stayed that way while Peter pried off her helmet. When she regained consciousness, she went into convulsions for five minutes. When the convulsions stopped, Peter and Norman ran some tests on her. She was blind now and could not control her arms sufficiently to grasp and carry to her mouth some food placed in her hands. Finally, she was wheeled into another room, where she was given an injection. Jennie died in less than a minute, and Norman and Peter began the work of decapitating her, describing the condition of her brain, and preparing slices of her brain tissue for microscopic analysis.*

Of course, Jennie was not human; let's say she was a squirrel monkey. However, the sort of equipment used in the experiment on Jennie could be manufactured to be used on humans. Also, since the ultimate purpose, presumably, of this sort of head injury test is to help develop therapies for human head injury victims, the information to be gained from running these experiments on human subjects would be even more useful than the information from animal tests. However, such tests are not done on humans; it would be grossly immoral to run such tests on humans. Why is it not also immoral to run such tests on squirrel monkeys? That, I think, is an instance of the core moral question concerning animal research: why is it not immoral to exploit animals in research, when so exploiting humans would be morally intolerable?

In this paper, I want to discuss three possible, and not uncommon, answers to that question. However, before listing these three answers, I want to emphasize that the most commonly offered justification for animal research is *irrelevant* to the question just posed. That justification is that animal research has produced numerous benefits for human beings. Many animal rights advocates question, and with good reason, how much animal research actually does result in benefits for human beings. For example, Roger Caras, an accomplished journalist who has spent decades investigating stories about animals, has written:

> Perhaps none of us knows quite as much as we should. But I do know this from long association with the scientific community (not as an adversary but as a friend): about 80 percent of what goes on in the laboratory has nothing whatsoever to do with the good of mankind. Only 20 percent can be exalted to that level. That remaining 80 percent is for the fun, profit, reputation, or other benefit of the experimenters.[1]

While this is a serious issue, economically and politically as well as morally, it is a separate one from that raised above. Even if all animal research issued in human benefits, the question of whether the research was morally justified would remain. That this is so can be quickly indicated by two comparisons. If research like that described above were done on humans rather than animals and issued in benefits for other humans, it would, like Nazi research on humans, still be morally deplored. Second, we may presume that slavery (generally) benefitted slave owners, yet it is considered a morally unacceptable institution nonetheless. Thus, that one group benefits by exploiting another does not show that that exploitation is morally justified. It follows that the

* This vignette was suggested by the head injury experiments carried on at the University of Pennsylvania for over a decade.

fact that we have benefitted by exploiting animals in research is beside the point when the question is whether our exploiting animals in research is morally justified.

Turning to relevant responses to this question, I propose discussing the following three common justifications of animal research:

1. People would have to be coerced into participating in such experiments — rather like the Nazi concentration camp experiments — since no one would freely and with understanding consent to have such experiments performed on him/her. It is such violations of the freedom and self-determination — what philosophers call "autonomy" — of rational beings which makes actions immoral. However, free, informed consent cannot be an issue where animals are concerned, since they are not rational, autonomous beings. Consequently, as long as these experiments are conducted humanely, there cannot be anything immoral about the treatment of animals in them.
2. We have special obligations to humans, due to our participation in the human community, which we do not have to animals, since they cannot participate in this community. We are thereby obligated to protect and to further human interests in ways we are not obligated to protect and further animal interests, and this justifies our sacrificing animal interests — humanely, of course — for human benefit. These obligations to protect and further the interests of humans also prohibit our exploiting humans in experiments.
3. Human life is morally more worthy than animal life; consequently, we are justified in (humanely) sacrificing these lower life forms in order to protect and enhance (the quality of) human life. Sacrificing humans in experiments would, of course, not result in this exchange of lesser for greater worth and would, consequently, not be justified.

Before discussing these claims, I need to clarify some terminology. First, I use "animal" to refer to all non-human beings who have interests, and I understand "S has an interest in x" to mean that x can affect S's feelings of well-being, which include pleasure and pain, feeling fit and feeling ill, feelings of fulfillment and of frustration, amusement and boredom, elation and depression, and so forth. Second, I use "suffering" to refer not only to experiencing negative feelings, but also to refer to losing opportunities for experiencing positive feelings. When an animal loses the remainder of what could have been a decent quality of life, it has "suffered a loss", even if it is killed painlessly. Finally, "exploit" should always be understood here in its moral sense of "to use selfishly", and I will use "exploitive animal research" to refer to research in which the suffering of the animals — which includes their capture, confinement, social deprivation, etc., as well as what is done to them during the experiment itself — is not adequately compensated by (likely or even just hoped for) benefits *for the research subjects themselves.* I do not think that all research with animals need be exploitive, although I suspect that virtually all laboratory research with animals currently is exploitive.

II. INFORMED CONSENT AND THE MORALITY OF ANIMAL RESEARCH

There probably are as many people who would claim that the fact that animals are incapable of freely and with understanding giving or withholding consent to participate in experiments makes it morally impermissible to do any research with animals as there are people who would claim that this inability removes humanely conducted experiments with animals from the moral arena. However, both extremes are mistaken. (Henceforth, "consent" will be used for "freely and with understanding give or withhold consent".)

Young children cannot consent to participate in experiments, but this is not viewed either as an obstacle to doing *any* research with them or as a justification for humanely sacrificing their interests in research. Research which is innocuous to the children involved, e.g., Piaget tests of the conceptual development of children, are routinely conducted. Sometimes, the idea of appropriate compensation is even introduced into such research, as when children are given free medical care in exchange for their participation in studies which require following a novel diet or periodically giving blood samples. Hopefully therapeutic research, e.g., experimental heart reconstructive surgery, is not as common, but is also considered morally acceptable when the hoped-for beneficiaries of the research include the children who are the subjects of the research. Of course, in all three kinds of cases consent must be obtained, but this may be given by someone who understands the interests of the children and is intent on protecting and furthering those interests.

Arguing by analogy, the conclusion to be drawn from animals' (supposed) inability to consent to participate in experiments is that guardians who understand the animals' interests and are intent on protecting and furthering those interests ought to be assigned to look after the animals. It would be the responsibility of such guardians to insure that research with their animals either was innocuous to the animals, was intended to be therapeutic for them, or provided them adequate compensation.

Some philosophers might object here that animals (employing the common meaning of the term now) cannot have interests or that we cannot know what their interests are.[2] I will not spend much time here answering such objections. This is because the philosophical bases for them are both technical and unconvincing and have been adequately refuted elsewhere.[3–5] Some schools of philosophy are famous for trying to "argue the hind leg off a donkey" and otherwise to convince us that things are not at all as they appear to be. Philosophers who try to convince us that an animal cannot want to get out of a cage because it cannot say "I want to get out of this cage," cannot foresee the months it will be locked in the cage, cannot understand why it has been locked in a cage, or lacks other, similar intellectual accomplishments are firmly in that tradition. And just as we refute those schools by going out into the field and taking a look at the donkey for ourselves, so our immediate experience of animals as feeling beings refutes those who claim that animals have no interests. Even the practice of animal research itself refutes them, e.g., the practice of doing pain and despair research on animals. Just where among the different species of animals the ability to feel pleasure and pain, fulfillment and frustration, etc. begins is a difficult question to answer, especially since there may be a gradual continuum here rather than a sharp dividing line. However, our uncertainty concerning whether shrimp can feel pain does not cast doubt on our experience of primates, pigs, dogs, cats, rabbits, rats, mice, and other laboratory animals having that capacity. If I step on a shrimp, I may wonder whether it felt pain, but if I step on the cat's tail, I have no more doubt that it felt pain than I do that a person feels pain when I step on his/her toe. Thus, questioning whether animals have interests is in the same category with questioning whether other people have minds: these are questions that philosophers can exercise their wits about, but if they come up with negative answers to them, that only proves the inadequacy of their philosophy.

Thus, it cannot seriously be questioned that being imprisoned in psychologically sterile surroundings, being "stressed" with shocks and through deprivation, having their nerves cut or their eyes sewn shut, being killed, etc., run counter to the best interests of the healthy animals commonly employed in research. We know that all these things are adverse for animals as readily and in the same ways that we know that they would be adverse for young children.*

So, that animals are incapable of consenting to participate in experiments does not indicate that the only moral issues concerning the interests of the animals employed in research are issues of humane treatment. If exploiting animals in research is to be morally justified, while exploiting

* See Marian Stamp Dawkins' *Animal Suffering* (Chapman and Hall, New York, 1980) for a discussion of some of the practical issues involved in determining what is and is not in the animals' interest. She concludes that giving animals the opportunity to *choose* for themselves which environments they prefer and to show what they find positively or negatively *reinforcing* is the closest we can come to being able to ask an animal what it is feeling (p. 111).

humans is not, then the justification for this difference must lie elsewhere than in differences between their and our abilities to consent.*

We may also conclude here that if we did appoint responsible guardians for research animals, virtually all laboratory animal research would probably come to a halt, because no responsible guardian would agree to have his/her ward dealt with as virtually all laboratory research currently deals with animals. For example, no responsible guardian would agree to Jennie's participation in the above head injury experiment. Most laboratory animal research is (supposedly) directed toward human benefit, but the experiments may not be conducted on humans because they involve exploiting, often severely, the subjects of the research. However, one of the most important functions of guardians is to prevent the exploitation of those who are not in a position to protect themselves. It follows that a primary reason for employing animals in laboratory research would also be a reason for severely inhibiting that research, if animals had guardians.

III. "THE BROTHERHOOD OF MAN" AND THE MORALITY OF ANIMAL RESEARCH

The second of the justifications of exploitive animal research proposed above is that we have obligations to human beings that we cannot have to animals because we are members of a human community of psychological, social, economic, and political interests in which animals cannot share.[7]

While some animal rights advocates project a world in which every sentient being is treated "equally", our everyday morality indicates that even among human beings there is no overriding egalitarian imperative. In our everyday moral practice we continue to recognize distinctions in the obligations we have to different groups of people. For example, parents have many obligations to their own children that they do not have to other children, and citizens have obligations to their own country that they do not have to other countries. As the philosophers who contend that animals have no interests are out of touch with and, consequently, refuted by our everyday experience with animal psychology, so militantly egalitarian moralities are out of touch with the structure of morality, and their claims that, for example, "everyone is to count for one and no one for more than one" (Jeremy Bentham) and that "inherent value, which requires respect, is something that does not admit of degrees; all who have it have it equally" (Tom Regan) are refuted by our everyday moral experience. A world in which we treated each other not as children, parents, siblings, friends, neighbors, colleagues, fellow citizens, etc., but merely and indiscriminately as persons is not an ideal of our actual moral practice. Such a world would not be a morally better one; it would be a sadly impoverished place, devoid of many of the emotional relations, opportunities, fulfillments, responsibilities, and reassurances that make life worth living for social animals like ourselves.

Militantly egalitarian moralities may also be out of touch with human psychology, since it is doubtful that many, if any, of us can really regard everyone — let alone every animal — just as individuals meriting equal consideration and concern. Thus, it is the variegated pattern of obligations in everyday moral practice and the undesirability, even impossibility, of doing away with this variety which gives some credence to this second defense of exploitive animal research, especially when that defense is offered as part of a critique of an abstract, militantly egalitarian defense of animal rights.[8]

However, emphasizing this variegated pattern of our obligations can provide as distorted an image of everyday morality as do abstract theories of equality. One of the primary functions of morality, especially of moral rights, is to inhibit favoritism. To accomplish this, we have

* I would also question the common assumption that animals cannot consent, but since I have discussed that issue elsewhere, I will not enter into it here.[5,6]

developed a "Justice is blind" family of moral imperatives: "Take a disinterested viewpoint," "Give equal consideration to all concerned," "Try to universalize your actions," "Disregard individual differences," and so forth. This family of moral imperatives is as much a part of everyday morality as are the obligations we have only to family, neighbors, colleagues, or fellow citizens. For example, while a father is morally justified in giving priority to fulfilling the interests of his child, he would not be justified in doing so by taking food from another, needy child, by enslaving a stranger, or by killing a business competitor. A father can be morally criticized even for using his discretionary income to purchase luxuries for his child while contributing nothing to help starving children in other parts of the world.

Thus, in addition to familial, collegial, etc. obligations, everyday morality also contains egalitarian rights and responsibilities. These rights and responsibilities counterbalance those obligations by restricting the ways in which they may be fulfilled. They do this by imposing impartial obligations on us, by requiring us to add a disinterested appraisal to our judgments of what is right, and so forth. It must be recognized that the fabric of everyday moral practice is a weave of contrary forces.[5,9]

It follows that this special-obligations defense of exploitive animal research cannot suffice. That we have obligations to protect and further the interests of P which we do not have to Q does not provide sufficient grounds for concluding that we would be justified in exploiting Q in order to fulfill our obligations to P. For example, the special obligations we have to family, friends, fellow citizens, etc. would not justify our exploiting strangers in order to fulfill those obligations. Therefore, even if we postulate humanistic obligations we have to all humans but to no animals, e.g., the obligation to help those who are ill, it does not follow that we are justified in exploiting animals to fulfill those obligations. Some argument showing that we are justified in fulfilling these special, humanistic obligations *by exploiting those to whom we are not thus obligated* must be provided before this second defense of exploitive animal research can fulfill its purpose.

To put the matter formally, this second defense of exploitive animal research looks (roughly) like this:

A1. We have obligations to humans that we do not have to animals.
A3. Therefore, we may exploit animals in order to fulfill our obligations to humans.

Thus stated, the argument is elliptical, since it has no premise concerning when we are and are not entitled to exploit others. The missing premise must be something like the following:

A2. When we have obligations to P that we do not have to Q, we may exploit Q in order to fulfill those obligations to P.

Inserting this premise into the above argument yields the following formally valid defense of exploitive animal research:

A1. We have obligations to humans that we do not have to animals.
A2. When we have obligations to P that we do not have to Q, we may exploit Q in order to fulfill those obligations to P.
A3. Therefore, we may exploit animals in order to fulfill our obligations to humans.

Although the argument now satisfies logical requirements, it is also now clear that it rests on a highly questionable premise, *viz.,* A2. Concentrating on A1, critics of militantly egalitarian defenses of animal rights have overlooked their own Achilles heel and failed to provide any defense of A2. And, given the kinds of restrictions which our "Justice is blind" family of principles places on fulfilling our other special obligations, it is severely doubtful that a convincing argument in favor of A2 could be found. Consequently, this "brotherhood of Man" defense of exploitive animal research is unsound.

This conclusion should not be surprising. Although extending moral rights to animals would require a considerable change in our way of life, doing so has been proposed, by Peter Singer, Bernard Rollin, and myself, for example, as the next logical step in the development of our everyday moral concepts and practice.[5,10–12] Briefly, Western moral history has been and remains "the story of liberty," to borrow a phrase from the 19th century German philosopher G. W. F. Hegel. In this story the scope of our moral concern is gradually increased as we overcome the ethnic, religious, racial, and sexual biases which have limited that concern and served as justifications for all sorts of iniquitous and exploitive practices, from slavery through the Inquisition and the Holocaust to many expressions of economic discrimination against women. The next stage in this story, we argue, is recognizing that (1) just as our basic moral principles are not logically tied to differences of national origin, race, religion, and sex, so they are not logically tied to differences of species, and (2) as a result of this, we need next to overcome our species bias and to reject it as a basis for justifying our exploitation of animals. Now, just as these previous liberations from ethnic, religious, racial, and sexual biases have been accomplished without denying that we have special obligations to our families, friends, neighbors, etc., so it is to be expected that liberation from our species bias can also be accomplished without denying these special obligations. Consequently, while some defenders of animal rights may have failed to give sufficient attention to the variety and complexity of our moral principles and their interrelations in actual practice, there is no reason to believe that defending animal rights requires oversimplifying our everyday moral concepts and practice or to believe that that variety and complexity in themselves constitute a refutation of animal rights and an adequate defense of our continued exploitation of animals.

IV. HUMAN SUPERIORITY AND THE MORALITY OF ANIMAL RESEARCH

Ultimately, it is the belief that human life is morally more worthy than animal life and the belief that this justifies us in sacrificing animals for our benefit explains why everyday morality does not protect animals against exploitation. However, both of these beliefs are highly questionable, not only morally, but epistemically and logically as well.

A. ARE HUMANS A MORALLY SUPERIOR LIFE FORM?

While acknowledging that animals have talents we lack, e.g., the bat's sensitivity to sound and the acute vision possessed by many birds, we still consider ourselves a superior life form because of our "rationality". Specifically, it has been our ability to employ reason to control our lives and surroundings which has been heralded in our traditional moral literature, both philosophical and religious, as the mark of human superiority. This ability has even been called "the image of God," the creator and controller of the universe, in us. However, even if humans are, ordinarily, capable of flights of reason of which animals are not, it does not immediately follow that we are *morally* more worthy than animals.

For example, if we think of reason in terms of the traditional analogy between human reason and God the creator, it seems that what we are talking about when we talk about human superiority is our great ability to dominate and control. Citing that ability as our morally crucial superiority to animals suggests that this justification of exploitive animal research is based on the belief that those who are strong enough to exploit others are *for that very reason* morally justified in doing so. If that is what is being argued here — and two prominent medical researchers recently explicitly subscribed to this "king of the jungle" position* — then this justification is an instance of the "might makes right" philosophy. Given our common moral principles and belief that we have made moral progress by abolishing feudalism, slavery, and other forms of subjugation, citing the great power our reason can create and command as justification for our employing that power to exploit those weaker than ourselves cannot be a morally credible argument.

Of course, there are other, more credible interpretations of the moral significance of superior reasoning ability. According to the school of moral philosophy, best exemplified in the works of the 18th century German philosopher, Immanuel Kant (and, consequently, commonly called "Kantianism"), a moral agent is supposed to be one who can act our of respect for impersonal laws, and it supposedly requires something like normal, adult, human intelligence to be able to recognize such laws, to counterbalance selfish feelings, and to do the morally right thing consistently.

However, while this Kantian view of the moral significance of reason is morally superior to the previous, Machiavellian account, it is exaggerated. As I have argued elsewhere, the Kantian claim that acting from a sense of duty is the *only* moral motive is mistaken.**[5,13] Generous sentiments are commonly considered to be at least as morally admirable as motivation through a sense of duty. For example, loving parents are commonly considered morally admirable; indeed, they are commonly considered morally more admirable than "dutiful parents". To describe someone as a "dutiful parent" is to damn him or her with faint praise. Rather than being the highest achievement of moral sensitivity, our sense of duty often has only the value of something which we must fall back on when our generous sentiments fail us. In many, perhaps all, cases, our sense of duty is to morality what a police force is to society: something we have to rely on because, regrettably, we are not sufficiently considerate, compassionate, altruistic, and otherwise virtuous to get along together without it.

However, once we allow sentiment-inspired actions to enter the moral arena, not only reasoned, but also instinctual and conditioned actions which are intentional, not ulteriorly motivated responses to the needs of others must count as moral actions and those who do them as moral agents. For example, a mother bird who feigns a broken wing and risks her life to distract a fox from her nest is a moral agent, even if she acts on "maternal instinct" and is not motivated by judging her action to be one that could serve as a law obligating everyone to act as she is acting (which is how the Kantians would have good mothers be motivated). The same is true of a human mother who instinctively rushes into a burning house to save her baby. In both cases, use of the word "instinct" does not preclude the mother recognizing and being motivated by the danger to her young, and it is that recognition and response which are morally crucial in such situations. Thus, whether bird or human, being a devoted mother is a moral virtue, and this is an evaluation based on the mother's self-sacrificing devotion to her youngsters. A mother's capacity for abstract practical reasoning is not what is at issue when she is being morally evaluated as a mother. As I have discussed elsewhere, that ability becomes morally crucial only in a limited family of situations, e.g., those requiring the evaluating and balancing of conflicting claims in a property dispute.[5,14] Good judges need to be good reasoners, but moral concern and action are neither limited to nor patterned exclusively after what goes on in court.

Consequently, we cannot so blithely dismiss loyal dogs, courageous lions, responsible wolves, industrious beavers, self-sacrificing parents of a wide variety of species, etc. as not being

* Dr. A. K. Ommaya, of George Washington University, on the Spring 1984 "Frontline" program (PBS Television) devoted to animal research, and Dr. Norman Schumway, of Stanford University, on the Fall 1984 segment of "60 Minutes" (CBS Television) devoted to the same topic.

** In the article and book referred to here, I distinguish two kinds of moral agents, virtuous agents and fully moral agents, concluding that animals can be the former, even if not the latter. Here, it is not germane to go into this distinction; so, I shall simply refer to animals as "moral agents".

moral agents because they are "merely creatures of instinct." Seeing virtue in other animals is not anthropomorphizing, unless we presume that they do their virtuous deeds as a result of the same reasoning process that we have to employ when our own generous sentiments, parental instincts, social instincts, moral training, etc. fail us and we have to combat our lesser, selfish selves in order to do what we ought to do. However, just as there is no reason to suppose that other animals (routinely) go through this process, so there is no moral need for making such a presumption. What makes an action an expression of moral agency is not that the agent had to put his or her internal house in order in order to overcome temptation and do what is right. An action is an expression of moral agency if it is done intentionally and straightforwardly (i.e., without ulterior motives) in response to what the agent perceives to be the need of another. The actions of many animals appear to be such expressions, which is why it is anthropocentric to insist that animals cannot be moral agents.

It follows that if we, as moral agents, are superior to animals, it must be a question of degree. Given our particularly bloody, destructive, exploitive habits and history, an impartial survey of the relevant facts would not obviously lead one to the conclusion that we are the most moral of species. Furthermore, how we would even go about trying to determine which species is/are the most moral is itself a staggering question. As long as we think that, on conceptual grounds, animals cannot be moral agents at all, we do not have a methodological problem here. But once we recognize that animals are moral agents, there are severe methodological problems in trying to determine which species is/are superior moral agents. For example, do we evaluate capacity for doing what is right or moral deeds actually performed? If we evaluate deeds actually performed, do we just count them, or do we try to evaluate them against the capacity available to be actualized, so that although P has done fewer moral deeds than Q, he/she might still be considered superior to Q, since he/she has actualized a greater part of his/her capacity to do moral deeds than has Q? What allowance do we make for the different living conditions and traditions of different species? Are certain kinds of moral capacities or deeds to be weighted over other kinds, and if so, on what basis? Answering such fundamental, methodological questions nonarbitrarily would seem to be impossible. Consequently, answering questions about which species is/are superior moral agents would seem to be impossible, once we acknowledge that we are not the only moral agents around.

Another school of moral philosophy which provides a family of interpretations of the moral significance of our superior intellectual capacity is utilitarianism. Utilitarianism originated in Britain near the end of the 18th century with the writings and reform activities of Jeremy Bentham. It is based on the principle, called "the principle of utility", that the morally right thing to do is that which will lead to the greatest happiness of the greatest number of those who will be affected by one's action. Now, in connection with evaluating what will maximize happiness, it is sometimes claimed that humans are capable of experiencing enjoyment, fulfillment, distress, and frustration of a greater subtlety and variety than other animals. The animals' range of experience is (supposedly) limited by their limited intellects to matters of sensation, digestion, and reproduction, while, thanks to our superior intellect, we are capable of appreciating fine art, conceptual matters, moral fulfillment, flights of imagination, remembrance, and anticipation, etc. in addition to what the animals can experience. Since the principle of utility commands us to maximize the excess of enjoyment and fulfillment over distress and frustration in life, it can then be argued that we are justified in sacrificing less sensitive beings, i.e., animals, in order to benefit more sensitive beings, i.e., humans. (Henceforth, "feelings" will be used in place of "enjoyment, fulfillment, distress, and frustration".)

In response to this, it must be acknowledged that we can experience some things, e.g., the satisfaction of solving a symbolic logic problem, which animals cannot. However, it does not follow from that that our feelings are superior to those of animals, either in a qualitative or a quantitative sense of "superior".

First, how could it be shown that human feelings are qualitatively superior to those of animals? Perhaps the staunchest advocate of qualitative differences among feelings is John Stuart Mill, the great 19th century defender of utilitarianism and champion of the slogan, "Better to be Socrates dissatisfied than a pig satisfied!" Being a staunch empiricist, Mill acknowledges that the only way to determine which are the qualitatively superior feelings is to find someone who can appreciate the lot and ask him/her which ones he/she prefers.[15] This is the only testing procedure available because the only way to determine the quality of a feeling is to feel it. However, while this sort of test might be run in order to compare the different qualities of human feelings, it is impossible to run such a test when the issue is the (supposed) differences in quality among the feelings experienced by members of different species. Even the multitalented John Stuart could not experience the enjoyments of bats and gulls and the fulfillments of dogs and dolphins. Consequently, although we can experience things that animals cannot, and although we (or at least some of us) place an especially high value on the things that only those with something like normal, adult, human intelligence can experience, we cannot justify a claim that our feelings are qualitatively superior to those of other species. Claims that cannot be justified should, of course, not be made and cannot provide a sound basis for moral action.

Turning to the issue of whether human feelings are quantitatively superior to those of animals, some utilitarians, such as Peter Singer, the contemporary Australian animal liberationist, believe that humans are capable of a greater quantity of feeling, especially of fulfillment, than are animals. This is because we are capable of projecting the future and remembering the past to a vastly greater extent than are animals. These abilities make it possible for us to experience such feelings as hope and regret, while animals, with their limited temporal capacities, cannot (supposedly) experience such feelings.[16]

However, even if this is true — and I have argued elsewhere that there is, once again, more of a continuum than a dichotomy between our and their temporal abilities[3–5] — it may also be true, and for the same reason, that the feelings of animals are more intense than ours. This is because their feelings are not diluted by recollections of the (far) past or hopes for the (distant) future.[11] For example, the distress and frustration experienced by a human prisoner of war can be alleviated by his/her recollections of past freedom and his/her hope for future release, while a dog trapped in a laboratory cage (supposedly) has no such recollection or hope to ease its distress and frustration. Similarly, humans are notorious for not getting full enjoyment from present pleasures because they have fixated on past sorrows or are fretting about future difficulties, while animals, like dogs playing on the beach, do not seem to have their present enjoyment thus diluted. Now, if animal feelings are more intense than ours, this extra intensity could counterbalance the extra feelings our extensive temporal capacity provides us. Thus, on quantitative, utilitarian grounds, our superior temporal capacity can be as much of a moral liability as an asset; it could even be the case that our more dispersed way of life is morally less valuable than animals' more intense way of life, on these grounds.

Of course, once again, there is no way to tell whether the feelings of animals are more intense than ours and, if so, whether that greater intensity is sufficient to counterbalance or even to outweigh the greater extent of our feelings. Again, this is because the only way to make such comparisons would be to find subjects who could each experience all of these feelings and ask them to draw these comparisons, and there are no such subjects. Consequently, this quantitative account of the moral superiority of human feelings encounters the same problem as the qualitative account and must suffer the same fate: since they cannot be justified (or falsified), they are not significant claims and cannot contribute to a justification of our continued exploitation of animals, in research or elsewhere.

In listening to utilitarian discussions of feelings, one can get the impression that feelings are like jelly beans in a jar, the jar being an individual life, and that we can evaluate the individual's life by seeing how many and what sort of mix of beans he/she has in his/her jar. Human lives are supposed to be more valuable because they have a quantity and variety of beans not found in animal lives.

But feelings are not like jelly beans, and they are not related to the one who has them as beans are to their receptacle. Feelings cannot exist by themselves, and they are what the one who has them makes of them. When it comes to feelings, "to be is to be perceived," as the 18th century idealist George Berkeley says. That is why we cannot simply survey a life to decide its utilitarian value; the one who lives that life must show us what its value is for him or her. A life which looks like it would be a hard and dull one for us to lead, e.g., the life of a beaver, may be experienced by the one who lives it to be enjoyable and fulfilling. And it is *the value that life has for the one who lives it* that counts in the utilitarian calculus, since it is only by his or her living it that it exists and actually comes to have value. The value that others place on S's life through imagining what it would be like for them to live that life is merely imaginary and, consequently, is irrelevant to the value S's life really has. Only insofar as S's life actually impacts the feelings of others, e.g., through S's cutting down their trees or providing them entertainment, does the value of S's life for others enter into the utilitarian calculus.

It follows that if we consider the lives of a dog and a human, for example, which are very different lives, but which, let us presume, are both experienced by those living them to be enjoyable and fulfilling, it is impossible to determine which life has greater utilitarian value. The human life may have greater variety to it, and the human might feel frustrated if he or she were given the dog's life to lead, but that is irrelevant to the question of whether the dog's life is more or less enjoyable and fulfilling *for the dog* than the human's life is enjoyable and fulfilling *for the human*. Since we cannot feel what the animal feels, we cannot determine whether a human gets more or less enjoyment and fulfillment from his or her enjoyable and fulfilling life than an animal gets from its enjoyable and fulfilling life.

Since we cannot make such comparisons, utilitarian evaluations must be limited to determining whether way of life x is more or less enjoyable and fulfilling than way of life y for a particular subject or type of subject which could actually live x or y, e.g., whether squirrel monkeys prefer to live in laboratory cages or in the jungle. Utilitarian evaluations must not involve comparing way of life x for p with way of life y for Q, e.g., comparing an enjoyable, fulfilling way of life for a monkey with an enjoyable, fulfilling way of life for a human. It further follows that our efforts to fulfill our utilitarian goals ought to be directed to improving the quality of life by finding better lives for sentient beings, e.g., by developing more enjoyable, healthful diets for people, rather than by substituting supposedly superior subjects for supposedly inferior ones, e.g., by killing baby baboons to save the lives of human infants.

A third utilitarian argument for the moral superiority of human life would emphasize that our reasoning ability makes it possible for us effectively to combat distress and frustration and to otherwise increase the enjoyment and fulfillment in life. Our greater intelligence allows us to do far more than other animals to fight diseases, produce food, construct shelter, and to develop technologies which otherwise prolong and enhance (the quality of) life. While the above qualitative and quantitative arguments focus on the ability to experience utilitarian goods, this argument focuses on the ability to produce those goods and sees human moral superiority in our superior ability to be utilitarian agents.

This moral agency argument encounters the same methodological difficulties as its Kantian cousin discussed above. Furthermore, when the effects of our technology on animals are taken seriously, it is doubtful that we are producing more enjoyment and fulfillment than distress and frustration. We destroy the quality and take the lives of hundreds of millions of laboratory animals every year. It is hard to imagine that our research produces utilitarian goods sufficient to outweigh that massive, annual evil. In the areas of farming, ranching, fishing, hunting, and trapping, the number of animals exploited annually is even greater, and, consequently, it is even harder in these areas to imagine that we are producing a better utilitarian world. This is especially true when we remember that utilitarian reasoning requires that we compare what we are doing not only with what things would be like if we stopped doing what we are doing, but also with what things would be like if we substituted non-animal-exploiting alternatives for what we are doing. For example, in the area of research, the comparison is not between doing invasive research on animals and not doing any research, but between doing invasive research and doing noninvasive research, in the sense both of doing research on problems that do not require invasive procedures and of developing noninvasive procedures for problems that are currently being attacked invasively.

So, applying this utilitarian agency test to human history and "accomplishments", we might have to conclude that to date our willingness to pursue our own interests at the expense of others has dominated our ability to produce a better world for all and has made of us utilitarian liabilities rather than assets. Like promising young men and women, we cannot continue to be valued for our potential forever; eventually, we must prove our mettle by performing, and to date we have not, as a species, performed as superior, impartial, utilitarian agents.*

There could, of course, be other suggestions for how we are a morally superior form of life. However, we shall not examine any further candidates here. The Kantian and utilitarian arguments we have examined are the ones suggested by our philosophical tradition and everyday morality. If they cannot justify our claim to superiority, then it is doubtful that it can be morally justified, and, consequently, it is reasonable to conclude that that claim ought to be rejected until, if ever, it is provided a moral justification.

B. WOULD SUPERIORITY JUSTIFY EXPLOITATION?

For the purposes of argument, let us presume to be true what we just determined there is not good reason to believe, i.e., that humans (ordinarily) are superior beings in a morally significant way. Many people presume that if we are superior beings, it immediately follows that we are justified in exploiting lesser beings for our benefit. However, as it stands, the following is clearly not a valid argument:

B1. P is superior to Q in a morally significant way.
B3. Therefore, P is justified in exploiting Q for his or her benefit.

There is a missing premise here which must be something like:

B2. Those who are morally superior to others are justified in exploiting those others for their benefit.

This general principle is not obviously correct. Morally superior beings sometimes have extensive obligations to and only circumscribed privileges over their inferiors, as in the case of human adults dealing with young children. Similarly, matters of life and death are morally more important than matters of convenience, but, as Judith Jarvis Thompson has argued, even if Isaac would die unless he were hooked up to Susan for several months, so that he could use her kidneys to cleanse his blood until he recuperated from some renal disease, that would not justify him forcing Susan to sacrifice her lifestyle to provide him this service.[17] Just what follows from P being morally superior to Q is an open question. Consequently, B2 needs to find support in some general, moral framework, if it is to be morally acceptable. Can it do so?

Approaching the matter from a Kantian perspective, what would make us a morally superior life form is our (supposedly) greater ability to set aside self-interest and to make disinterested appraisals of what ought to be done. However, it would be at least a cruel irony to cite this ability as justification for our disregarding the interests of weaker, (supposedly) morally inferior animals and exploiting them as mere means to our ends. Just as the philosopher-kings who would

* A popular version of this third utilitarian defense of human superiority contends that we produce goods of immensely greater intrinsic value than do animals, e.g., fine art and science. However, this contention obviously encounters the problems which bedevil the other utilitarian positions just discussed; so we need not spend time on it here.

rule the ideally just society imagined by the ancient Greek philosopher Plato would not be justified in exploiting those whom their superiority entitles them to govern, so our (supposedly) greater ability to act impartially does not provide a basis for our acting selfishly in our dealings with those of (supposedly) less moral ability. Indeed, to the extent that we exploit those weaker than ourselves, we bring into disrepute our claim to moral superiority. Therefore, Kantianism does not support B2.

The same is true of utilitarianism. Utilitarian concern is directed (1) to maximizing the excess of enjoyment and fulfillment over distress and frustration in life and (2) to doing so without prejudice, i.e., without favoring the interests of one group over those of another. This second utilitarian concern is sometimes overlooked, but it was his antipathy to aristocratic privilege that led Bentham to inaugurate utilitarianism. Replacing a hierarchical worldview, where the interests of "the better classes" are considered more worthy of fulfillment than those of their "inferiors", with a presumption in favor of equal consideration for all has always been a basic part of utilitarianism.

Now, B2 is an expression of such a hierarchical worldview and is, therefore, not supported by the principle of utility. That principle does support preferring a life capable of a greater utilitarian good over another when we are forced to choose between them, e.g., when we must choose between using scarce medical supplies to save an older or a younger person. However, the principle of utility does not support viewing the world in terms of classes of superior and inferior beings, with the latter being routinely exploited as a resource for fulfilling the interests of the former. That is, of course, exactly what B2 proposes.

Thus, neither Kantianism nor utilitarianism supports the contention that we are morally justified in exploiting animals in our efforts to fulfill our interests because our lives are (supposedly) of greater moral worth than theirs. Both Kantian and utilitarian concerns indicate that research with animals — as well as all our other dealings with animals — ought to be designed in ways which insure fairness and which contribute to minimizing suffering and otherwise maximizing the well-being of *all* concerned. Securing our own well-being by exploiting the hundreds of millions of healthy animals used annually in research is neither designed nor likely to help us attain these moral goals. Rather, it increases the distress and frustration in life and insures that these evils are distributed unfairly.

V. CONCLUSION

Perhaps the morally most important thing to recognize about contemporary animal research practices is that they are expressions of an aristocratic worldview in which we humans, the "better class" in this hierarchy, are entitled to treat those of inferior status as resources for the preservation and enhancement of our "superior way of life." Of course, we ought to be humane in our use of these inferiors — *noblesse oblige!* — but this humane obligation remains secondary to and circumscribed by the presumption and exercise of aristocratic privilege. For example, our traditional, humane ethic prohibits tormenting animals, but is overruled when it comes to performing painful and lethal experiments on animals, like the one described at the beginning of this essay, if those experiments could benefit us.

Such aristocratic worldviews have been the target of what we commonly consider moral progress. Our rejection of slavery, feudalism, aristocracies of birth, sexual and racial discrimination, etc. all have in common the discrediting of supposed "natural hierarchies" — e.g., the supposed superiority of men to women and of whites to blacks — in favor of a presumption of equal consideration for all. Consequently, it is not inappropriate to see contemporary animal research practices as vestiges of feudal mentality on the contemporary scene and as prime candidates for rejection as the next step forward in our moral progress.

In order to better accomplish that progress, I would suggest that research with animals be governed by the following principle.

Animals may be employed in research only when a guardian who understands the animals' interests and is intent on protecting and furthering those interests consents for them to participate in the experiment after determining that either (1) participating in the experiment would (likely or even just hopefully) be either innocuous to or beneficial for the research subjects, or (2) conducting this research on these subjects is the only available way to attain a *clear and present, massive, desperately needed* good which *greatly* outweighs the sacrifice involved in the experiment and that this sacrifice is *minimized and fairly distributed among those likely to benefit from the research.*

Adhering to this principle would be to apply to animals employed in research the same sort of moral protection against exploitation currently enjoyed by human research subjects.

Concretely, what would follow from adhering to this principle? Consider a case much in the news of late, research directed toward using animals as organ donors for humans. It might be thought that this research could be justified under heading (2) of the above principle, where the sacrifice of individuals in research is permitted. However, that is not the case, for the following pair of reasons.

First, the burdens of this research are not distributed fairly among those likely to benefit from it. Since it is not contemplated that heart, liver, etc. transplants will be used to save animals' lives, animals as a group have nothing to gain from this research. And since the individual animals employed in this research are healthy and will be killed in the research, they have nothing to gain here. Since they bear all the burdens and receive none of the benefits, this research is grossly unfair.

Second, the good to be obtained through this research does not outweigh the evil it generates. Many animals will be killed in this research, and if the transplant techniques prove successful, then the number of animals killed to provide donor organs will be approximately the same as the number of human recipients of these organs. Perhaps some animals will be killed to donate more than one organ, but some people will need to receive more than one donation. Consequently, such research is not directed toward minimizing suffering; it aims merely at shifting the suffering from "us" to "them".

Other forms of animal research, such as mother deprivation studies and psychosexual studies, are even more obviously incapable of meeting the moral standards we have put forward here, i.e., the moral standards currently employed in research with human subjects. So, although stopping exploitive animal research would not entail stopping all animal research — any more than respecting human rights entails stopping all research with human subjects — if we were to apply common principles of fairness and utilitarian moral concerns impartially to animal research, virtually all of that research would have to be radically restructured or terminated.*

* An earlier, shorter version of this paper, entitled "On Justifying the Exploitation of Animals in Research," was published in *J. Med. Philos.*, 13, 177, 1988.

REFERENCES

1. **Caras, R.,** Are we right in demanding an end to animal cruelty?, in *On the Fifth Day,* Morris, R. K. and Fox, M. W., Eds., Acropolis Books, Washington, DC, 1978, 130.
2. **Frey, R. G.,** *Interests and Rights: The Case Against the Animals,* Clarendon Press, Oxford, 1980.
3. **Sapontzis, S. F.,** The moral significance of interests, *Environ. Ethics,* 4, 145, 1982.
4. **Sapontzis, S. F.,** Interests and animals, needs and language, *Ethics Anim.,* 4, 38, 1983.
5. **Sapontzis, S. F.,** *Morals, Reason, and Animals,* Temple University Press, Philadelphia, 1987.
6. **Sapontzis, S. F.,** Some reflections on animal research, *Bet. Spec.,* 1, 18, 1985.

7. **Francis, L. and Norman, R.,** Some animals are more equal than others, *Philosophy,* 53, 507, 1978.
8. **Diamond, C.,** Eating meat and eating people, *Philosophy,* 53, 465, 1978.
9. **Sapontzis, S. F.,** Moral community and animal rights, *Am. Philos. Q.,* 22, 251, 1985.
10. **Singer, P.,** *Animal Liberation: A New Ethic for Our Treatment of Animals,* Avon, New York, 1975, chap. 1.
11. **Rollin, B.,** *Animal Rights and Human Morality, Part 1,* Prometheus Books, Buffalo, 1981.
12. **Sapontzis, S. F.,** The evolution of animals in moral philosophy, *Bet. Spec.,* 3, 61, 1987.
13. **Sapontzis, S. F.,** Are animals moral beings?, *Am. Philos. Q.,* 17, 45, 1980.
14. **Sapontzis, S. F.,** Moral value and reason, *Monist,* 66, 146, 1983.
15. **Midgley, M.,** *Beast and Man,* Cornell University Press, Ithaca, NY 1978.
16. **Singer, P.,** *Practical Ethics,* Cambridge University Press, New York, 1979, chap. 4.
17. **Thompson, J. J.,** A defense of abortion, *Philos. Public Affairs,* 1, 47, 1971.

Chapter 2

ETHICS AND RESEARCH ANIMALS — THEORY AND PRACTICE

Bernard E. Rollin

EDITOR'S PROEM

The issue of an ideal ethical theory for animals and its practical application to the care and treatment of laboratory animals has emerged as a major question for society in general and for the biomedical research community in particular. In his essay, Dr. Rollin articulates a rational ethic for the treatment of animals which he extracts from our consensus moral theory and practice for humans and shows how this ethic can ramify in the practice of science. He also explores and criticizes unnoticed philosophical assumptions in science which have become part of what scientists take for granted, yet which are, in his view, significantly flawed.

TABLE OF CONTENTS

I. INTRODUCTION

Over the past decade, scientists across the world have felt the pressure of increasing restrictions on the acquisition and use of animals in research.[1] New laws imposing constraints on all aspects of animal research have been promulgated in a variety of countries, including the U.S., Britain, Holland, and Germany, and the inevitability of legislation acknowledged in numerous others, for example, Canada and Australia.[2,3] To many researchers affected by these new laws, their presence is odious, for they are often viewed as politically motivated, irrationally based impediments to the progress of knowledge or to the battle against disease, cynically promulgated by politicians who must cater to the vagaries of public opinion to survive. Older, established scientists tend to chafe most under such regulations, for throughout their careers they have become accustomed to carte blanche in their use of animals. Further, many scientists are inclined to see socially imposed constraints on animal use as dangerous encroachments upon hard-won academic freedom, encroachments which fundamentally threaten the very integrity and fabric of scientific inquiry.

Indeed, prior to the 1985 Amendments to the Animal Welfare Act and the contemporaneous passage of law which turned National Institutes of Health (NIH) guidelines and policy into federal law,[1,3] scientists were virtually autonomous in their use of animals in research. (In many ways they still are, even in the face of the new regulations.) Although the original Animal Welfare Act was passed in 1966, it had relatively little effect on scientific autonomy, having been promulgated in response to well-publicized accounts of dog-napping of pets by animal dealers, which pets ended up in research laboratories. Thus, it did not reflect reasoned moral concern for laboratory animals. A major purpose of the Act was to allay fears of the pet-owning public that their animals might suffer similar fates, and thus, for purposes of the interpreted Act, well over 90% of the animals used in research were (and still are) not animals. From its inception, rodents and farm animals used in research were excluded from any coverage. For purposes of the Act and the inspectors who enforce it, a dead dog is an animal, a live mouse or rat is not. Furthermore, the Animal Welfare Act, from its beginning and through all of its amendments, has always statutorily denied any concern with the design or the conduct of research.

By the same token, though NIH regulations for animal care and use have been promulgated since the early 1960s and have allegedly been binding on all institutions receiving federal funding for biomedical research, it is well known that prior to the 1985 law giving them status based in statute, these regulations were cavalierly ignored, as NIH had no mechanism for enforcing them nor any desire to do so. Even though it could theoretically seize all federal funding at an institution failing to honor its contractual agreements to abide by the regulations, this never occurred prior to 1982.

Thus, one can see that despite the existence of a paper trail of regulations and restrictions regarding laboratory animals, which could be (and was) pointed to if it was necessary to infrequently assure the public that protection for laboratory animals did in fact exist, researchers enjoyed a laissez faire situation. To most researchers, this was as it should be. The free use of animals was seen as essential to progress in biomedical and other research; research was seen as by its nature a free and untrammeled creative process; thus, meaningful restrictions on animal use were seen not only as unnecessary but as undesirable. Furthermore, researchers reasoned, freedom in the use of animals had produced a wide variety of beneficial results; new drugs, new operative procedures, new therapies and so on, all of which were of direct and patent advantage to the public, so surely there was no source of or grounds for complaint. Thus, many researchers could not comprehend what appeared to them to be a sudden, faddish, and dangerous social movement to constrain and curtail what had traditionally enjoyed salutary neglect.

What had in fact occurred was burgeoning social concern for the care and treatment of animals.[4] Although such concern had surfaced sporadically throughout the 20th century, it had tended to be sentimentally and emotionally based, directed at particular issues (for example, the use of pound animals in research), and heavily oriented towards select animals which were perceived as cute and cuddly. It was this same spirit, encapsulated in the humane societies across the country, which tended to schematize animal issues in a sentimental and emotional way in terms of kindness and cruelty and, thus, tended to obscure the deeper moral issues and institutional and social bases for animal abuse and neglect.

Beginning roughly in the 1970s, however, new elements entered into the growing social interest in animals. Unquestionably, traditional emotional elements loomed large in public concern. But for the first time, these factors were augmented and rationalized by reflective ethical thinking, representing a serious attempt to articulate our obligations to other sentient life forms. The seminal work of philosophers like Peter Singer, Tom Regan, Bernard Rollin, Steven Sapontzis and others gave rational voice to what had traditionally been confused emotions and intuitions. This new thinking reflected a logical and inevitable extension of social thought which dominated the 1960s — concern for disenfranchised and exploited groups such as minorities and women. At the same time, increasing public awareness of environmental concern lent further support to extending ethical concepts to animals, though there are well-characterized tensions between environmental ethical thought and moral thought which focuses on the welfare of individual animals. Public disenchantment with science and technology doubtless also fueled new suspicions of animal use in research.

At any rate, by the late 1970s, concern for animals became an articulate and forceful political voice to be reckoned with. Correlatively, accountability was demanded in all areas of social life, and research use of animals was not exempted. Clearly, significant tensions prevailed between the traditional autonomy scientists enjoyed in all areas of research, including animal use, and changing public mores with regard to animals. Unfortunately, these tensions were exacerbated by a significant but often overlooked feature of science in the 20th century, what I have elsewhere called the ideology or common sense of science.

II. ETHICS AND THE COMMON SENSE OF SCIENCE

Although 20th century science has tended, quite intentionally, to separate itself from philosophical concerns, it is patent that no area of human activity can avoid making philosophical commitments, for all disciplines must rest on concepts and assumptions taken for granted by practitioners of the discipline. Twentieth century science, too, has its philosophy, though that philosophy is typically invisible to its practitioners, who tend to see the assumptions of science not as debatable philosophical precepts, but as self-evident truths. Thus, the philosophical assumptions made by science include an aversion to philosophical examination of these assumptions, and, in part for that reason, they have tended to harden into an ideology virtually universally pervasive among scientists, which I have termed the common sense of science, for it is to scientific activity what ordinary common sense is to daily life.

One major component of scientific common sense directly relevant to the issue of animal use in biomedicine is the belief that science is value free, ought to make no valuational commitments, and, thus, *a fortiori,* has no truck with ethics. This notion, like many other components of scientific common sense, is rooted in the logical positivism of the early 20th century, which stressed the need for objectivity, empiricism, and verification in science. Since value claims in general and ethical claims in particular are not subject to empirical test and verification, they have no place in science. They are at best to scientific ideology emotional predilections and cannot be dealt with objectively. It is for this reason that otherwise cool and rational scientists are often every bit as emotional on such ethical issues as animal use as their opponents are — their training and ideology has led them to the view that ethical issues are in fact nothing but emotional issues, where rational thought has no place, and they thus believe that battles are won by manipulating emotions and tugging at heartstrings. The possibility of a rational ethic on anything is instinctively seen as an oxymoron or solecism.

It is not difficult to determine that science has indeed viewed itself as value free. Introductory textbooks, such as Keeton's[5] or Mader's[6] recent college biology texts, stress in their preliminary discussions of science and scientific methodology that science is value free and unable to make ethical pronouncements. Such a position is reflected in the teaching of science, where science educators typically make such a doctrine explicit or implicitly communicate it by their failure to discuss ethical issues occasioned by the material they are teaching. Leading scientists, in public pronouncements, promulgate the value-free view of scientific inquiry. An excellent example of this may be found in a recent PBS television documentary dealing with the Manhattan Project, the development of the atomic bomb during World War II. When queried as to their ethical stance on the development of the bomb, most of the scientists replied that they left such questions to the politicians, since ethics is not in the purview of scientists. Scientific journals rarely articulate ethical issues occasioned by their subject matter, and scientific conferences consider them only when the issues are galvanizing significant concern among society at large.

In the face of this ideological rejection of ethics in science, it is not surprising that ethical issues have been neglected. Thus, many abuses of human subjects in medical research occurred in the 20th century and were not seen as morally problematic, but as scientific necessities — atrocious research on prisoners, patients, indigents, third-world citizens, and other uninformed or disenfranchised people has regularly been documented throughout this century. Yet the U.S. scientific community strongly resisted formalized constraints on the use of human subjects, reluctantly accepting self-regulation in the mid-1960s only in the face of threats of congressional action.[7]

It is not surprising, therefore, that it has been widespread among scientists to assume that the use of animals in research is simply a scientific question to which ethics is irrelevant. An excellent paradigmatic example of this view may be found in a recent textbook of abnormal psychology authored by two distinguished researchers. In the book, there is a photograph of a laboratory rat, accompanied by the following caption: "For moral reasons, animals are used in psychological research."[8] This statement typifies the idea that, in working with animals, one has somehow circumvented the need for dealing with moral issues.

A moment's reflection reveals that the ideological view of science as value free in general, or ethics free in particular, must be wrong. Science rests on a multiplicity of value judgments — it is a valuational presupposition, of course, to prefer replicable data to nonreplicable, the verifiable to the unverifiable, experiments to anecdotes. Indeed, the scientific revolution of Newton, Galileo, Descartes, and others, which replaced Aristotelian medieval science with "modern science", was based in a value judgment, namely, that it was better or more desirable to explain the world in mathematical terms, by one set of laws at the expense of qualitative distinctions, than to persist in a science that did justice to the fact that the world contained many different kinds of things, and insisted that each thing be explained according to its own kind. Predictability was valued more highly than diversity, surely a judgment of value, not logic.

The concepts of health and disease, key concepts of biomedical science, rest on value notions. One need only consider the World Health Organization definition of health as a "complete state of physical, mental and social well-being" in order to realize this. The recent definitions of alcoholism and wife-beating as sickness rather than badness again shows the value basis of medicine, as does the American Medical Association's declaring obesity a *disease,* not a cause thereof, and the waxing, waning, and waxing of PMS as a disease.

Indeed, one need look no further than the issue at hand, the use of animals in research, to find a paradigmatic example of a moral judgment built into the very foundation of scientific activity. Insofar as we forebear from doing biomedical research on unwanted children or political prisoners, even though they are scientifically a far higher fidelity model for the rest of us than rats or other animals are, we have a moral judgment standing at the very basis of biomedical science.

For that matter, let us recall that many scientists argue that modern biomedicine is essentially connected to and dependent on the invasive use of animals in order to function and progress. Connected with this claim, then, is a series of assumptions made by researchers, namely that the advance of applied knowledge that benefits humans, or for that matter, of pure knowledge with no obvious use, licenses the invasive use of animals or that the knowledge or control gained through research is worth more than the pain or death of the animals used in that research process. Such judgments are widely shared, perhaps universally shared among researchers, but they are unquestionably moral judgments. It is therefore appropriate to examine the validity of the implicit moral judgments which scientists committed to in their justification of the traditional carte blanche use of animals in research.

III. CONSTRUCTING AN IDEAL FOR ANIMALS

This, indeed, is what philosophers writing on this issue over the past decade have demanded — that scientists confront the issue of the morality of animal use in a rational and dialectical way. Although, as we have said, the common sense of science denies that moral issues are rationally adjudicable, this too is patently false. If, for example, someone claims to be a Christian and a moral relativist, we are licensed to dismiss that moral claim as absurd because it is patently self-contradictory. If someone claims to be a libertarian and demands censorship of literature, we can again rationally dismiss that position. And if someone refuses to hire a qualified person only because of gender or skin color, that too is not only unjust, but rationally indefensible. We can demand coherence, consistency, universalizability, and uniform application of moral principles and strongly criticize those who fail in these respects. If we could not do this, we could make no moral progress nor even have a coherent society.

As Sapontzis shows in this volume,[9] a rational case against invasive use of animals can be made, one which cannot simply be dismissed because it seems paradoxical or inconvenient. Similarly, one can construct a rational case demonstrating that our theoretical duties to animals are far more significant than our practice has demonstrated. Our society in general, and research in particular, has suffered from the lack of any articulated moral ideal for the treatment of animals. As we all know, we do share a consensus ideal ethic for the treatment of humans in our society; that consensus enforms our laws, our policies, and our behavior. To be sure, we don't always live up to our ideals, both individually and collectively, but we can be called to account on such failures — and progress beyond them — precisely because we do have a moral ideal concerning the value of individual humans. Lacking such an ideal in the case of animals, we have tended to take the way things are as the way they ought to be.

One of the major benefits of the social stirrings of the past decade regarding animal welfare is to force the articulation of just such an ideal for the treatment of animals, one which can serve as a social consensus to guide and constrain our treatment of animals. The need for such an ideal is patent, for technological progress has provided us with the potential for major and dramatic manipulation of animal lives. The development of confinement agriculture[10] and the rise of genetic engineering[11] provide two clear and dramatic examples of the need for moral guidance regarding animal treatment. Whereas traditional agriculture required husbandry which more or less fit the animals' biological natures, else the animals would not remain healthy and the agricultural operation would not be profitable or productive, the use of technological agriculture allows us to fit square pegs into round holes and put animals into environments for which they are biologically unsuitable. Whereas crowding of chickens one hundred years ago would have resulted in decimation of flocks through disease, this can now be held in check by vaccines and antibiotics, though the animals are still living in fundamentally uncongenial environments. By the same token, we can now genetically engineer animals to our benefit, such as the Harvard mouse which is highly susceptible to mammary tumors, yet we may do so at the expense of the animals' quality of life.

Such developments, coupled with increased use of animals in scientific research, have made it clear that whatever traditional consensus ethic for animals did exist was quite inadequate for providing guidance on treating animals in society today. In fact, the only thing resembling a consensus ethic which hitherto obtained in society is the prohibition against cruelty encapsulated in anticruelty legislation. Such laws basically exist to ferret out malicious, willful, "unnecessary" abuse — torturing animals for fun — and do not at all address the far more profound abuses growing out of accepted practices like trapping, confinement agriculture, or research. In fact, such laws embody and perpetuate the humane society stereotype that all harm done to animals is derived from cruelty, something patently not the case.

Hence, as we mentioned earlier, the recent attempt of philosophers to articulate the ideal which the *zeitgeist* demands. Such a task is understandably difficult, for one is unclear as to how to proceed. One can certainly create a moral ideal for animals *de novo,* but why should such a creation be compelling to others? In my own work, therefore, I attempted to extract the ideal for animals from moral principles which are socially accepted for the treatment of people, *mutatis mutandis*. In other words, granted that we do have a consensus moral theory for the treatment of people, what implications can be drawn from this for our treatment of animals? Thus, I do not attempt to impose my ideal on others; I rather attempt to extract it from principles to which those I am dealing with already acquiesce.[12]

Thus, although the common sense of science eschews any truck with morality, we have seen that science in fact does make moral assumptions. Second, scientists are also citizens and members of society, and in that capacity at least, they do inevitably share in our consensus ethic about humans. So it is not unreasonable to use the method outlined above to extract a moral ideal for animals by logical extension of our extant social ethic.

It is important to briefly summarize the socially accepted ethical ideal for humans which pervades our thinking and practice before we attempt to apply it to animals. In democratic societies, we accept the notion that individual humans are the basic objects of moral concern, not the state, the *Reich,* the *Volk,* the Church, or some other abstract entity. We attempt to cash out this insight in part by generally making our social decisions in terms of what would benefit the majority, the preponderance of individuals, i.e., in utilitarian terms of greatest benefit to the greatest number. In such calculations, each individual is counted as one, and thus no one's interests are ignored. But such decision-making presents the risk of riding roughshod over the minority in any given case. So democratic societies have developed the notion of individual rights, protective fences built around the individual which guard him or her in certain ways from encroachment by the interests of the majority.

These rights are based upon plausible hypotheses about *human nature,* i.e., about the interests or needs of human beings which are central to people, and whose infringement or thwarting *matters most* to people (or, we feel, *ought* to matter). So, for example, we protect freedom of speech, even when virtually no one wishes to hear the speaker's ideas, say in the case of a Nazi, regardless of cost. Similarly, we protect the right of assembly, choosing one's own companions, one's own beliefs, and also the individual's right not to be tortured even if it is in the general interest to torture, as in the case of a criminal who has stolen and hidden vast amounts of public money. And all of these rights are not simply abstract moral notions, but are built into the legal system. Thus, the notion of human nature is pivotal to our ethic — we feel obliged to protect the set of needs and desires which we hypothesize as being at the core of what it means to be human.

What we have so far outlined is not difficult to extract from most people in our society. In fact, calling attention to the moral principles people unconsciously accept has traditionally been a major way of effecting social change. Arguably, something of this sort occurred when (thinking) segregationists accepted integration, or when occupations such as veterinary medicine, which had traditionally barred women, began to admit them. In both of these cases, presumably no change in moral principles is required. What is demanded is a realization that a moral commitment to equality of opportunity, justice, fairness, and so on, which the segregationist or person who barred women from veterinary school himself had as a fundamental commitment, entails a change in practice. In other words, to put it simply, such people had readily accepted democratic moral principles as applying to all persons. What they had ignored was the fact that the class of persons was far greater than what they acknowledged and included blacks and women. It is this same sharpening of the applicability of accepted moral principles, rather than the adoption of new ones, which led to the steady augmentation of the class of full rights-bearers in the U.S. Constitution beyond the original limited group of white, adult, native-born, male, property owners.

As we all know, the history of western civilization in general, and Western democracies in particular, has been one of ever-increasing extension of moral concern to previously disenfranchised individuals. Whereas at various points in history individuals were excluded from moral and legal protection, or from having their interests compete in the moral arena, on the basis of characteristics like ancestry, gender, color, religion, age, race, citizenry, nationality, such traits have gradually come to be seen as lacking *moral relevance*. In other words, while there are, of course, differences between whites and blacks, men and women, citizens and foreigners, the key question has been — do these differences justify a difference in how they ought morally be treated, and the answer has been negative.

The next crucial step is to show how the above ethic can inexorably be applied to animals as well. If one can show that there are no rationally defensible grounds for differentiating animals from humans as candidates for moral concern, we must logically bring to bear upon questions of animal treatment the entire moral machinery we use to deal with human questions, and it turns out that none of the standard reasons offered up in the history of thought to exclude animals from the moral arena will stand up to rational scrutiny and meet the test of moral relevance.

For example, it has often been suggested that animals are not fully worthy of moral concern because they don't have immortal souls. (Incidentally, one is surprised at how often scientists say something like this; witness the letters in U.S. veterinary journals.) Aside from our obvious inability to state with any certainty who does or doesn't have a soul or even what it is, this view is open to a much more striking response, first enunciated by Cardinal Bellarmine. "True", said Bellarmine, "as a Catholic I must accept Church doctrine that animals have no souls, but this has nothing to do with their being excluded from moral concern. In fact," he argued, "they ought to be treated better than people, since this is their only chance at existence, whereas wrongs to humans will be redressed in the afterlife!" So the really interesting point in this example is not theological, but rather the dramatic way it enjoins us to be clear about the moral relevance of the differences we point to.

Other alleged differences between people and animals fare no better. Some have said we can do as we wish to animals because we are superior; but what does "superior" mean? Some have said that it means that we are "at the top of the evolutionary ladder," but, as we all know, there is no evolutionary ladder, only a branching tree. And if it does make sense to talk about species superiority, it is only in terms of differential reproduction, species longevity, and adaptability, in which case we share top billing with many other species, like the rat, and both lose hands-down to the cockroach. If "superior" means that we are more powerful than other creatures and can in fact do as we wish with them, this is surely true, but has no moral relevance. To say that it does is to affirm that might makes right and to destroy morality altogether, to confuse *de facto* authority with *de jure* authority. If one accepts this position, one is forced to say that the government has the right to kill people as it sees fit, since it is after all more powerful than any of us, or that the mugger or rapist is perfectly morally justified in exploiting his victim.

Still other alleged differences turn out to be equally irrelevant from a moral point of view. "Man is rational, animals are not," is a favorite justification for the exclusion of animals from moral concern. But what is the moral relevance of rationality? Doubtless one needs to be a rational being to be a moral agent or actor, to be held morally responsible for what one does; but one surely doesn't need to be rational to be an object of moral attention and concern — consider children, infants, the insane, the senile, the comatose, the retarded, etc. Furthermore, if rationality is the key feature of what makes something worthy of moral attention, why is so much of our moral concern devoted to aspects of human life which have nothing to do with rationality? Suppose I discovered that I could make my students more rational by wiring their seats and shocking them when their attention wandered. If rationality were the key feature relevant to moral concern, such behavior would not only be permissible, but obligatory. Yet we would rightly condemn such behavior as monstrous, showing that rationality is not the only thing involved in being an object of moral concern. There are many other features involved, as indeed this hypothetical case shows. It is wrong to shock the students because it causes pain and infringes on their freedom. Various other differences, such as the claim that ethics is based in social contracts and animals can't enter into contracts, turn out to be either false or to lack the requisite degree of moral relevance which would justify not considering them morally.

Equally important, a moment's reflection makes it patent that not only are there no morally relevant differences for excluding animals from moral concern as we in society define it, there are in fact significant morally relevant similarities which animals share with humans. The same sorts of features which we find in people which give rise to our talking about right and wrong actions with regard to people are also to be found in animals. The features I am talking about which are common to people and to at least "higher" animals (and possibly many "lower" ones as well) are *interests* — needs, desires, predilections, the fulfillment and thwarting of what *matters* to the person or animal in question. Cars have needs — for gas, oil, and so on — but they do not have interests, since we have absolutely no reason to believe that it matters to the car itself whether or not it gets its oil. That is why it is impossible to behave immorally towards cars in themselves — they are merely tools for human benefit. However, animals with interests cannot be looked at as mere tools, for they have lives which matter *to them.*

There are, of course, categories of interests and interests which are common to all animals (including humans) — food, reproduction, avoidance of pain. But even more significant are the unique variations on these general interests, and the particular interests, which arise in different species. Even as we talk of human nature, as defined by the particular set of interests constitutive of and fundamental to the human animal, we can also talk of animal natures as well — the "pigness" of the pig, the "dogness" of the dog. Following Aristotle, I like to talk of the *telos* of different species of animals as being the distinctive set of needs and interests, physical and behavioral, genetically determined and environmentally expressed, which determine the sort of life it is suited to live. This is not a mystical notion — it follows directly from modern biology and genetics and is certainly obvious to any one who is around animals and, indeed, to common sense; hence, the song which tells us that "fish gotta swim and birds gotta fly."

Recall that we have argued that our consensus ethic for humans protects certain aspects of human nature deemed to be essential to the human *telos* and shields them from infringement by the majority and by the general welfare. If it is the case that one can find no morally relevant grounds for excluding animals from the application of that ethic and if animals too have a *telos,* it follows inexorably that animals too should have their fundamental interests encoded in and protected by rights which enjoy both a legal and moral status. In this way, we indeed illustrate that the motion of animal rights is implicit (albeit unrecognized) in our consensus social ethics.

IV. THE COMMON SENSE OF SCIENCE AND ANIMAL CONSCIOUSNESS

One possible way of forestalling the foregoing surprising conclusion has unfortunately been central to the common sense of science discussed above. As I have discussed at length elsewhere,[13–16] one component of scientific ideology has been the claim that one cannot assert, legitimately, that animals are conscious in the sense of enjoying subjective experiences, feeling pain, fear, anxiety, loneliness, boredom, joy, happiness, pleasure, and the other noxious and positive mental states which figure so significantly in our moral concern for humans. This skepticism about attributing thought to animals enjoys a long history and was most famously promulgated by Descartes, who declared that animals were simply machines, driven by clockwork. Such a position, of course, provided justification for experiments in the burgeoning science of physiology in Descartes' time which required dissection of living animals without anesthesia. While Descartes' position was hotly contested by many philosophers and scientists, most notably by Darwin, who stated that if physiological and morphological traits were phylogenetically continuous, so too were mental and psychological ones, agnosticism about animal minds resurfaced in the early 20th century, receiving succor from non-Cartesian sources. We mentioned earlier that the positivism which shaped scientific ideology denied the validity of talking about ethics, since moral claims were not verifiable. The same positivistic tendency nurtured the development of psychological behaviorism, which denied the studiability of mind and consciousness, and affirmed that only overt behavior was open to scientific inquiry. This methodological aversion to treating mental states as real was enormously influential, shaping the thinking of psychologists, zoologists, biomedical scientists, and even the European ethologists who otherwise rejected behaviorism.[15,16]

It is clear that the denial of mentation to animals did have untoward moral consequences in science. Scientific books and papers routinely stopped short of attributing felt pain, fear, etc. to animals, and any such extrapolations beyond overt behavior were seen as pernicious "anthropomorphism", this despite the fact that much animal research, for example, pain research, presupposed that animals could feel pain. Though all analgesics in the U.S. were routinely tested on laboratory animals, these animals virtually never received analgesics in the course of research, and one searched in vain for a literature on laboratory animal analgesia. Incredibly, the first conference on animal pain ever held in the U.S. was only convened in 1983 and, even then, dealt almost exclusively with the machinery or "plumbing" of pain, ignoring the subjective and morally relevant aspects.[17] The scientific literature never discussed suffering in animals, and in its zeal to avoid "unverifiable" talk about mental states like fear, anxiety, loneliness, boredom, the research community talked blanketly in terms of mechanical, physiological "stress responses", which tended to be simplistically defined in terms of Cannon's alarm reaction for short-term stress, or Selye's activation of the pituitary-adrenal axis for long-term stress.

Thus, it became imperative for any moralist, like myself, attempting to effect changes in the scientific gestalt on animals, to refute the common sense of science's agnosticism about animal mentation and attempt to press forward a "reappropriation of common sense" — ordinary common sense accepts without question that animals possess thoughts and feelings.[13–16]

It is my contention that science cannot avoid talking about mental states in animals, that failure to do so makes for inadequate explanation, and that such talk is no more unscientific than a whole host of other accepted scientific notions. We cannot verify any of the following directly experientially: the existence of an external world independent of our perceptions; the reality of the past; or the existence of quanta; yet all of these are presupposed in science because they provide us with vehicles of explanation which generate testable consequences. The same point holds of mental states in animals — they help us to explain how the animals behave. Recent research has shown, for example, that talking about stress in animals cannot be done in purely mechanical terms and cannot circumvent reference to their states of awareness, for the same "stressor" can have very different physiological effects in an animal, depending on the animal's emotional state, or on how it has been treated prior to the noxious stimulus, or on whether it can anticipate or control the stimulus, etc.[18] Furthermore, purely psychological stressors, like putting an animal into an unfamiliar environment, can have greater physiological effects than such "physical" stressors as heat.[19] By the same token, pain as an experienced state in animals must be postulated for a variety of compelling reasons — for one thing, to make pain research on animals a coherent project! The mechanisms of pain activation and pain mitigation are virtually the same in all vertebrates (e.g., endorphins and enkephalins, serotonin, substance P, stimulation-produced analgesia),[20] and if pain were purely mechanical and unfelt in animals, why would the experiential dimension suddenly emerge evolutionarily in humans? (We know, in fact, that humans who have congenital or acquired inability to *feel* pain do not do well at all.)[16]

Incidentally, it is clear that seeing animals as Cartesian machines was not only morally pernicious, but scientifically pernicious as well. As I have discussed elsewhere, much scientific research has been put *en prise* by a failure to see how various forms of pain, suffering, disease, and unhappiness can skew relevant physiological and metabolic variables.[12] Riley showed that, not surprisingly, animals living under unpleasant conditions experienced higher incidence of tumors;[21] the implications of this for cancer research is obvious, especially in light of numerous scandals which have come to light regarding the husbandry of laboratory animals used in such research.[12] And far too few researchers do appreciate fully just how vulnerable animals are to all sorts of variables which do arise.[22]

Interestingly enough, cavalier disregard of animals as anything other than (in most cases) cheap tools, leads, as we mentioned, to compromised or even meaningless research results. Researchers who would never dream of dragging a microtome to work behind a pick-up truck will treat animals in ways which bespeak ignorance of or disregard for the extraordinary sensitivity of animals to noxious experiences. Unimaginable amounts of data are jeopardized by a failure to acknowledge, let alone control for, widely overlooked variables. Few researchers worry about such things as noise levels, technician personality and handling of the animal, restraint, housing congenial to the animals' natures, opportunities for play and social interaction for the animals, yet all of these have demonstrable physiological and metabolic effects. To cite some eloquent examples: Nerein et al. showed in 1980 that given two groups of rabbits fed a 2% cholesterol diet, the group treated with tender loving care developed 60% fewer atherosclerotic lesions.[23] Barnett et al. studied the effects on pigs of being gently stroked vs. slapped or shocked. The pigs which were handled unpleasantly showed significantly lower growth performance, feed conversion, and reproductive success. Boars unpleasantly handled developed smaller testicles at 160 d of age and showed later attainment of behavioral puberty.[24] Seabrook has shown what successful dairymen always knew, namely, that how herdsmen treat cows is a major factor in milk production.[25] Ingram has shown that variation in environmental conditions during the neonatal period can change brain chemistry in piglets.[26] Riley has shown, as we mentioned, that higher stress environments lead to greater incidence of certain cancers.[21] Isaac has discussed the incredible range of experiences which can serve as stressors for laboratory rats. In fact, he tells us, any stimulus can serve as a stressor.[27]

Few scientists control for anything like this range of variables — in the case of research on swine, an up and coming research animal, there aren't even standards for housing, care, and husbandry — one can easily imagine the incommensurability of data across laboratories. When I spoke before the Shock Society, an international group of circulatory shock researchers, they were bitterly divided on the legitimacy of using anesthetized animals in traumatic or burn or other shock research. Many wished to use anesthesia for ethical reasons, yet others declared that anesthesia would skew results. Interestingly enough, not a single researcher had heard of Gärtner's monumental results, which showed that having a familiar technician simply move rats in a cage 3 ft was enough, 100 s later, to produce major changes in 25 plasma variables, all of which changes demonstrated a microcirculatory shock profile, and which persisted for a considerable period of time.[28] I pointed out that until they controlled for such morally neutral stress variables, they could hardly piously withhold anesthesia on scientific grounds.

What is equally mischievous, both morally and scientifically in the light of our preceding discussion, is the widespread lack of knowledge of research animals' needs and natures, physical and psychological. Thus, one can get an M.D.-Ph.D. degree in an animal-using area of biomedicine and never learn anything about the animals one uses other than that they model some particular disease or syndrome. In the face of this remarkable fact, it is obvious that the vast majority of animal abuse in research flows from researcher ignorance, coupled with the common sense of science's denial of concern with ethics and with subjective states in animals.

V. THE IDEAL AND ITS APPLICATION

Thus, if the denial of consciousness and feeling to animals is as implausible as we have argued, there is indeed no way to forestall the earlier conclusion we drew from our social consensus ethic regarding humans, namely, that animals theoretically ought to enjoy a far more significant moral status than they have traditionally been accorded and that they should have rights legally codified which protect fundamental interests and aspects of their natures from encroachment.

Though, as we said earlier, scientists tend not to be highly trained in philosophy, many have indeed responded positively to the rational articulation of a moral ideal for animals, though many others have responded with the same sort of vituperative emotion they officially deplore in their opponents.[29] In my extensive interactions with researchers all over the world, I have found very few who, upon reflection, would fail to admit that animals ideally are not simply expendable tools. Most realize that as soon as one has admitted that animals can be hurt in ways which matter to them, or admitted even that animals are entitled to humane care and treatment, or that unnecessary animal suffering is wrong, one has implicitly but inescapably presupposed that animals are in the moral arena, that one can be morally wrong in how one uses or treats animals, none of which one would say of inanimate objects, such as chairs and wheelbarrows. Once weaned away from the distancing of moral issues which scientific common sense provides, virtually all researchers with whom I have discussed these issues will eloquently recount examples of animal "abuses" (i.e., moral wrongs) systematically perpetuated in some areas of research! Admitting the existence of abuses, of course, entails recognizing some moral status for animals. And in a startling response to an article by Rowan and myself,[4] Dr. Theodore Cooper, a board member of the Association for Biomedical Research, chartered to protect research from intrusion, asserted that "intelligent men and women no longer dispute that animals used in research have rights,"[30] and that "every scientist who properly carries the designation recognizes that animals have rights."[30]

At the same time, however, most researchers continue to believe that one can justify hurting animals for human (and animal) benefit in research. A statement by Dr. Cooper in the above-mentioned article typifies the position of many researchers:

> The main purpose of...research...is to improve or protect the health of people...Scientists make a distinction between the value of human life and the value of animal life...Most modern societies (not all) place a higher value on human life than on nonhuman animal life. Therefore, in the absence of the perfect information or the currently needed information, the scientist is willing to experiment on the animals to make progress.[30]

Most defenses of invasive animal use make a similar argument; but is this argument cogent? Let us assume without question both that a great deal of human benefit has flowed from animal research and that animal life is valued less than human life. That in itself, of course, would not *ipso facto* morally justify the invasive use of animals. The last 100 years have witnessed a great deal of documented exploitation of humans for research — humans who were or are valued less than first-line humans — blacks, primitives, criminal and political prisoners, women, retarded and insane persons, indigents, derelicts, and the elderly. Even if great benefit flowed from these activities, that would not make them right. (If we felt it did, we presumably would do all of our research on humans of this sort for scientific reasons.) What this argument comes down to, I think, is the assertion that given perceived human need and self-interest, and given the benefits which flow from animal research, and given animal impotence, invasive research will continue to be done. (Most scientists I know do not feel comfortable in asserting that they have a *right* to hurt animals for our benefit, but they nonetheless believe it will and should be done for the benefit it brings.)

This point was articulated well to me by members of the Australian biomedical research community at a recent conference on ethics and animal experimentation. Philosophers and others who have articulated the ideal for animals, they explained, have indeed developed cogent and hitherto ignored points; but, they continued, do you really expect us to immediately give up our careers, our research activities, and the benefits to humans which flow from them? Such an expectation, they argued, is naive. What they wanted from philosophers, and others concerned with animal rights, is some program for improving the lot of animals, of approximating to the ideal, short of abandoning the entire enterprise. This response is, it appears, typical of how researchers do react as they begin to understand the moral dimensions we have outlined. (*Some* researchers do, of course, move out of using animal research when they understand the moral issues at stake, but these are not a majority.)

So the task for those of us committed to bettering the lot of animals in research becomes, in the short run, one of making viable suggestions for improving the situation of research animals — for coming closer to the ideal in a world where some animal use in science is currently seen as inevitable. Many moralists are understandably uncomfortable with halfway measures, but the majority recognize that social change is invariably incremental and that, as I have put it elsewhere, a dog being used for experimental surgical purposes is better off with a moralist pressing for morphine than for immediate abolition. In short, a morally concerned individual working in this area should always ask oneself one key question: "Are the animals any better off in virtue of my efforts?"

Ways in which the situation of animals used in research could come closer to the ideal are easy to formulate, given that we are only just emerging from a situation which tended to see animals as cheap and expendable tools, with no moral status. If one takes seriously the argument we have developed, and if one continues to justify the use of animals by appeal to human benefit, *it at least follows that one ought to maximize the interests of the animals one uses consonant with that use,* something which has hitherto never been the case. This general dictum can be cashed out in terms of two practical principles for research:

1. **The Utilitarian Principle** — If the only justification that researchers themselves can bring forth in defense of invasive animal use invokes the tangible benefits which flow from such research, it follows logically that *the only invasive research which ought to be pursued is research where the benefit to humans and/or animals likely to emerge from the research outweighs the cost in suffering to the animals.*

   Admittedly such calculation is fraught with difficulties — how does one weigh one parameter against a disparate one? But the crucial point to remember is that we do currently make such cost/benefit decisions in a variety of areas, including research on humans. All that needs be done is that such calculations be exported to the area of animal use. Certainly there will be hard cases, but at least extreme cases will be clear. Invasive research aimed at developing a new weapon, a new nail polish, or at discovering knowledge of no clear benefit to humans and/or animals, for example, territorial aggression studies, would clearly not be permitted.

   One standard researcher response to this principle is to invoke the serendipity argument. It is argued that though it may not appear that a particular piece of research will produce foreseeable benefit, one never knows what will arise adventitiously. The response to that is simple: by definition, one cannot plan for serendipity. Society does not fund a great deal of research for a wide variety of reasons. Much research is turned down by the granting agencies because it is perceived as poorly designed, less important than other things, and so on. If the serendipity argument were valid, one could not make such discriminations, and one would be logically compelled to fund everything.

In sum, the Utilitarian Principle decrees that only patently beneficial invasive research ought to be done.

2. **The Rights Principle** — If a piece of research is permitted by the Utilitarian Principle, it should still be tested by the Rights Principle, which asserts that *in the context of research, all research should be conducted in such a way as to maximize the animal's potential for living its life according to its nature or telos, and certain fundamental rights should be preserved regardless of considerations of cost.* In other words, if we are embarking on a piece of research which meets the Utilitarian Principle, we by no means have carte blanche, we must attend to the animal's rights following from its nature — the right to be free from pain, to be housed and fed in accordance with its nature, to exercise, to companionship if it is a social animal, etc. The animal used in research should thus be treated, in Kant's terminology, as an end in itself, not merely as a means or tool, since it is not a test tube, but a being with its own moral status as defined by the ideal.[12]

How close are we in our research use of animals to living by these principles? Unfortunately, not very close. At present, animal research is not typically judged by the Utilitarian Principle by funding agencies; it is typically screened by reference to scientific merit or human benefit alone, with cost to animals in themselves not functioning as a significant vector in decision making. One small exception to this can be found in relatively infrequent decisions by federally mandated animal care committees in the course of protocol review to forbid certain research protocols on the grounds of cost to animals. Typically this occurs where a project is funded by small grants or by the researchers themselves, and no peer review system judges scientific merit. Such decisions are not explicitly chartered by federal law or by the *NIH Guide,* and, thus, committees which make them are in something of a grey area.

The Rights Principle, while still not actualized to any great extent, at least has a firmer basis in law. The recent federal legislation passed in 1985 to protect laboratory animals is based on the notion of ensuring certain basic animal interests, most notably, control of pain and suffering not essential to the research process. This was an explicit intention on the part of those of us who were involved in drafting the concept of that legislation.[12] The 1985 amendment to the Animal Welfare Act also represents a beginning in legislating respect for fundamental aspects of the natures of at least dogs and non-human primates when it mandates exercise for dogs, and environments for non-human primates which enhance their psychological well-being. Thus, the law, while only a beginning, at the same time is conceptually revolutionary insofar as it mandates, as against the common sense of science, that animals do feel pain and do suffer; that they have interests beyond food and water; and that at least primates enjoy mental lives.

As far as they go, these laws have been very salubrious; they have, as we intended in drafting their concept, made researchers begin to think about animals in moral terms, forced recognition of pain and suffering, and assured protection of some very basic rights, such as control of pain, thereby launching a major attack on the ideology of science's skepticism about both ethics and animal consciousness.[16] This, in turn, has ramified in far greater scientific concern with identifying and controlling pain and suffering in animals and has led to a proliferation of articles and seminars on pain and its control where until recently one found nothing.[16] Insofar as these laws demand training of all personnel using animals, we begin to get a remedy for one major source of animal abuse mentioned earlier — the fact that researchers could freely use animals in all areas of science without knowing anything whatsoever about the animals they use, except that they model a particular disease, syndrome, or process. Hence, the legendary stories rife among laboratory animal veterinarians of researchers who think that dogs with body temperatures higher than 98.6°F are sick, who wonder at what age white mice become rats, who think that mice are induced ovulators, etc.

By the same token, insofar as at least one of the new laws requires training in ethics, and mandated protocol review inevitably forces upon researchers rumination on ethical questions, the new laws again launch a frontal assault on the common sense of science's tendency to see animal research and animal use as merely a technical, scientific question, and not a moral one. One hopes that increasing numbers of researchers are slowly beginning to progress beyond the attitude we characterized at the beginning of our discussion, which sees the new laws and policies simply as irrational impediments to science and progress. The reader who has followed our discussion can begin to understand the moral basis for legislation.

One lingering doubt which might conceivably remain even in the face of our discussion is the question of whether the same results of increasing researcher sensitivity could not have been accomplished otherwise than by burdensome legislation. Science, it is asserted, should not be tainted by legislative regulation. The simple response to this is that scientists are as human as everyone else. In society, we simply do not trust others to do the right thing where objects of moral concern can be harmed by a failure to do the right thing. We ensure that people will do the right thing by formalizing the requirements. For example, we have laws against driving on the wrong side of the road even though, presumably, only a lunatic would do such a thing, since it is in one's serious interest not to risk death. Scientists are indeed human and cut corners like everyone else under pressures of career advancement, budgets, time pressure, and so on, as increasing numbers of reports of data falsification eloquently document. The fact that the U.S. government felt compelled to pass the Good Laboratory Practices Act of 1978, an act which mandates practices like record keeping and disease control, which are essential to science anyway, yet which were found to be flagrantly ignored, eloquently demonstrates that researchers, like everyone else, need to be nudged to do what their calling naturally demands. It is well known that animal care and treatment did historically suffer from being unregulated, as the case of laboratory animal analgesia alluded to earlier demonstrates.

VI. SUMMARY AND CONCLUSION

One can see in the face of the argument presented that new legislated restrictions on scientific use of animals do not represent the irrational Luddism and antiscientific spirit that many scientists have claimed is their source, nor do they simply reflect politicians' catering to irrational, emotionally based social hysteria. At the core of new social concern about animals and the constraints on animal use which follow in its wake is a reasoned moral position clearly demonstrating that animals deserve a moral status considerably higher than they have traditionally been accorded. We have tried to summarize this moral position in this paper.

Researchers should therefore not expect that concern for the welfare of laboratory animals is a transient, ephemeral fad. It is likely to grow, not diminish, as all social movements recognizing an unnoticed set of moral questions have done. Members of the research community therefore ought not expect moral questions to go away, nor should they perpetuate the ideology which asserts that ethical questions are of no concern to science. Instead, the biomedical community should actively pursue and anticipate and define the moral questions associated with animal research and attempt to develop their own answers to them, rather than wait for society as a whole to do so.

In my view, two sets of moral concerns related to animal research are of greatest significance at the moment. First, some mechanism needs to be developed which will exclude invasive research which produces no benefit, but simply advances knowledge or careers. Some types of psychological research, for example, are very vulnerable to this criticism. The current mechanism of peer review, whereby experts in the field judge the value and fundability of research plainly does not address these concerns. Researchers who throughout their whole careers have taken a particular sort of invasive animal use for granted in their field are not the best source for eliminating such a use from the field. A better alternative, perhaps, would be to allow local committees with greater representation from the citizenry at large to pass on the value of a piece of animal research. Society pays for animal research; researchers ought to be able to successfully defend their need to spend public money to hurt animals to a set of citizens. Such an approach works for our justice system; perhaps researchers need to convince something comparable to a jury of their need to hurt animals for the sake of research.

The second major set of concerns which the animal research community ought to deal with in an anticipatory manner is the way in which animals are maintained, a concern already implicit in the new amendment to the Animal Welfare Act. Laboratory animals have traditionally been kept in husbandry conditions which are convenient to researchers in terms of ease of cleaning, maintenance, etc. The standard stainless steel or polycarbonate rodent cage was developed for human convenience and efficiency, not out of regard for the animals' natures. Sophisticated laboratory animal scientists are increasingly coming to realize that life under conditions which do not accommodate the animal's *telos* is a major source of suffering for animals — some have suggested that being forced to live under conditions for which they are not suited is a far greater source of suffering for laboratory animals than are the experimental manipulations performed on them which have elicited the bulk of social concern thus far.

In my own thinking, I have found it useful to employ the theoretical notion of "happiness" as a basis for designing husbandry systems for animals. An ideal for design of such systems ought to be the satisfaction of the complex set of mental and physical interests which make up the animals nature or *telos*.[12,16,22] It is reasonable to call such a state of satisfaction *happiness* for animals, just as Aristotle long ago pointed out that the meaning of human happiness was the satisfaction of the human rational *telos*.[16,22] Fortunately, there is a growing body of ethological literature which points the way towards scientifically based design of such systems.[31-33]

The implementation of such an approach to animal research represents a viable place wherein science can begin to forthrightly deal with the moral issues it carries in its wake.

REFERENCES

1. **Newcomer, C.,** Laws, regulations and policies pertaining to the welfare of animals, in *The Experimental Animal in Biomedical Research,* Vol. 1, Rollin, B. E., Ed., CRC Press, Boca Raton, FL, 1990, chap. 3.
2. **Hampson, J.,** Laws relating to animal experimentation, in *Laboratory Animals: An Introduction for New Experimenters,* Tuffery, A. A., Ed., John Wiley & Sons, London, 1987, chap. 2.
3. **Rollin, B. E.,** Laws relevant to animal experimentation in the United States, in *Laboratory Animals: An Introduction for New Experimenters,* Tuffery, A. A., Ed., John Wiley & Sons, London, 1987, chap. 16.
4. **Rowan, A. and Rollin, B. E.,** Animal research for and against: a philosophical, social, and historical perspective, *Perspect. Biol. Med.,* 27, 1, 1983.
5. **Keeton, W. T. and Gould, J. H.,** *Biological Science,* Norton, New York, 1986, chap. 1.
6. **Mader, S. S.,** *Biology,* Wm. C. Brown, Dubuque, Iowa, 1987, chap. 1.
7. **Katz, J.,** The regulation of human experimentation in the United States: a personal odyssey, *IRB,* 9, 1, 1987.
8. **Rosenhan, D. and Seligman, M.,** *Abnormal Psychology,* Norton, New York, 1984.
9. **Sapontzis, S.,** The case against invasive research with aniamls, in *The Experimental Animal in Biomedical Research,* Vol. 1, Rollin, B. E., Ed., CRC Press, Boca Raton, FL, 1990, chap. 1.
10. **Rollin, B. E.,** Social ethics, animal rights, and agriculture, forthcoming, in *Ethics and Agriculture,* Blatz, C., Ed., University of Idaho Press, Moscow, 1990.
11. **Rollin, B. E.,** The Frankenstein thing, in *Genetic Engineering of Animals,* Evans, J. W. and Hollaender, A., Eds., Plenum Press, New York, 1986.
12. **Rollin, B. E.,** *Animal Rights and Human Morality,* Prometheus Books, Buffalo, 1981.
13. **Rollin, B. E.,** Animal pain, in *Advances in Animal Welfare Science 1985–1986,* Fox, M. W. and Mickley, L., Eds., Martinus Nijhoff, The Hague, 1986.

14. **Rollin, B. E.,** Animal consciousness and scientific change, *New Ideas Psychol.,* 4, 141, 1986.
15. **Rollin, B. E.,** Animal pain, scientific ideology, and the reappropriation of common sense, *JAVMA,* 191, 1222, 1987.
16. **Rollin, B. E.,** *The Unheeded Cry: Animal Consciousness, Animal Pain, and Science,* Oxford University Press, Oxford, 1989.
17. **Kitchell, R. L. and Erickson, H. H., Eds.,** *Animal Pain: Perception and Alleviation,* American Physiological Society, Bethesda, MD, 1983.
18. **Mason, J. W.,** A re-evaluation of the concept of "non-specificity" in stress theory, *J. Psychiatr. Res.,* 8, 323, 1971.
19. **Kilgour, R.,** The application of animal behavior and the humane care of farm animals, *J. Anim. Sci.,* 46, 1478, 1978.
20. **Kitchell, R. L.,** The nature of pain in animals, in *The Experimental Animal in Biomedical Research,* Vol. 1, Rollin, B. E., Ed., CRC Press, Boca Raton, FL, 1990, chap. 12.
21. **Riley, V.,** Mouse mammary tumors: alteration of incidence as apparent function of stress, *Science,* 189, 465, 1975.
22. **Rollin, B. E.,** Moral, social, and scientific perspectives on the use of swine in biomedical research, in *Swine in Biomedical Research,* Vol. 1, Tumbleson, M., Ed., Alan R. Liss, New York, 1986.
23. **Nerein, R. M. et al.,** Social environment as a factor in diet-induced atherosclersis, *Science,* 208, 1475, 1980.
24. **Barnett, J. L. et al.,** The welfare of confined sows: psychological, behavioral, and production responses to contrasting housing systems and handler attitudes, *Ann. Rech. Vet.,* 15, 217, 1984.
25. **Seabrook, M. F.,** The psychological relationship between dairy cows and dairy cowmen and its implications for animal welfare, *Int. J. Study Anim. Prob.,* 1, 295, 1980.
26. **Ingram, W. H.,** Physiology and behavior of young pigs in relation to the environment, in *The Welfare of Pigs,* Sybesma, W., Ed., Martinus Nijhoff, The Hague, 1981.
27. **Isaac, W.,** Causes and consequences of stress in the rat and mouse, American Association for Laboratory Animal Science Seminar on Stress, Atlanta, 1979.
28. **Gärtner, D. et al.,** Stress response of rats to handling and experimental procedures, *Lab. Anim.,* 14, 267, 1980.
29. **Rollin, B. E.,** Seven pillars of folly: barriers to reason in animal welfare, *ATLA,* 12, 243, 1985.
30. **Cooper, T. and Stucki, J.,** Commentary, *Perspect. Biol. Med.,* 27, 20, 1983.
31. **Markowitz, H. and Line, S.,** The need for responsive environments, in *The Experimental Animal in Biomedical Research,* Vol. 1, Rollin, B. E., Ed., CRC Press, Boca Raton, FL, 1990, chap. 10.
32. **Wemelsfelder, F.,** Boredom and laboratory animal welfare, in *The Experimental Animal in Biomedical Research,* Vol. 1, Rollin, B. E., Ed., Boca Raton, FL, chap. 16.
33. **Dawkins, M.,** *Animal Suffering: The Science of Animal Welfare,* Chapman and Hall, London, 1980.

Part II
Legal and Regulatory Issues

Chapter 3

LAWS, REGULATIONS, AND POLICIES PERTAINING TO THE WELFARE OF LABORATORY ANIMALS

Christian E. Newcomer

EDITOR'S PROEM

Rollin's discussion emphasizes the moral basis for encoding in law a set of fundamental protections for the interests of animals used in research. Society has accepted this argument and has, through recent legislation, already imposed significant regulations governing animal care and use upon the biomedical community, and is likely to continue to do so. In his essay, Dr. Newcomer summarizes the history and content of the complex set of laws, regulations, and guidelines, many of which are still evolving and forthcoming, which structure the stage whereupon biomedical research is played out.

TABLE OF CONTENTS

I. INTRODUCTION

The social imperative to regulate the use of animals in biomedical research was established and brought to worldwide attention with the passage of the British "Cruelty to Animals Act of 1876".[1,2] Subsequently, other nations have enacted laws or implemented other regulatory mechanisms providing for the protection of research animal subjects including most of the European countries, the U.S., Canada, Japan, and Australia.[1,2] In addition, international nongovernmental organizations, such as the World Health Organization and UNESCO cosponsored Council for International Organizations of Medical Sciences (CIOMS)[3] and the International Council for Laboratory Animal Sciences (ICLAS),[4] have adopted resolutions which promote appropriate standards for the scientific utilization of laboratory animals. Public sentiment, frequently crystallized around specific incidents or issues related to animal use within the scientific community, has provided the impetus for a stricter regulatory environment for research animal utilization while supporting the scientific need for continued animal research.[5] The approaches taken to the regulation of laboratory animal welfare are varied among nations, reflecting the diverse complexion of public sentiment and political trends from an international perspective. A review of this information is beyond the scope of this chapter, but readers may wish to explore this material to speculate on the prospect and potential effect of policy importation in this area.

In the U.S., the Laboratory Animal Welfare Act of 1966 (P.L. 89-544) and subsequent amendments (P.L. 91-579, P.L. 94-279, and 99-198),[6] and the Health Research Extension Act of 1985 (P.L. 99-158)[7] are the principal laws concerned with the welfare of laboratory animals used in research, testing, or instructional exercises. The Good Laboratory Practices Regulations[8] also have had some impact on the welfare of laboratory animals. This chapter focuses on these federal laws and regulations promoting laboratory animal welfare, with a particular emphasis on the areas of recent regulatory convergence which have enhanced the welfare of laboratory animals significantly. Although these regulatory developments have necessitated the modification of institutional mechanisms of accountability, resulting in higher administrative costs and increased investigator participation in the process, they also have fostered institutional credibility from a public relations standpoint and allowed a new level of animal welfare assurance to be achieved.

Consideration also is given in this chapter to the recent publication of guidelines and principles regarding research animals by numerous professional organizations and scientific societies which attests to the renascent interest of scientists to invite public discussion while asserting the viability of internal evaluation and regulation. Finally, several state and local laws and ordinances which impede particular types of animal use in scientific studies also warrant discussion.

II. THE ANIMAL WELFARE ACT AND AMENDMENTS

A. DEVELOPMENT OF THE ANIMAL WELFARE ACT: 1966—1979

Although federal laws had been in place since 1873 for the humane treatment of farm animals during exportation, it was not until the passage of the Laboratory Animal Welfare Act of 1966 (P.L. 89-544) that protection was afforded to non-farm animals under the law. The focus of the Laboratory Animal Welfare Act of 1966 was prevention of illegal transfer of family pets to research institutions by requiring individuals or institutions that bought or sold dogs or cats to be licensed and by establishing a record-keeping system to document the legal acquisition and transfer of animals. In addition, institutions which used dogs or cats in their biomedical research programs were required to be registered, and standards were promulgated for the care of domestic dogs and cats as well as rabbits, guinea pigs, hamsters, and non-human primates, excluding the time interval when these animals were actually undergoing experimental procedures.

The types of enterprises covered, the species of animals regulated, and the assurance of minimal research animal care were expanded in the Animal Welfare Act of 1970 (P.L. 91-597). All warm-blooded animals were included; however, based on the discretion of the Secretary of Agriculture in the implementation of regulations, marine mammals, farm animals, laboratory rats and mice, and birds were excluded. The standards for animal care were to be applied continuously throughout the animals stay at the research institution, including the period of experimentation according to this amendment, but the actual design or method of experimental animal utilization remained inaccessible to regulation. Another important development was the annual reporting requirement for research facilities to record that the care, treatment, and use of animals, including appropriate use of anesthetics, analgesics, and tranquilizers, was in conformance with professionally acceptable standards. Animal exhibitors such as zoos and circuses, operators of auction sales in which domestic dogs or cats were sold, and any dealer who sold regulated animals on a wholesale basis were required to be licensed.

The Animal Welfare Acts of 1976 (P.L. 94-279) enlarged the scope and authority of the original Act. Carriers and intermediate handlers involved in the transportation of regulated animals were brought under jurisdiction, and the interstate promotion or shipment of animals for animal fighting ventures became illegal. Standards for the containers and the conditions of shipment of animals also were developed subsequently. Regulations and standards were promulgated in 1979 by the Secretary of Agriculture which brought marine mammals under the protection of the Act.

B. PRESENT REGULATIONS AND FUTURE DIRECTIONS OF THE ACT

Despite the progressive enlargement over the past two decades of the scope and authority of the Animal Welfare Act regulations[9] and their instrument of implementation, the U.S. Department of Agriculture (USDA), Animal and Plant Health Inspection Service (APHIS), and animal welfare and animal rights organizations frequently have asserted that the Act still imposes only minimum standards and have chided APHIS for its ineffective enforcement and insufficient monitoring of regulatory compliance.[10] This perspective has been shared also by other authors on this topic.[1] Furthermore, because the actual conditions of research animal use have remained inscrutable and immutable in the face of well-publicized instances of research animal abuse, congressional support mounted for legislation providing a mechanism to begin to penetrate the arena of research animal utilization resulting in the 1985 amendment known as the "Improved Standards for Laboratory Animals Act" (P.L. 99-198). With this amendment Congress has not rescinded its position that nothing in the Act should be construed as authorizing the Secretary of Agriculture to promulgate rules, regulations, or orders with regard to the performance of actual research or experimentation by a research facility. Because the regulatory relationship in animal research in the U.S. occurs between the government and the institution rather than the government and the individual, this amendment charges research institutions to develop an internal system of research animal proposal review and program monitoring to ensure compliance with the Act. It also introduces two measures significantly modifying existing animal care standards and prohibits certain experimental situations which were detrimental to animal welfare.

At the time of this writing, only speculation can be offered on the regulatory implementation of this important legislative step represented by P.L. 99-198. The first publication[11] of the proposed regulations for only a portion of the amendment evoked more than 8000 responses from which APHIS has yet to rebound.[12] Among the many points of contention between APHIS and the reviewers of the regulations on behalf of the scientific contingent was the cost of implementing the regulations: APHIS projected no significant financial impact, whereas the Association for Biomedical Research estimated a $500 million cost nationwide. Others have emphasized the numerous inconsistencies between the regulations proposed and the regulations already in effect at most institutions in order to be eligible for research funding by agencies within the Public Health Service. Nevertheless, the amendment is broadly consonant with the Health Research Extension Act of 1985 (P.L. 99-158) and brings considerable unity to the institutional review process and accountability for animal research, regardless of the funding source when animal species regulated by the Animal Welfare Act are being used.

One of the blind spots of the Animal Welfare Act frequently noted by animal welfare and animal rights organizations is the omission of rats, mice, and native farm animals from coverage under the regulations,[9] especially since mice and rats are believed to constitute 75 to 90% of the number of animals used in research. In many institutions, however, compliance with the regulations of the Public Health Service Policy would be in effect to ensure the humane care and welfare of these animals. Many of these rodents also may be used in facilities complying with the Good Laboratory Practice Regulations, but these regulations serve primarily to protect the integrity of the study data generated and assign little intrinsic value to animal welfare. For example, although extensive documentation is required for the procedures used in laboratory animal management, the Good Laboratory Practice Regulations only allude to veterinary care and do not include any consideration of animal euthanasia practices. Accordingly, it must be acknowledged that the use of rodents is largely unregulated and that, at present, it is conceivable that even large domestic animals could be used by a privately funded research operation under similar circumstances. Thus, the unrelenting concern of the animal welfare and animal rights activists over the unregulated use of rodents in LD_{50} testing and in instructional exercises and research programs at small institutions of higher education not receiving federal funds is likely to keep these issues in the public eye.

The essential network of licensing and registration imposed by the Animal Welfare Act regulations has been of paramount importance to the enforcement of minimum standards of animal care during transit or when held by dealers, exhibitors, or research institutions and has aided in the prevention of stolen dogs and cats being used in research. In conjunction with the licensing and registration procedures, a system of animal identification and individual records is required for dogs and cats, enabling the source, description, and disposition of animal to be ascertained. These records must be retained for 1 year following the euthanasia of the animal.

Licensure and registration apparently have not eliminated the acquisition of dogs and cats by dealers by fraudulent means from individuals who had stolen animals and offered them for sale at flea markets or trade-day type sales. Consquently, APHIS has proposed that dealers not be permitted to obtain random source dogs and cats from nongovernment, contract or humane pounds or shelters, or from individuals who did not breed and raise the dogs on their own premises.[11] The proposed regulations sought to further deter the problem of the sale of stolen animals by requiring that the seller's vehicle license number and driver's license number be recorded to enable the source of the animals to be determined.

The transportation standards include the standards for primary enclosure construction and space allotment, availabilty of food and water, care in transit, terminal facilities, and handling. Temperature ranges are established for terminal facilities where animals are held for an intermediate stop or before or after shipment. Temperatures are not to exceed 85°F or fall below 45°F unless the animals are acclimated to lower temperatures. A USDA-accredited veterinarian must provide a certificate signed not more than 10 days prior stating that the animals are acclimated to lower temperatures. Shipping delays, due to the uncertainty common carriers may have about being able to maintain compliance with the provisions of the Act during inclement weather, occasionally disrupt animal delivery schedules. Also, dogs, cats, and non-human primates must be accompanied by a health certificate during transportation by intermediate handlers or carriers.

The specific standards promulgated for the care of animals by licensees, registrants, and U.S. goverment agencies cannot be covered in detail in this chapter, but the reader should be aware that standards are available for dogs, cats, rabbits, guinea pigs, hamsters, non-human primates, and marine mammals. General standards also are provided for the other animals covered by the Act. Minimum requirements are given for facilities and operating standards including the elimination of environmental extremes, provision of a sound working environment for personnel, and construction features conducive to the maintenance of sanitation and the exclusion of vermin. Space requirements are established for the various species as are husbandry standards concerning the feeding, watering, and grouping of animals and the maintenance of animal cage and room sanitation. The present space requirements for animal housing allow each animal to make normal postural adjustments and to move freely within the enclosure.

The recent amendment (P.L. 99-198) to the Act has directed facilities to provide a physical environment suitable to promote the psychological well-being of non-human primates. The scientific community has stressed the complexity of this issue given the paucity of information on psychological well-being for each of the non-human primates species used in biomedical research.[12] The regulatory implementation of this amendment may entail increased space requirements for non-human primates as well as a period of exercise. Exercise for dogs as determined by the attending veterinarian in accordance with the regulations also has been mandated in P.L. 99-198; however, the regulations on this issue have not yet been available for review and comment. These amendments reflect the intent of Congress to advance animal welfare, but also demonstrate the congressional naïveté about the ease of implementing regulations in the absence of a supporting intellectual base on these issues.

Veterinary care, programs of disease control and prevention, and euthanasia must be established and maintained under the supervision and assistance of a doctor of veterinary medicine according to the Act. Dogs, cats, non-human primates and the other mammals covered but not specifically named in the Act are to be observed daily by an animal caretaker, and the observation requirement for rabbits, guinea pigs, and hamsters is every 48 h. Animals which are injured or diseased must be given veterinary care or be euthanatized unless such action is inconsistent with the research purposes for which the animal was obtained and is being held.

The recent amendment has reaffirmed the importance of palliating pain and distress in laboratory animals through the use of anesthetics, analgesics, tranquilizers, or euthanasia. It requires that the principal investigator consider alternatives to painful or distressful procedures in experimental animals and that the veterinarian be consulted in the planning of these procedures to avert experimental designs that are detrimental to animal welfare. During experimental procedures eliciting pain or stress in which these drugs are withheld for scientific reasons, the study should be limited to only the necessary time to accomplish the research objectives. The use of paralytic agents in the absence of anesthetics during painful procedures in animals is prohibited. An animal may not be used in more than one major operative experiment from which it is allowed to recover except when scientifically necessary, and laboratory personnel must provide pre- and postsurgical care in accordance with established veterinary medical and nursing procedures. This new amendment has amplified the veterinary care aspect of the Act in important areas and has delineated critical issues for veterinary-investigator interaction. As established in the earlier amendments to the Act, annual reports continue to be required providing the common name and numbers of animals used in activities involving no pain or distress; the numbers of animals used in activities that otherwise would have involved pain or distress and for which appropriate anesthetic, analgesic, or tranquilizer was administered; and the numbers of animals used in activities involving pain or distress for which the use of anesthetic, analgesic, or tranquilizer was withheld because these would adversely affect the procedure or data. A detailed explanation is required for animals in the latter category.

The implications of P.L. 99-198 in regards to the development of and charge given to the Institutional Animal Committee in the areas of program development and monitoring, review, and oversight of research animal use and investigative training, and accountability for fulfilling the institution's reporting requirements, are far reaching. The amendment requires the chief executive officer of each research institution to appoint an Institutional Animal Committee of not less than three members possessing sufficient ability to assess animal care, treatment, and experimental practices as dictated by the needs of the facility. Further, these individuals should represent the concerns of society regarding the welfare of animal subjects used in the facility. The following information is only a brief synopsis of Committee composition and functions as defined in P.L. 99-198. Readers should refer to the proposed rules[11] by APHIS for a more exhaustive approximation of the eventual regulations.

The Committee composition must include at least one veterinarian and one individual who is not affiliated with the animal facility by means other than Committee membership nor who has an immediate family member affiliated with the facility. This individual is intended to represent the general community interests in the proper care and treatment of animals. In cases where the Committee is composed of more than three members, not more than three members are to be from the same administrative unit of the facility.

The Committee is required to conduct semiannual reviews of animal facilities and animal studies areas to ascertain that the control of pain and suffering in experimental animals is occurring as described in the reports filed by investigators with the Committee. This requirement is accomplished by the review of the procedure by the Committee as well as the review of the condition of the experimental animals. The Committee is required to file a report of each facility inspection which includes reports of any violations of the standards promulgated or of assurances required by the Secretary of Agriculture. The necessary areas of consideration include deficiencies of animal care or treatment, deviations of research practices from originally approved protocols that adversely affect animal welfare, and any notification to the facility regarding deficient conditions, and the corrections made thereafter. The minority views of Committee members must be recorded in the report.

The facility report is to be kept on file for a period of 3 years for review by APHIS and any funding federal agency. This report also must be circulated to an administrative official of the facility to initiate corrective action on any deficiencies or deviations from the provisions of the Act. If corrective measures do not ensue as a result of this process, the Committee is obligated to notify in writing APHIS and the funding federal agency of these deficiencies or deviations.

One final area of Committee responsibility is to assure that the facility is providing training to investigators, animal technicians, and other personnel involved with the care and treatment of animals. The training program should include instruction on humane practice of animal care and experimentation; methods of reporting deficiencies of animal care and treatment; research and testing methods that minimize or eliminate the use of animals or limit animal pain or distress; and utilization of the information service at the National Agriculture Library established by this amendment. The intent of this library is to prevent research duplication and to promote the adoption of techniques by scientists which refine, reduce, or replace the use of animals in experimentation.

III. PUBLIC HEALTH SERVICE POLICY ON HUMANE CARE AND USE OF LABORATORY ANIMALS AND THE HEALTH RESEARCH EXTENSION ACT OF 1985

A. HISTORY OF DEVELOPMENT OF PUBLIC HEALTH SERVICE POLICY

Since its inception in 1971, the policy of the National Institutes of Health (NIH) and the Public Health Service (PHS) on the humane care and use of laboratory animals has emerged as the most encompassing regulatory consideration of animal welfare issues by a federal agency. The present PHS policy resulted from the passage of the Health Research Extension Act (P.L. 99-158) and contains important conceptual parallels with the recent Animal Welfare Act Amendment (P.L. 99-198). Readers interested in the details of the evolution of the policy should consult the excellent review on this subject.[13] The following discussion highlights only the key developments in this process.

In the 1971 NIH policy, institutions and organizations using warm-blooded animals in projects or demonstrations supported by NIH grants were required to assure the NIH that they would evaluate their animal facilities with regard to the care, use, and treatment of animals by one of two mechanisms: accreditation by a recognized professional laboratory animal accrediting body, i.e., American Association for Accreditation of Laboratory Animal Care (AAALAC), or an internal institutional committee without a specified membership other than the requirement that it include a veterinarian. The institution was obligated to conform to the guidelines in the *NIH Guide* and to follow applicable portions of the Animal Welfare Act. Thus, since this policy always has subsumed compliance with the Animal Welfare Act and has included all warm-blooded animals, it has represented the most comprehensive assurance of animal welfare required by a federal agency. Annual reports were to be submitted to the responsible institutional officials and to be kept on file for review by NIH officials. Facilities accredited by AAALAC were not required to have an evaluation committee.

In 1973 the NIH policy was replaced by the first PHS policy, and the dual mechanism of compliance with the policy was retained. Partially accredited facilities or unaccredited facilities were required to have committees of at least three members, and the requirement to have a veterinarian on the committee was waived in institutions not using a large number of animals when a scientist with demonstrated expertise in the care and use of laboratory animals served on the committee. Institutions were now required to submit assurance statements to the NIH listing the facilities and components either covered by accreditation or through the institutional review mechanism.

The PHS policy of 1979 required all institutions to establish committees to oversee the animal care program regardless of a program's active and successful participation in the AAALAC accreditation process. Assurances to the NIH had to state the institutional commitment to compliance with the principles set forth in the *NIH Guide,* and AAALAC accreditation was encouraged as the best means of demonstrating conformance with these principles. Records of committee activities, findings, and recommendations were to be maintained for review by PHS officials as was information pertaining to institutional AAALAC review and accreditation. Grants were required to provide specific information relative to animal welfare issues, and the review of proposals or projects by the committee was encouraged but not required. The committee was to have five members who were expected to have appropriate education and experience to discharge their duty to oversee the animal care program.

B. PUBLIC HEALTH SERVICE POLICY UNDER THE HEALTH RESEARCH EXTENSION ACT OF 1985 (P.L. 99-158)

Working upon the backdrop of major revisions which had been incorporated into the PHS Policy in 1985, the current version of the policy was acheived in 1986 following the passage of the Health Research Extension Act of 1985 (P.L. 99-158).[14] The provisions of the PHS Policy apply to both intramural and extramural PHS-sponsored research. The PHS Policy adopts the *Guide for the Care and Use of Laboratory Animals*[15] *(Guide)* as its regulations for the evaluation of institutional animal care and use program compliance and endorses the *U.S. Government Principles for the Utilization and Care of Vertebrate Animals Used in Testing, Research and Training,* an appendix to the *Guide* developed by the Interagency Research Animal Committee. The principal features of the policy are submission of an assurance, establishment of an Institutional Animal Care and Use Committee (IACUC), review of PHS-conducted or -supported research projects by the IACUC, provision of specific information in applications and proposals to the PHS, record keeping, and reporting requirements. The following discussion highlights areas of the policy which have an influence on the welfare of research animals. It should be noted that all vetebrate animals are covered under this policy.

A central purpose of the written assurance which must be submitted to the Office for Protection from Research Risks (OPRR) is to define the scope of the program including the qualifications, lines of authority, and areas of responsibility for the key personnel in the administration of the institutional animal care and use program. The assurance contains a list of the subdivisions of the institution covered under the assurance, animal inventory data and facility information, IACUC membership and operating guidelines, a description of the occupational health program for individuals having significant contact with laboratory animals, and a synopsis of the provisions for the training of personnel. Programs also must report their institutional status: Category 1 institutions maintain AAALAC accreditation and need not file a complete program evaluation, whereas Category 2 institutions must submit a complete program evaluation conducted by the IACUC. In both categories, institutions are required to perform a biannual program review according to the criteria of the *Guide* by the IACUC and submit a letter to the OPRR annually describing program deficiencies and the plan for corrective action. On the basis of this evaluation OPRR may approve or disapprove the assurance or negotiate an approvable assurance. Research activities supported or conducted by the PHS are not permitted in programs which have failed to secure an approved assurance.

The composition, responsibilities, and authority of the IACUC according to the policy are consonant generally with those defined for the Institution Animal Committee in the Animal Welfare Act Amendment (P.L. 99-198). Although certain functions must be fulfilled by the IACUC according to the policy, nothing in the policy should be construed to dictate the philosophical approach of an institution to resolving the spectrum of difficult issues related to research animal utilization. The emergence of the IACUC as the appropriate stage for program development and the arena for an ongoing and progressive effort to define, evaluate, and implement new animal welfare initiatives has received discussion beyond the scope of this chapter. Readers interested in a broad overview of the vital activities of the IACUC in various institutional settings from the perspectives of different interest groups within the scientific community are encouraged to review the volume dedicated to this subject.[16]

Members of the IACUC are to be appointed by the chief executive officer of the institution. The PHS elected to require the IACUC to have at least five members. The policy specifies that the IACUC must include a veterinarian with training and experience in laboratory animal science and medicine who has direct or delegated program responsibility for activities involving animals at the institution. Other membership profiles must be fulfilled also: one practicing scientist experienced in research involving animals, one member whose primary concerns are in a nonscientific area, and one member who is unaffiliated with the institution.

The biannual review of facilities and program by the IACUC is comprehensive and relies on the *Guide* as the basis of evaluation. The deficiencies noted must be relayed to the responsible institutional official in a report which distinguishes minor deficiencies from major deficiencies, i.e., deficiencies which may pose a threat to the health and safety of animals. When deficiencies are noted, the reports must delineate the plan and schedule of correction. The IACUC also is obliged to make recommendations to the institutional official regarding any aspect of the program, facilities, and personnel training of the institution and to review concerns involving the care and use of animals at the institution.

The review of proposed research projects or proposed modifications to ongoing research projects to determine if these proposals are in conformance with the PHS Policy and the assurance of the institution rests with the IACUC, and the approval status assigned by the IACUC is inviolable. However, written notification of the IACUC action on a proposal must be given to the investigator, and if approval is withheld for an activity, the reasons for the decision must be given and the investigator afforded the opportunity to respond in writing or in person to the IACUC. The IACUC also has the authority to suspend an ongoing activity. Due to the gravity of this situation, the institutional official in consultation with the IACUC must review the reasons for the suspension, take appropriate corrective action, and report the action with full explanation to the OPRR.

The policy rules for the proposal review do not require full consideration of each proposal if the IACUC members unanimously consent to waive this fundamental right; it is required that each IACUC member receive a description of the proposal in order to make this decision. Under these circumstances, the chairperson may designate a member qualified to conduct the review and given the authority to approve, require modifications to secure approval, or request full committee review of a proposal. A convened meeting of a quorum of the IACUC is essential for full committee review, and a proposal must carry a majority of the quorum to become approved. Consultants may be useful in the resolution of complex issues by the IACUC, but may only participate in offical actions if they are also IACUC members.

The proposal review process should include a consideration of the discomfort, distress, or pain to animals during experimental procedures and ensure that anesthetics, analgesics, and/or tranquilizers are used for their control whenever possible. The IACUC also must review the investigator's reasons for withholding these agents if more than slight or momentary pain or distress is likely to occur. Animals that would otherwise experience severe or chronic pain or distress to the animals that cannot be relieved must be euthanatized at the end of the procedure, or if warranted, during the procedure. The euthanasia methods utilized must be consistent with the recommendations of the American Veterinary Medical Association Panel on Euthanasia,[17] and deviations for scientific reasons must be justified in writing by the investigator. The qualifications and training of the personnel conducting the experimentation or maintaining the animals should be determined to be appropriate. The well-being of animals also must be ensured by the provision of appropriate living conditions and the availability of consistent medical care by a qualified veterinarian in accordance with the *Guide*.

C. THE GUIDE FOR THE CARE AND USE OF LABORATORY ANIMALS

The present *Guide* evolved from the 1963 publication originally entitled, *Guide for Laboratory Animal Facilities and Care*, which was revised in 1965, 1968, 1972, and 1978, reflecting the growth and development of laboratory animal science and medicine in the U.S. throughout this period. Readers will benefit from reviewing this panoramic document and may find that it provides a useful point of entry to the literature on animal-related issues germane to their area of investigation. The *Guide* considers many facets of the care and use of laboratory animals in a generally prescriptive manner, but does not usurp the authority and responsibility of an institution for developing the criteria, approach, or plans for specific program implementation. Although recent highly publicized episodes of animal welfare violations in several major U.S. research centers have led to the pointed criticism of the *Guide* on this accord,[1] the approach taken in the *Guide* to encourage the application of professional judgement and expertise at the institutional level has endured and contributed to program excellence in diverse and numerous institutions. Consequently, from the perspective of the PHS, flagrant animal welfare violations are considered to be indicative of an institutional failure and not a failure of the *Guide*.

Several areas of the *Guide* encompass broad animal welfare concerns. The housing or caging system should provide a comfortable environment and adequate space to the animal and facilitate social interactions and an appropriate level of physical activity. Consideration also is given to the selection and delivery of food, water, and bedding and to the sanitation practices of the facility. The use of animal restraint devices is to be evaluated and approved by the IACUC, with strict limitations to prevent unnecessary restraint and injury to the animal.

The provisions for veterinary care include the development of preventive medicine programs and measures for disease surveillance, diagnosis, treatment, and control. Guidance is to be provided by the veterinarian to animal users regarding animal handling, immobilization, anesthesia, analgesia, and euthanasia; the choice and use of the most appropriate drugs are matters for the attending veterinarian's professional judgement. The provisions for surgery and postsurgical care include the requirement for appropriate training of personnel active in this area, the need for equipment to optimize the animal's recovery, and the maintenance of medical records. The use of aseptic technique is required in most animals, including rabbits, undergoing major survival surgery and is strongly suggested in all other survival surgery. Multiple major survival surgery, i.e., the invasion of a body cavity, on a single animal is permitted only after IACUC approval and is generally restricted to situations where it is necessary to perform the related components of a research project in stages. Cost savings alone is not an adequate reason for performing multiple major survival procedures.

Many scientific societies and consortiums have produced guidelines modeled after the *Guide* for the use of the specific type of research animals including fishes,[18] amphibians and reptiles,[19] wild birds,[20] wild mammals,[21] and agricultural animals.[22] Typically, institutions have chosen not to exempt field studies or studies conducted with wild animals from review, albeit technically it may not be required under the existing regulatory provisions. These societies have recognized the quandary that this poses for many IACUCs without experience in these matters, and these informative documents should assist IACUCs in their deliberations.

IV. SURVEY OF OTHER LAWS CONCERNED WITH ANIMAL WELFARE

Recurrent themes are evident in the review of legislative proposals regarding research animal welfare at the national, state, and local levels.[23] Members of the scientific community need to remain aware of the legislative initiatives at all levels to play a viable role in the formulation of a rational and articulate public discussion. Mounting public support for the elimination of the Draize eye irritancy test and the LD_{50} test resulted in the passage of an ordinance in Cambridge, MA preventing these studies in 1987. The introduction of similar proposals into state legislatures in Pennsylvania, New Jersey, and Maryland and into the federal legislature suggests that this issue will continue to be a focus of attention.

The prohibition of the use of pound animals in biomedical research already has widespread precedent at the state level and has now been introduced, without action, at the federal level. An additional 13 states saw the introduction of restrictive legislation in this area during 1987, but only New York enacted legislation, bringing the number of states preventing the release of pound animals to 12.

Several other important measures are under consideration at the federal level. Efforts to reinforce the Animal Welfare Act include a provision to protect farm animals used in non-agricultural research by deleting the USDA Secretary's discretion to determine animal species to be covered. In addition, an effort to compel the enforcement of the Act by allowing anyone to sue the federal government on his own behalf or on behalf of any animal has been proposed. Thus, although the 1985 amendment strengthened the Animal Welfare Act in some areas, there remains considerable public skepticism about the Act due its ineffective enforcement and selective view of animal pain and suffering.

V. CONCLUSION

The recent regulatory provisions in the U.S. should foster public confidence in the institutional accountability for the use of research animals. A recent poll commissioned by the American Medical Association demonstrating the solid public support for the use of animals in research suggests that this may already be occurring.[24] However, the scientific community must continue to be engaged in the ongoing evaluation of animal welfare concerns through public communication and cooperation with the institutional review process on sensitive issues if it expects to play a role in the direction of future regulation.

REFERENCES

1. **Hampson, J.,** Law relating to animal experimentation, in *Laboratory Animals: An Introduction for New Experimenters,* John Wiley & Sons, London, 1987, 21.
2. **Rowsell, H. C.,** A Comparative Overview of International Regulations, Guidelines and Policies on the Care and Use of Experimental Animals, Canadian Council on Animal Care, Ottawa, 1985.
3. Council for International Organizations of Medical Sciences, International Guiding Principles for Biomedical Research Involving Animals, CIOMS, Geneva, 1985.
4. **Rowsell, H. C. and McWilliam, A. A.,** *Laboratory Animal Science: A Global View,* Gustav Fischer Verlag, Stuttgart, 1983, 139.
5. The Principal Findings on a National Survey on the Use of Animals in Research and Testing, Foundation of Biomedical Research, Washington, DC, 1985.
6. Animal Welfare Act, Title 7, United States Code, 2131 et. seq., as amended by P.L. 99-198, December 23, 1986.
7. Health Research Extensions Act of 1985, Public Law 99-158, Title 5, United States Code, November 20, 1985.
8. U.S. Department of Health and Human Services, Food and Drug Administration, Good Laboratory Practice Regulations: Nonclinical Laboratory Studies, Title 21, Code of Federal Regulations, Part 58.
9. U.S. Department of Agriculture, Animal and Plant Health Inspection Service, Animal Welfare, Title 9, Subchapter A, Code of Federal Regulations, Parts 1, 2, 3, and 4.
10. *Beyond the Laboratory Door,* Animal Welfare Institute, Washington, DC, 1985.
11. U.S. Department of Agriculture, Animal and Plant Health Inspection Agency, 9 CFR Parts 1 and 2, Animal Welfare; Proposed Rules, *Fed. Regist.,* 52, 10292, 1987.
12. **Holden, C.,** Animal regulations: so far, so good, *Science,* 238, 880, 1987.
13. **Whitney, R. A., Jr.,** Animal care and use committees: history and current national policies in the United States, *Lab. Anim. Sci.,* 37 (Spec. Issue), 19, 1987.
14. Public Health Service Policy on Humane Care and Use of Laboratory Animals, Office for Protection from Research Risks, U.S. Department of Health and Human Services, Washington, DC, 1986.
15. Committee on Care and Use of Laboratory Animals of the Institute of Laboratory Animal Resources, Guide for the Care and Use of Laboratory Animals, U.S. Department of Health and Human Services, Public Health Service, National Institutes of Health, NIH Publication No. 85-23, Bethesda, MD, 1985.
16. Scientists Center for Animal Welfare, Effective animal care and use committees, *Lab. Anim. Sci.,* 37, 1987.
17. AVMA Panel on Euthanasia, *J. Am. Vet. Med. Assoc.,* 188, 252, 1986.
18. American Society of Ichthyologists and Herpetologists, American Fisheries Society, American Institute of Fisheries Research Biologists, Guidelines for Use of Fishes in Field Research, 1987.
19. American Society of Ichthyologists and Herpetologists, The Herpetologists League, Society for the Study of Amphibians and Reptiles, Guidelines for Use of Live Amphibians and Reptiles in Field Research, 1987.
20. Ad Hoc Committee on the Use of Wild Birds in Research, Guidelines for use of wild birds in research, *Auk,* 105(Suppl.), 1987.
21. Ad Hoc Committee on Acceptable Field Methods in Mammalogy, Acceptable field methods in mammalogy: preliminary guidelines approved by the American Society of Mammalogists, *J. Mammal.,* 68 (Suppl.), 1987.
22. Consortium for Developing a Guide for the Care and Use of Agricultural Animals in Agriculture Research and Teaching, Guide for the Care and Use of Agricultural Animals in Agriculture Research and Teaching, Division of Agriculture, National Association of State Universities and Land-Grant Colleges, Washington, DC, 1988.
23. Annual Report, National Association for Biomedical Research, 1987.

Chapter 4

INSTITUTIONAL ANIMAL COMMITTEES: THEORY AND PRACTICE

Rebecca Dresser

EDITOR'S PROEM

It is clear that the heart of the new legislation, summarized in the preceding chapter by Newcomer, is what one sociologist has called "enforced self-regulation". The biomedical community is legally mandated to consider the needs and interests of animals, and the major vehicle for achieving this result is the local animal care and use committee. Charged with a variety of tasks from protocol review to facilities inspection, the animal care committee carries the major burden of articulating an ethic for animals and putting it into practice. In this essay, Professor Dresser outlines some of the specific duties and problems confronting institutional animal care committees.

TABLE OF CONTENTS

I. INTRODUCTION

1985 is a year unlikely to be soon forgotten by anyone involved in the care and use of laboratory animals. During that year, the two major federal regulatory systems governing scientific uses of animals were substantially revised. Both the Animal Welfare Act (AWA) and the Public Health Service (PHS) funding policy were modified to intensify the responsibilities of investigators and institutions using animals in research, testing, or education. Although federal agencies assumed a portion of the enforcement duties, the revised AWA and PHS Policy each assigned significant oversight duties to institutional review committees composed of institutional employees and community members.

As a result of the federal revisions, nearly every U.S. research facility must establish an Institutional Animal Care and Use Committee (IACUC) to perform a variety of functions related to laboratory animal housing and treatment. While the AWA and PHS Policy set forth specific requirements on the committee's duties, membership, recordkeeping, and other matters, many of their provisions are general and remain subject to widely varying interpretations. Consequently, institutional committee members now confront a number of substantive and procedural questions related to their roles and responsibilities. In this chapter, the federal provisions on institutional animal committees are described and the issues that the federal authorities left for committees to resolve on their own are discussed.

II. FEDERAL PROVISIONS

A. ORIGINS OF THE INSTITUTIONAL COMMITTEE SYSTEM

Resorting to local committees for decision making on complex and controversial ethical issues is far from a novel approach. The recent federal action requiring IACUCs has several precedents in regulatory policy.

The first wide-scale adoption of local committee oversight in the U.S. arose in response to discoveries of abuses in the treatment of human subjects in biomedical research. These abuses were most egregious in the case of the Nazi physicians' experiments on concentration camp prisoners, but they were documented as well in a number of U.S. research projects.[1] In the 1970s, federal regulatory procedures closely resembling the provisions now in effect for animal research emerged to guard against future mistreatment of human research subjects. Institutional review boards composed of affiliated and community members were assigned oversight responsibilities for all federally sponsored research on human subjects.[2]

Since then, the institutional committee has become popular in the treatment setting as well. A number of hospitals have established such committees to address problematic decisions on life-sustaining care for disabled newborns and other patients in severely compromised states.[3]

The concept of local committee oversight is not completely alien to the laboratory animal area either. For several years, the National Institutes of Health (NIH) required institutions receiving agency funds to establish animal care committees. Committee responsibilities were relatively unspecified, however, and no unaffiliated member was required. In a series of NIH sites visits conducted in 1984, site visitors were disappointed to find that the committees they studied generally failed to meet the responsibilities described in their organizational charters.[4] This finding resulted in the decision of the federal agency to define more precisely the role and duties of the animal care committee.

Swedish law governing the performance of animal research also supplied guidance to drafters of the 1985 revisions in U.S. policy. Since 1979, Swedish law has mandated review of commercial and nonprofit animal research proposals by a regional ethics committee composed of an equal proportion of scientists, laboratory or animal technicians, and laypersons. All proposals that will confer more than minimal pain or distress on animals are reviewed by a subcommittee whose members work with researchers to assess the justification of a project and determine the least harmful approach to the scientific question.[5]

In sum, institutional committees have a history that made them a logical choice for playing a role in the recent federal effort to reform laboratory animal use. Ideally, a local committee of diverse membership stimulates a thorough review of difficult ethical issues from a range of professional and social perspectives. It enables government agencies to avoid the need to draft comprehensive rules to govern the complex and varied activities associated with biomedical research. The local committee can also be an effective means of educating its members and their colleagues, thereby encouraging the integration of ethical and regulatory standards.

Yet the local committee has its detractors as well. Scientists complain that the added bureaucracy entailed in committee review hampers their efforts. Some commentators criticize the inconsistency a decentralized system permits. Inconsistent decision making among different committees and within the same committee has been documented in one study of institutional review boards.[6] Other critics argue that the true appeal of the committee lies in its ability to diffuse the responsibility for difficult ethical choices. Still others are dubious about whether committees composed primarily of institutional employees are likely to render decisions detrimental to their colleagues' interests.[7]

Despite this criticism, the institutional committee remains a popular device to resolve contemporary biomedical dilemmas. With the new federal policy on animal use in research, testing, and education, the institutional committee has become an integral component of the modern biomedical science endeavor.

B. THE 1985 ANIMAL WELFARE ACT AMENDMENTS

The Animal Welfare Act[8] governs all animal research activity in or substantially affecting interstate or foreign commerce. The U.S. Department of Agriculture (USDA) is charged with administering and enforcing the law. The original law was enacted in 1966, and it has been amended three times since then. On the one hand, the influence of the law is wide reaching, for it applies to both public and private research facilities. However, an existing regulatory exemption drastically narrows the scope of the statute. A directive from the Secretary of Agriculture currently limits the coverage of the law to dogs, cats, non-human primates, guinea pigs, hamsters, and rabbits. Thus, none of the AWA's provisions, including those pertaining to institutional committees, presently apply to the numerous rats, mice, farm animals, birds, and other species used in scientific projects.[9]

In December of 1985, Congress passed the Food Security Act of 1985, which contained several significant amendments to the AWA. Among them were provisions requiring research facilities to establish a committee composed of not fewer than three members having "sufficient ability to assess animal care, treatment, and practices in experimental research," who "shall represent society's concerns regarding the welfare of animal subjects." The committee must include at least one veterinarian and one member otherwise unaffiliated with the facility who is "intended to provide representation for general community interests in the proper care and treatment of animals."

The amendments create specific duties for the institutional committee. A majority of the committee is to "inspect at least semiannually all animal study areas and animal facilities," and to include in the inspection a review of "practices involving pain to animals, and the condition of animals" in the facility. The research institution must maintain on file for three years and make available to federal officials a committee report on each inspection that documents any violation of the legal standards for animal care and treatment, any notification to the facility of the violation, and any subsequent correction of the violation. This report must include minority views of committee members. If a violation persists after the facility administrator has been notified and given an opportunity for corrections, the law instructs the committee to notify the USDA and any relevant federal funding agency.

The AWA amendments also include standards for the committee to apply in reviewing laboratory animal care and use. Animal pain and distress must be minimized, through the provision of "adequate veterinary care" and "appropriate use of anesthetic, analgesic, tranquilizing drugs, or euthanasia." Investigators must consider alternatives to any procedure likely to cause pain and distress in animals. They must also consult a veterinarian when they plan any potentially painful practices on animals. Adequate pre- and postsurgical care must be provided, and pain- or distress-relieving drugs or euthanasia may be withheld only when "scientifically necessary." No single animal may be subjected to major survival surgery more than once except in the presence of scientific justification or other special circumstances determined by the Secretary of Agriculture. Finally, facilities are required to arrange for exercise for dogs and "a physical environment adequate to promote the psychological well-being of primates."

In August of 1989, the Secretary of Agriculture issued final regulations to implement the AWA.[10] The proposed regulations add further specificity to the institutional review committee's role. The regulations provide a more detailed description of the procedures the IACUC should adopt in conducting facility inspections. They also require the IACUC to review the facility's program for humane care and use of animals, and to make recommendations for improvements to the institutional official responsible for compliance with USDA standards.

Other regulations supplement the committee's standards for protocol review. All survival surgery must be performed under aseptic conditions. Multiple major survival surgery is permitted if necessary for scientific reasons, to maintain an animal's health, or in other special circumstances as permitted by the Secretary of Agriculture. Cost savings alone is thus rejected as a justification for multiple major survival surgery. Unless there is a legitimate scientific reason to withhold them, pain- or stress-relieving drugs must be used whenever an animal is subjected to any procedure that would cause more than momentary or slight pain or distress and "would reasonably be expected to cause pain or distress in a human being."

Committees are also directed to obtain certain written assurances from investigators using animals in research. These assurances must state that the investigator has considered alternatives to painful or distressing procedures on animals, and that the experiment is not unnecessarily duplicative of prior research. They must also document the methods and sources used to determine that alternatives were not available. Investigators must also provide a detailed explanation to the committee whenever drugs are withheld from an animal in pain or distress. The USDA is now in the process of preparing more specific instructions on the special provisions on dogs and primates, which should aid committees in addressing the AWA requirements.

The regulations also assign to committees the responsibility to evaluate the training and qualifications of researchers. The AWA amendments include a mandate for facilities to supply training to personnel involved in animal care and treatment. The training must include instructions on

1. The humane practice of animal maintenance and experimentation.
2. Research or testing methods that minimize or eliminate the use of animals or limit animal pain or distress.
3. Utilization of the information service at the National Agricultural Library.
4. Methods whereby deficiencies in animal care and treatment should be reported.

According to the regulations, the institutional committee shall review the training program and establish a reporting system to effectuate item 4, above. The regulations also supplement the list of mandatory training topics to include instruction on the care and basic needs of individual laboratory animal species, proper pre- and postsurgical care, use of pain- and distress-relieving drugs, aseptic surgical techniques, and the AWA itself. Finally, the authorized institutional official must sign a statement on the annual report of the facility to the USDA, assuring the agency that the facility has fulfilled its regulatory duties and complies with AWA standards. The annual report, which includes additional information on a number of topics, must also indicate any exceptions to the standards set by the AWA and its implementing regulations that were approved by the IACUC.

C. THE REVISED PHS POLICY

The PHS Policy on Humane Care and Use of Laboratory Animals[11] constitutes the second set of federal rules that governs a substantial portion of laboratory animal use in this country. The policy applies explicitly to all activities involving animals supported or conducted by the following agencies: the NIH, Food and Drug Administration, Centers for Disease Control, Health Resources and Services Administration, and the Alcohol, Drug Abuse, and Mental Health Administration. Other agencies, such as the National Science Foundation, have also adopted the policy.[12] Moreover, the effect of the policy is enhanced by the common practice of institutions to apply its provisions to all animal projects conducted under their auspices. Furthermore, the policy applies to all vertebrate animals, which extends its effect beyond that of the AWA.

The PHS Policy was extensively revised in 1985. In 1986, Congress enacted some provisions of the policy into law, with a few additional changes.[13] Once again, the institutional review committee comprises an integral component of the federal rules. The policy requires every institution receiving funds from the relevant federal agencies to establish an IACUC. The policy requires a larger committee than the AWA does. The IACUC must have at least five members, including one veterinarian with training or experience in laboratory animal science and medicine who has program responsibility for the activities of the institution involving animals, one practicing scientist with experience in animal research, one nonscientist, and one unaffiliated member. A qualified individual may satisfy more than one category; for example, an unaffiliated member may also fill the nonscientist position.

The policy assigns the institutional committee duties similar to those described in the amended AWA. At least once every 6 months, the IACUC must review the animal program of the institution and inspect all animal facilities, using the *Guide for the Care and Use of Laboratory Animals*[14] *(Guide)* as the basis for its evaluation. The *Guide* contains recommended standards for laboratory animal use, management, quality, health, institutional monitoring, veterinary care, physical plant conditions, and personnel policies. The PHS provisions also direct the committee to review at least once every 3 years all proposals related to the care and use of animals and authorizes the committee to require modifications, withhold approval from, or suspend any research activities that fail to comply with the substantive requirements of the policy. The committee must also consider any other "concerns involving the care and use of animals at the institution," and recommend to a designated institutional official any changes it deems necessary in the program, facilities, and personnel training of the institution.

The policy also creates specific record keeping and reporting requirements for the committee. Every 6 months, the committee must submit reports (including minority views) to the designated institutional official, describing the degree of compliance of the institution with the policy and *Guide,* any major or minor deficiencies, and a plan for correcting such deficiencies. Institutions must keep this report, together with committee meeting minutes and protocol review records, for at least 3 years. All of this material must be available to PHS representatives for their review. Committees must also annually report certain information to the NIH Office of Protection from Research Risks and promptly notify the agency of any serious noncompliance with the policy or *Guide,* and of any committee suspension of research activity.

The PHS Policy includes several substantive principles governing animal care and use. Among them are mandates to avoid or minimize laboratory animal discomfort, pain, and distress, to relieve pain and distress through the appropriate use of sedation, analgesia, anesthesia, and humane killing, and to adopt the euthanasia techniques recommended by the American Veterinary Medical Association Panel on Euthanasia. Scientific necessity is the sole acceptable basis for departing from these principles. Other policy provisions require facilities to provide animals with appropriate species-specific living conditions and medical care and to ensure that animal care and use are performed by qualified personnel.

The policy also states that animal use must be consistent with the AWA and the *Guide*. The *Guide* strongly discourages prolonged physical restraint and multiple major survival surgery on a single animal. The *Guide* also includes the U.S. Government Principles for the Utilization and Care of Vertebrate Animals Used in Testing, Research, and Training. While some of these principles closely resemble those contained in the policy itself, others supplement the policy. Particularly noteworthy is the principle stating, "Procedures involving animals should be designed and performed with due consideration of their relevance to human or animal health, the advancement of knowledge, or the good of society." A second principle instructs investigators to consider, in the absence of evidence to the contrary, that "procedures that cause pain or distress in human beings may cause pain or distress in animals." Another principle prohibits the performance of surgical or other painful procedures on unanesthetized, paralyzed animals, although a second principle leaves open the possibility of some exceptions to this general rule. Last, investigators are directed to select the appropriate species, quality, and minimum number of animals needed to obtain valid results and to consider replacing animal use with non-animal alternatives.

In sum, the revised AWA and PHS Policy create numerous responsibilities for institutional committees. Committees face three broad challenges as they take steps to implement the federal revisions. Perhaps the easiest to cope with is the array of administrative and procedural decisions to be made as the full institutional oversight system is constructed. More politically charged is the process of achieving institutional legitimacy among the researchers and other personnel whose activities are now subject to committee scrutiny. The remaining task centers on the committee's value choices. Because many of the substantive principles of the federal policy are quite general, the committee must determine how stringently to apply the principles in specific animal care and experimental settings. Particularly crucial will be the level of proof the committee demands from investigators claiming scientific justification for exceptions to the AWA and PHS humane treatment standards. In subsequent sections of this chapter, the committee's role in light of these three major challenges are discussed.

III. BEYOND THE FEDERAL REQUIREMENTS

A. PROTOCOL REVIEW

1. Substantive Issues

Protocol review is one of the institutional committee's primary responsibilities. The process presents both substantive and procedural issues for committee members. Substantive questions arise as the committee attempts to apply the federal principles to the specific protocols it reviews. Committee members will undoubtedly confront much uncertainty and conflict as they seek to comply with the broad federal directives to alleviate and reduce animal pain and distress.

One possible approach to the problem is to formulate committee policies to govern common experimental procedures conducted on laboratory animals.[15] Such policies may be adopted for interventions imposing low, moderate, and high levels of pain or distress on animals. Thus, for example, the committee can adopt a policy setting forth the requisite measures to ameliorate laboratory animal discomfort in such relatively unintrusive interventions as blood sampling and inoculation. In addition, it can promulgate rules to govern projects imposing heavier burdens on animals, such as the use of complete Freund's adjuvant, creation of ascites fluid in antibody production, food and water deprivation, behavioral conditioning involving negative reinforcement, lengthy physical restraint, cancer, infectious disease, and toxicity studies, and research involving the deliberate imposition of pain and distress. In their policies, committees can require modifications or alternatives to specific experimental interventions to lessen the harm imposed on research animals. Committees can also take the more controversial step of demanding especially compelling justification for performance of procedures inflicting a high degree of pain or distress on animals. Finally, some committees may go even further, and decide to prohibit certain extremely harmful interventions on laboratory animals. Thus, for instance, the University of Southern California Animal Ethics Review Board has adopted a policy that forbids the use of prolonged physical restraint in any experimental procedure on animals.[16]

The process of hammering out such policies will always entail substantial discussion and debate among committee members; but eventually most committees will find the time and effort worthwhile. Formulating policies on specific interventions will increase the committee's long-term efficiency by reducing the need to discuss the issues surrounding such interventions every time they are proposed by an individual investigator. Committee policies also contribute to the consistency of committee decisions, which in turn enhances the fair treatment of investigators and the protection of animal subjects. As new knowledge and techniques emerge offering a means to reduce further the animal harm inflicted in various procedures, the committee's policies can be revised.[17]

A related approach to protocol review involves the classification of common research interventions into categories of severity. Classification helps investigators and committees identify the research proposals presenting the greatest anticipated pain or distress to animal subjects. The approach allows the committee to devote special effort to reducing animal harm in the most severe protocols. Classification systems can also be used to delineate procedures that require compelling justification, or are absolutely prohibited, because of the extreme pain or distress they inflict on laboratory animals.

Several institutional committees have voluntarily incorporated into their review process a protocol classification system created by the Scientists Center for Animal Welfare (SCAW) (Table 1). The system was originally based on one used in the animal research review process mandated by Swedish law. SCAW has refined and expanded the classification system, and promoted its use as a means of assessing the harm an intervention inflicts on animals. Besides assisting the committee in protocol review, the system helps define the optimal committee composition by determining the areas of expertise most needed from committee members. The system also helps to educate investigators and other institutional personnel on the nature of animal protocols performed in their institution.[17]

A third method of enhancing protocol review is to formulate general guidelines on recognition of animal pain and distress. In many protocols, monitoring animals for signs of these states is crucial to minimizing animal harm. Besides requiring investigators to ensure that their research animals are observed frequently, committees can assemble material on and distribute to investigators and technicians the common changes in appearance and behavior that signal animal pain and distress, as well as the appropriate action to take when evidence of such states is observed. Two recent publications supply concise and valuable information on these topics.[18]

The most difficult substantive issue the committee will encounter in protocol review concerns assessing the merit of animal research projects. This issue unavoidably arises whenever an investigator claims a scientific need to subject animals to painful or distressing procedures. It also arises when investigators assert that the number or species of animals they propose to use is required to obtain valid scientific results, and when an investigator argues that adopting a nonanimal method would compromise the scientific validity of a project. The possibility of unnecessary duplication of existing research also can necessitate merit judgments of proposed projects. Last, to apply the *Guide* directive to give "due consideration" to the "relevance to human or animal health, the advancement of knowledge, or the good of society" of a proposed project, the committee members must exercise some judgments on the merit of the research they review.

TABLE 1
Categories of Biomedical Experiments Based on Increasing Ethical Concerns for Non-Human Species

Category	Examples and comments
Category A Experiments involving either no living materials or use of plants, bacteria, protozoa, or invertebrate animal species	Biochemical, botanic, bacteriologic, microbiologic, or invertebrate animal studies, tissue cultures, studies on tissues obtained from autopsy or from slaughter-houses, or studies on embryonated eggs; invertebrate animals have nervous systems and respond to noxious stimuli, and therefore also must be treated humanely.
Category B Experiments on vertebrate animal species that are expected to produce little or no discomfort	Mere holding of animals captive for experimental purposes; simple procedures such as injections of relatively harmless substances and blood sampling; physical examinations; experiments on completely anesthetized animals that do not regain consciousness; food/water deprivation for short periods (a few hours); standard methods of euthanasia that induce rapid unconsciousness, such as anesthetic overdose or decapitation preceded by sedation or light anesthesia.
Category C Experiments that involve some minor stress or pain (short-duration pain) to vertebrate animal species	With anesthesia, exposure of blood vessels or implantation of chronic catheters; behavioral experiments on awake animals that involve short-term stressful restraint; immunization employing Freund's adjuvant; noxious stimuli from which escape is possible; surgical procedures under anesthesia that may result in some minor postsurgical discomfort; Category C procedures incur additional concern in proportion to the degree and duration of unavoidable stress or discomfort.
Category D Experiments that involve significant but unavoidable stress or pain to vertebrate animal species	Deliberate induction of behavioral stress in order to test its effect; major surgical procedures under anesthesia that result in significant postoperative discomfort; induction of an anatomic or physiologic deficit that will result in pain or distress; application of noxious stimuli from which escape is impossible; prolonged periods (up to several hours or more) of physical restraint; maternal deprivation with substitution of punitive surrogates; induction of aggressive behavior leading to self-mutilation or intraspecies aggression; procedures that produce pain in which anesthetics are not used, such as toxicity testing with death as an end point, production of radiation sickness, certain injections, and stress and shock research that would result in pain approaching the pain tolerance threshold, i.e., the point at which intense emotional reactions occur; Category D experiments present an explicit responsibility on the investigator to explore alternative designs to ensure that animal distress is minimized or eliminated.

TABLE 1 (continued)
Categories of Biomedical Experiments Based on Increasing Ethical Concerns for Non-Human Species

Category	Examples and comments
Category E Procedures that involve inflicting severe pain near, at, or above the pain tolerance threshold off unanesthetized, conscious animals	Use of muscle relaxants or paralytic drugs such as succinyl choline or other curariform drugs used alone for surgical restraint without the use of anesthetics; severe burn or trauma infliction on unanesthetized animals; attempts to induce psychotic-like behavior; killing by use of microwave ovens designed for domestic kitchens or by strychnine; inescapably severe stress or terminal stress; category E experiments are considered highly questionable or unacceptable irrespective of the significance of anticipated results. Many of these procedures are specifically prohibited in national policies and therefore may result in withdrawal of federal funds and/or institutional USDA registration.

The federal provisions furnish committees with little guidance on conducting merit evaluation. As a result, this is an issue the committee will have to handle on its own, at least for the time being. Merit questions fall into three primary categories. The first concerns project design. In this category are such questions as the following. Is the proposed animal study necessary to investigate the research question, or could a nonanimal or less harmful animal method be used? Is the project design the one most likely to yield valid and reliable results? The second category of questions concerns the ability of the researcher and technicians to perform the proposed study. Here the relevant inquiry centers on the skill, training, and facilities of those who will conduct the proposed animal project. Project design and personnel qualifications are crucial elements for reviewing the merit of animal research, for a poorly designed or performed study will never produce important scientific knowledge. The third major component of merit assessment concerns the significance of the goal of the project. In this area, social and moral judgments on the importance of various animal projects become relevant. It is this aspect of merit that has triggered debate over the need for using animals in activities such as product safety testing, studies of addiction and maternal deprivation, and cardiovascular and cancer research.

Merit evaluation is obviously a complex, sensitive, and value-laden process. How can an institutional committee of three or five people possibly make the judgments entailed in such a process? Although there are definite limits on the committee's capacities to perform merit evaluation, several measures are available to institute a rough form of merit evaluation at the committee level.

As a rule, the institutional committee is unequipped to engage in precise merit assessment of the design of a proposed project, for the committee's limited membership makes it unlikely that any member will have sufficient expertise in the investigator's precise field of inquiry. Yet the committee can still raise basic questions on design and demand some documentation to support the investigator's responses. This step should be effective to screen out some unmeritorious proposals, for it will at a minimum put investigators on notice that they must consider and respond to these questions as they plan their projects.

Other areas the institutional committee can examine are the training and skills of investigators and their assistants. The committee can ask about the experience these individuals have had with a specific animal species, model, and technique. It can also inquire about the adequacy of the facility and equipment available to the investigator. Such an examination will be politically delicate in some cases, but the AWA regulations clearly assign the committee this responsibility.

The committee can take additional measures to enhance the merit review process. The AWA regulations and the PHS Policy explicitly state that the IACUC may enlist outside consultants "to assist in the review of complex issues." Such consultants could provide a valuable contribution, particularly when the committee has questions on the merit of a project imposing high costs on research animals. Furthermore, when proposed projects are to be submitted to a government or major private funding entity, the committee can rely partially on these funding entities to award support to the most meritorious projects. Internally funded projects present a more troubling group, however, because merit review is often perfunctory or nonexistent in this context. To address this problem, the committee can work with institutional administrators and staff to create an effective institutional merit review system for all proposals that will not be evaluated by outside experts.[20]

2. Procedural Questions

Substantive issues are not the only challenges that the committee must confront in protocol review. Other important issues pertain to the committee's operating procedures. The committee must design its protocol review form and determine its procedures on expedited protocol review, voting requirements, monitoring research projects, and investigator appeals of committee decisions.

Anyone who has served on an IACUC is aware of the importance of the protocol review form. When carefully designed, this form encourages thorough and efficient committee review. Conversely, a form that fails to elicit a comprehensive protocol description from the investigator delays and complicates committee functioning. Thus, the time committee members devote to constructing their protocol review form will be rewarded by a subsequent increase in committee efficiency. Besides the need to elicit through the form information essential to reviewing the substantive issues raised by a protocol, it is also crucial to obtain this information in terms comprehensible to reviewers unfamiliar with the investigator's field of research. Committees devising their forms can benefit from studying forms used at other institutions, which most committees are willing to share. Forms usually require periodic updating as well to incorporate evolving professional and regulatory standards.

Expedited protocol review is a means for the committee to increase its efficiency and to focus its attention on the protocols imposing the greatest costs on animals. In this system, minimally intrusive proposals are reviewed by a committee veterinarian and usually one or two other members. The full committee is thus left with more time to consider the remaining protocols, which entail a higher level of pain or distress. The SCAW classification system discussed earlier supplies a framework for dividing proposals into the expedited and full review categories. The PHS Policy permits expedited review, as long as the full committee receives a list of all proposals and can obtain a description of each proposal if they so desire. The AWA regulations also allow this approach.

A second operational issue concerns voting requirements. The PHS Policy states that a quorum (majority) of members must be present at protocol review meetings. Decisions on protocols must then be approved by a majority of the members at the meeting. The AWA amendments require a quorum for all formal committee actions. The regulations also define quorum as a majority of the committee members present at the meeting.

The federal provisions create minimum standards for voting requirements. Individual committees may want to be more stringent, however, by demanding more than simple majority approval. Thus, a committee can require protocol review decisions to have the support of a 2/3 or 3/4 majority, or even a consensus. The latter standard ensures that the views of unaffiliated committee members are taken into account. The federal requirements to include minority views in the committee's inspection reports to USDA indicate the importance federal officials assign to dissenting opinions.

Monitoring ongoing research is a third important issue for institutional animal committees. Many committee members resist playing "watchdog" over their colleagues' conduct. Yet some oversight is necessary to ensure that humane animal use is more than simply a paper requirement. Ideally, the veterinary care staff of the institution can assume much of this responsibility through conducting daily rounds on the animal population. Problems with animals used in specific projects can be reported to the committee, if necessary. The reporting system mandated in the AWA amendments should enhance this process. In addition, committee members may decide to observe particular interventions that raise questions about animal pain or distress. Announced or unannounced committee site visits are a more controversial approach. Another option is to appoint one committee member to monitor projects that involve an unusual degree of animal pain or distress.[17]

Committees must also address the matter of an appeals process for investigators whose protocols have been modified, disapproved, or suspended by committee vote. Under the PHS Policy, no institutional official can override a committee's failure to approve an activity involving the care and use of animals. The AWA regulations adopt the same position. The PHS Policy does require the committee to allow an investigator to respond in writing to a committee decision to withhold approval from a proposal. The committee could decide, however, to go further and establish an appeal mechanism in which investigators are given the opportunity to appear before the committee to reargue their cases. Adopting this course will counter investigator perceptions of arbitrary and summary behavior by the committee.

B. INSPECTIONS

Inspections constitute the other primary task the federal provisions assign to institutional committees. Inspections raise several issues for committee resolution. These include determining the areas for inspection, the standards to adopt in evaluating animal areas, which members shall conduct an inspection, and when it will occur. The committee must also confront the necessity to report precisely and candidly to federal agencies the deficiencies it discovers in its own institution.

The PHS Policy requires committees to inspect "all of the institution's facilities (including satellite facilities)." The AWA amendments broaden the mandated inspections to include "all animal study areas and animal facilities," with the possible exception of natural environments prohibitive to easy access. The difficulties committees encounter in fulfilling these responsibilities will vary among institutions. Research facilities can have many satellitte facilities scattered throughout the premises. Indeed, a major task for some committees is to determine every location where animals are housed in their institutions. Other committees must cope with the logistics of inspecting animal housing that is miles away from the primary research facility. Germ-free, barrier, and biohazard containment facilities also present practical impediments to inspection. The AWA demand for inspection of animal study areas adds to these problems. The laboratory inspection requirement is insignificant for committees in institutions in which most research is performed on rats, mice, and other species exempt from AWA coverage. However, it could confer substantial burdens on committees serving in institutions in which there is heavy research use of primates, dogs, cats, rabbits, guinea pigs, and hamsters.

Committee members performing inspections must have standards upon which to base their judgments on the condition of a facility. The PHS Policy instructs committees to use the *Guide* as the basis for their evaluations. The AWA regulations state that committees conducting inspections should refer to the USDA regulations on housing and care standards. The *Guide* and USDA regulations are relatively lengthy, detailed, and at times unclear. Another challenge for committee members is to gain at least a rough familiarity with these federal standards. The knowledge and expertise of the veterinarians of the facility is absolutely crucial to assisting the committee in formulating defensible conclusions in this area.

Another relevant question concerns who must conduct committee inspections. The PHS currently gives committees the discretion to assign inspection duties to designated subcommittees of the full committee. The subcomittee's report must then be reviewed and approved by the full committee. The AWA regulations allow the use of subcommittees composed of at least two committee members. Consultants may also be invited to assist in the inspections. Again, however, the full IACUC must review and approve subcommittee reports.

Finally, institutional committees cannot avoid the matter of honesty. The federal system puts the institutional committee in the awkward position of reporting deficiencies in their colleagues' animal care. Sometimes this responsibility furnishes a welcome means of stimulating the institution to take action to make much-needed improvements. In other instances, the committee may encounter pressure to downplay institutional shortcomings. Yielding to this pressure not only compromises laboratory animal welfare, it also exposes the institution to an increased risk of agency penalties and adverse publicity if unreported problems are discovered by federal officials. Frank and forthright reporting is thus the most prudent and defensible strategy for the committee to adopt. In tense situations, committee members should seek to convince institutional personnel that this strategy best serves the long-term interests of the institution as well.

C. PROGRAM OVERSIGHT AND EVALUATION

Federal policy now assigns institutional committees a variety of additional responsibilities. For example, the PHS Policy directs the committee to review the employee health plan, veterinary care arrangements, and other program requirements of an institution as discussed in the *Guide*. Both the PHS Policy and the AWA regulations direct the committee to review the training program of an institution for personnel involved in animal care and use. In many institutions, the committee may find itself heavily involved in designing institutional programs in these and other areas.

D. SELECTING COMMITTEE MEMBERS

The AWA and the PHS Policy both allow membership appointment procedures to be determined largely by individual committees and institutions. The sole federal constraint is for the chief executive officer to make the appointments. Because this individual generally will follow committee recommendations on membership, the committee will usually have the power to determine the identity of and frequency with which new members are appointed.

The committee's success in managing its diverse and extensive responsibilities strongly depends on the energy, knowledge, and commitment of its members. As a result, the qualifications and interests of potential new members should always be examined before appointments are made. Good communication skills, combined with the ability to be forceful but diplomatic, are ideal characteristics for all committee members, particularly the chairperson.

Although it is important to consider the qualifications of every candidate for committee membership, the need is especially compelling in the case of the community member. In the AWA amendments, Congress explicitly designated this individual to represent "general community interests in the proper care and treatment of animals." This is admittedly a broad directive, and one could argue about the content of such interests. Yet the provision does indicate that the unaffiliated member has a special role in representing animal welfare and other public concerns that might not be voiced by institutional employees. Committees should choose their outside members accordingly.

Some commentators argue that public members should be appointed from animal advocacy groups. Concerned advocates willing to work within the system may be well-equipped to serve as the committee's "community conscience."[21] At the very least, public members should be free of preexisting biases in favor of laboratory animal use, such as "large donors to local medical facilities or individuals affiliated with groups serving victims of particular diseases."[22] In a widely publicized incident, the executive director of a lobbying group representing the views of the medical community on laboratory animal use was appointed the public member of an institutional animal committee.[23] As one writer has warned, the failure to avoid such obvious conflicts of interest defeats the intent of the new federal policy and simply invites more restrictive federal regulations.[22]

A related matter is the number of outside and nonscientist members on committees. The lone public member of a large committee whose remaining members are all individuals involved in laboratory animal use is likely to feel isolated and overwhelmed.[25] Committees interested in obtaining a genuine portrayal of public concerns should appoint other additional unaffiliated or affiliated members willing to voice these concerns.

E. CONFIDENTIALITY AND LIABILITY

The current political climate surrounding laboratory animal use has triggered anxiety over public accessibility to committee records and meetings, and potential legal liability of committee members. In this section, these issues are discussed in light of federal and state law.

The Freedom of Information Act (FOIA)[26] provides that government agency records are open for public inspection unless the records fall under one of the legal exemptions to the provisions of the statute. In general, the reports of the institution to the USDA and approved assurances describing institutional compliance with the PHS Policy are available to the public through the FOIA. Some questions have been raised about the confidentiality of committee records on protocol review and institutional inspections. As long as this material remains within the institution, existing court decisions indicate that it is not subject to federal FOIA disclosure provisions.[27] There is presently some concern, however, about the accessibility of institutional records photocopied by USDA inspectors, particularly of protocols containing the investigator's scientific work product before it is prepared for publication.

Although committee records on file at the institution are not available to the public through FOIA provisions, some state laws could require public disclosure of such records. State public records or freedom of information laws could be construed to require public institutions to release some or all IACUC records. Precise resolution of this issue will depend on the legislation, regulations, and court decisions in the particular jurisdiction.

The public's right to attend institutional committee meetings is similarly dependent on state law. In some states, authorities have interpreted the law to require public admission to IACUC meetings at state-supported institutions.[27] Other state authorities could make the same determination if a member of the public or media representative raises this issue. Many states exempt certain hospital and medical committee meetings from public accessibility rules, but it is unclear whether these exemptions would be extended to include IACUC meetings.

Finally, there is the question of committee member liability. Although some committee members are quite concerned about their potential legal liability, the actual risks in this area are remote. Because the committee handles no human patient care or research, the possibility of malpractice or other tort liability lawsuits of this nature is nonexistent.

An investigator who loses a grant or suffers other legally recognized damages as a result of committee action could claim a violation of a contractual right to conduct research that meets institutional and federal standards. The committee could be held responsible if its decisions was not made in good faith, with reasonable care, and in a timely fashion. If the committee action was based on the investigator's failure to adhere to institutional and federal policy, however, there would be no legal support for the investigator's claim. Moreover, the odds of such a suit being filed, much less succeeding in court, are very low, based on the human subjects research experience.[28] It is also conceivable that hospital-based IACUCs would be protected by state statutes granting immunity from liability to hospital or medical peer review committees. Many of these state statutes, however, explicitly grant immunity only to committees whose membership is limited to professional and administrative staff.[29] Because institutional animal committees have a broader membership, they could be excluded from this statutory protection.

In theory, committee members could be vulnerable to suits filed by animal activists on behalf of animals used in activities approved by the committee. Although some writers have argued that animal protection groups should be given standing to assert legal claims for nonhuman animals,[30] as yet no court has agreed with this argument. Most recently, a federal appeals court ruled against such a claim made on behalf of the Silver Spring monkeys removed from Dr. Edward Taub's laboratory. There is always the possibility of future change in the courts' rulings, however. If this type of lawsuit became a realistic threat to committee members, state and institutional officials would probably act to protect members through enacting immunity legislation or providing members with adequate liability insurance coverage.

IV. CONCLUSION

In this chapter, I have attempted to discuss many of the issues members of institutional animal committees will encounter as they seek to implement the revised federal policy governing animal use in research, testing, and education. In certain areas, the committee's role and responsibilities are clear. It should also be obvious, however, that in several areas, federal officials have placed difficult ethical and political decisions squarely on the committee's shoulders. In some respects, one could legitimately argue that government officials aware of the public's ambivalence about laboratory animal use have simply written a few general principles and platitudes into law and policy, then handed to institutional committees the bulk of the responsibility for responding to the current societal unease.

On a more positive note, however, committee members do have the opportunity to effect real change in the norms and practices of investigators and institutions. Although institutional committees are unequipped to resolve the broader ethical and social questions surrounding contemporary animal use, they can substantially contribute to a system in which humane animal care and treatment is the accepted rule in the laboratory.

REFERENCES

1. **Beecher, H. K.,** Ethics and clinical research, *N. Engl. J. Med.,* 274, 1354, 1966.
2. **Greewald, R. A., Ryan, M. K., and Mulvihill, J. E., Eds.,** *Human Subjects Research: A Handbook for Institutional Review Boards,* Plenum Press, New York, 1982.
3. **Cranford, R. E. and Doudera, A. E., Eds.,** *Institutional Ethics Committees and Health Care Decision Making,* Health Administration Press, Ann Arbor, MI, 1984.
4. U.S. Department of Health and Human Services, Laboratory Animal Welfare, *NIH Guide for Grants and Contracts,* 1, April 5, 1984.
5. **Obrink, K.,** Swedish law on laboratory animals, in *Scientific Perspectives on Animal Welfare,* Dodds, W. J. and Orlans, F. B., Eds., Academic Press, New York, 1982, 55.
6. **Goldman, J. and Katz, M. D.,** Inconsistency and institutional review boards, *J. Am. Med. Assoc.,* 248, 197, 1982.
7. **Dresser, R.,** Research on animals: values, politics, and regulatory reform, *South. Calif. Law Rev.,* 58, 1147, 1985.
8. U.S. Code 7, Sec. 2131-2157, 1982 and 1985 Supplement.
9. Code of Federal Regulations, Section 1.1 (n), 1986.
10. U.S. Department of Agriculture, Animal welfare, Final Rule, *Fed. Regist.,* 54, 36112, 1989.
11. Office of Protection from Research Risks, National Institutes of Health, Public Health Service Policy on Humane Care and Use of Laboratory Animals, Bethesda, MD, 1986.
12. U.S. Congress, Office of Technology Assessment, Alternatives to Animal Use in Research, Testing, and Education, U.S. Government Printing Office, Washington, DC, 1986, 392.
13. Health Research Extension Act of 1985, Public Law 99-158, Section 495, November 20, 1985.

14. U.S. Department of Health and Human Services, *Guide for the Care and Use of Laboratory Animals*, NIH Publ. No. 85-23, National Institutes of Health, Bethesda, MD, 1985.
15. Pioneering policy guidelines issued by two committees, *Scientists Center for Animal Welfare Newsletter*, 9, 3, Winter 1987.
16. University of Southern California, Animal Ethics Review Board, Revised Code of Ethics, 1984.
17. **Dresser, R.**, Refining the ACUC process: policies and procedures, *Scientists Center for Animal Welfare Newsletter*, 9, 3, Winter 1987.
18. **Sanford, J., Ewbank, R., Molony, V., Tavernor, W. D., and Uvarov, O.**, Guidelines for the recognition and assessment of pain in animals, *Vet. Rec.*, 118, 334, 1986; **Morton, D. B. and Griffiths, P. H. M.**, Guidelines on the recognition and assessment of pain in animals, *Vet. Rec.*, 116, 431, 1986.
19. **Orlans, F. B.**, Research protocol review for animal welfare, *Invest. Radiol.*, 22, 253, 1987.
20. **Dresser, R.**, Measuring merit in animal research, *Theor. Med.*, 10, 21, 1989.
21. **Hutchison, J.**, The role of lay members on animal care committees, *New Paths*, 1, 5, Fall 1986.
22. **Sapolsky, H.**, Assuring the effectiveness of animal research committees, *Scientists Center for Animal Welfare Newsletter*, 9, 7, Winter 1987.
23. **Keller, J.**, Stacked deck?, *Cambridge Tab*, January 13, 1987.
24. Experimental Animals Committee, *Guidelines for Lay Members of Animal Care Committees*, Canadian Federation of Humane Societies, Nepean, Ontario, 1986, 51.
25. U.S. Code, 5, Sec. 552, 1976.
26. **Pardue, L.**, PAWS vs. U. of Washington, *Anim. Agenda*, 7, 9, July/August, 1987; Committee in the Sunshine, *Scientists Center for Animal Welfare Newsletter*, 9, 5, Spring 1987.
27. **Robertson, J.**, The law of institutional review boards, *UCLA Law Review*, 26, 484, 1979.
28. **Robertson, J.**, Committees as decision makers: alternative structures and responsibilities, in *Institutional Ethics Committees and Health Care Decision Making*, Cranford, R.E. and Doudera, A.E., Eds., Health Administration Press, Ann Arbor, MI, 1984, 85.
29. **Stone, C.**, Should trees have standing? Toward legal rights for natural objects, *South. Calif. Law Rev.*, 45, 450, 1972.
30. *International Primate Protection League v. Institute for Behavioral Research*, No. 86-1508, U.S. Court of Appeals for the Fourth Circuit, September 4, 1986.

Part III
Methodological Considerations

Chapter 5

CHOOSING THE ANIMAL — REASONS, EXCUSES, AND WELFARE

Charles W. Leathers

EDITOR'S PROEM

The passage of the new legislation described by Dr. Newcomer has placed a great onus on both researchers and local review committees to scrutinize methodological aspects of animal research with far greater care than ever before. One fundamental aspect of such scrutiny concerns the species of animal to be employed. Traditionally, as Dr. Leathers indicates, animals have too often been chosen for research protocols in terms of convenience and previous acquaintance of researchers with the species, rather than in terms of proper fit with the scientific question being asked. As a result, the validity of the research is put *en prise* and the animals used may have in fact suffered and died needlessly. In his discussion, Dr. Leathers focuses on a number of the scientific and ethical concerns which arise relative to choosing the correct animal.

TABLE OF CONTENTS

I. INTRODUCTION

Individual investigators have the responsibility and the authority for designing and controlling their research. The complexity and diversity of biomedical research has lead to astounding development of sophisticated instrumentation, reagents, and technical manipulations. Likewise, fascinating varieties of research animals are continually being identified and characterized. The choice of research animal to be used is one of the most pivotal decisions to be made by an investigator and encompasses the deliberation of whether non-animal methods would be appropriate. The validity of any information derived from the animal is dependent on its appropriateness, health, and welfare. The most brilliant design, the most elegant procedure, the purest reagents, along with investigator talent, public money, and animal life are all wasted if the choice of animal is incorrect.

Despite the diversity of research questions, only a fraction of the diversity of the animal kingdom is represented in research laboratories.[1] Relatively few species are now used in the majority of animal studies, and acceptance of a new ("exotic") species is often slow.[2] Knowledge of the normative biology, diseases, and research application of an animal is often derived painstakingly, and often by chance observation.[3] Once established, however, a species may be used habitually, for a variety of studies, without consideration of other options.[4] Choice of research animal may become a matter of convenience for the investigator; they have used the same species in the past, it is available inexpensively from a local supplier, and the cages are already here. Yet familiarity often fosters sloppy assumptions. The familiar species may not represent the genetic, microbiologic, physiologic, or psychologic facets needed (or wanted) for the proposed research design. The erroneous choice of animal escalates beyond just sloppiness; it is poor science that often does not consider the well-being of the individual animal when assessing the validity of research results.[5] More than one investigator has remarked that they do their research up in the lab, and the animals are down in the vivarium. This artifactual separation of research technology from animal care and use is a common misunderstanding.[6]

Investigators should expect and encourage more rigor in selection, verification, and quality of an animal subject than for a pH meter, chromatograph, or isotope. The sensory and behavioral repertoire of the individual animal is a crucial consideration for assuring its welfare and is a part of good animal research design. Understanding the spectrum of animal-related information likely will require collaboration among principal investigators, zoologists, behavioral scientists, animal scientists, and laboratory animal veterinarians.

The best characterized laboratory mammal is the mouse. Its genetic diversity is reflected in the number of specific traits and defects that are still being recognized and described.[7] Strains and substrains of mice are being developed continually, and the value of the laboratory mouse is unquestioned.[8,9] However, appropriate use of laboratory mice necessitates careful understanding of genetic characteristics, microbiologic background, and behavioral variations. Differences among selected mouse strains or stocks beg the question of proper identification. Yet how many principal investigators, post-doctoral fellows, or journal editors can specify what a so-called Swiss mouse is or isn't?[10] Maintenance of genetic purity within a specific group of mice cannot rely on casual observation. Biochemical testing and disciplined surveillance are proper components in contemporary mouse colony production. Genetic contamination will not only interfere with research validity, but also necessitate elimination of selected mouse populations.

Mice of selected strains are now available commercially without detectable viral antibodies. These mice are appropriate for many types of research, yet the housing requirements necessary to maintain these mice antibody free exceed the capacity of conventional vivaria. Investigators need to recognize the environmental requirements and constraints of using certain mice and should understand the controls necessary to purchase, ship, quarantine, and utilize such exacting animals.[11–13]

The proper control of genetic and microbiologic variation among laboratory mice is a specialized and expensive undertaking. Not every investigator will have accessible, local expertise and facilities to accommodate these defined colonies. Commercial or other production units may become more widely used as centralized research laboratories, with visiting investigators having access to the appropriately monitored animals. Such an arrangement underscores the link between excellent animal care and valid research.[14]

Rats, guinea pigs, hamsters, and rabbits have followed the same steps of recognition, characterization, and refinement as mentioned for mice. Numerous strains and stocks of these species exhibit characteristics that have expanded the research understanding of biologic phenomena. The choice of animal can rarely end with species alone. The exact genotype, gender, age, microbiologic status, and source are all basic to defining the particular individual.

Albino mammals (principally rodents and rabbits) are the most common varieties used in research laboratories. These established strains and stocks have provided fundamental insights for biomedical scientists. Yet the choice of albino animals has been challenged.[15,16] Investigators have been reminded that albino animals are mutants, with distorted biochemical and physiologic features compared to nonalbino individuals. Albino mammals lack normal optic systems and are susceptible to retinal damage from ambient lighting within animal rooms. Melanin and its intermediary metabolism are integral to production of catecholamines and thyroxine and proper functioning of neuroreceptors in the eye and ear. Albinism or other syndromes of hypopigmentation are associated with functional deficits in not only rodents, but also cats, dogs, mink, and human beings.[17] The unwitting use of an albino animal may not only invalidate the experimental design, but also needlessly relegate mutant animals to unacceptable environmental and experimental stresses.

It is inappropriate to consider any particular mouse, rat, guinea pig, hamster, rabbit, dog, cat, or non-human primate (or any other species) as a "generic" laboratory animal.[18–20] Similarly, it is unreasonable to consider any particular research question or method as simply "generic". Implicit in the choice of research animal is an intimate match between the protocol and the animal. This is especially true for the naturally occurring animal models of human disease. Many of the intriguing glimpses into biologic function have come unexpectedly from the atypical or unusual animal.[21,22] There is probably no valid reason for considering any species an "exotic" laboratory animal.[23–26] This variety of animals presents not only new opportunities for advancement of our understanding, but also new challenges for the care and use of these species.[27]

Well-documented animal models of human disease reflect not only the variety of species that have been studied, but also their valuable contributions to the understanding of biomedical problems. These models are summarized in Table 1.

Dogs are only a small percentage of the animals used in research, yet they receive considerable attention because of their position as pet animals. Numerous canine diseases have been identified and characterized as valuable models for human disorders: diabetes mellitus, systemic lupus erythematosis, cyclic neutropenia, glaucoma, retinal degeneration, and hemophilia. Other studies involving experimental surgery, organ transplantation, radiobiology, trauma research, nutrition, pharmacologic screening, and behavior have, and continue to be, benefited by the use of dogs.[28]

Random-source dogs and more strictly defined dogs from commercial breeding facilities each fill a niche in various research fields. Consideration of the defined needs of the dog and of the specific protocol are integral to the selection of an individual dog. The socialization of a dog intended for research is similar to that required of an acceptable pet, and several commercial suppliers of laboratory dogs emphasize the proper social development of their animals. The physical health of research dogs is critical to their usefulness and to the validity of experimental results. Initial costs in treatment or preventive measures are eclipsed by the cost of lost time and effort in a research project and in lost animal resources due to clinical or subclinical diseases that invalidate the research results. The physical and psychological well-being of research dogs is

TABLE 1
Animal Models of Human Disease

Animal group	Biomedical problem	Specific disease
Rodents		
Mouse	Genetic/developmental defect	Anemia, hereditary
		Athymic
		Autosomal trisomies
		Chediak-Higashi syndrome
		Copper malabsorption, X-linked
		Exencephaly
		Hereditary asplenia
		L cell mutant
		Megacolon, aganglionic
		Megaloblastic anemia
		Polycystic kidney disease
		Testicular feminization
	Neoplastic disease	Adenocarcinoma, DES
		Adenoma, salivary
		Angiosarcoma, liver
		Carcinoma, cervix
		Carcinoma, embryonal
		Hodgkin's disease
		Leukemia, myelogenous
		Malignant tumor transplant
		Mammary tumor
		Ovarian tumor
		Preneoplastic lymphoid hyperplasia
		Teratoma and teratocarcinoma
	Metabolic/nutritional disease	Amyloidosis
		Diabetes mellitus
		Gammopathies, monoclonal
		Gestational diabetes
		Globoid cell leukodystrophy
		Glucose-6-phosphate dehydrogenase deficiency
		Histidinemia
		Hypervitaminosis A
		Hypophosphatemia (rickets)
		Mast cell deficiency
		Methylmercury poisoning
		Niemann-Pick disease
		Nonobese diabetic
		Ochratoxicosis
		Ornithine transcarbamylase deficiency
		Paraproteinemia, idiopathic
		Thalassemia, alpha
	Degenerative disease	Adenosis, vagina/cervix
		Autoimmune disease
		Biliary obstruction
		Diverticulosis, oviduct
		Dysbaric osteonecrosis
		Proliferative glomerulonephritis
		Immunosuppression
		Macroglobulinemic neuropathy
		Menke's disease
		Motor neuron disease
		Pulmonary fibrosis, bleomycin
		Pulmonary fibrosis, solvents/oxygen

TABLE 1 (continued)
Animal Models of Human Disease

Animal group	Biomedical problem	Specific disease
		Reye's syndrome
		Salpingitis
		Vitiligo
	Infectious disease	Avian reovirus
		Capillaria hepatica
		Cytomegalovirus
		Encephalomyocarditis
		Giardiasis
		Hepatitis, reovirus
		Influenza B
		Lymphocytic choriomeningitis
		Meningoencephalitis, amoebic
		Meningoencephalitis, *Angiostrongylus*
		Scrapie
		Theiler's encephalomyelitis
		Trypanosomiasis
		Yersinia infection
Rat	Genetic/developmental defect	Amnionic fluid deficiency
		Fetal colon implants
		Fetal lung growth
		Hereditary hyperbilirubinemia
		Hydrocephalus
		Hydronephrosis
		Intrauterine growth retardation
		Megacolon, aganglionic
	Neoplastic disease	Adenocarcinoma, colon
		Adenocarcinoma, intestine
		Adenocarcinoma, prostate
		Aflatoxin carcinogenesis
		Angiosarcoma, hepatic
		Carcinoma, bladder
		Carcinoma, esophagus
		Carcinoma, pancreas
		Carcinoma, kidney
		Carcinoma, squamous cell, lung
		Carcinoma, yolk sac
		Interstitial cell tumor
		Lymphoblastic leukemia
		Malignant histiocytoma
		Medullary carcinoma, thyroid
		Neurogenic tumors, N-nitrosourea
		Osteosarcoma, Moloney sarcoma virus
		Pituitary tumors
		Urothelial tumors
	Metabolic/nutritional disease	Adrenal apoplexy
		Alcoholic fatty liver
		Anemia
		Cirrhosis
		Diabetes insipidus
		Diabetes mellitus
		Ethanol dependence
		Fructose-induced lesions
		Hepatic necrosis, halothane induced
		Hypervitaminosis A
		Hypothyroidism

TABLE 1 (continued)
Animal Models of Human Disease

Animal group	Biomedical problem	Specific disease
		Lead encephalopathy
		Lipotrope deficiency
		Mucopolysaccharidosis
		Obesity
		Ochratoxicosis
		Osteopetrosis
		Phenylketonuria
		Skeletal muscle, defective glucose/glycogen
		Striatal lesions, kainic acid induced
		Urolithiasis
		Vasculitis, pulmonary, glucan induced
	Degenerative disease	Aneurysm, cerebral
		Arthritis
		Autoimmune thyroiditis
		Duodenal ulcer
		Hypertension, induced
		Hypertension, spontaneous
		Hypertrophy, right ventricle
		Immunosuppression
		Ligation, cerebral artery
		Myocardial infarction
		Optic disc swelling
		Periodontitis
		Retinal degeneration
		Silica-induced pulmonary lipoproteinosis
		Thromboembolism
		Uterine vessel ligation
	Infectious disease	*Pneumocystis* pneumonia
		Venezuelan equine encephalitis
Guinea pig	Neoplastic disease	Transplantable leukemia
	Metabolic/nutritional disease	Hypervitaminosis A
		Hypovitaminosis C
		Mannosidosis
		Ulcerative colitis
	Degenerative disease	Allergic optic neuritis
		Antitubular BM nephritis
		Inflammatory bowel disease
		Optic disc swelling
	Infectious disease	Genital herpes
		Entamoeba histolytica
		Pichinde virus
		Tuberculosis
Hamster	Genetic/developmental defect	Autoimmunity
	Neoplastic disease	Benzo(a)pyrene-induced tumors
		Carcinoma, larynx
		Cholangiocarcinoma
		Pancreatic tumors
		Spontaneous carcinoma, lung
		Tumors of respiratory tract
	Metabolic/nutritional disease	Hypervitaminosis A
		Diabetes mellitus
	Degenerative disease	Cardiomyopathy
		Thrombosis, atrial
	Infectious disease	Besnoitiosis, chronic
		Scrapie

TABLE 1 (continued)
Animal Models of Human Disease

Animal group	Biomedical problem	Specific disease
		Syphilis
		Transmissible mink encephalopathy
Gerbil		Aural cholesteatoma
		Lead neuropathy
		Stroke
Mystromys		Diabetes mellitus
Praomys (Mastomys)		Autoimmune thyroiditis
		Carcinoids and adenocarcinoma
		Gastric ulcer
Squirrel		Prophyria
Woodchuck		Hepatocellular carcinoma
		Viral hepatitis
Rabbits	Genetic/developmental defect	Hydrocephalus
		Pelger-Huet anomaly
	Neoplastic disease	Fibroma
		Nephroblastoma
		Uterine adenocarcinoma
		VX-2 carcinoma
	Metabolic/nutritional disease	Diabetes mellitus
		Hypervitaminosis A
		Hypovitaminosis A
		Monoclonal gammopathies
		Methylmercury poisoning
	Degenerative disease	Acute respiratory distress syndrome
		Cholecystitis
		Inflammatory bowel disease
	Infectious disease	*Campylobacter* enteritis
		Chaga's disease
		Cryptococcal meningitis
Non-human primates	Genetic/developmental defect	Biliary atresia, neonatal
		Congenital hydrocephalus
		Down's syndrome
		Thalidomide syndrome
	Neoplastic disease	Colonic adenocarcinoma
		Malignant lymphoma
	Metabolic/nutritional disease	Atherosclerosis
		Calcium pyrophosphate deposition disease
		Cholelithiasis
		Diabetes mellitus
		Hemolytic anemia
		Hyperbilirubinemia
		Hypervitaminosis A
		Lead poisoning
		Monocrotaline poisoning
		Oxygen toxicity
		Toxemia of pregnancy
	Degenerative disease	Adenosis of vagina/cervix
		Baldness
		Cardiopulmonary disease
		Cor pulmonale
		Endocardial fibroelastosis
		Optic disc swelling
		Pulmonary hypertension
	Infectious disease	Acquired immune deficiency syndrome
		Adenovirus SV-20 pneumonia

TABLE 1 (continued)
Animal Models of Human Disease

Animal group	Biomedical problem	Specific disease
		Adult respiratory distress syndrome
		Chlamydial conjunctivitis
		Chlamydial genital tract infection
		Enterpathogenic *E. coli* infection
		Fetal pneumonia
		Gonorrhea
		Hepatitis A
		Hepatitis B
		Hepatitis non-A, non-B
		Malaria
		Periodontitis
		Streptococcal pneumonia
		Schistosoma haematobium infection
		Shigellosis
		Tetanus
		Varicella-like disease
		Wuchereria bancrofti infection
Ruminants		
Cattle	Genetic/developmental defect	Chediak-Higashi syndrome
		Hereditary parakeratosis
		Hereditary syndactyly
		Hereditary thymic hypoplasia, zinc deficiency, lethal trait A46
		Hydrocephalus
		Tibial hemimelia
	Neoplastic disease	Lymphosarcoma
		Ultimobranchial thyroid tumors
	Metabolic/nutritional disease	Glycogenosis, Type II
		GM1 gangliosidosis
		Mannosidosis, induced
		Mannosidosis, spontaneous
		Osteopetrosis
	Degenerative disease	Pancreatolithiasis
	Infectious disease	Ostertagiasis
		Rotaviral enteritis
		Venereal vibriosis
Sheep	Genetic/developmental defect	Muscular dystrophy
		Dubin-Johnson syndrome
		Gilbert's syndrome
	Neoplastic disease	Adenocarcinoma, intestine
		Lymphosarcoma
		Pulmonary carcinoma
		Squamous cell carcinoma, aural
	Metabolic/nutritional disease	Congenital goiter
		Copper poisoning, chronic
		Glucose-6-phosphate dehydrogenase deficiency
		Glutathione deficiency
		Mannosidosis, induced
		Photosensitivity
	Degenerative disease	Prosthetic cardiac valves
		Ceroid-lipofuscinosis
		Anti-BM glomerulonephritis
	Infectious disease	Bluetongue
		Jaagsiekte

TABLE 1 (continued)
Animal Models of Human Disease

Animal group	Biomedical problem	Specific disease
Goat		Hereditary myotonia
Carnivores		
Dog	Genetic/developmental defect	Abiotrophy, neuronal
		Anemia, hemolytic
		Cyclic hematopoiesis
		Cyclic neutropenia
		Dermatomyositis, familial
		Ehlers-Danlos syndrome
		Esophageal achalasia
		Glaucoma, hereditary
		Hemophilia
		Nephritis, hereditary
		Pelger-Huet anomaly
		Sensory neuropathy, hereditary
		Spinal dysraphism
		Spinal muscular atrophy
		Syringomyelia
		von Willebrand's disease
	Neoplastic disease	Mammary neoplasia
		Non-beta-cell pancreatic tumors
		Osteosarcoma
		Squamous cell carcinoma
	Metabolic/nutritional disease	Acromegaly
		Alpha-L-induronidase deficiency
		Ceroid-lipofuscinosis
		Copper toxicosis, inherited
		Cushing's-like syndrome
		Diabetes mellitus
		Gammopathies, monoclonal
		GM_2 gangliosidosis
		Gaucher's disease
		Globoid cell leukodystrophy
		Glycogenosis, Type II
		Glycoproteinosis, neuronal
		Hypercalcemia with malignancy
		Hypervitaminosis A
		Methemoglobin reductase deficiency
		Mucopolysaccharidosis
		Primary polycythemia
		Pseudohyperparathyroidism
		Pyruvate kinase deficiency
		Renal osteodystrophy
	Degenerative disease	Amyloidosis
		Arthritis, rheumatoid
		Asthma
		Atrioventricular block
		Craniomandibular osteopathy
		Dental calculus
		Endocardiosis
		Gastrointestinal polyps
		Hip dysplasia
		Immune-mediated thrombocytopenia
		Lafora's disease
		Lymphocytic thyroiditis
		Optic disc swelling

TABLE 1 (continued)
Animal Models of Human Disease

Animal group	Biomedical problem	Specific disease
		Pemphigus vulgaris
		Polyneuritis and polyradiculoneuritis
		Solar dermatosis
		Systemic lupus erythematosis
		Urolithiasis, sturvite
	Infectious disease	Bronchitis, chronic
		Coronaviral infection
		Distemper-associated demyelinating encephalomyelitis
		Herpesvirus infection
		Myocarditis, parvoviral
		Strongyloidiasis
Cat	Genetic/developmental defect	Chediak-Higashi syndrome
		Congenital deafness
		Esophageal achalasia
		Factor XII deficiency
		Klinefelter's syndrome
		Prophyria
		Sacrococcygeal agenesis and dysgenesis
		Spina bifida
	Neoplastic disease	Leukemia, acute, lymphoblastic
		Mammary carcinoma
	Metabolic/nutritional disease	Alpha-l-iduronidase deficiency
		Arylsulfate B deficiency
		Diabetes mellitus
		Gammopathies, monoclonal
		GM gangliosidosis
		Hypervitaminosis A, skeletal
		Methylmercury poisoning
		Mucopolysaccharidosis
		Spingomyelinosis
		T-2 toxicosis
	Degenerative disease	Aplastic anemia
		Endocardial fibroelastosis
		Gyrate atrophy
		Linear granuloma
		Retinal degeneration
	Infectious disease	Acquired immunodeficiency syndrome
		Amoebic dysentary
		Brugia malayi infection
		Feline leukemia virus disease
		Platynosomum infection
		Sporotrichosis
Mink		Aleutian disease
		Chediak-Higashi syndrome
		Ehlers-Danlos syndrome
		Hemivertebra
		Muscular dystrophy
		Scrapie
		Transmissible mink encephalopathy
Ferret		Brain malformations, induced
		Hyperammonemia
		Reye's syndrome
		Subacute sclerosing panencephalitis, measles

TABLE 1 (continued)
Animal Models of Human Disease

Animal group	Biomedical problem	Specific disease
Other mammals and marsupials		
Horse		Agammaglobulinemia, X-linked
		Anti-GMB nephritis
		Combined immunodeficiency (severe)
		Exostosis, multiple, hereditary
		Infectious anemia
		Lymphosarcoma
		Mannosidosis, induced
		Selective IgM deficiency
		Thrombocytopenia purpura
		Vitiligo
Pig		Arthritis
		Cerebrospinal lipodystrophy
		GM_2 gangliosidosis
		Hypervitaminosis A
		Lactational osteoporosis
		Lymphosarcoma
		Malignant hyperthermia
		Melanoma
		Ochratoxicosis
		Vitiligo
		von Willebrand's disease
Armadillo		Lepromatous leprosy
Opossum		Endocarditis, infectious
Koala		*Mycobacterium ulcerans* infection
Birds		
Chicken		Atherosclerosis
		Muscular dystrophy
		Myopathy, deep pectoral
		Osteomyelitis
		Osteopetrosis, viral
		Scleroderma
		Scoliosis
		Thyroiditis, lymphocytic
		Vitiligo
Duck		Amyloidosis
		Muscular dystrophy
		Torticollis
Turkey		Myopathy, deep pectoral
		Round heart disease
Fish		Diabetes, nutritional

fundamental, and investigators should understand the high standards for health that are mandated in the use of research dogs, regardless of source.

Effective and humane use of animals in research requires rapid acquisition of new information regarding behavior, nutrition, housing, reproduction, disease identification and prevention, analgesia, and anesthesia. Choosing an animal based upon one unique feature, without consideration of its individual biologic requirements, is myopic and contrary to proper experimental design and proper animal use.[29] New information is accumulating rapidly on the relatively new laboratory animal species.[30] Retesting of old assumptions regarding established laboratory animals is also yielding an expanded understanding of what is normal and what is artifactual. Investigators should remain receptive to this information base and be prepared to analyze and reevaluate their choice of animal.

Social interactions among experimental animals, and with investigators, technicians, or caretakers, are important factors explaining some experimental variation. Studies of a wide range of species have substantiated the impact of human contact on animal response to experimental situations.[31–33] These impacts have varied from differences in postoperative mortality to lesion development and disease resistance and emphasize consideration of individual animal behavior and welfare as important components in experimental design.[34]

The psychological well-being of non-human primate used in research is a topic that has stimulated wide discussion, especially following a recent amendment to the Federal Animal Welfare Act. The order Primates is so diverse that appropriate guidelines for providing psychological well-being will require tailoring to not only individual species, but also the various origins of individual animals (colony born, wild born, maternally deprived, group housed).[35]

Progressive and innovative professionals at zoological parks have grappled with how to avoid behavioral, social, and other stresses upon captive animals and have made remarkable progress. This information is adaptable and transferable to the biomedical research laboratory.[36–39] Less well understood are non-mammalian behavioral and social considerations. These factors will achieve greater prominence as model systems are developed in more non-mammalian vertebrates and invertebrates.[40,41]

II. CONCLUSION

Biomedical advances notwithstanding, we should be humbled by how much remains unknown. The pain, infirmity, and death due to undiagnosed and untreated disease are borne by all animal life. The health and well-being of people share many biologic foundations with other animals. Research animals will continue to yield basic, exciting, and prodigious information that can enrich the future of people and all other animals.[42] The identification, characterization, and use of research animals is an evolving process; a complex interplay among animal, environment, and investigator that is tested, revised, and refined continually. Each investigator choosing animals must accept the challenge of understanding the uniqueness of each laboratory animal and avoid blissful contentment with what is already known. Welfare of any experimental animal is integral to its research use and deserves careful consideration in quality experimental design.

REFERENCES

1. **Prichard, R. W.,** Animal models in human medicine, in Animal Models of Thrombosis and Hemorrhagic Disease, U.S. Department of Health, Education, and Welfare, Washington, DC, 1976, 169.
2. **Schmidt-Nielsen, B.,** Research animals in experimental medicine, *Exp. Biol. Med.,* 7, 46, 1982.
3. **Schmidt-Nielsen, K.,** The unusual animal, or to expect the unexpected, *Fed. Proc. Fed. Am. Soc. Exp. Biol.,* 26, 981, 1967.
4. **Bustad, L. K.,** The experimental subject — a choice, not an echo, *Perspect. Biol. Med.,* 14, 1, 1970.
5. Office for Protection from Research Risks, National Symposium on Imperatives in Research Animal Use: Scientific Needs and Animal Welfare, National Institutes of Health, Bethesda, MD, 1985.
6. **Gamble, M. R.,** The design of experiments, in *Laboratory Animals: An Introduction for New Experimenters,* Tuffery, A. A., Ed., John Wiley & Sons, Chichester, 1987, chap. 3.
7. **Staats, J.,** Standardized nomenclature for inbred strains of mice: eighth listing, *Cancer Res.,* 45, 945, 1985.
8. **Festing, M. F. W.,** A case for using inbred strains of laboratory animals in evaluating the safety of drugs, *Food Cosmet. Toxicol.,* 13, 369, 1975.
9. **Moriwaki, K.,** Genetic significance of laboratory mice in biomedical research, in *Animal Models: Assessing the Scope of Their Use in Biomedical Research,* Kawamata, J. and Melby, E. C., Jr., Eds., Alan R. Liss, New York, 1987, 53.

10. **Lynch, C. J.,** The so-called Swiss mouse, *Lab. Anim. Care,* 19, 214, 1969.
11. **Lang, C. M. and Vesell, E. S.,** Environmental and genetic factors affecting laboratory animals: impact on biomedical research, *Fed. Proc. Fed. Am. Soc. Exp. Biol.,* 35, 1123, 1976.
12. **Hughes, H. C., Jr. and Lang, C. M.,** Basic principles in selecting animal species for research projects, *Clin. Toxicol.,* 13, 611, 1978.
13. **Held, J. R.,** Quality animals for biomedical research: lessons from the past, opportunities for the future, in *the Importance of Laboratory Animal Genetics, Health, and the Environment in Biomedical Research,* Melby, E. C., Jr. and Balk, M. W., Eds., Academic Press, New York, 1983, 259.
14. **Clough, G.,** Quality in laboratory animals, in *Laboratory Animals: An Introduction for New Experimenters,* Tuffey, A. A., Ed., John Wiley & Sons, Chichester, 1987, chap. 5.
15. **Creel, D.,** Inappropriate use of albino animals as models in research, *Pharmacol. Biochem. Behav.,* 12, 969, 1980.
16. **Lockard, R. B.,** The albino rat: a defensible choice or a bad habit?, *Am. Psychol.,* 23, 734, 1968.
17. **Bolande, R. P.,** The neurocristopathies; a unifying concept of disease arising in neural crest maldevelopment, *Hum. Pathol.,* 5, 409, 1975.
18. **Burch, G. E.,** Of the normal dog, *Am. Heart J.,* 58, 805, 1959.
19. **Stevens, C.,** Humane considerations for animal models, in Animal Models of Thrombosis and Hemorrhagic Disease, U.S. Department of Health, Education, and Welfare, Washington, DC, 1976, 151.
20. **Malone, T. E.,** Toward refinement, replacement, and reduction in the care and use of laboratory animals, in *Scientific Perspectives on Animal Welfare,* Dodds, W. J. and Orlans, F. B., Eds., Academic Press, New York, 1982, 7.
21. **Leader R. W. and Padgett, G. A.,** The genesis and validation of animal models, *Am. J. Pathol.,* 101, S11, 1980.
22. **Van Citters, R.,** The role of animal research in clinical medicine, in *Research Animals in Medicine,* Harmison, L. T., Ed., Department of Health, Education, and Welfare, Washington, DC, 1973, 3.
23. **Capen, C. C., Jones, T. C., and Migaki, G., Eds.,** *Handbook: Animal Models of Human Disease,* Registry of Comparative Pathology, Armed Forces Institute of Pathology, Washington, DC, 1972—1989, fascicles 1—17.
24. **Jones, T. C.,** The value of animal models, *Am. J. Pathol.,* 101, S3, 1980.
25. **Mitruka, B. M., Rawsley, H. M., and Vadehra, D. V.,** *Animals for Medical Research,* John Wiley & Sons, New York, 1976.
26. **Balk, M. W.,** Emerging models in the U.S.A.; swine, woodchucks, and the hairless guinea pig, in *Animal Models: Assessing the Scope of Their Use in Biomedical Research,* Kawamata, J. and Melby, E. C., Jr., Eds., Alan R. Liss, New York, 1987, 311.
27. **Melby, E. C., Jr.,** Overview of the state of the art in development and utilization of animal models in the U.S.A., in *Animal Models: Assessing the Scope of Their Use in Biomedical Research,* Kawamata, J. and Melby, E. C., Jr., Eds., Alan R. Liss, New York, 1987, 1.
28. **Gay, W. I., Ed.,** The dog as a research subject, in *Health Benefits of Animal Research,* Foundation for Biomedical Research, Washington, DC, 1986, chap. 5.
29. **Bustad, L. K. and Rosenblatt, L. S.,** Choosing the experimental subject and its environment, presented at Symp. Dose Rate Mammalian Radiation Biology, Oak Ridge, TN, April 29 to May 1, 1968, 1.
30. **Nomura, T., Katsuki, M., Yokoyama, M., and Tajima, Y.,** Future perspectives in the development of new animal models, in *Animal Models: Assessing the Scope of Their Use in Biomedical Research,* Kawamata, J. and Melby, E. C., Jr., Eds., Alan R. Liss, 1987, 337.
31. **Hammet, F. S.,** Studies of thyroid apparatus, *Am. J. Physiol.,* 56, 196, 1921.
32. **Gross, W. B. and Siegal, P. B.,** Adaptations of chickens to their handler, and experimental results, *Avian Dis.,* 23, 708, 1979.
33. **Nerein, R. M.,** Social environment as a factor in diet-induced atherosclerosis, *Science,* 208, 1475, 1980.
34. **Dodds, W. J. and Abelseth, M. K.,** Criteria for selecting the animal to meet the research need, *Lab. Anim. Sci.,* 30, 460, 1980.
35. **Rosenblum, L. A.,** Behavioral factors and social needs of various species, presented at ACLAM Forum: The ACLAM Diplomate's Approach to Laboratory Animal Welfare, Columbia, MD, April 27 to 30, 1986.
36. **Hediger, H.,** *The Psychology and Behavior of Animals in Zoos and Circuses,* Dover, New York, 1968 (republication of 1955 translation), chap. 1.
37. **Wallach, J. D.,** Gauntlet of the cage, *J. Zoo Anim. Med.,* 3, 30, 1972.
38. **Snyder, R. L.,** Behavioral stress in captive animals, in *Research in Zoos and Aquariums,* Institute of Laboratory Animal Resources, National Academy of Sciences, Washington, DC, 1975, 41.
39. **Markowitz, H.,** *Behavioral Enrichment in the Zoo,* Van Nostrand Reinhold, New York, 1982.
40. **Umminger, B. L. and Pang, P. K. T.,** Fish as animal models in biomedical research, *ILAR News,* 22, 12, 1979.
41. Committee on Models for Biomedical Research, *Models for Biomedical Research — A New Perspective,* National Academy Press, Washington, DC, 1985, chap. 4.
42. **Weihe, W. H.,** Use and misuse of an imprecise concept: alternative methods in animal experiments, *Lab. Anim.,* 19, 19, 1985.

Chapter 6

BASIC PRINCIPLES OF EXPERIMENTAL DESIGN

George E. Seidel, Jr.

EDITOR'S PROEM

One of the most frequent and avoidable sources of wasted animal lives and unnecessary animal suffering in biomedical research arises out of improperly designed experiments, the results of which are meaningless. All too often, researchers are insufficiently trained in the subtleties (or even the basics) of statistics and experimental design and, thus, do a poor job of constructing experiments. Statistical consultants are sought after the fact in a frantic effort to salvage something of value from the experiment, but in many cases it is too late. Choice of experimental design is often dictated by a researcher's experience or habituation rather than by sound reasoning. Historically, funding agencies often have not looked carefully at statistical considerations and design, and much of this responsibility has been shifted to animal care committees, charged with ascertaining that a researcher has employed the proper species and number. In this chapter, Dr. Seidel forcefully illustrates some of the major components and errors in experimental design, pressing the point that it is far better to seek statistical expertise in advance than to attempt to salvage something of value from the wreckage of a bad experiment.

TABLE OF CONTENTS

I. STEPS BEFORE DESIGNING EXPERIMENTS

This chapter concerns wise use of animals in experiments. The first question in experimental design always should be: "Is this the best way of doing the experiment?" To answer this question, one must consider the purpose of the experiment and whether alternative approaches (Table 1) are preferable.

Often there are multiple, or at least ancillary, reasons for experiments, such as training physicians or scientists, establishing a scientific reputation, or complying with laws concerning safety and efficacy of products. There always are constraints to experiments (Table 2). For example, in many cases protocols must be reviewed by institutional animal use committees to insure humane care of animals. In most cases the quality and usefulness of an experiment improve as one improves quality of animal care. In this regard, training and attitude of personnel usually are vastly more important than physical facilities.

The main purpose of this chapter is to provide insight into experimental design and analysis. The experimental design phase of a project really begins after one has wrestled with issues described in the preceding paragraphs and Tables 1 and 2. For purposes of further discussion, it is assumed that the project is worthwhile, the animal model is appropriate, the correct measurements will be taken, animals will be cared for properly, etc.

II. THE PRELIMINARY EXPERIMENT

The vast majority of experiments with animals evolve in stages something like the following: (1) the thinking phase, in which ideas and approaches to a problem are considered; (2) the "checking out the system" phase, in which the characteristics of a technique are explored, technical problems are worked out, animal responses are observed, etc.; (3) the preliminary experimental phase (in which limited numbers of animals are used) to insure that the proposed experiment is in fact feasible and to point out obvious omissions or problems; and (4) the formal experiment.

Not all animal experiments evolve in this way, and not all phases are appropriate in each case. Evidence that the second phase just described has been carried out is crucial when one is seeking funding from many granting agencies. The preliminary experiment frequently is skipped when it should not be; in many cases the planned experiment turns into a preliminary experiment due to unforeseen aspects of the work. Administrators and animal use committees may seek to discourage the "checking out the system" and "preliminary experiment" phases. This would be a serious mistake; the "checking out the system" phase must exist in science, and to pretend it does not exist is counterproductive. Preliminary experimentation usually improves the definitive experiment, and, in the long run, reduces numbers of animals used because unproductive approaches are eliminated.

TABLE 1
Examples of Alternatives to a Particular Experiment

1. Use a different species
2. Make different measurements
3. Change the frequency of measurements
4. Do a smaller, preliminary experiment first to test feasibility
5. Do the experiment at a different location, time of the day or season of the year
6. Consider alternative analgesia, anesthesia, tranquilization or methods of euthanasia
7. Decide not to do the experiment, or replace it with two experiments
8. Change the design or method of statistical analysis
9. Use cells *in vitro* instead of animals or the converse
10. Use an epidemiological study design

TABLE 2
Examples of Frequent Constraints on Experiments

1. Amount of funding
2. Number of observations or measurements that can be taken per unit of time
3. Numbers of a particular kind of animal available
4. Laws and institutional regulations
5. Attitudes and opinions of superiors, colleagues, regulatory officials, and others
6. Availability of suitable animal care facilities
7. Measurement of A may invalidate measurement of B
8. Myopia

III. NEED FOR AN EXPERIMENTAL DESIGN AND STATISTICAL PROCEDURES

With rare exceptions, valid experiments with animals require a well-thought-out experimental design that can be evaluated with statistical procedures. A major reason for statistical analysis is that most people, including scientists and clinicians, greatly overestimate their abilities to determine if one treatment is superior to another from perusing data. Statistical procedures take into account all of the information in an experiment so that the best evaluation can be made.

An experimental design may be as simple as a treatment and a control group and rarely may not require statistical analysis. For example, many molecular biology experiments proceed in a logical sequence by evaluation of presence or absence of molecules. However, even in these kinds of experiments, statistical analyses should be used in many cases.

Clinical experiments are done constantly by veterinarians in practice as they seek to improve methods of treating animals. In most cases, no particular design is used, and improved procedures are transmitted to colleagues by word of mouth, round tables at meetings, etc. In some cases the improvements are so logical or obvious that experimental design is irrelevant. However, often it is impossible to be sure that the improvements are superior without information from designed experiments. In fact, the randomized clinical experiment has proven to be exceptionally powerful in all areas of medicine. For example, it took such an experiment to determine that treatment of women with diethylstilbestrol (DES) in fact increased rather than decreased abortion rates. Millions of women were treated with this drug because therapy was based on erroneous impressions rather than experimental evidence.

IV. ATTRIBUTES OF A GOOD EXPERIMENT

A good experiment usually has three important attributes: (1) a testable hypothesis, (2) a random sample of the population of interest allocated randomly to each treatment, and (3) enough animals per group for validity. These will be discussed in detail. Other attributes such as honesty of the investigators, accuracy of recording data, and proper training of personnel are assumed.

A. TESTABLE HYPOTHESES

For a testable hypothesis, one must formulate a clear question or objective and be able to obtain data to answer the question or satisfy the objective. It is best to state hypotheses or objectives formally so that the experimental design can be evaluated. Frequently, hypotheses are formulated to compare alternate and null modes; for example, chickens will grow more rapidly on Diet A than on Diet B may be the alternate hypothesis, and Diet A will result in the same growth rate in chickens as the control Diet B may be the null hypothesis. Another approach: the objective of the experiment is to determine if chickens grow faster with Diet A than with the control diet. Each of these variants is a reasonably clear, acceptable statement of a hypothesis. A subtle but important aspect of these hypotheses is that they imply that Diet A is expected to be an improvement (in statistical jargon, a 1-tail test).

Many details are not included in these hypotheses, such as the age of the chickens, how chickens will be fed (e.g., *ad libitum),* whether the chickens will be weighed at regular intervals, the length of their bones measured, their circumference recorded (or all of these), but it is clear that the hypothesis is testable. An example of an untestable hypothesis would be: "Chickens will do better on Diet A than the control Diet B." Better could be less sickness, increased reproductive rate, increased longevity, increased feather production, or many other measurements. An example of a testable hypothesis of marginal value would be: "Chickens will gain weight on Diet A." Since the data cannot be put into context of other treatments given under similar circumstances, the information from this testable hypothesis will be of little value.

Untestable hypotheses are common. Sometimes they can be made testable by rewording, but the reworded hypothesis may no longer be of interest. An example might be having to change the word "testosterone" to "androgen" because the measurement system does not discriminate between testosterone and other androgens. Hypotheses also may be untestable due to inappropriate experimental approaches or designs. A common problem is comparing cortisol or prolactin concentrations in blood of two groups of animals. Often excitement due to the method of blood collection elevates concentrations of these hormones acutely to much higher levels than those of physiological interest. An example in which sensitivity of a technique limits hypothesis testing could be comparing estradiol-17-beta concentrations in muscle of male and female sheep. The available assay may give zero readings for both when, in fact, there is a small but physiologically meaningful difference.

B. RANDOM SAMPLING AND RANDOMIZATION OF ANIMALS TO TREATMENTS

The second attribute of good experiments, random sampling from the population of interest, can be very difficult. A subset of this, randomization of animals to treatments, is especially important. There are numerous books on sampling, and a full semester university course on this subject can only deal with some of the many problems and approaches to sampling. On the other hand, random sampling is primarily common sense once the population of interest has been defined. The population might be relatively homogeneous (e.g., 2-year-old Hereford bulls, or 3-week-old female rats of the Long-Evans strain) or fairly heterogeneous (e.g., dairy cows). The more defined, the better for internal validity, but the less generalizable to broader populations. Once the population is defined and the animals acquired, it is essential to assign the animals to treatment at random; otherwise the experiment will be invalid. A good way to do this is to identify the animals by giving each a number, and then put each number on bits of paper or pebbles, mix them in a container, and pull out the numbers randomly, assigning the first to treatment 1, second to treatment 2, etc. A common error is to take the first 10 animals in a pen for treatment 1, the next 10 for treatment 2, etc. This results in biases due to shyness, ease of catching and size.

In most cases, the results of the experiment can only be extrapolated to the population sampled. For example, it usually is invalid to extrapolate from one breed to another, from one age to another, from one species to another, or one sex to another. An example of such a problem occurred recently when a drug had been tested in cows, but approved for both cows and bulls. It turned out that the drug frequently resulted in death of bulls, even though no toxicity was seen in cows. Huge damages had to be paid by the drug company.

In rare cases, valid extrapolations to other populations can be made if enough ancillary data are available. An interesting example is that extrapolation among breeds of dairy bulls has been found to be valid when experiments concern spermatogenesis and seminal characteristics, whereas such extrapolations frequently would be inappropriate for rates of body growth. If in doubt, do not extrapolate outside of the population sampled.

A recurring question is the validity of experimental use of "random-source animals" such as dogs that otherwise would be destroyed at pounds compared to use of "purpose-bred" animals. Frequently, more information is available for purpose-bred animals, for example, age and health status, so they are preferred for most experiments. In a few cases, however, random-source animals are more representative of the population of interest and are thus a better choice; examples might include studies on contraception in pets or vaccination for diseases, particularly if initial experiments were promising with purpose-bred animals, and the next step was to extrapolate to a wider population. Frequently, the reason for using random-source animals is that they cost much less than purpose-bred animals. For some experiments, this is false economy because it is necessary to keep animals for a longer preexperimental phase while they are vaccinated, checked for specific diseases, etc. Also, more animals may be required for valid experiments because random-source animals will likely be more heterogeneous (see next section). An example of another kind of problem occurred using wild monkeys for malaria research; because they already had malaria, they were inappropriate. This sort of problem occurs less frequently with purpose-bred animals, but good experimental protocols require screening for such obvious potential pitfalls, whatever the source of animals.

In many cases, the question of "random-source" vs. "purpose-bred" animals is an emotional rather than a scientific issue. For those few experiments for which either is a valid model, one obviously should use purpose-bred animals if the alternative is wild animals; but if the alternative is pound animals about to be euthanized, then they usually should be the choice because fewer total animals will be sacrificed.

C. REPLICATION

The third important attribute of a good experiment is the use of an appropriate number of animals per treatment. A common serious flaw in animal experiments is insufficient replication, and, ironically, it is an easy attribute to deal with properly. Replication must be done for two reasons: (1) to estimate the background noise in experiments (in statistical jargon, error variance, s^2_e), and (2) to make experiments sufficiently powerful to determine if true differences exist between treatments. It is wasteful to use more animals per treatment than necessary for statistical validity; it is even more wasteful to use too few. In the first case, at least the experiment is valid and the information can be put to use. However, if too few animals are used, the effort and animals are completely wasted, and the work will have to be redone. Even worse, incorrect conclusions may be drawn, possibly leading to animal suffering, economic loss, and ruined reputations.

The problem of background noise is illustrated for weight gain in rats in Table 3. Even though the average gain for both populations was 18 g, the inbred strain had less variability than the outbred strain. Note that no treatment was given to either strain. Only one in ten of the inbred rats deviated from the mean by 2 or more g, whereas seven of ten of the outbred rats deviated

TABLE 3
Weight Gains (g) between 3 and 4 Weeks in 20 Male Rats of 2 Strains

Inbred strain			Outbred strain		
Rat no.	Weight	Difference from mean	Rat no.	Weight	Difference from mean
1	17	–1	11	17	–1
2	19	+1	12	20	+2
3	18	0	13	14	–4
4	16	–2	14	21	+3
5	19	+1	15	18	0
6	19	+1	16	20	+2
7	18	0	17	15	–3
8	17	–1	18	22	+4
9	18	0	19	16	–2
10	19	+1	20	17	–1
Absolute average (g)	18	0.8		18	2.2

by 2 or more g. Because there is more intrinsic variability in weight gain of the outbred strain, more animals will be required per treatment for statistical validity than with the inbred strain. Frequently, there is less variability with inbred animals (or identical twins), which makes them good subjects for some kinds of experiments. Estimated variability can be quantified mathematically by squaring deviations from the mean. This is called variance; exact formulas for variance can be obtained in any standard statistical textbook.[2,3]

To illustrate the problem with background noise, consider the following experiment. Rats normally are fed a 10% fat diet. Suppose that we want to determine if rats gain more weight between 3 and 4 weeks of age if given a 20% fat diet. Let's further assume that the true treatment effect is a gain of 2 extra grams each, with the 20% fat diet compared to the standard 10% fat diet, and that the outbred strain of rats in Table 3 will be used for the experiment with three rats per treatment. If by chance we had chosen rats 11, 13, and 15 to receive 20% fat, and rats 12, 14, and 16 to serve as controls, the new weight gains for the treated rats would be 17, 16, and 20 g (values in Table 3 plus 2 g) and for the controls 20, 21, and 20 g (same values as in Table 3). The means would be 17.7 g for the treated group and 20.3 g for the controls, and our conclusion would be completely erroneous; instead of the correct conclusion that a 20% fat diet resulted in an average additional gain in weight of 2 g, we would conclude that a loss of 2.6 g occurred. This is due to unfortunate sampling of the population, which can easily happen. As sample size is increased, such errors in sampling are much less likely to occur, that is, the power of the experiment increases. Also note that such an incorrect conclusion would have been very unlikely for rats selected from the inbred strain because the intrinsic variability is much lower. The important point is that the number of animals required for statistically valid experiments depends on the intrinsic variability (error variance) of the response being measured. There are formulas for calculating this number under various circumstances. This will be dealt with in more detail shortly.

V. MINIMIZING ERRORS IN EXPERIMENTAL CONCLUSIONS

Table 4 illustrates mistakes that are possible with experimental conclusions. One can minimize Type I errors (conclusions that a treatment difference occurred when in fact there was none) by choosing stringent levels of statistical significance from statistical tables (for example,

TABLE 4
Truth and Possible Conclusions for Experiments

Truth about experiment	Possible conclusions	End result
No treatment difference	No treatment difference	Correct conclusion
	Treatment difference	Type I error
Is a treatment difference	No treatment difference	Type II error
	Treatment difference	Correct conclusion

critical values for t-tests, F-tests, or Chi-square). This commonly is termed the alpha level of significance. Typically accepted levels of alpha for animal experiments are 0.05 or 0.01. With alpha of 0.01, on the average, there will be 1 wrong Type I conclusion per 100 statistical comparisons. Frequently, experiments are published and conclusions made with alpha $\leq$0.05. One must be more cautious about accepting conclusions when $p = 0.05$, but also realize that in 95% of comparisons, these conclusions will be correct.

The approach to minimizing Type II errors is to have sufficient numbers of animals per treatment to find a treatment difference if one in fact exists. The concept of treatment difference is complicated because observed treatment means are only estimates of true treatment effects; an infinite number of animals would be required to estimate a true treatment effect. Another problem is that one can argue persuasively that there always will be a difference between two (or more) treatments if enough observations are made. To deal with this problem, it is necessary to establish how large a treatment difference is expected or important. For example, if 10% more dogs survive distemper when immunized with product A compared to product B, one could argue that it is an important treatment difference worth detecting. On the other hand, if cows give 2 kg more milk per month on ration A than ration B, such a small difference (about $.50 worth of milk) probably is not important if the rations cost the same. Perhaps 5 or 10 kg/month would be important because of increasing profit to the dairyman, decreasing environmental costs of milk production, or decreasing the price of milk to consumers. Selecting the magnitude of an important treatment difference depends on costs of the experiment and value of the difference scientifically, economically, and socially.

To get to the heart of the matter, to decide rationally on how many animals to use per treatment, four quantities are required. These are d, the minimal treatment difference that is important to detect; s^2_e, an estimate of the error variance (s^2_e) in the experimental material obtained from a previous experiment or observing a group of untreated animals; alpha, the probability of making a Type I error (typically 0.05); and beta, the probability of making a Type II error (typically 0.1). With a beta of 0.1, the experiment has a 90% chance (1-beta) of detecting a treatment difference, d, if a true treatment difference of magnitude d or greater in fact exists. The ability of experiments to detect treatment differences (1-beta) is called the power of the experiment.

Formulas for determining the proper number of animals per treatment from alpha, beta, d, and s^2_e are found in standard statistical texts; usually a statistician should be consulted to assist with the calculations. For binomial responses (e.g., pregnant or not, dead or alive, male or female, etc.) an estimate of s^2_e is not required.

Frequently so few animals are used per treatment that there is no possibility of finding an important treatment difference, even if one exists. In fact, the difference found may even be in the wrong direction, as just illustrated for weight gain in rats. Such experiments waste animals, time, money, and other resources. Similarly, if more animals are used than the number required to detect an important difference, d, there is a similar waste of resources. Procedures to ascertain

TABLE 5
Illustration of Blocking by Weight in Experimental Design

Body weight	Parasite treatment (no. of gilts)	Control (no. of gilts)
Light	3	3
Medium	3	3
Heavy	3	3

appropriate numbers of animals per treatment are easy with help of a statistical consultant. A common occurrence after making these calculations is to find that the number of animals required to do a meaningful experiment exceeds the resources of the investigator, and the experiment is modified to become a meaningful experiment, or not done at all.

A very important point is that an experiment in which no treatment difference is detected is not a failure if the experiment was well designed and included sufficient replication. "Negative results" are frequently very useful.

VI. REDUCING ANIMAL NEEDS BY REDUCING ERROR VARIANCE

In the previous section, it was pointed out that for useful experiments, the number of animals needed per treatment increased as variance (s^2_e) increased. Two sources of variance (differences) between animals were discussed, those due to treatments and those due to background noise. Often there are additional identifiable sources of variation in experiments. One example is animal differences such as age, weight, or breed; e.g., animals never will be exactly the same weight. Another source of variation occurs due to methodological constraints. Since the entire experiment cannot be done instantaneously, parts may be done on different days or times of the day or different times of the year. These differences may or may not affect the responses being measured; if they do, they contribute to increased variance and increased animal requirements.

A common approach to dealing with these sources of variation is to work them into the experimental design, determine the amount of variation due to such factors, and subtract it from the background noise, thus lowering error variance and decreasing animal needs. The most common procedure for doing this is called blocking.

An example is given in Table 5. We may want to test effects of treatment for parasites on litter size in gilts, which are young female pigs. Usually, larger gilts tend to have larger first litters than smaller gilts. For this experiment, the pigs are divided into three groups by weight. Then, within these groups (blocks, in statistical jargon) gilts are randomly assigned to treatment and control groups. When these data are analyzed, three sources of variation can be studied: treatment variation, weight variation, and random noise s^2_e. The latter term will be smaller if variation in weight is removed, which makes the experiment more powerful to determine treatment differences and thus requires fewer animals. Two other points will be mentioned: first, an additional source of variation, the weight x treatment interaction, could be considered. This is beyond the scope of this chapter. Second, in small experiments there are some situations in which blocking may be inappropriate because of reducing degrees of freedom in error variance. Consulting someone knowledgable in these matters is very important so that the experiment can be designed optimally.

There are many other ways of reducing error variance by design and analysis strategies. For example, the previous experiment could also be approached by a technique called analysis of covariance, in which the weight of each pig is considered directly rather than by blocking gilts into light, medium, or heavy groups. Some of these approaches are fairly complicated. However, a few hours of consultation with a statistician is a small investment in improving experiments and decreasing resources required.

VII. BLINDNESS

Biases frequently creep into experiments because someone knows which animals are receiving which treatments. Biases can occur in the way animals are fed or handled or when estimating responses to treatments. These biases usually are unintentional, but can ruin an experiment nonetheless. For some treatments, unintentional biases are virtually impossible, for example, an injection either is or is not given. Similarly, for some responses, one simply records a result, e.g., weight. On the other hand, in some cases complete blindness is essential for valid experiments. A good example would be a clinical trial for new treatments of sick animals. In a so-called double-blind experiment, neither the veterinarian nor the client knows which treatment the animal is given because the treatments are coded. Only after the experiment is completed is the code broken so results can be analyzed. This prevents assigning animals from good clients to one treatment or the control, or sicker animals to one of the groups. It assures unbiasedness. It is not unethical if the owners involved have been informed and given their consent. Such clinical trials are done frequently with human patients when it is not known if a new treatment is better or worse than the standard treatment (or has no effect). Of course, there usually is some basis for thinking that the new therapy might be an improvement. It has been proven over and over that such approaches are one of the most efficient procedures for improving human and animal health.

Examples of evaluating responses blindly are reading histological slides for evidence of tumor cells, or examining a sample of sperm and estimating the percentage that are motile. It is essential to make observations blindly for any response that has a component of subjectivity. The slides or samples should be coded by someone other than the person making the observations. Blind measurements are not required for strictly objective criteria, but in some situations are still a good idea to prevent tinkering with the data.

VIII. THE EXPERIMENTAL UNIT

The entity that receives an experimental treatment is called the experimental unit. It may be a pregnant rat or a ram or a bowl of goldfish. Commonly, more than one measurement is made per experimental unit. For example, one might examine each of the fetuses of the pregnant rat or take multiple blood samples from the ram or weigh the individual goldfish. For purposes of estimating error variance for statistical tests, it is the experimental unit that must be considered. Thus, for example, total litter weight or average weight of a litter (and not the weight of each fetus) should be used in statistical analyses when the treatment is applied to the mother rather than the individual fetuses. Similarly, taking ten blood samples from a ram is very different from taking one blood sample from each of ten rams. Repeated sampling over time results in very complex statistical analyses,[1] and usually requires professional statistical help. In the third example, one bowl with ten goldfish is quite different from ten bowls with one goldfish each, at least if the bowl of fish is handled as a group (e.g., in feeding). Variability due to repeated sampling of the same experimental unit is called sampling variance. It usually is useful to make these kinds of repeated sampling measurements in experiments; the problem comes in confusing sampling variance with experimental error variance. Since sampling variance generally is considerably lower than error variance, substituting the former for the latter increases the likelihood of Type I errors, that is, concluding that there is a treatment difference when none exists. Substituting sampling variance for error variance is a very common mistake in animal experiments.

IX. ERRORS IN LOGIC

An example of an error in logic will be illustrated with an experiment on superovulation of cows. The response measured was number of embryos obtained per cow. This response is extremely variable; with the current standard treatment, nearly 30% of cows do not respond at all, i.e., produce no embryo, the majority produce between 3 and 15 embryos, and a few produce more than 15 embryos. The reason that some cows do not respond is partly due to chance and partly due to something about those particular cows. Perhaps the standard treatment is not appropriate for all cows. In any case, a group of cows was given the standard treatment, and those that did not respond, i.e., had no embryos, were treated for superovulation with a new treatment. The average number of embryos recovered with the new treatment was 3.2 compared to 0 with the standard treatment, and the experimenters concluded that the new treatment was superior for cows that responded poorly with the standard treatment.

There are several problems with this experiment. First, the two treatments were done at different times, which precluded assigning animals to the two treatments at random. Even worse, responses of the animals given the new treatment were compared to the same animals when they did not respond to the standard treatment. The error in logic is that part (perhaps most or all) of the reason the cows responded poorly to the standard treatment was chance or background noise; some had to be at the bottom of the distribution. A clear way of seeing the problem is that the experiment was rigged from the beginning because no treatment could possibly be worse than 0, which was the mean response for the standard treatment for these **selected** animals. We cannot conclude anything from this experiment. The new treatment could be worse, the same as, or better than the standard treatment. Obviously, the correct way of doing the experiment on this population (cows that respond poorly to standard superovulation) is to randomly assign half of them to be superovulated with the new treatment and the other half with the standard treatment (again).

A related error in logic that is exploited commercially in various guises is as follows. A company is set up to predict sex of babies to be born. The fee is $10, and if an error is made, the fee is refunded. This of course is a difficult offer for a customer to refuse. Knowing the sex ahead of birth is important for many people, and if a mistake is made (which the customer is led to believe is unlikely), it is of little consequence, was going to happen anyway, and the service cost nothing since the $10 is refunded. The customer is satisfied in either case. The method of sexing is to flip coins (or if the clientele are sufficiently scattered, to predict boys in all cases, both because this population usually prefers boys and because about 53% of babies are boys). Thus, the business makes $10 on half of the clients and nothing on the other half, or on the average, $5 each for doing nothing.

This trick sometimes is used commercially in sexing semen of domestic animals. Numbers of semen doses in a lot are chosen carefully to make refunds minimal. Frequently, more than one incorrect sex is required before a refund is given, and the refund does not include semen resulting in the correct sex or resulting in no pregnancy. Similarly, experimental data can be manipulated by excluding animals that do not respond or by excluding farms from which no treatment effect was observed. To evaluate claims, one needs to understand how experiments were done and determine if all data were used in analyses.

X. PROBLEMS WITH COMPUTER ANALYSES OF DATA

Computers have improved procedures for statistical analysis immensely, but often distance the researcher from the data; data frequently are collected and given to someone else to analyze. This can lead to disaster; the person in charge of the experiment must bear the ultimate responsibility for its correctness. It is one thing to administer treatments or evaluate responses

TABLE 6
Latin Square Design for Superovulating Cows

	Superovulation treatments (cow number)			
Season of year	**A**	**B**	**C**	**D**
Winter	1	2	3	4
Spring	2	1	4	3
Summer	4	3	1	2
Fall	3	4	2	1

Note: Each cow receives each treatment and each cow is superovulated in each season.

blindly, but entirely different to be so detached that no one really knows what was done or how data were analyzed or even if the correct data were analyzed. The following examples illustrate this problem.

Each of 20 cows was given 4 superovulation treatments. Furthermore, each cow was superovulated in each of the four seasons such that one fourth of the cows received each treatment in each season. This is called a Latin Square design and is illustrated for four of the cows in Table 6. The value of the design is being able to separate out variances for individual cows, seasons, and treatments so they can all be studied or controlled in the experiment.

The data were collected and given to another person for statistical analysis, and the results published as an abstract. Nearly a year later, a third person by chance perused the raw data and noted that the published means seemed incorrect, so the data were reanalyzed. It evolved that means were identical on reanalysis, but attributed to different factors. In the first analysis, means labeled as winter, spring, summer, and fall were actually means for superovulation treatments D, C, B, and A and vice versa. Apparently, during communication between researcher and computer programmer, rows became columns and vice versa, and to add further confusion, means were in reverse order — D to A instead of A to D. Two problems are illustrated, failure to label data adequately and failure to check the computer output independently. In most cases, at least one of the means of any computer output should be checked independently with a calculator; this eliminates such systematic errors.

A second example of communication problems between people and computers occurred as follows. A researcher sent samples of blood serum to two laboratories for analysis of cortisol. In one case, the computer software gave concentrations in nanograms per milliliter unless concentrations were less than 1 ng/ml, in which case concentrations were given in picograms per milliliter (1 ng = 1000 pg). Most of the concentrations were in the 2- to 40-ng/ml category, but one sample had little cortisol, about 100 pg/ml. From the second laboratory all concentrations were given in nanograms per milliliter. When the two sets of data were received, the researcher asked his technician to determine if the values from the two laboratories were in agreement. The technician, who had little training in these matters performed a simple linear regression/correlation analysis, but failed to change the 100-pg value to 0.1 ng. As a result, the regression line changed from a positive slope to a negative one (Figure 1). The researcher noted these parameters without examining the raw data and concluded that the assay systems of the two laboratories were not measuring the same molecule, arbitrarily decided that one laboratory was right and the other wrong, and from then on had all samples analyzed at the first laboratory. Fortunately, the information from this exercise lay dormant for 4 years, when a second researcher discovered the error. On reanalysis, the relationship between the two assay systems was shown to be highly positively correlated.

This again illustrates the need to examine the raw data. Simply graphing the data (and examining the graph) would have pointed out the error in this case.

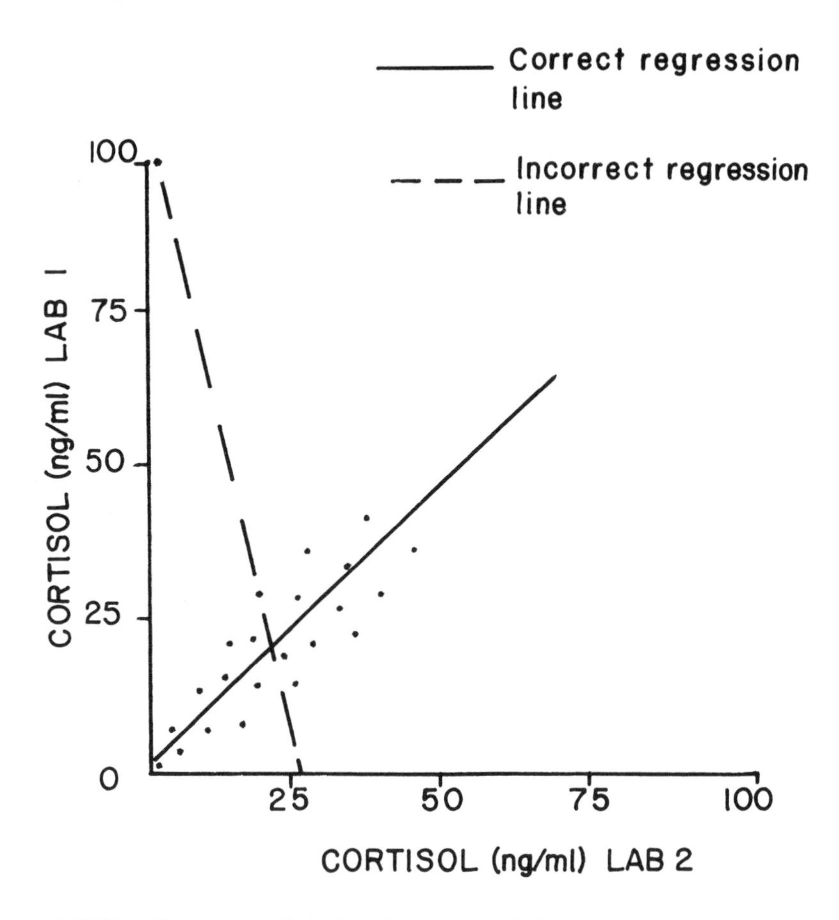

FIGURE 1. Consequences of mistaking picograms per milliliter for nanograms per milliliter for one data point.

XI. OUTLIERS

Often when an experiment is done, an observation is recorded that is inconsistent with the rest of the data. Such aberrant values are termed outliers. When such a data point is found, the first step should be to examine records to see if an error occurred, as illustrated in the previous section, or if something unusual occurred when the sample was collected. An example might be that urine was noted in a sample of semen collected from a rabbit. If no unusual occurrence was recorded, it is possible to apply certain statistical procedures to determine if the value could be considered an outlier on statistical grounds. If so, there is justification for ignoring that data point.

Such discarding of data that seem not to fit should be done only rarely, and when done, usually should be noted in published results. Another alternative is to analyze the data with and without the aberrant value and present both versions so the reader can decide if discarding the value is justified. An admittedly arbitrary rule of thumb that seems appropriate is that in typical animal experiments, one should not discard more than one data point per ten experiments because of suspecting it to be an unexplained outlier. If more data than this need to be discarded, usually very sloppy work is being done. It is easy to get into the trap of discarding data points that are not outliers.

XII. FATAL FLAWS IN EXPERIMENTS

Statistical consultants often are asked to analyze data after an experiment has been completed. Often this is a salvage operation which requires making the best of a poorly designed experiment. Sometimes there is no design at all, and not even a clear hypothesis. In some cases it is possible to do meaningful analyses of such data, but frequently one cannot address the original question directly.

In other cases, however, there are fatal flaws, and the hypothesis intended for testing cannot be tested. Probably the most common fatal flaw is bad controls. If one is testing a new treatment, in most cases there must be contemporary controls or another contemporary treatment for comparison. No amount of statistical prowess can rescue an experiment in which valid comparisons cannot be made. In some cases, historical data can be used as valid controls, but in most cases this is risky. A second common fatal flaw is biased assignment of animals to treatments. Frequently, small animals will be assigned to one treatment, and large ones to another, or sick animals to one treatment and healthy ones to another. These clearly lead to flawed experiments that are not salvageable. A third flaw that is highly suspect is failure to evaluate subjective responses blindly.

XIII. OTHER CONSIDERATIONS

Planning experiments to test simple hypotheses is a vastly superior approach to collecting lots of data to see what falls out upon statistical analysis. For example, one could make hundreds of measurements on prairie dogs to determine if there are correlations among the measurements. The first problem is that correlation and causation are not synonymous. The second problem is that if one observes 100 correlations at $p = 0.05$, there will be 5 due to chance alone. If there are 10 significant correlations at $p = 0.05$, about half (5) will be due to chance and half will be real, and it will be impossible to tell which is which, especially in the absence of reasonable hypotheses. Such correlation analyses are of value in many instances, but frequently significant correlations at best only suggest hypotheses for further experiments.

A final point is that there is no perfect experiment. Some researchers play around so much in attempting to design the perfect experiment that they rarely do a formal experiment, and even if they do, they become frustrated because this or that is not ideal. Standard statistical procedures are quite robust, and if a modicum of effort is invested in proper design and analysis, one will rarely be misled by experiments.

REFERENCES

1. **Gill, J. L. and Hafs, H. D.,** Analysis of repeated measurements of animals, *J. Anim. Sci.,* 33, 331, 1971.
2. **Snedecor, G. W. and Cochran, W. G.,** *Statistical Methods,* 7th ed., Iowa State Univ. Press, Ames, 1980.
3. **Steel, R. G. D. and Torrie, J. H.,** *Principles and Procedures of Statistics,* 2nd ed. McGraw-Hill, New York, 1980.

Part IV
Alternatives to the Use of Live Animals

Chapter 7

ALTERNATIVES TO ANIMALS: PHILOSOPHY, HISTORY, AND PRACTICE

Andrew N. Rowan

EDITOR'S PROEM

The demand for "alternatives to animals" is one of the most emotional aspects of the debate about animal research. Animal advocates demand the replacement of animals by non-animal alternatives, and accuse scientists of foot-dragging, while researchers often dismiss such demands as symptomatic of naivete about science. At the same time, new law and policy focus attention on seeking alternatives. In this chapter, Dr. Rowan attempts to disambiguate the various senses in which one can speak of alternatives, and also discusses some reasons why their development and acceptance may be forestalled. Using the field of toxicology as his major example, Rowan illustrates the multiplicity of modifications which count as alternatives, and also exemplifies the complex factors in society in general and in scientific culture which constrain rapid advances in alternatives. Implicit in Rowan's discussion is the suggestion that development of alternatives is as much a matter of mind-set as of technological break-throughs.

TABLE OF CONTENTS

I. INTRODUCTION

For 150 years the use of animals in medical research has aroused impassioned and bitter protest. In the 19th century, these forces were relatively powerful,[1,2] but their influence decreased steadily in the first half of the 20th century. As medical research went from one triumph to another, antivivisection groups lost even the support of humane societies. In 1950, however, with antivivisection societies reduced to relatively powerless rhetoric, the animal welfare movement underwent a reformation and became increasingly outspoken against at least some animal research. Initially, their protests did not have much general support. Opinion polls indicated that the public overwhelmingly approved of medical research, even if it used animals.[3] Today, public support for science in general has declined[4] and people appear to be more negative than positive about animal research,[5] although a majority of the public (77%) still say they accept some use of animals in research. The shift in attitude that has occurred over the last 30 years has strengthened the humane movement in its opposition to animal research. In the last 10 years, membership of the Humane Society of the U.S. has increased more than 10-fold to 550,000 and revenues have increased 4-fold to $10 million per annum. There has been a similar dramatic rise in the number of animal rights groups. (By and large, animal welfare groups campaign against animal pain and suffering, animal rights groups oppose animal exploitation and killing whether or not suffering occurs). Most such animal rights groups are still small (under 1000 members) but People for the Ethical Treatment of Animals (PETA) has, according to its claims, grown from 20 to over 250,000 members in 6 years.

The postwar change and invigoration of the animal protection movement has been prompted, in part, by the tremendous increase in biomedical research after World War II and also by the legally mandated surrender of unclaimed shelter animals to licensed research institutions that occurred between 1946 and 1955. The increase in research money available led to a rapid increase in the number of animals used in laboratories in the 1950s and 1960s. At the same time, the National Society for Medical Research orchestrated the passage of a number of state laws and city ordinances which required pounds and shelters to turn over unclaimed animals to research institutions. Release of these animals posed a threat to the operators of animal welfare societies and seemed to undermine their basic goal — namely, the provision of a suffering-free sanctuary and, when necessary, a humane death for animals. This threat, together with the obvious increase in animal research, jarred the humane movement out of its uneasy apathy on the question of the use of animals as laboratory tools. Increased public pressure resulting from protests organized by the animal movement led to the passage of the first act regulating research animal use in 1966. Strengthening amendments were passed in 1970 and 1985. The animal welfare movement now has more political acumen and power than at any time in this century and is kept on its toes by a growing band of young and active militants who equate speciesism (i.e., discrimination against species other than our own) with sexism and racism.[6] While there are many issues that stimulate the concern of the modern animal welfare activist (e.g., seal clubbing, whaling, coyote control, and intensive animal husbandry), animal research arguably leads the list. In particular, there is widespread support for federal legislation to curb *painful* research involving animals and to promote the development and use of alternatives to animals.

The concept of alternatives is, in fact, a relative newcomer to the debate over animal research, being no more than 30 years old. Nevertheless, it has grabbed the imagination of those who protest against the use of laboratory animals. It also seems to offer an ideal solution to those large numbers of the public who are uneasy about animal suffering in experiments, but who do not wish to forego the possible future benefits of biomedical research. Thus, they perceive that biomedical research may be able to continue *without* animals because we can use existing alternatives and/or develop new ones. However, such simplistic arguments, combined with the exaggerated rhetoric of some segments of the animal welfare movement, made biomedical research organizations very suspicious of the whole notion of alternatives. There also was, and still is, confusion over just what the term "alternatives" really means and whether the concept has any merit in biomedical research and testing.

Some biomedical researchers argue that the term "alternatives" is misleading because it gives the public the mistaken idea that we can immediately replace all use of animals. As a result, some scientists argue that we should refer to "adjunct" or supplementary methods that might be used to *complement* animal research. Nonetheless, it is also likely that some of the scientific opposition to the notion of alternatives (or the three R's) is based on the fact that alternatives is a concept developed and promoted by animal protection groups and is, therefore, presumed to be inimical to the best interests of biomedical and behavioral research. If the strong emotional overtones surrounding the "alternatives" controversy can be removed, it should become clear that science has much to gain and little to lose by incorporating the alternatives concept as part of a coordinated approach to new technique development, review, and implementation.

II. HISTORICAL BACKGROUND AND DEFINITION

Most people date the concept of alternatives back to the 1959 book by Russell and Burch,[7] who enunciated several principles for humane experimental technique. Their central thesis was that researchers should follow the "three R's" of replacement, reduction, and refinement. That is, the investigator should seek, where possible, to replace the use of living animals with nonsentient material, to reduce the number of animals used, and to refine techniques so as to reduce animal pain and suffering. It should be noted that Russell and Burch were both practicing scientists who recognized that only science would produce the advances which would lead to the development and use of more alternatives. Doubtless they would have enthusiastically endorsed Medawar's comment that "we must grapple with the paradox that nothing but research on animals will provide us with the knowledge that will make it possible for us, one day, to dispense with the use of them altogether."[8]

While the three R's of Russell and Burch provided the roots for the concept of alternatives, the term has different meanings for different people. There are some who use the term to refer only to total replacement of laboratory animal use, but most protagonists use the term to refer not only to replacement, but also to reduction and refinement. Therefore, the development of a new statistical technique which reduced the number of animals required for a toxicity test would normally qualify as an alternative.

Russell and Burch[7] also discussed the concepts of fidelity and discrimination in animal models. For example, a mammal is a high-fidelity model of the human. However, one may be able to conduct research or testing in a low-fidelity but high discrimination model, such as microorganisms for nutrition studies. Russell[9] argued that progress towards replacement methods is "hindered by an insidious and widespread assumption" which may be called the "hi-fi fallacy". The major premise is that high fidelity is desirable in general. The minor premise is that mammals are of exceptionally high fidelity as models of the human organism and therefore should be used as much as possible. This attitude has not changed much in the last 25 years. Nevertheless, the fact is that, as we learn more about a biological phenomenon, the more likely it is that we can develop adequate low-fidelity models to study it — as in, for example, the use of bacterial and mammalian cells to screen for carcinogens and mutagens.[10]

III. ALTERNATIVES AS TECHNIQUES

In the sometimes heated debate over alternatives and their merits, we frequently appear to forget that the term "alternative" refers specifically to a technique or method that results in a modified use of animals and that the development of a new technique could be of benefit to a wide range of fields, including immunology, cancer research, and toxicology. Funds and resources can therefore be applied to support such technique development and related resource needs with fairly predictable results. In toxicology, if more money is applied to methods development, evaluation, and validation, we can be reasonably confident that in 10 years we will have better test methods than if serendipity and the unsupported interest of individual scientists remained the only guiding principle.

"New laws and theories mark the great forward strides in science. A jumble of confusing and chaotic facts is suddenly made to make sense, as Mendeleev did for chemistry in divining the periodic table of elements. But how did all the facts get established in the first place? Because someone invented a new technique for querying nature. *There's a case to be made that technique, not theory, is the principal bottleneck in scientific progress.* Each new technique makes accessible a new field of discovery, which will be fully explored sooner or later by those who rush in." (From *N.Y. Times,* Editorial, 17 October, 1986 — emphasis added).

Bernard Dixon, a well-known science journalist and currently the European editor of *The Scientist,* has commented that scientific advance is based on four factors — the intuitive leap (imagination), rigorous self-criticism (usually including experimental verification), luck, and technique availability.[11] Of these, the first two are well accepted as important in the process of scientific discovery,[12] and few people would quibble with the idea that luck also plays a role. However, the idea that technique development and implementation is also important is likely to raise a few eyebrows. For example, few scientists would announce that they are working on the development of a better technique to measure, for instance, brain peptides. They would more likely describe the project as a search to understand memory, or pain modulation, or whatever. Some scientists do, nevertheless, recognize the place of technique development. Sir Hans Krebs bemoaned the fact that editors gave technique papers short shrift and went on to emphasize the importance of methods development to scientific advance.[13] The noted science historian, Derek de Solla Price, also talked of the importance of techniques and remarked that "it is often techniques and technologies, rather than theories or hypotheses, that originate scientific breakthroughs."[14] It is also noteworthy that many Nobel prizes have been awarded specifically for the development of new techniques — the prizes for radioimmunoassay (1977) and monoclonal antibody (1984) development being two recent examples.

Few studies have been undertaken to evaluate the relative importance of technique development. Comroe and Dripps comment tangentially on the topic in their retrospective analysis of the major developments in the care and treatment of cardiovascular and pulmonary diseases.[15] With the help of a panel of experts, they selected the top ten advances that were the most important for patient care since the 1940s. For these 10 advances, they identified 137 essential bodies of knowledge (e.g., electrocardiography, cardiac parameters, development of anticoagulants, etc.). They then identified 2500 specific scientific reports that were particularly important in the development of 1 or more of the 137 bodies of knowledge, and, finally, with the aid of numerous consultants, they narrowed these down to 529 key articles. Of these, 61.7% reported results from basic research and 15.3% were concerned solely with the development of new apparatus, techniques, or procedures. This is a high proportion when one considers that the importance of technique papers was probably considerably underestimated by Comroe and Dripps. The general attitude among scientists that the development of a new method is a rather mundane activity most likely would have biased the consultants against the selection of a "methods" paper as a key article. Furthermore, some of the key articles classified as basic research were also critically dependent on the development of a new technique. For example, in one of the key papers described the authors use an intracellular electrode to measure transmembrane potentials of heart muscle cells.[16] However, such a study required the prior development of such an electrode and subsequent improvements in the technology.

A recent paper by Frank on the history of neurophysiology specifically extolls the importance of technique availability.[17] The author cites Adrian (a 1932 Nobel Laureate and neurophysiologist) as arguing that techniques are the rate-limiting factor in scientific advance. Frank comments that while "experimental techniques are often seen as the necessary but rather prosaic way of getting from here to there — from a brilliant insight to a logical and effective publication," they are, in fact, intimately bound up with both the conceptual substance and the social process of biomedical experimentation.[17] Therefore, even the relatively few reports that point to the importance of technique availability constitute a strong argument supporting Adrian's thesis that technique development is the rate-limiting factor in science.

Frank adds another point that is relevant to discussions of alternatives. He notes that techniques have inertia and comments that the "investment of money or — more often — time and effort in the development of a technique will encourage an individual or a laboratory to continue to use it, or some relatively simple modifications or extensions of it, until an unusual circumstance — a significant crisis or special opportunity — arises".[17] The relevance of this is that it is often argued by those who seek to resist any adoption of an "alternatives" policy that scientists are constantly seeking new approaches and methods and will use them as soon as they are available. Anybody who has worked in a research laboratory knows that this is nonsense. Each laboratory has its own culture and habits, and many of them tend to stick with approaches that are familiar. This "technique inertia" is many-fold greater in toxicity testing where one is dealing not with the culture of a single laboratory, but with the resistance to change of a large bureaucracy.

There is plenty of room for a more active promotion of alternatives by funding agencies and government regulators. For example, Public Law 99-158, the Health Research Extension Act, contained a section requiring the National Institutes of Health to establish a plan for the development and validation of alternatives. On January 12, 1987, the Health Grants and Contracts weekly carried a notice announcing the availability of grants for studies of research methods that limit the use of vertebrate animals. As is well known, there is nothing that stimulates scientific creativity and interest quite so effectively as the availability of research funding.

IV. ALTERNATIVES IN TOXICITY TESTING

The alternatives concept has made its strongest showing in the field of toxicity testing, where a large number of animals are used in relatively few tests employing well-established techniques and methods. In Great Britain, 20% of the animal experiments (874,000) are conducted to evaluate toxic effects, especially acute toxicity.[18] Therefore, the animal welfare movement has tended to focus on toxicity testing as the most promising area for the development and application of alternatives.

The initial pressures from animal welfare and antivivisection groups were relatively unfocused and resulted in vague proposals to use and/or develop cell cultures and computer models instead of animals. However, in the mid-1970s, animal welfare groups began to learn to use criticisms and suggestions that had appeared in the scientific literature. Thus, the following proposal by Muul and colleagues[19] was widely quoted and cited in support of the need to develop alternatives.

> A team of expert toxicologists and scientists from related fields should be assembled to evaluate existing technology and identify a battery of the most predictive screening tests, including in vitro systems, animal models, and chemical behavior. A combination of quick tests could replace the conventional protocols, whereas any single test might not. This team could also perform cost-benefit analyses and estimate how much, if any, sacrifice of confidence would result from using a battery of screening tests at this time. *Use of a combination of screening tests might allow a tenfold reduction in cost and fivefold reduction in time for toxicological testing.* We would expect little or no sacrifice of safety, since most of the tests tend to err on the side of false positives. Standard methods could still be employed if indicated by the screening results. (Emphasis added.)

Another tactic involved support for practicing scientists to examine the issue. In 1980, for example, a panel of toxicologists convened under the sponsorship of the Canadian SPCA recommended that

> an increasing effort is necessary in the development of alternatives to the use of laboratory animals in toxicological testing. It is therefore recommended that the federal and provincial government departments and agencies and other organizations and foundations supporting toxicological research initiate and fund research programs with the specific objective of developing and validating non-animal models for use in the safety evaluation process.[20]

The support provided to animal welfare arguments by the above recommendations was bolstered by reference to research reports dealing with specific techniques and studies which could possibly lead to reductions in animal use. Tissue culture has come in for particular attention. For example, in 1954 Pomerat and Leake[21] suggested that tissue culture could be a useful tool in toxicology studies. Nevertheless, it remained largely a laboratory curiosity; 17 years later, Rofe[22] suggested that animal studies were unlikely to help overcome the difficulty of evaluating human risk from animal data and proposed "a new kind of multilateral study" which exploited the advantages of tissue and organ culture. But the traditions of toxicology and tissue culture remained separate. The barrier between the two were reinforced by such factors as inadequate toxicological training of cell biologists, ignorance of advances in cell culture techniques among animal toxicologists, and by simplistic animal welfare claims that viewed *in vitro* toxicology as *the* answer to all animal testing.

Ekwall[23] notes that tissue culture studies have suffered from a number of very specific problems. First, the wide variety of cell cultures and media that are available means that there is little standardization from one laboratory to another. As a result, each set of studies stands in isolation and cannot be directly compared with another. Second, the questions asked of tissue culture techniques are generally posed in terms restricted to concepts from whole-animal toxicology. This is directly related to the lack of information on chemical injury to basic cell functions — a deficiency which must be corrected before *in vitro* toxicity testing can come into its own.

The use of continuous cell lines which have lost the specialized functions of the parent cells has also been a major disadvantage in *in vitro* toxicity studies. However, this either has been or is being overcome by the expanded use of primary or early passage cells and by improved media formulations which extend the periods during which cell cultures exhibit the properties of the original differentiated tissue. In addition, tissues and cells of human origin, and of a quality suitable for toxicity studies, are becoming available. This is very important since the use of human tissue overcomes the problem of extrapolating from one species to another, although one still has the considerable task of predicting *in vivo* effects from *in vitro* data.

The development of alternatives involves both scientific and political issues. The following examples give some idea of the varying approaches being pursued and the tensions involved in changing from one system to another.

V. THE LD_{50} TEST — REDUCTION, NOT REPLACEMENT

The calculation of the median lethal dose (LD_{50}) for the measurement of toxicity was introduced by Trevan in 1927.[24] Initially, the LD_{50} was used to standardize such potent biologicals as digitalis and insulin. As the practice of food, drug, and chemical toxicity testing grew, the LD_{50} was pressed into service as a standard measure of the toxicity of a substance and it soon became a general screen that began to be viewed as a fundamental toxicological property of a chemical, rather like specific gravity. In fact, the LD_{50} represented a very shakey foundation for the development of the science of toxicology, and a number of concerned scientists pointed this out.

In 1960, Barnes stated that too much effort was being spent on producing accurate LD_{50} figures and not enough on defining acute toxicity effects.[25]

In 1973, Zbinden, a World Health Organization (WHO) toxicology consultant, called the LD_{50} test "a ritual mass execution of animals."[26]

On a later occasion he added that "most experts considered the modern toxicological routine procedure a wasteful endeavor in which scientific inventiveness and common sense have been replaced by a thoughtless completion of standard protocols.[27]

In 1969, Baker stated that acute toxicity studies "are of little use and are expensive in animals. The main information they give is an indication of the size of dose required to commit suicide."[28]

In 1968, three British toxicologists reviewed the usefulness of the LD_{50} and presented convincing arguments that the LD_{50} figure obtained was unnecessarily precise and that, for nearly every purpose, an approximate lethal dose test was more than adequate.[29] They showed how small changes in test protocol (animal genotype, animal environment, etc.) could have a very large effect on the resulting LD_{50} and that the precise LD_{50} was not more useful to toxicologists, pharmacologists, clinicians, or regulatory scientists than an approximate lethal dose figure. Despite these criticisms, the standard LD_{50} statistic, determined in a procedure using 40 to 200 animals, was unquestioningly adopted by a wide range of regulatory agencies as a standard requirement for the registration and toxicity classification of household products, pesticides, drugs, and other chemicals.

In the mid-1970s public pressure resulted in the British Home Office requesting its advisory committee on animal experimentation to review the use and relevance of the LD_{50} test. The committee stopped accepting evidence in mid-1978 and the report was published in mid-1979.[30] The main recommendations of the Home Office report included the following: first, LD_{50} tests should be allowed to continue, but investigators should bear in mind that the use of large numbers of animals was not necessary for safety evaluation purposes and that wherever practicable, a limit test (single large dose) should be used in preference to the LD_{50}; second, the establishment of a uniform code of laboratory practice in relation to toxicity testing should be considered; and third, LD_{50} tests should not be started at a time that will not ensure adequate supervision of the animal during expected period of maximum effect.

The conclusions of the report did not please the animal activists and tended to be much more conservative than many of the recommendations submitted by industry and scientific organizations. Interestingly, those associations with practical experience of animal testing tended to be critical, while the more academically inclined (and less knowledgeable on practical toxicology) Medical Research Council (MRC) tended to defend the use of the LD_{50}. For example, the MRC stated that "the LD_{50} test is the only reliable measure of acute toxicity," but then argued that only a small number of animals should be used.

However, the Home Office and the MRC were not attuned to developing scientific and toxicological opinion. In 1981, another review of the usefulness of the formal LD_{50} appeared that was even more damning than the 1968 paper.[31] The only really new information, however, were the results of an interlaboratory study in which the LD_{50}s of 5 standard chemicals were determined in over 60 reputable research laboratories using a standard protocol. The LD_{50} values for the chemicals varied by a factor of 2.4 to 11.9 among the laboratories.

A flurry of papers on the LD_{50} then appeared in the scientific literature (see Reference 32), most of which were critical. On October 21, 1982, the Pharmaceutical Manufacturers Association (U.S.) called for a revision of government regulations so that fewer animals are used in drug safety evaluation. They specifically noted that "the classical LD_{50} test which utilizes many animals to determine an LD_{50} value with mathematical precision lacks justification..." They proposed that (1) the precise determination of an LD_{50} should be limited to those rare cases where it is necessary; (2) an approximate lethal dose plus qualitative data usually represents adequate information on the acute toxicity of drugs; and (3) there should be an international effort to reach agreement among regulatory agencies that, for drugs, a precise LD_{50} determination is not necessary. Scientific meeting after scientific meeting criticized the formal LD_{50} and called for a looser standard requiring one tenth the number of animals. For example, at a 1982 conference in London on alternatives, those present in the audience (largely toxicologists and industry scientists) voted 20 to 1 for the abolition of the formal LD_{50}.[60]

Faced with such a broad consensus that the classical LD_{50} was not useful, it seems surprising that it continued to be required, but the inertia of the regulatory system is not easy to overcome. In November of 1983, the U.S. Food and Drug Administration (FDA) convened a meeting in Washington, DC, to set out its position on the classical LD_{50} test. As this meeting, the FDA stated that it "does not have any regulation specifying the need for LD_{50} testing"[33] and seemed surprised that there would even be any question about this. However, a representative of the Pharmaceutical Manufacturer's Association said that there had been considerable confusion about the FDA's requirements and stated that this was the first time that the FDA had stated its position clearly and unequivocally.[33] Despite this, there are still reports from industry that federal officials are requiring the performance of a classic LD_{50}. There is no doubt that the number of formal LD_{50} tests performed today has fallen considerably, but regulatory agencies are still being pressured to publish formal statements that will provide the regulated industry with better guidelines and fewer confused signals.[34, 35]

VI. THE DRAIZE TEST — SEARCH FOR A REPLACEMENT

In the mid-1970s, the animal protection movement began to focus its attention on the safety testing of cosmetics and toiletries using animals. The Fund for the Replacement of Animals in Medical Experiments (FRAME) produced a leaflet titled "What Price Vanity?" that proved to be very popular and backed up the leaflet with various technical materials on the safety testing of cosmetics. In the U.S., the Society for Animal Rights (now the International Society for Animal Rights) produced a lengthy report on cosmetic testing on animals and organized some of the first demonstrations against the cosmetic industry. The animal movement was given some encouragement when the author of a book examining the alternatives issues for the British Research Defense Society commented that the Draize Test was one of the few areas where a search for a non-animal alternative test had a real chance of success.[36]

Early in 1980, Henry Spira, a New York animal activist, launched his Coalition for the Abolition of the Draize Rabbit Blinding Test. Using a mixture of hard-ball politics, street demonstrations, and print advertisements, Spira brought Revlon (picked as a target because it was an industry flagship) to the negotiating table. When the dust had cleared, Revlon had committed to supporting a program at Rockefeller University that would search for an *in vitro* alternative to the Draize, and the rest of the cosmetics industry was scrambling to put together its own *in vitro* program. In the following 8 years, industry committed over $5 million to the search for a non-animal alternative to the Draize eye irritancy test, and a number of preliminary validation projects are now underway. However, there are signs that progress has been too slow and a new round of political pressure from the animal movement is starting to gather steam.

There are a number of good reviews that describe the classical Draize test and its shortcomings.[37–39] The main criticisms have been that the test does not provide a good prediction of human eye irritancy and that the subjective nature of the scoring raises questions about the the reliability of the test as a regulatory tool. There had been several attempts to modify the test or to improve it; however, any initiative was severely constrained by the inertia of the regulatory bodies. It was likely that no major revision of eye irritancy testing procedures could succeed without some powerful external political support. The animal protection movement has now provided the necessary external impetus, and a number of interesting new methods are being developed and explored. At this time, no single test is likely to replace the Draize, but the search for an alternative has had the effect of focusing attention on the mechanisms of the irritant response with the hope that a particular step (or steps) in the irritant reaction might be useful as an indicator of irritant potential.

Several modifications of the Draize test or of the general eye irritancy test protocol that reduce animal use or suffering have already been instituted with no noticeable deterioration in the protection of human health. These include reducing the number of animals required in the test, the use of local anesthetics, and the exclusion of certain compounds from testing on the basis of physical properties (e.g., pH) that are normally associated with eye irritancy.[32] One of the more interesting modifications that has been proposed is the use of the exfoliative cytology test.[40] In this technique, the eye is exposed to the test substance (usually in diluted form), and then exfoliated cells are retrieved from the eye at standard intervals after exposure via a distilled water rinse. The number of cells retrieved is a very sensitive index of irritancy and correlates well with published Draize irritancy scores. The test can be classified as a refinement alternative because the test is so sensitive that there is no need to subject the rabbit eye to anything more than mild irritation.

While refinements to the Draize test have been developed and implemented, this was one area where the animal protection movement hoped for a replacement test. More than 30 organ or cell culture methods have been proposed,[39] but none have been adequately validated yet. (A brief but comprehensive review of the Draize test and the development of alternatives to it with a European perspective has just been published.[41]) Variations on standard cytotoxicity and cell morphology are the most common types of alternative being investigated. Provided the tests are reasonably well designed and competently carried out, they give satisfactory but not exceptional correlation with Draize data. While many of the tests use corneal cells, it has been reported that other cell types appear to function as well in discriminating between nonirritants and irritants.[42] Some have criticized the cell culture systems because they provide no data on recovery from injury, but at least two cell systems have been developed that could provide such information.[39]

One of the more interesting test systems that has been developed in the search for a Draize alternative is the chorioallantoic membrane (CAM) assay. This test exposes the CAM of the developing chick embryo and assesses the effects of test substances on the CAM. In its initial versions, any substance that caused a necrotic lesion of a certain size was deemed to be an irritant.[43] The funding for the development of the CAM assay in the U.S. was provided by Colgate-Palmolive, but they then decided to bring the test in-house and develop it further.

While the CAM test performed satisfactorily in the hands of the Colgate-Palmolive scientists, it did give several false-positive (but no false-negative) results.[44] While investigating the time course of the response, it was found that irritant chemicals produced small hemorrhages (blebs) from the network of blood vessels in the CAM within 30 min of dosing, and Colgate-Palmolive now uses this as the endpoint.[61] Apart from the fact that it shortens the test considerably, it also improved the results (fewer false positives), and the endpoint appears to be more functionally significant than cell necrosis and death. The new endpoint takes advantage of the fact that one is dosing an organ system and observing an organotypic response (vascular permeability increase) that probably has some significance in eye irritation.

The CAM test has come in for some criticism. For example, it is technically not a non-animal alternative. The test is carried out between days 11 and 14 of the chick embryo development (the egg takes 21 d to hatch) and, in the U.K., a vertebrate embryo more than 50% through its development is defined as an animal. However, the CAM has no sensory neurons and appears to be incapable of sensing painful stimuli. Chicks in which the CAM has suffered some damage as a result of the test are still capable of normal development. There have also been claims that the CAM measures an inflammatory response, but the actual endpoint (prior to the modification by Colgate-Palmolive) assessed little more than cell death. Despite the many caveats about the CAM assay, it is an intriguing system and its potential is far from being fully explored.

Several preliminary investigations have also been undertaken to explore the significance of the release of prostaglandins, leukotrienes, and other factors that might be important in the development of an irritant response to chemical insult.[39] However, none of these tests have been taken beyond very preliminary stages of development, so it is not clear how useful they might be.

For the moment, one could argue that there are some promising alternative test systems on the horizon, but that much more development and validation must be done before they could be pressed into service as a replacement to the Draize. It should also be noted that different batteries of tests are likely to be required for the different industries. For example, the detergent industry knows that most of its products are irritating, and their goal is to produce a substance that will not trigger a strong enough irritant response to require labeling as an irritant. By contrast, pharmaceutical companies that produce topical ophthalmic therapeutics are constantly seeking for substances and drug vehicles that produce no eye irritation. It is likely that one battery of tests might be adequate to discriminate between moderate and severe eye irritants, but that another battery of tests will be required to discriminate between nonirritants and very mild irritants.

The use of the Draize test has been reduced by perhaps more than 50% according to a recent report, but its total elimination will take a long time. There is no evidence that any single non-whole animal test will ever be developed to replace the Draize test, but much progress has been made on the development of a range of different *in vitro* alternatives.[45]

VII. MONOCLONAL ANTIBODY PRODUCTION — A CASE FOR REFINEMENT

The LD_{50} and the Draize tests were used as examples of the search for alternatives that reduce or replace animal use, respectively. This leaves a third "R" of the alternatives trio still to be discussed — namely, refinement. There are many examples of experimental techniques that have been refined, i.e., altered, to reduce the distress and suffering caused to experimental animals. However, there are also many examples of techniques that could be refined further, and the use of mice as monoclonal antibody producing "factories" is one such area. (It should be noted that monoclonal antibodies may also be produced in cell culture,[46] but there are a number of reasons, including convenience, the expense of setting up a tissue culture laboratory, the relatively low titre of antibody in spent culture fluid, and problems with the culture of some clones of antibody-producing cells, why an investigator may decide to use ascites tumors in mice as the means of producing the antibodies. Eventually, it is likely that all monoclonal antibodies will be produced in culture, and firms that now use mice to produce antibodies for commercial purposes are beginning to invest in cell culture technology.[47])

In the production of monoclonal antibodies, one first has to produce a suitable clone of antibody-producing cells. Once one has such a cell clone, the cells may either be grown in cell culture or injected into the peritoneal cavity of mice where, under suitable conditions, they will produce an ascites tumor. A single mouse may produce as much antibody as can be obtained from multiple bleedings of a well-immunized goat.[48] However, the ascites tumor is by no means a benign growth, and it will kill the mouse in a relatively short period of time. The survival time is a reflection of the initial quantity of cells injected into the peritoneal cavity. In one series of experiments, mice injected with 3.5×10^7 cells survived an average of 8.5 d after the injection. Mice injected with one tenth this number of cells survived an average of 12.7 d.[49] Prior to death, the ascites tumor will cause distension of the abdomen and discomfort. In humans, patients with massive tense ascites tumors are frequently unable to walk and experience abdominal discomfort, indigestion, and heartburn.[50]

In order to get the monoclonal antibody cells to grow and produce an ascites tumor, the peritoneal cavity first has to be prepared or "primed" with an agent that causes irritation and inflammation. The usual agent used as a primer is pristane, although several other agents, including complete Freund's adjuvant, mineral oil, and thioglycolate, have been evaluated for their effectiveness as primers. In verbal comments at a 1987 colloquium on the alleviation of animal pain and distress, Amyx noted that large volumes of pristane injected into the peritoneal cavities of mice cause weight loss, a hunched appearance, and reduced activity. He noted that pristane is thought to cause granulomatous reactions and to interfere with peritoneal draining.[51] He suggested that the usual volume of pristane used (0.5 ml) was too large and that 0.2 ml of pristane resulted in adequate priming, but caused reduced symptoms of distress. While production of monoclonal antibodies is more of an art form than a science, some of the studies on priming indicate that there is no advantage to using more than 0.2 ml of pristane as a primer.[49] One group found that 0.1 ml was adequate, although below this volume there is a decline in the incidence of ascites tumors formed after injection of the cells.[52] Of other priming agents investigated, only incomplete Freund's adjuvant produced results comparable to pristane.[53]

The success of antibody production from ascites tumors is measured by the volume of ascites fluid obtained ("tapped") per mouse and the titre of the antibody in the fluid. Recently, it has been reported that cross-bred mice yielded four times as much ascitic fluid (about 28 ml or 3.2 ml/d) as the inbred mice used previously.[54] While the use of cross-breeds would reduce the number of mice needed to produce a given quantity of antibody, the paper does not address possible problems associated with such a rapid build-up of ascites fluid nor with the removal of large volumes of liquid in repeated samplings. Many institutions are now limiting investigators to a single survival tap of ascites fluid (there may be a second, nonsurvival tap) because of the sense that multiple taps on the same animal causes undue suffering. Of course, it also means that more animals have to be used to produce a given quantity of antibody. In human medicine, paracentesis of ascitic fluid is discouraged because of the fear that it might produce a hypovolemic reaction.[55] In mice, it is not uncommon to remove a volume of ascitic fluid larger than the animal's total blood volume, but the possibility of physiological stress due to hypovolemia is not considered in the literature on monoclonal antibody production.

The techniques now employed in monoclonal antibody production were developed in the past 10 years, at a time of increasing public concern over the welfare and use of experimental animals. It is thus surprising not to find more signs in the literature of a search for ways to produce monoclonal antibodies while minimizing animal distress. Clearly more can be done, and should be done, to refine the existing *in vivo* techniques to limit animal suffering while we wait for the technical improvements in *in vitro* technology that will allow most, if not all, monoclonal antibodies to be harvested from spent cell culture fluid.

VIII. CONCLUSION

Remarkable changes have occurred in the debate on alternatives since 1976 (when the present author first became involved). In 1976, it was difficult to find a toxicologist who would criticize the classic LD_{50} test in public. Today, it is hard to find one who will defend it. Many toxicologists have become interested, or even enthusiastic, proponents of the idea that *in vitro* toxicology is worth pursuing and promoting. For example, one industrial toxicologist commented to me that his company originally became involved in the search for alternatives out of political necessity and that he reluctantly went along with the initiative. However, today, he genuinely believes that the search has merit not only because it is attempting to reduce animal use in safety testing, but also because it makes sense from a scientific and hazard evaluation viewpoint. The funding that has been made available has certainly helped to promote interest in alternatives, but there are other forces at work as well.

For example, it was interesting to note that the Industrial Biotechnology Association press release of April 12, 1988, on the first animal patent (issued on a transgenic mouse that could be used in cancer research) noted that use of the mouse would lead to a lower requirement for animals and, hence, would be "a big step toward the goal of limited and humane use of animals."

Producers of various cell culture equipment have also on occasion advertised their products as alternatives to animals. Colgate-Palmolive has provided funds to the Society of Toxicology to support a post-doctoral fellowship in *in vitro* toxicology. The U.S. Pharmacopeial Convention held a conference in January 1988 on alternative methods for toxicity testing that is introduced with a logo showing a rabbit with a line through it.[56] In fact, considerable advances have been made in reducing the requirements for animals in biologics testing. The change in insulin testing has been described by Trethewey,[57] while the development of replacement, reduction, and refinement techniques in vaccine testing has been extensively reviewed in a recent book.[58]

Less enthusiasm and interest for the concept of alternatives (or the three R's) has been shown outside the pharmaceutical and toxicology laboratories. This is, in part, because there are significant differences between most university laboratories and laboratories engaged in drug discovery and in toxicity testing. While the methods used in drug discovery and in toxicity testing are *relatively* limited in number, every university laboratory tends to have its own particular laboratory culture and methodology to answer unique basic and applied research questions. There is no single way to examine the mechanism of metastasis, and each different experimental approach will probably require the use of numerous different laboratory techniques, each of which will be associated with a particular set of advantages and disadvantages. However, there are some standard techniques that are widely used in basic research, such as the production of monoclonal antibodies, or the sampling of blood, or the use of death as an endpoint, that could be the focus of research to develop alternatives that reduce animal use and animal suffering. The new institutional protocol review committees face these questions all the time and are beginning to press for answers. These pressures, as well as the more conventional public pressures that led to a Congressional requirement that the NIH establish an alternatives program,[59] are focusing attention on alternatives in basic biomedical research. Barring a major economic or military crisis, it is probable that more money will become available to develop and validate alternatives and that this will lead to a continuing growth in interest in the topic. Some scientists may view such a prediction with alarm, but it is more likely that more attention to alternative techniques will benefit both science and medicine (through improvements in research techniques) and also reduce animal use and suffering in the research laboratory.

REFERENCES

1. **French, R. D.,** *Antivivisection and Medical Science in Victorian Society,* Princeton University Press, Princeton, NJ, 1975.
2. **Turner, J.,** *Reckoning with the Beast: Animals, Pain and Humanity in the Victorian Mind,* Johns Hopkins University Press, Baltimore, 1980.
3. **Grafton, T. S.,** The founding and early history of the National Society for Medical Research, *Lab. Anim. Sci.,* 30, 759, 1980.
4. **Walsh, J.,** Public attitudes towards Science is yes, but..., *Science,* 215, 270, 1982.
5. **Braithwaite, J. and Braithwaite, V.,** Attitudes toward animal suffering: an exploratory study, *Int. J. Study Anim. Probl.,* 3, 42, 1982.
6. **Mason, J.,** The politics of animal rights: making the human connection, *Int. J. Study Anim. Probl.,* 2, 198, 1981.
7. **Russell, W. M. S. and Burch, R. L.,** *The Principles of Humane Experimental Technique,* Methuen, London, 1959.
8. **Medawar, P. B.,** *The Hope of Progress,* Methuen, London, 1972, 86.
9. **Russell, W. M. S.,** The increase of humanity in experimentation: replacement, reduction and refinement, *Lab. Anim. Bur. Collect. Pap.,* 6, 23, 1957.
10. **Hollstein, M., McCann, J., Angelosanto, F. A., and Nichols, W. W.,** Short-term tests for carcinogens and mutagens, *Mutat. Res.,* 65, 133, 1979.
11. **Dixon, B. D.,** *What is Science for?,* Penguin Books, London, 1976.
12. **Medawar, P. B.,** *The Art of the Soluble,* Pelican Books, London, 1969.

13. **Garfield, E.,** To remember Hans Krebs — Nobelist, friend and adviser, *Curr. Contents,* 13(30), 5, 1982.
14. **Price, D. J. D.,** Of sealing wax and string, *Nat. Hist.,* 84:49, 1984.
15. **Comroe, J. H. and Dripps, R. D.,** Scientific basis for support of biomedical science, *Science,* 192, 105, 1976.
16. **Draper, M. H. and Weidmann, S.,** Cardiac resting and action potentials recorded with an intracellular electrode, *J. Physiol.,* 115, 74, 1951.
17. **Frank, R. G.,** The Columbian exchange: American physiologists and neuroscience techniques, *Fed. Proc. Fed. Am. Soc. Exp. Biol.,* 45, 2665, 1986.
18. Home Office, Statistics of Experiments on Living Animals — Great Britain, 1981, Her Majesty's Stationery Office, London, 1982.
19. **Muul, I., Hegyeli, A. F., Dacre, J. C., and Woodard, G.,** Toxicological testing dilemma, *Science,* 193, 834, 1976.
20. CSPCA, Report of a Workshop on Alternatives to the Use of Laboratory Animals in Biomedical Research and Testing, Canadian Society for the Prevention of Cruelty to Animals, Montreal, 1980.
21. **Pomerat, C. M. and Leake, C. D.,** Short-term cultures for drug assays, *Ann. N.Y. Acad. Sci.,* 58, 1110, 1954.
22. **Rofe, P. C.,** Tissue culture and toxicology, *Food Cosmet. Toxicol.,* 9, 683, 1971.
23. **Ekwall, B.,** Screening of toxic compounds in tissue culture, *Toxicology,* 17, 127, 1980.
24. **Trevan, J. W.,** The error of determination of toxicity, *Proc. R. Soc., London Ser. B,* 101, 483, 1927.
25. **Barnes, J. M.,** in *Toxicity Testing: Modern Trends in Occupational Health,* Schilling, R. S. F., Ed., Butterworths, London, 1960,20.
26. **Zbinden, G.,** *Progress in Toxicology,* Springer-Verlag, Berlin, 1973.
27. **Zbinden, G.,** A look at the World from inside the toxicologist's cage, *Eur. J. Toxicol.,* 9, 33, 1976.
28. **Baker, S. B. de C.,** The study of the toxicity of potential drugs in laboratory animals, in *The Use of Animals in Toxicological Studies,* Universities Federation for the Welfare of Animals, Potters Bar, U.K., 1969, 23.
29. **Morrison, J. K., Quinton, R. M., and Reinert, H.,** The purpose and value of LD_{50} determinations, in *Modern Trends in Toxicology,* Vol. 1, Boyland, E. and Goulding R., Eds., Butterworths, London, 1968, 1.
30. Advisory Committee on the Administration of the Cruelty to Animals Act — 1976, Report on the LD50 Test, Home Office, London, 1979.
31. **Zbinden, G. and Flury-Roversi, M.,** Significance of the LD50 test for the toxicological evaluation of chemical substances, *Arch. Toxicol.,* 47, 77, 1981.
32. **Rowan, A. N. and Goldberg, A. M.,** Perspectives on alternatives to current animal testing techniques in preclinical toxicology, *Annu. Rev. Pharmacol. Toxicol.,* 25, 225, 1985.
33. **Sun, M.,** Lost of talk about the LD_{50}, *Science,* 222, 1106, 1983.
34. **Anonymous,** The FDA and trends in animal testing, *U.S. Regulat. Rep.,* 4(11), 3, 1988.
35. **Weiss R.,** Test tube toxicology, *Sci. News,* 133(3), 42, 1988.
36. **Smyth, D. H.,** *Alternatives to Animal Experiments,* Scolar Press, London, 1978.
37. **Griffith, J. F., Nixon, G. A., Bruce, R. D., Peer, P. J., and Bannan, E. A.,** Dose-response studies with chemical irritants in the albino rabbit eye as a basis for selecting optimum testing conditions for predicting hazard to the human eye, *Toxicol. Appl. Pharmacol.,* 55, 501, 1980.
38. **Falahee, K. J., Rose, C., Olin, S. S., and Seifried, H. E.,** Eye Irritation Testing: An Assessment of Methods and Guidelines for Testing Materials for Eye Irritance, Office of Pesticides and Toxic Substances, U.S. Environmental Protection Agency (EPA-560/11-82-001), Washington, DC, 1981.
39. **Frazier, J. M., Gad, S., Goldberg, A. M., and McCully, J. P., Eds.,** A critical evaluation of alternatives to acute ocular irritation testing, in *Alternative Methods in Toxicology,* Vol. 4, Mary Ann Liebert, New York, 1987.
40. **Walberg, J.,** Exfoliative cytology as a refinement of the Draize eye irritancy test, *Toxicol. Lett.,* 18, 49, 1983.
41. **Oliver, G. J. A., Kaestner, W., Simpson, B. J. E., Walker, A. P., York, M., and Bontinck, W. J.,** Eye Irritation Testing, European Chemical Industry Ecology and Toxicology Centre (ECETOC), Monogr. 11, Brussels, 1988.
42. **Borenfreund, E. and Borrero, O.,** In vitro cytotoxicity assays: potential alternatives to the Draize ocular irritancy test, *Toxicology,* 29, 195, 1984.
43. **Leighton, J., Nassauer, J. Tchao, R., and Verdone, J.,** Development of a procedure using the chick egg as an alternative to the Draize rabbit test, in *Alternative Methods in Toxicology,* Vol. 1, Goldberg, A. M., Ed., Mary Ann Liebert, New York, 1983, 163.
44. **Kong, B. M.,** The evolution of an alternative to the Draize eye test, *Soap Cosmet. Chem. Spec.,* July, 40, 1987.
45. **Holden, C.,** Much work but slow going on an alternative to the Draize test, *Science,* 242, 185, 1988.
46. **de St Groth, S. F.,** Automated production of monoclonal antibodies in in a cytostat, *Methods Enzymol.,* 121, 360, 1986.
47. **Anon.,** Publication of OTA Report, Charles River Biotechnical Services purchased Opticell hybridoma technology to complement its ability to produce monoclonal antibodies in mice on a contract basis, *ATLA,* 13, 156, 1986.
48. **Velton, D. E., Roberts, S. B., and Scharff, M. D.,** Hybridomas and monoclonal antibodies, *Lab. Manage.,* 19(1), 19, 1981.
49. **Brodeur, B. R., Tsang, P., and Larose, V.,** Parameters affecting ascites tumour formation in mice monoclonal antibody formation, *Immunol. Methods,* 71, 265, 1984.

50. **Mauch, P. M. and Ultmann, J. E.,** Treatment of malignant ascites, in *Cancer, Principles and Practice of Oncology,* Vol. 2, 2nd ed., DeVita, V. T., Hellman, S., and Rosenbergs, S., Eds., 1985, 2150.
51. **Amyx, H. L.,** Control of animal pain and distress in antibody production and infectious disease studies, *J. Am. Vet. Med. Assoc.,* 191, 1287, 1987.
52. **Hoogenraad, N. J. and Wraight, C. J.,** The effect of pristane on ascites tumor formation and monoclonal antibody production, *Methods Enzymol.,* 121, 375, 1986.
53. **Gillette, R. W.,** Alternatives to pristane priming for ascitic fluid and monoclonal antibody production, *J. Immunol. Methods,* 99, 21, 1987.
54. **Brodeur, B. R. and Tsang, P. S.,** High yield monoclonal antibody production in ascites, *J. Immunol. Methods,* 86, 239, 1986.
55. **Boyer, T. D.,** Removal of ascites: what's the rush?, *Gastroenterology,* 90, 2022, 1986.
56. **Dabbah, R. and Ryan, S., Eds.,** USP Open Conference on Alternative Methods for Toxicity Testing — The In Vitro Option, U.S. Pharmacopeial Convention, Rockville, MD, 1988.
57. **Trethewey, J.,** Moving away from animal tests for the standardization of insulin formulations, *Trends Pharmacol. Sci.,* 8, 287, 1987.
58. **Hendricksen, C. F. M.,** *Laboratory Animals in Vaccine Production and Control: Replacement, Reduction and Refinement,* Kluwer Academic, Boston, 1988.
59. According to a letter from J.D. Willett of the Division of Research Resources, NIH, dated April 12, 1988, 41 grant applications had been received responding to the program announcement "Research into Methods of Research that do not Use Vertebrate Animals Used in Research." Of these, 26 had been through the complete review process and 7 had been funded.
60. **Balls, M.,** Personal communication, 1983.
61. **Kong, B.,** Personal communication, 1988.

Part V
Animal Care and Husbandry

Chapter 8

ENVIRONMENTAL VARIABLES AND ANIMAL NEEDS

Emerson L. Besch

EDITOR'S PROEM

All too few investigators are aware of the extraordinary range of environmental variables which can affect research animals and thereby confound research results as well. Failure to control for such variables, and the "noise" which such failure generates, perhaps represents the major methodological flaw pervading biomedical research on animals. Few researchers would purchase a microtome and fail to familiarize themselves with all of the details of its use and maintenance. Animals are incalculably more sensitive than microtomes, yet many researchers persist in seeing them as requiring food, water, and little else. In his essay, Dr. Besch delineates some of the major environmental considerations which are relevant to the husbandry and welfare of animals used in research and demonstrates the numerous ways in which environmental influences can affect all aspects of animals' physiology and metabolism.

TABLE OF CONTENTS

I. INTRODUCTION

Historically, the relationship between animals and humans has been interdependent: humans used animals for food, as beasts of burden, and as companions, and, with domestication, animals have become dependent upon humans, even for survival of some species.[1] Animals also have played a very important role in the development of science and medicine.[2] Many advances in medical science were made possible by the use of experimental animals and biomedical research will continue for the immediate or potential benefit to both humans and other animals.[3]

The usefulness of biomedical and behavioral research data depends upon their validity, and in order to obtain reliable and reproducible experimental results, all factors responsible for variation within and between experiments must be defined and controlled. Although animals possess a genetically controlled response pattern (genotype), which can be modified by selective breeding, biological responses are influenced by both genetic heritage and environment. The heredity potential of an animal has been widely accepted, but only relatively recently have environmental influences been realized.[4] To minimize variability, experimental animals should be maintained in environments that are appropriate to the species and its life history.[3]

It is common practice to define the quality of the animal's environment in terms of physical factors such as mass and energy, where mass factors include gaseous and particulate contaminants and water vapor, while energy factors include heat, sound, and light.[5,6] More appropriately, the variables important to the animal's environment are depicted by the ecosystem complex.[7] In this three-dimensional paradigm, the *physical* factors define the physical environment (temperature, noise, illumination), *organismic* factors define the organism (animal) within the environment, and *adaptive* factors enable the animal to physiologically adapt to the physical environment. Together, the physical, organismic, and adaptive factors not only interact to influence the physiology, affectivity, and behavior of the animal, but also must be defined when evaluating the effects of environment on the animal (Figure 1).

II. PHYSIOLOGICAL ADAPTATION

Organisms generally respond to environmental change through alterations influenced, in part, by the magnitude and duration of exposure to circumstances that induce adaptations.[8] Those alterations favoring survival are said to be adaptive, and adaptive variations may be genetically determined or environmentally induced. One biological assumption regarding adaptational physiology is that organisms are in dynamic equilibrium with their environment, and interactions with their environment are through continuous inputs and outputs.[9]

An animal adapts through an alteration of the internal state (physiological adjustment) with the environment or maintenance of internal constancy (physiological regulation) over a wide range of environmental change. Although acclimatization can shift the limits at which adjusting (i.e., poikilothermic or poikiloosmotic) animals can survive, their internal state fluctuates according to the environment. On the other hand, self-regulating (homeothermic or homeoosmotic) animals utilize negative feedback mechanisms to compensate for changes in the environment. Ultimately, at some environmental limit, these feedback mechanisms are inadequate and regulation fails.[9]

Environmental variables associated with animal housing relate not only to the type of research animal (e.g., poikilothermic or homeothermic), but also to the time course of adaptations to environmental change. For example, homeotherms tolerate a wider range of environmental fluctuations, but a narrower range of internal variation compared to poikilotherms.[9] Also, through acclimatization, homeotherms compensate for changes in their environment. Movement of animals from natural environments should be accomplished at an early age because the artificial environment, designed to meet the optimum requirements of the animal,

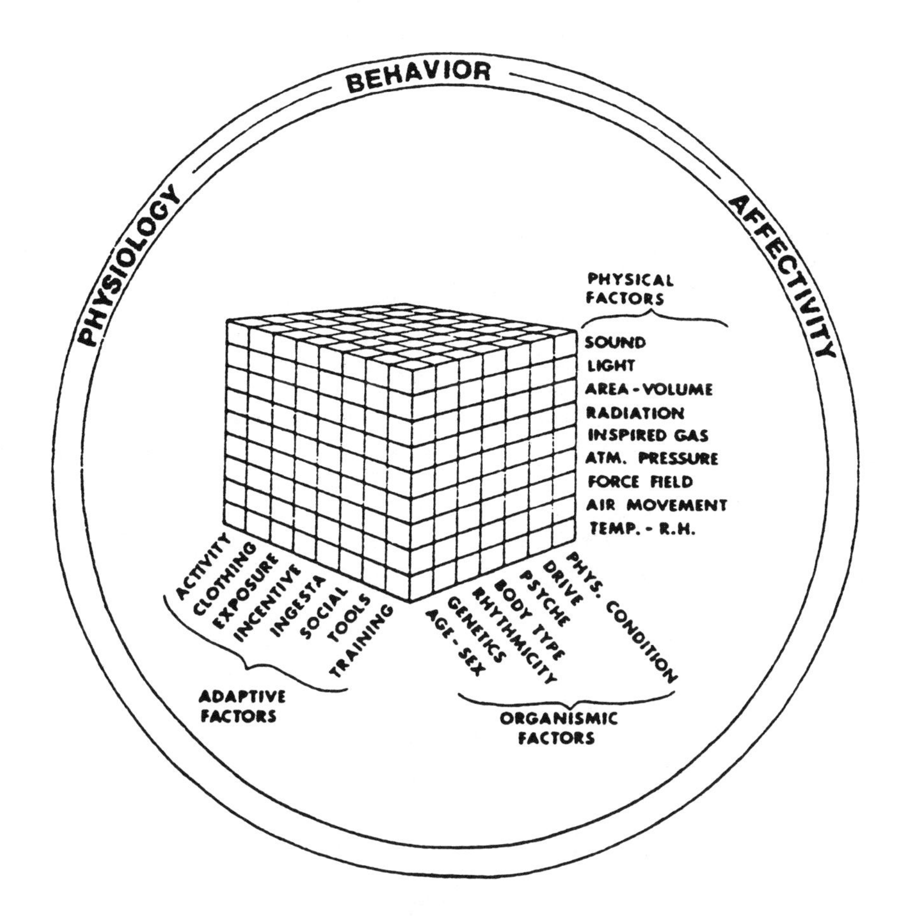

FIGURE 1. The ecosystem complex — a three dimensional representation of the variables which must be controlled or otherwise defined when studying the effects of the environment on physiology, behavior, and affectivity. (From Rohles, F. H., Jr., *ASHRAE Trans.*, 84, 725, 1978. With permission.)

influences the survival and reproduction through contributions to the gene pool or modifications in behavior and physiology.[10]

Although most research animals are purpose-bred and are adapted to their surroundings, their biological responses can be modified by environmental factors including methods and amounts of handling, confinement, restraint, population density, cage type, bedding material, environmental chemicals, noise, photoperiod, temperature, and humidity.[5,6] Controlling the laboratory animal environment prevents undesired responses,[11] assures health and comfort of the animal (Institute of Laboratory Animal Resources [ILAR], 1985), and minimizes bias in research results.[12]

III. PSYCHOLOGICAL FACTORS AND ENVIRONMENTAL STRESSORS

Altered environments elicit responses that operate to keep the internal environment of the animal constant (i.e., maintain homeostasis). The responses are categorized into two types: short latency reactions[13] or delayed responses characterized by the release of adrenocorticotropic hormone (ACTH) from the anterior pituitary gland.[14] Most research on the influence of physical events on animal physiology have emphasized the pituitary-adrenal system. Nonetheless, psychological factors are equally effective in increasing pituitary-adrenal activity, which may explain why corticosterone levels in blood are used as an index of sensitivity to environmental stimuli.[15]

The influence of psychological factors on pituitary-adrenal activity can be demonstrated in pigs, in which the observed responses following a 10-min exposure to a new environment were the same as exposure to inescapable electric shock.[15] Exposure of animals to a new environment (open-field testing), while not inducing pain, allows for and has been used to assess behavioral and pituitary-adrenal activity in pigs,[16,17] sheep,[18] and chicks.[19] The importance of social factors is exemplified by placing together two pigs from different social origins and observing their behavior. Plasma corticosteroids are higher in fighting than in nonfighting animals. Further, whether fighting or just pushing or biting each other, the increase in plasma corticosteroids is more pronounced in subordinate than in dominant animals.[16]

Increased concentrations of plasma corticosteroids have been observed in animals acutely exposed to heat or cold,[20,21] but part of this response appears to be more related to emotional reactions such as anxiety and restlessness[22] than to the thermal stimulus. In animals chronically exposed to heat, plasma corticosteroid levels are depressed,[20] but in animals chronically exposed to cold, these levels are increased.[15] The influence of psychological factors on the specific pituitary-adrenal response to heat or cold stressors is described in Figure 2.

There is other evidence to suggest that stressor induced pituitary-adrenal responses that occur in conscious individuals depend on the emotions experienced by the subjects.[23] Studies of neuroendocrine responses of monkeys to heat, cold, fasting, or exercise show that the pituitary-adrenal axis is not activated if emotional arousal is carefully avoided.[23] From other reports it appears that factors such as novelty, conflict, uncertainty, and relevant feedback mechanisms also may influence the response of animals exposed to stressors.[24] As an example, rats exposed to electric shocks showed a lowered severity of gastric ulcerations — as a measurement of stress — when the shocks were preceded by warning signals (i.e., shocks highly predictable).

IV. THERMAL ENVIRONMENT/HUMIDITY

Temperature and humidity are important environmental variables,[25] although there appear to have been few systematic investigations that contributed to the development of recommended temperature and humidity ranges for animal quarters. These recommendations have been developed over time, through practice and clinical judgment.[3] An often quoted source[26] of temperature and humidity values for laboratory animals is based on recommendations from "..... a few reliable sources ....." The American Society of Heating, Refrigerating, and Air Conditioning Engineers (ASHRAE) publishes temperature and humidity data used by architects and engineers in designing animal facilities;[27] those data are based on Institute of Laboratory Animal Resources (ILAR) recommendations.[28-34] A survey conducted in 1963[35] revealed that between 78 to 89% of the respondents utilized temperature and humidity values similar to those of Runkle.[26] Presumably, data from that survey were largely based on clinical judgment.

The recommended dry-bulb temperature (DBT) values (Table 1) result largely from experience regarding optimal growth, development, reactivity, and adaptability of the various species. Those temperatures, although generally below, are similar to the reported thermoneutral zones,[6] and animals housed within these ranges do not appear to have been adversely affected.[5] Animals exposed to microenvironmental temperatures near to or greater than the upper critical temperature display various responses (Table 2). From other observations of rectal temperature changes in dogs exposed to varying effective temperatures,[53] it is apparent that the predicted maximum acceptable microenvironmental temperature of enclosures used to house dogs is below the upper critical temperature for these animals.

Yamauchi et al.[54] studied the effect of room temperatures between 12 to 32°C on reproduction, body and organ weight, food and water intake, and hematology in two generations of rats

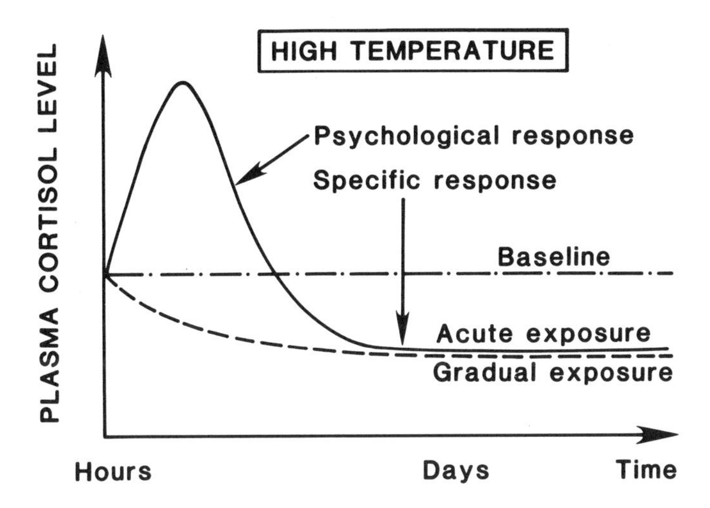

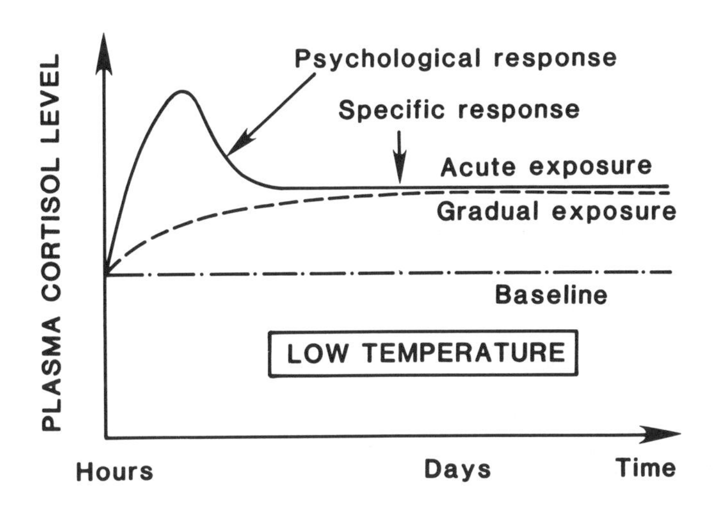

FIGURE 2. Comparisons of the influence of psychological factors on specific responses of cattle to heat or cold stressors. (From Dantzer, R. and Mormede, P., *J. Anim. Sci.*, 57, 6, 1983. With permission.)

and recommended that the optimum temperature range for rats is 20 to 26°C. Laboratory animals appear to have little need for metabolic adaptation and appear to adjust through behavioral adaptation to temperatures of 20 to 23°C and relative humidity of 40 to 60%.[25] Mice housed in wire mesh cages have been reported to display significant behavioral avoidance of ambient temperatures below 22°C[55] or above 30°C[56] during their inactive periods.

As with DBTs, recommended relative humidity values for laboratory animals depend upon the source of the data (Table 1). Further, while the moisture content of air is important, little has been reported on the independent effects of relative humidity on laboratory animals. It is known however, that variations in relative humidity appear to be less significant at low DBTs than at high ones.[57] When the environmental air temperature reaches the body temperature of the animal, evaporation is the only available means of heat loss. At high levels of relative humidity (i.e., at or near saturation), evaporative heat loss from the animal is either absent or severely impaired.

TABLE 1
Recommended Relative Humidity, Dry-Bulb Temperature, and Thermoneutral Temperature for Common Laboratory Animals

Species	Relative humidity (%)	Dry-bulb temperature (°C)	Thermoneutral temperature (°C)		Ref.
			Low	High	
Mouse	40—70	18—26	27	30	36
			28	35	37
Hamster	40—70	18—26			
Rat	40—70	18—26	28	30	38
			26	28	39
			25	29	36
Guinea pig	40—70	18—26	30	31	36
Chicken	45—70	16—27[a]	22	27[b]	40
			15	28	37
Rabbit	40—60	16—21	25	29	41
			28	32	42
Cat	30—70	18—29	24	27	43
Non-human primate	30—70[c]	18—29	25	31[d]	44
			25	35[e]	45
			20	29[f]	37
Dog	30—70	18—29[g]	26	29[h]	46
			22	26[i]	46
			23	27	47
			25	29	37

[a] For brooder chicks 6 weeks old, optimum temperature varies with age of bird.[29]
[b] Pullets 8 to 13 weeks old.
[c] Primate infants have poor thermostability and should be maintained at a constant ambient room or isolette temperature between 26.7 and 32°C.[34]
[d] *Macaca mulatta.*
[e] Thermoneutral range for *Saimira sciureus* in still air conditions.
[f] Chimpanzee.
[g] National Research Council[30] recommends 27 to 29°C in postoperative recovery and whelping cages.
[h] Measurements made in summer.
[i] Measurements made in winter.

From Besch, E. L., in *Animal Stress,* Moberg, G., Ed., American Psychological Society, Washington, DC, 1985, 297. With permission of the American Physiological Society.

It also is known that low humidity increases the levels of dust and upper respiratory tract infections,[11,57] and ringtail in rats,[58,59] mice,[60] and hamsters.[61] Ringtail is a scaliness, annulation, and reddening of tail tip and necrosis thought to cause increased heat loss through evaporation resulting in reduced blood flow to tail.[58] Although temperature and humidity can influence susceptibility to infectious disease, at the same temperature, low humidity usually is associated with high susceptibility.[11]

V. MICRO- AND MACROENVIRONMENTS

Although the importance of the animal's immediate surroundings (microenvironment) relative to its health and well-being has been known for many years,[62,63] it was only relatively recently that quantitative guidelines were suggested for maintaining environmental quality within the microenvironment.[64] It is well known that the health status of animals is related to the

TABLE 2
Responses of the Rat to Various Dry-Bulb Temperatures

Measurement	T_{DB} (°C)	Ref.
Upper critical temperature	32.0	48
Increased evaporative water loss	29.2	49
	32.0—34.0	50
Increased body water turnover	29.2	51
Elevated plasma corticosterone	29.2	51
	32.5	52
Decreased food and increased water intake	29.2	51

microenvironmental conditions and that exposure of animals to elevated temperature, humidity, and concentrations of ammonia can increase the animal's susceptibility to infectious, toxic, and other harmful agents.[11,65] Further, environmental control often is related to the secondary enclosure (room) not the primary enclosure (cage). Unless cages are individually ventilated, the temperature and humidity gradients[66-68] and carbon dioxide and ammonia levels[69] and, as a consequence, hepatic drug metabolism,[70] are impaired in rats exposed to a dirty environment. Further, histopathological changes occur in the tracheal epithelium,[71] and progression and severity of murine respiratory mycoplasmosis is enhanced[72,73] in rats exposed to ammonia levels commonly encountered in shoe-box cages.

Further, while control of environmental variables within primary enclosures is thought to be important, methods to achieve this objective are less than clear, particularly when the primary enclosure is a cage.[64] Controls for temperature, room ventilation, and illumination, for example, affect the animal room and, insofar as the room is the primary enclosure, microenvironmental control is achieved. However, when the animal is housed in a cage, this becomes the primary enclosure which is contained in a secondary enclosure (animal room). Because direct control of environmental variables is limited to the secondary enclosure (room), the primary enclosure (cage) is only indirectly (i.e., room-coupled or passively) controlled. Ventilation air quantity and quality of the cage can be assured only when it is actively ventilated (i.e., ventilation is supply coupled).

Because there is little available data in the literature,[74] space requirements for animals are based on successful experience and professional judgment.[3] It is generally accepted that animal cages should provide sufficient space for normal postural adjustments and freedom of movement for each animal.[3] Further, the enclosure must be adequately ventilated to provide for the health and comfort of the animal at all times.[75] Although federal regulations[75] specify the ventilation openings for containers used to transport live animals in public commerce, the guidelines[28-34] do not specify the ventilation methods for cages. In large measure, specifications for ventilation openings in cages and shipping containers have been derived through the consensus process, with little information regarding the description of criteria, if any, that were used to evaluate acceptability.[64]

Similarly, a protective filter apparatus has been used on shoe-box rodent cages to provide some microenvironmental control.[25] Nonetheless, the filter cap has been shown to alter the animal's microenvironment.[66,68] Because of the perceived value of cage filters in preventing cross-contamination, their use continues. Use of forced-air ventilation of rodent cages appears to significantly decrease levels of ammonia and moisture content of soiled bedding.[76,77]

The implications of lack of environmental control in some primary enclosures and the resultant influence on the physiological well-being of the animal are exemplified by the data presented in Figure 3. That is, because of the DBT and dew-point temperature (DPT) differentials resulting from design or room-coupled (passive) ventilation of the cage (primary

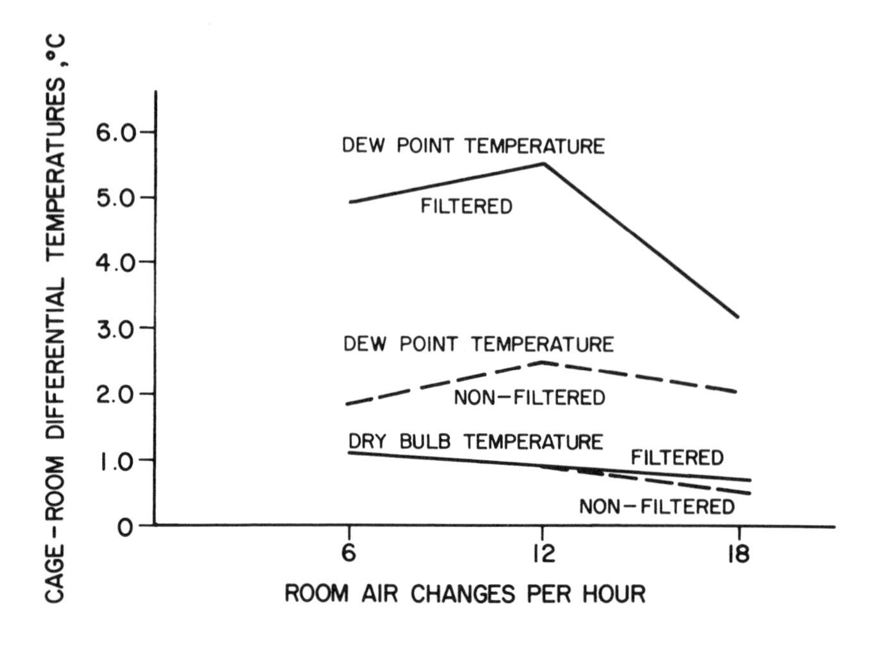

FIGURE 3. Temperature differentials between room and filtered and nonfiltered mouse cages. (From Besch, E. L., *ASHRAE Trans.*, 81, 549, 1975. With permission.)

enclosure), the animal actually may be exposed to a DBT or DPT higher than either that which is recommended (Table 1) or is observed in the room (secondary enclosure). The evaluation of the consequences of an accidental exposure of rats to a room temperature of 26.6 to 31.6°C for 29 h (temperature was never below 23°C) is instructive in this regard.[78] While the temperature of the room never exceeded 31.6°C, the cage temperature was estimated to be 37.7°C. As a consequence of the thermal conditions of the cage, the animals showed, 18 d later, bilaterally atrophied testicles in 25% of the surviving males; weights of the testicles were about one half the size of nonexposed but equal-age males.[78]

Because the concentration of ammonia in an animal cage is directly related to the generation rate and inversely related to the mass flow rate of odor-free air introduced into the space,[79] increasing room-air changes per hour does not necessarily greatly reduce the concentration of ammonia at equilibrium.[6] It also has been shown that the ammonia concentration is inversely related to ventilation, but positively correlated to exposure days between cage cleaning for rats and for rabbits.[5,80] Contact bedding has been shown to be useful in controlling ammonia generation, and raised floorwalk inserts significantly reduced the aerosolization of bedding particles that could be inhaled or ingested by rats.[81] Some strains of mice have displayed a high incidence, but others a low incidence of corneal opacities[82] thought to be affected by ammonia. Twice weekly cage cleaning reduces the incidence of corneal opacity to a low level.

VI. VENTILATION AIR QUALITY AND QUANTITY

Ventilation is defined as the movement of air to and from an occupied space through mechanical or natural means. In animal facilities, ventilation is an important factor in the production, maintenance, and use of laboratory animals[3] because it provides the basis for removing thermal loads, diluting gaseous and particulate contaminants, and controlling heat loss and gain to the enclosures. Animal rooms are considered to have the highest odor index of various spaces evaluated,[83] and early studies[79,84] recognized the importance of controlling air

TABLE 3
Induction of Drug-Metabolizing Enzymes in Liver Microsomes of Mice and Rats by Softwood Bedding

Bedding	Sleeping time (min)	Hexobarbital oxidase[a,b]	Aniline hydroxylase[a]	Ethyl morphine *N*-demethylase
Hardwood	35.3 ± 5.5 (12)[c]	0.46 ± 0.01 (6)	0.35 ± 0.06 (6)	3.8 ± 0.6 (6)
Red cedar	16.0 ± 3.1 (12)	0.95 ± 0.03 (6)	1.00 ± 0.20 (6)	7.0 ± 0.5 (6)

[a] Micromoles of substrates metabolized per gram of liver for 10 min.
[b] Metabolized for 15 min.
[c] Mean ± SE; number of animals in parentheses.

From Vesell, E. S., *Science*, 157, 1057, 1967. With permission from the American Association for the Advancement of Science.

quality and in providing "odor-free" environments through frequent changes of room air. In those studies, the required ventilation to remove animal odors to a satisfactory level was calculated on the basis of volumetric air exchange rate per animal, which was extrapolated to be about 11 room air changes per hour. Thus, the concept of ventilation of animal facilities based on room air changes per hour was established.

Although in theory room air exchange per hour (RAER) would appear to be the same as volumetric air exchange rate per animal, the former does not take into account the spatial dimensions of the room. For example, in an animal room 6.1 m × 6.1 m × 2.4 m (20 ft × 20 ft × 8 ft), ten room air changes per hour would require 893 m^3/h (32,000 ft^3/h) of outside air. Should the height of the same room be increased to 3.0 m (10 ft), there would be a 25% increase in room volume and room air changes per hour would decrease to eight, with no change in the cage air exchange rate. The latter is true because in the typical animal room, the cage ventilation results from passively coupling the cage with the room through a room-coupling coefficient.[85] Through use of the experimentally derived and the room air supply rate, $\dot{V}$, the cage air exchange can be calculated. The use of the cage air exchange not only allows for differences in the room size and fractional loads in animal cages, but also more effectively ventilates the enclosure. Further, over a range of room air exchange rates, the cage air exchange rates are relatively constant.[85]

VII. ENVIRONMENTAL CHEMICALS

Over the past 20 to 30 years, the use of chemicals has resulted not only in the maintenance of clean and sanitary facilities, but also healthier and longer-lived research animals.[86] Chemicals which have been described elsewhere[87] have resulted in some undesirable consequences. In particular, hepatic microsomal enzymes (HME) appear to be sensitive to environmental chemicals or gaseous contaminants such as ammonia. Chemicals reported to affect HME activity are contained in diets,[87] insecticides,[88] room deodorizers,[89] and bedding materials such as red cedar and pine.[90] In one study, five types of commercially available bedding were assessed for endogenous ureolytic activity. The results showed the highest levels of urease activity were found in heat-treated hardwood chips and a regular grade of crushed corncobs.[91] Pentobarbital sleeptimes (Table 3) for mice were significantly longer on hardwood than white pine or red cedar bedding.[92] Further, liver:body weight ratios of mice on red cedar bedding were significantly increased compared to mice on white pine, white spruce, or mixed hardwood beddings; autoclaving the bedding did not alter pentobarbital sleeptimes or liver:body weight ratios.[93]

TABLE 4
Plasma Corticosterone of Rats in Differing Population Densities

Housing	Rats per cage	Corticosterone (μg/100 ml plasma)	Source of blood
Animal house	20	8.5 ± 0.7[a]	Decapitation
Animal house	20	9.5 ± 1.1	Aortic puncture
Laboratory	20	27.4 ± 3.4	Aortic puncture
Animal house	1	5.8 ± 0.5	Decapitation
Animal house	1	5.8 ± 0.3	Aortic puncture
Laboratory	1	30.4 ± 2.2	Aortic puncture

[a] Mean ± SE

From Barrett, A. M. and Stockham, M. A., *J. Endocrinol.*, 26, 97, 1963. With permission.

VIII. SOCIAL ENVIRONMENT

Social environments for laboratory animals are unlike natural environments,[74] and information regarding care of animals with respect to their social environments is limited[74] or contradictory.[94] There have been few systematic studies on space requirements for animals, and caging systems have been based on experience or professional judgment. Nonetheless, some information is available on social behaviors in a laboratory environment.[74]

Factors to be considered in selecting an appropriate social environment include whether the animals are housed singly or in groups and whether they are territorial or communal.[3] Density is considered to be a principal feature of the animal's social environment,[74] and the distinction between density and crowding of laboratory animals has been described elsewhere.[95] Density involves the number of animals, whereas crowding is a perceived lack of space. The influence of differing population densities on plasma corticosterone in rats is summarized in Table 4.

Available evidence suggests that crowding has serious consequences on reproduction[97] and on behavioral responses[98] in mice. Population size and density of laying hens has been shown to influence egg production.[99] Bell et al.[100] report that group size and living space are significant determinants of behavior of mice. Female mice show a consistent repression of estrous cycling when grouped together in a room with no males compared to isolated females.[101] On the other hand, isolation stress has been shown to cause behavioral changes, lowered resistance to stress, changes in food consumption and weight gains, hematological abnormalities, and endocrinal changes;[102] isolated mice exposed to odors of mice in groups or in isolation had significantly larger adrenal glands (Table 5). Also, heat dissipation in individually caged dogs is not significantly different when the animals were in visual, aural, or olfactory contact with each other, but was significantly elevated in the isolated dog.[104]

IX. HANDLING, CONFINEMENT, AND RESTRAINT

Handling, confinement, and restraint are important variables in the animal's environment, and each may influence the animal's physiology, behavior, or affectivity. Further, the laboratory environment connotes restricted activity relative to that of the "natural state" of the animal. To some extent, the response of animals to confinement depends upon whether they are purpose-bred or random source. Although data regarding the quality or quantity of an animal's activity are sparse,[3] some information is available on the influence of handling on laboratory animals.

TABLE 5
Effect of Air from Crowded Mice on Adrenal Glands of Isolated Mice

Measurement	Crowded group (6 mice per cage)	Isolated group (1 male-female pair per cage)	Control group (1 male-female pair per cage)
Paired adrenal glands			
Mean fresh weight (mg)	4.8[a]	5.8[a]	3.9
Terminal body weight (g)	28.5	27.9	29.9
Number of offspring per female	1.8	6.7	7.6

[a] $p < 0.05$ compared to control group.

From Sattler, K. M., *Psychonom. Sci.*, 29, 294, 1972. With permission.

For example, it has been reported that the more albino rats (males, Wistar strain) are handled and petted, the better they thrive in a laboratory situation.[105] Results from studies in which weanling female rats were handled for 3 d prior to follicle-stimulating hormone (FSH) assay suggest that twice as many nonhandled rats would be required in an assay to obtain the same degree of precision as with handled rats.[106] Gentled rats, compared to nongentled rats, display more activity in an open-field situation, but less fearful behavior and less physiological damage to cardiovascular, endocrine, and gastrointestinal systems under prolonged emotional stress.[105] Gentling also has been shown to increase growth rate of pigs,[107] mice,[108] and albino rats.[109] Rats handled repeatedly withstand surgery better than rats which have not been handled.[110] Handling and duration of handling also significantly alter blood glucose and liver glycogen.[111]

Confinement implies a limitation in the range of movement, and various species of animals respond differently. It has been reported that long-term confinement of beagle dogs is not detrimental,[112] but confinement causes delayed puberty in pigs;[113] the latter is thought to be caused by factors other than chronic stress.[114] Domestic fowl purpose-bred for egg production typically are housed in cages that provide each bird about 0.19 to 0.28 m^2 (2 to 3 ft^2) with little adverse consequences to the animal; lowering the space to 0.09 m^2 (1 ft^2) per animal significantly depresses egg production.[99] In a study on population size and density, it was shown that egg production differences were not significant between 30-, 60-, and 100-bird flocks provided 1, 2, and 3 ft^2 per bird, although 30-bird flocks consistently had the best production regardless of floor space allocation.[99] Herring gulls *(Larus argentatus),* on the other hand, kept in captivity for a maximum of 28 d, developed marked heterophilia and nonregenerative anemia.[115] These hematological findings remained abnormal throughout captivity, although behavioral signs of stress were not seen after day 3 in most gulls. The conclusion is that gulls brought into captivity appear to adapt behaviorally, but maintain an abnormal hemogram.

While brief physical restraint may be necessary for examination of the animal, during collection of samples or other clinical or experimental manipulations, prolonged restraint should be avoided unless essential to research objectives.[3] Restraint stress has been observed in all animal studies, including man.[116] The most common and observable changes were behavioral, which included a period of anxiety and hyperactivity accompanied by reduced food intake and loss of body mass and followed by lethargy. Consistent findings include lymphopenia and neutrophilia.[116-117] Restrained animals also display organ tissue and body mass changes,[118] increased blood acidity, plasma protein, calcium and magnesium,[119] and increased plasma corticosterone.[120]

TABLE 6
Effects of Environmental Temperature on the Approximate Lethal Doses of Compounds Administered Intraperitoneally to Albino Rats

	Environmental temperature		
Compound	8°C (mg/kg)	26°C (mg/kg)	36°C (mg/kg)
Acetylsalicylic acid	80	420	55
Atropine	280	420	55
Heparin	180	420	120
Ethyl alcohol	1880	1225	800
Pentobarbital	55	80	10
Warfarin	180	420	120
Caffeine	280	280	55

From Keplinger, M. L., Lanier, G. E., and Deichmann, D. B., *Toxicol. Appl. Pharmacol.*, 1, 156, 1959. With permission.

X. DRUG RESPONSES

Many environmental factors have been shown to influence an animal's response to drugs; the magnitude of the influence varies according to the drug or the species being investigated.[121] Perhaps the most widely investigated physical factor has been temperature, which affects the duration of the response, the potency or toxicity, and the variance of responses of drugs.[122,123] Fluctuations of body temperature in homeotherms are slight, and it is difficult to measure the influence of temperature on pharmacological action.[124] Nonetheless, in mice and other small homeotherms, body temperature is partly determined by environmental temperature.[124] Because environmental temperature can affect body temperature, both should be measured in pharmacological experimentation.

Ambient temperature appears to be a major factor determining thermal responses to many drugs and may affect drug action and toxicity.[125] For example, in rats, reserpine produces a consistent hypothermia at 23°C, but inconsistent effects at temperatures between 24 to 39°C.[125] In mice, sympathomimetic amines display higher toxicities at 26.7°C than at 15.6°C and variance is influenced, but in the opposite direction from potency.[126] Factors such as sound and degree of confinement also affect toxicity of sympathomimetic amines.[126] Piglets, on the other hand, when exposed to mild stimuli such as change in environment, slight frustration, or changes in ambient temperature between 5 and 30°C only rarely display elevated plasma concentrations of ACTH or corticosteroids.[21] Regarding the duration of drug response, an increase in environmental temperature above about 26°C has been found to decrease duration of barbiturate-induced sleep in mice and rats;[127,128] the reason for the shorter sleeping time in mice is thought to be due to a reduction in barbiturate-induced hypothermia.[127]

There are reported to be various relationships between environmental temperature and drug toxicity.[123] The most common is U-shaped with minimum toxicity occurring between room remperature and thermal neutrality (i.e., about 17 to 30°C) and increasing at temperatures above and below this point. This type response has been reported for a number of compounds in rats.[129] Other responses (Table 6) include a continuously increasing toxicity with increasing temperature, such as with alcohol or ephedrine in rats,[129] constant toxicity over a wide range of temperature followed with an increase, such as with caffeine or procaine in rats,[129] and striking effects of temperature on drug toxicity, as with insulin which is 80 times as toxic in mice at 40°C as at 20°C.[122] Gaylord and Hodge[130] reported that in rats the duration of pentobarbital sodium

TABLE 7
Light Intensity Effects on Laboratory Animals

Measurement	Light intensity lx (fc)	Ref.
Retinal changes (rat)	1345—270 (25—125)	142
Retinal degeneration (rat)	194 (18)	143
Retinal damage (rat)	64 (6)	144

sleep time decreases as environmental temperature increases. The influence of environmental temperature on the nature of the drug response is exemplified by the action of reserpine in mice, which has been shown to produce sedation at 4 or 12°C, but not at 30 or 36°C.[131]

Many body systems display rhythmic cycles of varying lengths, those occurring at 24-h intervals are referred to as circadian cycles, while those at 1-year intervals are called circannual cycles. There is evidence to suggest that time of day and season affect drug responses. For example, the duration of pentobarbitone anesthesia in mice is shorter in the dark than in the light period;[127] the tranquilizing drug chlordiazepoxide (Librium®) caused maximum mortality midway during the dark period of the circadian cycle.[132] Also, Heller et al.[133] found that the diuretic action of pitressin in rats was greater in the spring and summer than in autumn and winter, with variability the least in winter.

XI. ILLUMINATION

Sufficient illumination in an animal room is considered to be essential for maintaining good housekeeping practices, observing animals, and providing safe working conditions for employees.[3] Although the precise lighting requirements for maintaining the physiological well-being and health of animals are not known, it is generally accepted that the biologic response of animals is related to either the photoperiod (i.e., the number of hours of light per 24-h day), intensity (i.e., lux or lumens/m^2), or spectral characteristics (i.e., wavelength or color). Of these, more has been reported on animals' responses to photoperiod than to intensity or wavelength.

Through stimulation of photoreceptors, visible light synchronizes biorhythms in animals.[134] Those rhythms that persist in the absence of environmental synchronizers are considered to be free running and have been termed circadian (i.e., about 1 d in length) rhythms. Photoperiodicity regulates both circadian and ultradian (i.e., greater than 1 d in length) rhythms in animals.[135] In particular, photoperiodicity influences reproductive cycles,[136] drug effectiveness and toxicity,[137] body temperature,[138] locomotor activity,[139] and other biochemical measurements.[140] Day lengths of 10 to 14 h appear to be critical to photoperiod reactions.[141] It also has been reported that mice injected with Librium· during the dark period have a higher mortality than mice injected during the light period.[132]

The synchronization of free-running cycles with the 24-h clock is influenced by light intensity. That is, with gradually increasing light intensity, the period length will increase (i.e., be greater than 24 h) in nocturnal and decrease (i.e., be less than 24 h) in diurnal animals.[136] Light intensity also appears to affect the amplitude of the circadian cycle. Nonetheless, optimal light intensity for laboratory animals has been the subject of debate in recent years. Until recently,[3] light levels of 807 to 1076 lx (75 to 100 fc) have been recommended for animal rooms (Table 7), although it has been reported that rats exposed to 4 d of continuous light at 194 lx display degeneration of retinal photoreceptor cells.[143] Severely damaged retinas have been reported in albino rats exposed to light intensities of 270 to 1345 lx for 3 to 7 d.[142] In a report on chronic studies, the cage shelf level was an important factor in the incidence of retinal atrophy in mice;

TABLE 8
Noise Effects on Laboratory Animals

Measurement	dB	Ref.
Recommended levels	85	147
Self-generated noise		
Dog rooms	110	148
Pig rooms	80	148
Rhesus monkey quarters	80	149
Decreased adrenal ascorbic acid — rat	83	150
Elevated serum lipids — rat	102	151
Elevated plasma corticosterone — rat	107	152
Increased adrenal size		
Rabbit	112	152
Rat	107	152
Decreased spleen and thymus	107	152
Eosinopenia — rat	107	152

19.7% of those on the top shelf had retinal atrophy compared to 0.2% of those on all other shelf levels.[146]

XII. NOISE

Noise has been described as an unwanted or undesired sound but, nonetheless, it is a commonly identified environmental problem in animal quarters. Noise results from mechanical equipment, cage washers, fan motors, and other animals (Table 8). The recommended noise level in animal rooms is 85 dB.[147] Sound pressure levels of 110 dB have been measured in dog rooms,[148] 80 dB in pig rooms,[148] and 80 dB in rhesus money quarters.[149] A review of the techniques and procedures available to reduce noise in the animal care facility has been reported elsewhere.[149]

Conflicting reports have been published on the nonauditory affects of noise on laboratory animals, particularly at low or moderate levels, but the preponderance of evidence suggests that exposure to intense noise can produce both auditory and nonauditory effects. For example, animals have been reported to display increases in plasma corticosterone[152] and serum lipids[151] and decreases in spleen and thymus size[152] and adrenal ascorbic acid.[152]

XIII. GOALS FOR ACHIEVING ENVIRONMENTAL QUALITY

In general, animals respond to environmental stimulation through hormonal and behavioral changes. These changes are important because they allow the animal to physiologically adjust to the environmental stressor. In order to prevent unwanted physiological, behavioral, and psychological responses, all individuals who use animals in research, education, and testing should understand the interplay between environ-mental variables and animal needs. In particular, environments must always be suitable for the species in terms of temperature, humidity, ventilation, cage size, lighting, noise, and any other factors which may be stressful. This requires that users of animals define, control, and maintain appropriate housing, anticipate needs that require investigation, and provide for the elaboration of new information regarding the interactions between animals and environmental variables.

REFERENCES

1. **Clark, J. D.,** Regulation of animal use: voluntary and involuntary, *J. Vet. Med. Educ.,* 6, 86, 1979.
2. **French, C. O.,** Trends in humane care legislation, *JAVMA,* 161, 1533, 1972.
3. **ILAR (Institute of Laboratory Animal Resources) Committee on the Care and Use of Laboratory Animals.** Guide for The Care and Use of Laboratory Animals, NIH Publ. 85-23, U.S. Department of Health and Human Services, Washington, DC, 1985.
4. **Conalty, M. L., Ed.,** *Husbandry of Laboratory Animals,* Academic Press, London, 1967.
5. **Besch, E. L.,** Environmental quality within animal facilities, *Lab. Anim. Sci.,* 30, 385, 1980.
6. **Besch, E. L.,** Definition of laboratory animal environmental conditions, in *Animal Stress,* Moberg G., Ed., American Physiological Society, Washington, DC, 1985, 297.
7. **Rohles, F. H., Jr.,** The empirical approach to thermal comfort, *ASHRAE Trans.,* 84, 725, 1978.
8. **Adolph, E. F.,** General and specific characteristics of physiological adaptations, *Am. J. Physiol.,* 184, 18, 1956.
9. **Prosser, C. L.,** Perspectives of adaptation: theoretical aspects, in *Handbook of Physiology, Adaptation to the Environment,* Sect. 4, Dill, D. B., Prosser, C. L., and Wilber, C. G., Eds., American Physiological Society, Washington, DC, 1964, 11.
10. **Price, E. O.,** The laboratory animal and its environment, in *Laboratory Animal Handbooks 7, Control of Animal House Environment,* Laboratory Animal Handbooks, McSheehy, T., Ed., Laboratory Animals, London, 1976, 7.
11. **Baetjer, A. M.,** Role of environmental temperature and humidity in susceptibility to disease, *Arch. Environ. Health,* 16, 565, 1968.
12. **Weihe, W. H.,** The significance of the physical environment for the health and state of adaptation of laboratory animals, in *Defining The Laboratory Animal,* National Academy of Sciences, Washington, DC, 1971, 353.
13. **Cannon, W. B.,** Stresses and strains of homeostasis, *Am. J. Med. Sci.,* 189, 1, 1935.
14. **Selye, H.,** A syndrome produced by diverse nocuous agents, *Nature (London),* 138, 32, 1936.
15. **Dantzer, R. and Mormede, P.,** Stress in farm animals: a need for reevaluation, *J. Anim. Sci.,* 57, 6, 1983.
16. **Arnone, M. and Dantzer, R.,** Does frustration induce aggression in pigs?, *Appl. Anim. Ethol.,* 6, 351, 1980.
17. **Fraser, D.,** The vocalizations and other behavior of young pigs in an "open field" test, *Appl. Anim. Ethol.,* 1, 3, 1974.
18. **Moberg, G. P., Anderson, C., O. and Underwood, T. R.,** Ontogeny of the adrenal and behavioral responses of lambs to emotional stress, *J. Anim. Sci.,* 51, 138, 1980.
19. **Jones, R. B.,** Sex and strain differences in the open-field responses of the domestic chick, *Appl. Anim. Ethol.,* 3, 255, 1977.
20. **Alvarez, M. B. and Johnson, H. D.,** Environmental heat exposure on cattle plasma catecholamine and glucocorticoids, *J. Dairy Sci.,* 56, 189, 1973.
21. **Blatchford, D., Holzbauer, M., Ingram, D. L., and Sharman, D. F.,** Responses of the pituitary-adrenal system of the pig to environmental changes and drugs, *Br. J. Pharmacol.,* 62, 241, 1978.
22. **Johnson, H. D. and Vanjonack, W. J.,** Effects of environmental and other stressors on blood hormone patterns in lactating animals, *J. Dairy Sci.,* 59, 1603, 1976.
23. **Mason, J. W.,** A re-evaluation of the concept of 'non-specificity' in stress theory, *J. Psychiatr. Res.,* 8, 323, 1971.
24. **Weiss, J. M.,** Psychological factors in stress and disease, *Sci. Am.,* 226, 104, 1972.
25. **Weihe, W. H.,** The effects on animals of change in ambient temperature and humidity, in *Laboratory Animal Handbooks, Control of The Animal House Environment,* McSheehy, T., Ed., Laboratory Animals, London, 1976, 41.
26. **Runkle, R. S.,** Laboratory animal housing. II, *J. Am. Inst. Arch.,* 41, 77, 1964.
27. **Anon.,** *ASHRAE Handbook and Product Directory, Applications,* American Society of Heating, Refrigerating, and Air Conditioning Engineers, Atlanta, 1987, 30.13.
28. **ILAR (Institute of Laboratory Animal Resources) Committee on Standards,** Standards for the Breeding, Care and Management of Laboratory Rabbits, National Academy of Sciences, Washington, DC, 1965.
29. **ILAR (Institute of Laboratory Animal Resources) Subcommittee on Avian Standards,** Committee on Standards. Standards and Guidelines for the Breeding, Care and Management of Laboratory Animals: Chickens, National Academy of Sciences, Washington, DC, 1966.
30. **ILAR (Institute of Laboratory Animal Resources) Subcommittee on Dog and Cat Standards,** Committee on Standards. Standards and Guidelines for the Breeding, Care and Management of Laboratory Animals: Dogs, National Academy of Sciences, Washington, DC, 1973.
31. **ILAR (Institute of Laboratory Animal Resources) Subcommittee on Revision of Non-human Primate Standards,** Committee on Standards. Nonhuman Primates: Standards and Guidelines for the Breeding, Care, and Management of Laboratory Animals, National Academy of Sciences, Washington, DC, 1973.
32. **ILAR (Institute of Laboratory Animal Resources) Committee on Rodents,** Laboratory animal managment — rodents, *ILAR News,* 20, L1, 1977.

33. **ILAR (Institute of Laboratory Animal Resources) Committee on Cats,** Laboratory animal management — cats, *ILAR News,* 21, C1, 1978.
34. **ILAR (Institute of Laboratory Animal Resources) Subcommittee on Care and Use, Committee on Nonhuman Primates,** Laboratory animal management — non-human primate, *ILAR News,* 23, P1, 1980.
35. **Anon.,** Questionnaire reveals animal room conditions, *Heat. Pip. Air Cond.,* 35,98, 1963.
36. **Herrington, L. P.,** The heat regulation of small animals at various environmental temperatures, *Am. J. Physiol.,* 129, 123, 1940.
37. **Benedict, F. G.,** *Vital Energetics: A Study in Comparative Basal Metabolism,* Carnegie Institute of Washington, Washington, DC, 1938, 64.
38. **Poole, S. and Stephenson, J. D.,** Body temperature regulation and thermoneutrality, *Q. J. Exp. Physiol.,* 62, 143, 1977.
39. **Clarkson, D. P., Schatte, C. L., and Jordan, J. P.,** Thermal neutral temperature of rats in helium-oxygen, argon-oxygen, and air, *Am. J. Physiol.,* 222, 1494, 1972.
40. **Meltzer, A., Goodman, G., and Fistool, J.,** Thermoneutral zone and resting metabolic rate for growing white leghorn-type chickens, *Br. J. Poult. Sci.,* 23, 383, 1982.
41. **Szelenyi, Z. and Moore, R. E.,** Thermal neutrality and the effect of intraventricular 5-hydroxytryptamine on oxygen consumption in the conscious neonatal rabbit, *Acta Physiol. Acad. Sci. Hung.,* 55, 135, 1980.
42. **Lee, R. C.,** Basal metabolism of the adult rabbit and prerequisites for its measurement, *J. Nutr.,* 18, 473, 1939.
43. **Forster, R. E., Jr. and Ferguson, T. B.,** Relationship between hypothalamic temperature and thermoregulatory effectors in unanesthetized cat, *Am. J. Physiol.,* 169, 255, 1952.
44. **Johnson, G. S. and Elizondo, R. S.,** Thermoregulation in *Macaca mulatta:* a thermal balance study, *J. Appl. Physiol.,* 46, 268, 1979.
45. **Stitt, J. T. and Hardy, J. D.,** Thermoregulation in the squirrel monkey *(Saimiri sciureus), J. Appl. Physiol.,* 31, 48, 1971.
46. **Sugano, Y.,** Seasonal changes in heat balance of dogs acclimatized to outdoor climate, *Jpn. J. Physiol.,* 32, 465, 1971.
47. **Hammel, H. T., Wyndham, C. H., and Hardy, J. D.,** Heat production and heat loss in the dog at 8—36°C environmental temperature, *Am. J. Physiol.,* 194, 99, 1958.
48. **Hamilton, C. L.,** Interactions of food intake and temperature regulation in the rat, *J. Comp. Physiol. Psychol.,* 56, 476, 1963.
49. **Gwosdow-Cohen, A. R.,** Influence of Preconditioning on Physiological Responses to Changing Thermal Environments, M.S. thesis, University of Florida, Gainesville, 1980.
50. **Hainsworth, F. R., Stricker, E. M., and Epstein, A. N.,** Water metabolism of rats in the heat: dehydration and drinking, *Am. J. Physiol.,* 214, 983, 1968.
51. **Attah, M. Y. and Besch, E. L.,** Estrous cycle variations of food and water intake in rats in the heat, *J. Appl. Physiol.,* 42, 874, 1977.
52. **Gwosdow-Cohen, A. R., Chen, C. L., and Besch, E. L.,** Radioimmunoassay (RIA) of serum corticosterone in rats, *Proc. Soc. Exp. Biol. Med.,* 170, 29, 1982.
53. **Besch, E. L., Kadono, H., and Brigmon, R. L.,** Body temperature changes in dogs exposed to varying effective temperatures, *Lab. Anim. Sci.,* 34, 177, 1984.
54. **Yamauchi, C., Fujita, S., Obara, T., and Ueda, T.,** Effects of room temperature on reproduction, body and organ weights, food and water intake, and hematology in rats, *Lab. Anim. Sci.,* 31, 251, 1981.
55. **Murakami, H. and Kinoshita, K.,** Temperature preference of adolescent mice, *Lab. Anim. Sci.,* 28, 277, 1978.
56. **Murakami, H. and Kinoshita, K.,** Spontaneous activity and heat avoidance of mice, *J. Appl. Physiol.,* 43, 573, 1977.
57. **Clough G. and Gamble, M. R.,** *Laboratory Animal Houses: A Guide to the Design and Planning of Animal Facilities,* LAC Manual Ser. No. 4, Environmental Physiology Department, Medical Research Council, Laboratory Animals Centre, Carshalton, England, 1976.
58. **Njaa, L. R., Utne, F., and Braekkan, O. R.,** Effect of relative humidity on rat breeding and ringtail, *Nature (London),* 180, 290, 1957.
59. **Flynn, R. J.,** Studies on the aetiology of ring tail of rats, *Proc. Anim. Care Panel,* 9, 155, 1959.
60. **Nelson, J. B.,** The problems of disease and quality in laboratory Animals, *J. Med. Educ.,* 35, 34, 1960.
61. **Stuhlman, R. A. and Wagner, J. E.,** Ringtail in *Mystromys albicandatus:* a case report, *Lab. Anim. Sci.,* 21, 585, 1971.
62. **Henriques, V. and Hansen, C.,** Uber Eiweissynthese im Tierkorper, *Hoppe-Seyler's Z. Physiol. Chem.,* 43, 418, 1904.
63. **Reyniers, J. A.,** Housing laboratory animals, *Mod. Hosp.,* 58, 64, 1942.
64. **Woods, J. E.,** The animal enclosure-a microenvironment, *Lab. Anim. Sci.,* 30, 407, 1980.
65. **Broderson, J. R., Lindsey, J. R., and Crawford, J. E.,** The role of environmental ammonia in respiratory mycoplasmosis of rats, *Am. J. Pathol.,* 85, 115, 1976.
66. **Simmons, M. L., Robie, D. M., Jones, J. B., and Serrano, L. J.,** Effect of a filter cover on temperature and humidity in a mouse cage, *Lab. Anim.,* 2, 113, 1968.

67. **Murakami, H.,** Differences between internal and external environments of the mouse cage, *Lab. Anim. Sci.,* 21, 680, 1971.
68. **Besch, E. L.,** Animal cage room dry-bulb and dew-point temperature differentials, *ASHRAE Trans.,* 81, 549, 1975.
69. **Serrano, L. J.,** Carbon dioxide and ammonia in mouse cages: effect of cage covers, population, and activity, *Lab. Anim. Sci.,* 21, 75, 1971.
70. **Vesell, E. S., Lang, C. M., White, W. J., Passananti, G. T., and Tripp, S. L.,** Hepatic drug metabolism in rats: impairment in a dirty environment, *Science,* 179, 896, 1973.
71. **Gamble, M. R. and Clough, G.,** Ammonia build-up in animal boxes and its effect on rat tracheal epithelium, *Lab. Anim.,* 10, 93, 1976.
72. **Lindsey, J. R. and Conner, M. W.,** Influence of cage sanitization frequency on intracage ammonia (NH_3) concentration and progression of murine respiratory mycoplasmosis in the rat, *Zentralbl. Bakteriol. I Abt. Orig.,* 241, 215, 1978.
73. **Schoeb, T. R., Davidson, M. K., and Lindsey, J. R.,** Intracage ammonia promotes growth of *Mycoplasma pulmonis* in the respiratory tract of rats, *Infect. Immun.,* 38, 212, 1982.
74. **Davis, D. E.,** Social behavior in a laboratory environment, in *Laboratory Animal Housing,* National Academy of Sciences, Washington, DC, 1978, 44.
75. **Anon.,** Laboratory Animal Welfare Act, 1966, (Public Law 89-544), Amended 1970 (Public Law 91-579), Amended 1976 (Public Law 94-279), Code of Federal Regulations, Title 9, Animals and Animal Products, Subchapter A-Animal Welfare, U.S. Government Printing Office, Washington, DC, January, 1982.
76. **Keller, G. L., Mattingly, S. F., and Knapke, F. B., Jr.,** A forced-air individually ventilated caging system for rodents, *Lab. Anim. Sci.,* 33, 580, 1983.
77. **Wu, D., Joiner, G. N., and McFarland, A. R.,** A forced-air ventilation system for rodent cages, *Lab. Anim. Sci.,* 35, 499, 1985.
78. **Pucak, G. J., Lee, C. S., and Zaino, A. S.,** Effects of prolonged high temperature on testicular development and fertility in the male rat, *Lab. Anim. Sci.,* 27, 76, 1977.
79. **Munkelt, F. H.,** Air purification and deodorization by use of activated carbon, *Refrig. Eng.,* 56, 222, 1948.
80. **White, W. J. and Mans, A. M.,** Effect of bedding changes and room ventilation rates on blood and brain ammonia levels in normal rats and rats with portacaval shunts, *Lab. Anim. Sci.,* 34, 49, 1984.
81. **Raynor, T. H., Steinhagen, W. H., and Harum, T. E., Jr.,** Differences in the microenvironment of a polycarbonate caging system: bedding vs. raised wire floors, *Lab. Anim.,* 17, 85, 1983.
82. **Van Winkle, T. J. and Balk, M. W.,** Spontaneous corneal opacities in laboratory mice, *Lab. Anim. Sci.,* 36, 248, 1986.
83. **Barnebey, H. L.,** Activated charcoal for air purification, *ASHRAE Trans.,* 64, 481, 1958.
84. **Munkelt, F. H.,** Odor control in animal laboratories, *Heat. Pip. Air Cond.,* 10, 289, 1938.
85. **Woods, J. E., Nevins, R. G., and Besch, E. L.,** Analysis of thermal and ventilation requirements for laboratory animal cage environments, *ASHRAE Trans.,* 81, 45, 1975.
86. **Burek, J. D. and Schwetz, B. A.,** Considerations in the selection and use of chemicals within the animal facility, *Lab. Anim. Sci.,* 30, 414, 1980.
87. **Newberne, P. M. and Fox, J. G.,** Chemicals and toxins in the animal facility, in *Laboratory Animal Housing,* National Academy of Science, Washington, DC, 1978, 118.
88. **Conney, A. H. and Burns, J. J.,** Metabolic interactions among environmental chemicals and drugs, Science, 178, 576, 1972.
89. **Cinti, D. L., Lemelin, M. A., and Christian, J.,** Induction of liver microsomal mixed-function oxidases by volatile hydrocarbons, *Biochem. Pharmacol.,* 25, 100, 1976.
90. **Vesell, E. S., Lang, C. M., White, W. J., Passananti, G. T., Hill, R. N., Clemens, T. L, Liu, D. K., and Johnson, W. D.,** Environmental and genetic factors affecting the response of laboratory animals to drugs, *Fed. Proc.,* 35, 1125, 1976.
91. **Gale, G. R. and Smith, A. B.,** Ureolytic and urease-activating properties of commercial laboratory animal breeding, *Lab. Anim. Sci.,* 31, 56, 1981.
92. **Vesell, E. S.,** Induction of drug-metabolizing enzymes in liver microsomes of mice and rats by softwood bedding, *Science,* 157, 1057, 1967.
93. **Cunliffe-Beamer, T. L., Freeman, L. C., and Meyers, D. D.,** Barbiturate sleeptime in mice exposed to autoclaved or unautoclaved wood beddings, *Lab. Anim. Sci.,* 31, 672, 1981.
94. **Brain, P. and Benton, D.,** The interpretation of physiological correlates of differential housing in laboratory rats, *Life Sci.,* 24, 99, 1979.
95. **Stokols, D.,** On the distinction between density and crowding; some implications for future research, *Psychol. Rev.,* 79, 275, 1972.
96. **Barrett, A. M. and Stockham, M. A.,** The effect of housing conditions and simple experimental procedures upon the corticosterone level in the plasma of rats, *J. Endocrinol.,* 26, 97, 1963.
97. **Christian, J. J. and Lemunyan, C. D.,** Adverse effects of crowding on lactation and reproduction of mice and two generations of their progeny, *Endocrinol.,* 63, 517, 1958.

98. **Welch, B. L. and Welch, A. S.,** Sustained effects of brief daily stress (fighting) upon brain and adrenal catecholamines and adrenal, spleen, and heart weights of mice, *Proc. Natl. Acad. Sci., U.S.A.,* 64, 100, 1969.
99. **Fox, T. W. and Clayton, J. T.,** Population size and density related to laying house performance, *Poult. Sci.,* 39, 896, 1960.
100. **Bell, R. W., Miller, C. E., and Ordy, J. M.,** Effects of population density and living space upon neuroanatomy, neurochemistry and behavior in the $C5_B1/10$ mouse, *J. Comp. Physiol. Psychol.,* 75, 258, 1971.
101. **Bronson, F. H. and Chapman, V. M.,** Adrenal-oestrous relationships in grouped or isolated female mice, *Nature (London),* 218, 483, 1968.
102. **Baer, H.,** Long-term isolation stress and its effects on drug response in rodents, *Lab. Anim. Sci.,* 21, 341, 1971.
103. **Sattler, K. M.,** Olfactory and auditory stress on mice *(Mus musculus), Psychonom. Sci.,* 29, 294, 1972.
104. **Woods, J. E. and Besch, E. L.,** Influence of group size on heat dissipation from dogs in a controlled environment, *Lab. Anim. Sci.,* 24, 72, 1974.
105. **Weininger, O.,** Physiological damage under emotional stress as a function of early experience, *Science,* 119, 285, 1954.
106. **Sharpe, R. M., Choudbury, S. A. R., and Brown, P. S.,** The effect of handling weanling rats on their usefulness in subsequent assays of follicle-stimulating hormone, *Lab. Anim.,* 7, 311, 1973.
107. **Hemsworth, P. H., Barnett, J. L., and Hansen, C.,** The influence of handling by humans on the behavior, growth, and corticosteroids in the juvenile female pig, *Horm. Behav.,* 15, 396, 1981.
108. **Porter, G. and Festing, M.,** Effects of daily handling and other factors on weight gain of mice from birth to six weeks of age, *Lab. Anim.,* 3, 7, 1969.
109. **Weininger, O.,** The effects of early experience on behavior and growth characteristics, *J. Comp. Physiol. Psychol.,* 49, 1, 1956.
110. **Hammett, F. S.,** Studies of the thyroid apparatus. V. The significance of the comparative mortality rates of parathyroidectomized wild Norway rats and excitable and non-excitable albino rats, *Endocrinology,* 6, 221, 1922.
111. **Besch, E. L. and Chou, B. J.,** Physiological responses to blood collection methods in rats, *Proc. Soc. Exp. Biol. Med.,* 138, 1019, 1971.
112. **Newton, W. M.,** An evaluation of the effects of various degrees of long-term confinement on adult Beagle dogs, *Lab. Anim. Sci.,* 22, 860, 1972.
113. **Rampacek, G. A., Kraeling, R. R., and Kiser, T. E.,** Delayed puberty in gilts in total confinement, *Theriogenology,* 15, 491, 1981.
114. **Rampacek, G. P., Kraeling, R. R., Fonda, E. S., and Barb, C. R.,** Comparison of physiological indicators of chronic stress in confined and nonconfined gilts, *J. Anim. Sci.,* 58, 401, 1984.
115. **Hoffman, A. M. and Leighton, F. A.,** Hemograms and microscopic lesions of herring gulls during captivity, *J. Am. Vet. Med. Assoc.,* 187, 1125, 1985.
116. **Smith, A. H.,** Response of animals to reduced acceleration fields, in *Principles of Gravitational Physiology,* NASA Contract NSR-09-010-027, 1970.
117. **Besch, E. L., Smith, A. H., Burton, R. R., and Sluka, S. J.,** Physiological limitations of animal restraint, *Aerosp. Med.,* 38, 1130, 1967.
118. **Besch, E. L., Burton, R. R., and Smith, A. H.,** Organ and body mass changes in restrained and fasted domestic fowl, *Proc. Soc. Exp. Biol. Med.,* 141, 456, 1972.
119. **Upton, P. K. and Morgan, D. J.,** The effect of sampling technique on some blood parameters in the rat, *Lab. Anim.,* 9, 85, 1975.
120. **Tache, Y., Ducharme, J. R., Charpenet, G., Haour, F., Saez, J., and Collu, R.,** Effect of chronic intermittent immobilization stress on hypophyso-gonadal function of rats, *Acta Endocrinol.,* 93, 168, 1980.
121. **Ellis, T. M.,** Environmental influences on drug responses in laboratory animals, in *Husbandry of Laboratory Animals,* Conalty, M. L., Ed., Academic Press, London, 1967, 569.
122. **Chen, K. K., Anderson, R. C., Steldt, F. A., and Mills, C. A.,** Environmental temperature and drug action in mice, *J. Pharmacol. Exp. Ther.,* 79, 127, 1943.
123. **Fuhrman, G. J. and Fuhrman, F. A.,** Effect of temperature on the action of drugs, *Annu. Rev. Pharmacol.,* 1, 65, 1961.
124. **Fuhrman, F. A.,** The effect of body temperature on drug action, *Physiol. Rev.,* 26, 247, 1946.
125. **Shemano, I. and Nickerson, M.,** Effect of ambient temperature on thermal responses to drugs, *Can. J. Biochem. Physiol.,* 36, 1243, 1958.
126. **Chance, M. R. A.,** Factors influencing the toxicity of sympathomimetic amines to solitary mice, *J. Pharmacol. Exp. Ther.,* 89, 289, 1947.
127. **Davis, W. M.,** Day-night periodicity in pentobarbital response of mice and the influence of socio-psychological conditions, *Experientia,* 18, 235, 1962.
128. **Komlos, E. and Foldes, I.,** The influence of external temperature on the narcotic potency of evipan and the stimulant action of amphetamine in the white rat, *Arch. Exp. Pathol. Pharmacol.,* 236, 335, 1959.

129. **Keplinger, M. L., Lanier, G. E., and Deichmann, W. B.,** Effects of environmental temperature on the acute toxicity of a number of compounds in rats, *Toxicol. Appl. Pharmacol.,* 1, 156, 1959.
130. **Gaylord, C. and Hodge, H. C.,** Duration of sleep produced by pentobarbital sodium in normal and castrate female mice, *Proc. Soc. Exp. Biol. Med.,* 55, 46, 1944.
131. **Dandiya, P. C., Johnson, G., and Sellers, E. A.,** Influence of variation in environmental temperature on the acute toxicity of reserpine and chlorpromazine in mice, *Can. J. Biochem. Physiol.,* 38, 591, 1960.
132. **Marte, E. and Halberg, F.,** Circadian susceptibility rhythm of mice to librium, *Fed. Proc. Fed. Am. Soc. Exp. Biol.,* 20 (Abstr.), 305, 1961
133. **Heller, H., Herdan, G., and Zaidi, S. M. A.,** Seasonal variations in the response of rats to the antidiuretic hormone, *Br. J. Pharmacol. Chemother.,* 12, 100, 1957.
134. **Weihe, W. H.,** The effect of light on animals, in *Laboratory Animal Handbooks, Control of The Animal House Environment,* McSheehy, T., Ed., Laboratory Animals, London, 1976, 63.
135. **Halberg, F.,** Chronobiology, *Annu. Rev. Physiol.,* 31, 675, 1969.
136. **Aschoff, J.,** Exogenous and endogenous components in circadian rhythms, *Cold Spring Harbor Symp. Quant. Biol.,* 25, 11, 1960.
137. **Haus, E. and Halberg, F.,** 24-Hour rhythm in susceptibility of C mice to a toxic dose of ethanol, *J. Appl. Physiol.,* 14, 878, 1959.
138. **Fioretti, M. C., Riccardi, C., Menconi, E., and Martini, L.,** Control of the circadian rhythm of the body temperature in the rat, *Life Sci.,* 14, 2111, 1974.
139. **Besch, E. L.,** Activity responses to altered photoperiods, *Aerosp. Med.,* 40, 1111, 1969.
140. **Lindsey, J. R., Conner, M. W., and Baker, H. J.,** Physical, chemical and microbial factors affecting biologic response, in *Laboratory Animal Housing,* National Academy of Sciences, Washington, DC, 1978, 31.
141. **Bunning, E.,** *The Physiological Clock,* 2nd ed., revised, Springer-Verlag, New York, 1967, 126.
142. **Semple-Rowland, S. L. and Dawson, W. W.,** Retinal cyclic light damage threshold for albino rats, *Lab. Anim. Sci.,* 37, 289, 1987.
143. **O'Steen, W. K.,** Retinal and optic nerve serotonin and retinal degeneration as influenced by photoperiod, *Exp. Neurol.,* 27, 194, 1970.
144. **Anderson, K. V., Coyle, F. P., and O'Steen, W. K.,** Retinal degeneration produced by low intensity of light, *Exp. Neurol.,* 35, 233, 1972.
145. **Kumar, M. S. A., Chen, C. L., Besch, E. L., Simpkins, J. W., and Estes, K. S.,** Altered hypothalamic dopamine depletion rate and LHRH content in noncyclic hamsters, *Brain Res. Bull.,* 8, 33, 1982.
146. **Greenman, D. L., Bryant, P., Kodell, R. L., and Sheldon, W.,** Influence of cage shelf level on retinal atrophy in mice, *Lab. Anim. Sci.,* 32, 353, 1982.
147. **Anthony, A.,** Criteria for acoustics in animal housing, *Lab. Anim. Care,* 13, 340, 1963.
148. **Sierens, S. E.,** The Design, Construction and Calibration of an Acoustical Reverberation Chamber for Measuring the Sound Power Levels of Laboratory Animals, M.S. thesis, University of Florida, Gainesville, 1976.
149. **Peterson, E. A.,** Noise and laboratory animals, *Lab. Anim. Sci.,* 30, 422, 1980.
150. **Geber, W. F., Anderson, T. A., and Van Dyne, B.,** Physiologic responses of the albino rat to chronic noise stress, *Arch. Environ. Health,* 12, 751, 1966.
151. **Friedman, M., Byers, S. O., and Brown, A. E.,** Plasma lipid responses of rats and rabbits to an auditory stimulus, *Am. J. Physiol.,* 212, 1174, 1967.
152. **Nayfield, K. C. and Besch, E. L.,** Comparative responses of rabbits and rats to elevated noise, *Lab. Anim. Sci.,* 31, 386, 1981.

Chapter 9

PREVENTION AND CONTROL OF ANIMAL DISEASE

Warren W. Frost and Thomas E. Hamm, Jr.

EDITOR'S PROEM

Of all branches of laboratory animal science, perhaps the best established is laboratory animal medicine. And the core of laboratory animal medicine has been the prevention of disease in research facilities. It is intuitively plain that adventitious disease benefits no one — the animals suffer and die for no purpose, and the research project is jeopardized or destroyed. (Indeed, as this essay indicates, conditions short of full-blown disease can also endanger both animals and research.) Unfortunately, too many people using animals are not aware that disease prevention is not simple, but involves a variety of complex considerations. In this paper, Drs. Frost and Hamm lay out some of the fundamental concepts relevant to the prevention of disease in research facilities.

TABLE OF CONTENTS

I. INTRODUCTION

When research animals are used as an integral part of an experimental procedure, prevention and control of animal disease is paramount to assuring confidence in the findings of such studies. It is widely recognized that overt clinical disease can destroy or diminish the credibility of research data. Diseases of laboratory animals associated with high mortalities such as canine distemper, feline panleukopenia, or Sendai virus in a highly susceptible mouse strain may necessitate the abandonment and perhaps repetition of an experiment. Diseases with a high morbidity but low mortality, such as many of the internal parasitic diseases of laboratory animals, may create havoc by altered physiological and/or biochemical responses, depressed growth rate, or compromised reproductive performance. Subclinical or latent infections of laboratory animals may pose an even greater threat to the validity of a study, since immunological responses, biochemical changes, or histopathological lesions associated with such asymptomatic infections may be mistakenly attributed to experimental procedures. As examples, mammalian microsporidia such as *Encephalitozoon cuniculi*, which occur in rabbits and rodent species, are capable of significantly altering the growth of transplantable tumors and changing the lifespan of the host animals.[1] Lactic dehydrogenase virus (LDV), an RNA virus sometimes found in clinically normal mice, markedly increases plasma enzyme activity of several enzymes including lactic dehydrogenase, isocitric dehydrogenase, malic dehydrogenase, and others. Since LDV is a common contaminant of tumors, studies which originally correlated tumor growth with dramatically increased plasma enzyme activity had to be reinterpreted when it was discovered that LDV was a contaminating virus present in the tumors.[2] The presence of lymphocytic choriomeningitis virus (LCM), a normally asymptomatic infection of mice and other laboratory species, can also impact on cancer research, and the presence of this organism as a contaminant of animal or tissue culture extracts modifies the response to experimental murine leukemia.[3] Sendai virus, a common respiratory infection found in mouse, rat, and hamster colonies, can have varying effects on the incidence of different experimental neoplasms.[4] Mouse hepatitis virus, an ubiquitous infection of laboratory mice, is a common contaminant of transplantable tumors or cell lines which produces spontaneous regression, abnormal invasion patterns, and altered passage times when such tissues are transplanted.[5]

Various microbial agents also have profound affects on the immune system. As reported by Hall et al.,[4] in summarizing studies which described alterations in immune function caused by Sendai virus, "...55 of 63 indices tested were abnormal as late as 8 months after disappearance of clinical signs of infection." Mouse hepatitis virus has been implicated in causing an initial immunostimulation and a more prolonged immunosuppression of the immune system. This virus also activates natural killer cells, a cell system independent of thymus-derived lymphocytes which can cause cytotoxic lysis of target cells without preimmunization of the host.[5] Subclinical infections with the organism *Mycoplasma arthritidis* can alter interferon production and immune responses in the rat and mouse.[6] The presence of the bacterial organism *Pseudomonas aeruginosa* or the protozoan *Hexamita muris* in the intestinal tract of mice alter their response to whole body X-irradiation, resulting in a higher mortality in infected than noninfected mice exposed to the same dose of X-irradiation.[7,8]

The presence of many of these adventitious infectious organisms in the asymptomatic form understandably confuses and undermines experimental studies with other infectious organisms. Several strains of *Mycoplasma* are synergistic with other seemingly innocuous infectious agents, producing overt clinical signs when both agents are present.[6] Experimentally, the intracerebral inoculation of mouse hepatitis virus and *M. pulmonis* produces an encephalitis, while no such inflammation occurs when either agent is inoculated individually. Sendai virus can interact with several bacterial organisms and *M. pulmonis* to produce increased morbidity and mortality.[4]

The few examples listed above are only a small representation of the many complications caused by adventitious agents when they are present in laboratory animals used in biomedical studies. There remain many subtle effects on cellular kinetics and biochemical responses which may never be fully understood because of the wide spectrum of virulence of different strains of organisms, varying organotropism, difference in susceptibilities between animal species and strains, and complex interactions of concomitant infections.

For humane as well as scientific reasons, the invalidation of a research study by the presence of an overt disease must obviously be avoided or minimized as much as possible to prevent subjecting the animal to the needless pain or discomfort which may be associated with such infection. It is never acceptable to knowingly compensate for an expected increase in disease-related morbidity or mortality if facilities, practices, or knowledge are adequate to prevent or minimize such disease. This is not only unsatisfactory from a scientific standpoint, but violates a basic humane principle of scientific experimentation which encourages the reduction in the numbers of animals used in any given experimental study when such reduction is compatible with sound experimental design.[9] This principle is compromised when a study is negated by the presence of an asymptomatic infection in the animal subjects, resulting in erroneous data or necessitating repetition of the study and consequent use of additional animals.

Fortunately, a greater appreciation for the devastating effects of such disease has resulted in improved facilities, equipment, and animal husbandry techniques and a decrease in the incidence of such problems. Nevertheless, these diseases still occur, and it is the intent of this discussion to offer general guidelines towards preventing and/or controlling diseases of laboratory animals.

II. ANIMAL SOURCE

A critical step in preventing disease is to obtain animals from disease-free sources. To accomplish this goal, the individuals responsible for animal health must have control over both commercial and noncommercial sources of animals. Some research institutions rely on the establishment of intramural breeding colonies, especially for rodent species, to provide for their own animal needs. Such colonies eliminate the need for frequent introduction of new animals, are convenient for the investigator, and may increase control over study-related requirements. Institutional colonies are primarily utilized when large numbers of pregnant or neonatal animals are required or when there is a requirement for specialized strains or stocks of animals not readily available from a commercial source, such as transgenic animals. Proper genetic and health control of such colonies within a research facility is expensive, labor-intensive, and occupies space that may be needed for housing animals on experiment. Also, animal production in institutional colonies is difficult to coordinate with investigator requirements for a particular age or sex of animal, and excess animals may be required to ensure that there are always animals on hand to meet sporadic demands. With the advances in barrier maintenance and emphasis on health and genetic monitoring programs, it is now possible to obtain a large number of strains and stocks of well-defined, disease-free animals which can be more efficiently and economically produced by one of several commercial suppliers. This is particularly true when the common rodent species are required for experimental studies or instructional needs. Most research institutions now purchase the majority of laboratory rodents from commercial sources. The directory, *Animals for Research*,[10] as compiled by the Institute of Laboratory Animal Resources, and the *Lab Animal Buyer's Guide*[11] list sources of research animals. The former directory provides information as to the source, strain, or stock designation and, when available, vendor-established microbiological status.

In selecting a source, it is helpful to become familiar with the terminology used to describe the common methods of rearing and maintaining laboratory animals, particularly as it relates to

TABLE 1
Classification of Animals Based on Microbiological Status

Classification	Criteria
Axenic animal	Derived by hysterectomy; reared and maintained in isolators with germ-free techniques; demonstrably free from all forms of associated life
Gnotobiotic animal	As above, except that any additionally acquired forms of life are fully known; these should be few in number and nonpathogenic
Defined microbially	An axenic animal that has been intentionally associated with one or more microorganisms
Barrier maintained	A defined microbially associated animal that has been removed from the isolator and placed in a barrier; such animals are repeatedly tested to monitor for (1) presence of the deliberately given organisms and (2) presence of accidentally acquired organisms
Monitored animal	An animal housed in a low-security barrier system and demonstrated by sequential monitoring to be free of major pathogens; other nonpathogenic associated components are largely unknown.
Conventional animal	An animal with an unknown and uncontrolled microbial burden. Generally reared under open animal room conditions

Reprinted from Institute of Laboratory Animal Resources, National Research Council Long-Term Holding of Laboratory Rodents, National Academy of Sciences, Washington, DC, 1976.

the presence or absence of microorganisms, helminths, and arthropods. The terminology generally used is defined in Table 1. While most commonly used when referring to the microbiological status of rodents, this terminology applies to all species. Terms such as "specific pathogen free", "disease free", "clean", and "caesarean derived", while less specific, are generally used when referring to barrier-maintained animals. In all cases, the presence or absence of potentially pathogenic organisms should be specified. The barrier referred to in these definitions is accepted to mean a physical barrier coupled with operational procedures designed to prevent the entrance of unwanted organisms into the barrier. Barriers may be classified in four categories ranging from maximum security (Type 1) to minimal security (Type 4), depending upon the methods employed for the introduction of animals, personnel and supplies, animal care and handling, air supply filtration, and the depth and breadth of monitoring procedures.[12]

Two additional terms used in describing sources of research animals are "random source" and "purpose-bred". The term "random source" generally refers to those animals which are used in research studies, though they have not been specifically bred for research purposes. As the term would indicate, "purpose-bred" animals refers to any species bred and raised specifically for use in research. Domestic species used in research, including dogs and cats obtained from pounds and larger domestic species obtained from farms as well as species caught from the wild, are considered to be random-source animals. Purpose-bred colonies of domestic and wildlife species have been established when random source animals were not readily available or when random source animals were considered to be inappropriate for specific studies. Rabbits used in research have been traditionally obtained from random source breeders that raise animals primarily for meat or pelts. Increasingly, some rabbitries utilize improved facilities and techniques to produce rabbits that are free of the usual infections ubiquitous in many meat-producing rabbitries and provide "purpose-bred" rabbits exclusively for research. The term "purpose-bred", however, does not attest to the health of the animal. Conversely, though random source animals may have wide genetic variability and are initially undefined as to their health status, they have proven invaluable in selected studies following appropriate health screening and adaptation to a laboratory environment. Through humane and appropriate socialization, handling, and varied environmental stimulation, both random source and purpose-bred animals can become well adapted to a laboratory environment.[13]

Increasingly, the larger commercial breeders of laboratory animals have their own health surveillance programs for detecting disease within production colonies. Though some of these programs are comprehensive and provide meaningful insight into the health status of commercial colonies, the information from other suppliers is incomplete and may represent biased health surveillance data from an inadequate number of animals sampled too infrequently to be of significance in evaluating the health status of the source colony. Animal health monitoring programs are most effective when they incorporate all of the various diagnostic disciplines, including clinical observation, serology, parasitology, bacteriology, and gross anatomic and, where indicated, histopathological evaluations. Information as to the numbers of animals sampled as well as the methods, time, and frequency of selecting animals for surveillance procedures should be provided by the suppliers in order to evaluate the significance of the monitoring results. Such information is most useful when compiled and distributed to user facilities on a regular basis.

Regardless of the availability of such information from suppliers, it is prudent for users of animals from commercial sources to perform their own evaluation by obtaining representative numbers of animals from the intended source at frequent intervals, preferably without disclosing the intended purpose, and performing an independent evaluation, using in-house institutional diagnostic services, one of several commercial diagnostic laboratories, or National Institutes of Health (NIH)-supported Animal Resource Diagnostic Laboratories.[14] Since the health status may change at any time, health surveillance of each arriving shipment is desirable. Based on such information, many institutions have chosen to establish quality assurance standards for animals based on the presence or absence of certain microbial agents in any given species of laboratory animal. By knowing the nature and objectives of the proposed studies and the potential impact of specific pathogens on laboratory species, the veterinary staff in concert with the research investigator can formulate standards for selecting animals suitable for the proposed study. A classification for rodent quality assurance, as utilized by one biomedical institution, is reproduced in Table 2. Animals in Class A are free of all rodent pathogens. Since the absence of disease organisms in this category would maximize life span and eliminate the danger of subclinical infection, these animals are suitable for long-term studies, studies requiring a normal immune system, and radiation studies. At the other extreme, animals in Class E harbor organisms such as lymphocytic choriomeningitis (LCM) and *Salmonella,* which are potential pathogens of man and should not generally be used in any experimental procedure unless the study of such pathogens is part of the experimental protocol. The intermediate categories represented by B, C, and D columns are used in experimental procedures as long as the presence of the organism does not jeopardize the health of any animals or the intended study. For this system to succeed, health care personnel and research investigators must coordinate efforts to match the animal to the project requirements during the selection process. More conventional animals, such as those with the microbial flora shown in Class C, are less expensive to purchase and maintain if they meet other experimental criteria. Animals within these categories should be separately housed to prevent cross-contamination. As a minimum criterion, Class A and B animals should never be housed in the same room as Class C, D, and E animals.[15]

III. TRANSPORTATION AND RECEIPT

Most suppliers and users of laboratory animals are well aware of the need to produce, maintain, and use only healthy animals in biomedical research. Unfortunately, animal health may still be placed in jeopardy when animals are transferred to the research institution, especially when the means of conveyance is not under the direct control of the supplier or user. Congress has addressed the humane treatment of animals and felt that the Secretary of Agriculture should "reevaluate the transportation, purchase, sale, housing, care, handling and

TABLE 2
Rodent Quality Assurance

Microbial agents	A	B	C	D	E
Serology					
Pneumonia virus of mice	–	–	+	+	+
Reovirus 3	–	+	+	+	+
GD–VII virus (Thieler's encephalomyelitis virus)	–	–	+	+	+
K virus	–	+	+	+	
Minute virus of mice	–	+	+	+	+
Polyoma virus	–	–	+	+	+
Ectromelia virus	–	–	–	–	+
Sendai virus	–	–	+	+	+
Rat coronavirus/sialodacryoadenitis virus	–	–	+	+	+
Mouse adenovirus	–	–	+	+	+
Mouse hepatitis virus	–	–	+	+	+
Lymphocytic choriomeningitis virus	–	–	–	–	+
Kilham rat virus	–	+	+	+	+
Lactic dehydrogenase-elevating virus	–	–	+	+	+
Toolan's H–1 virus	–	–	+	+	+
Mycoplasma spp.	–	–	+	+	+
Microbiology					
Salmonella spp.	–	–	–	–	+
Staphylococcus aureus	–	+	+	+	+
Klebsiella pneumoniae	–	+	+	+	+
Corynebacteria kutscheri	–	–	+	+	+
Pseudomonas aeruginosa	–	–	+	+	+
Pasteurella multocida	–	+	+	+	+
Pasteurella pneumotropica	–	+	+	+	+
Yersinia pseudotuberculosis	–	–	–	+	+
Yersinia intercolitica	–	–	–	+	+
Bordetella bronchiseptica	–	–	+	+	+
Streptococcus pneumoniae	–	–	–	+	+
Beta hemolytic *Streptococcus*	–	+	+	+	+
Mycoplasma spp.	–	–	+	+	+
Trichophyton spp.	–	–	–	+	+
Microsporum spp.	–	–	–	+	+
Ectoparasites					
Myobia spp.	–	–	–	+	+
Myocoptes spp.	–	–	–	–	+
Radfordia spp.	–	–	–	–	+
Ornithonyssus bacoti	–	–	–	–	+
Trichoecious romboutsi	–	–	–	–	+
Chirodiscoides caviae	–	–	–	–	+
Cliricola porcelli	–	–	–	–	+
Polyplax spp.	–	–	–	–	+
Endoparasites					
Syphacia/Aspicularis spp.	–	–	–	+	+
Trichosomoides crassicauda	–	–	–	+	+
Hexamita spp.	–	–	–	+	+
Hymenolepis diminuta	–	–	–	+	+
Hymenolepis nana	–	–	–	–	+

Contributed by the Department of Comparative Medicine, University of Washington, Seattle. With permission.

treatment of such animals by individuals or organizations engaged in using them in research or experimental purposes, or in transporting, buying or selling them for such use."[16] Federal laws therefore place stringent requirements on common carriers when they transport live animals.

Through several amendments and revisions of the Animal Welfare Act (P.L. 89-544 as amended by P.L. 91-579 and P.L. 94-279), transportation standards have been established for common carriers and are overseen by the U.S. Department of Agriculture. Jurisdiction is currently limited to those species covered by this Act which include dogs, cats, non-human primates, guinea pigs, rabbits, hamsters, and wild-caught species used or intended for use in research. The rules regulate many aspects of air and surface transportation of animals including the size of the container, number of animals per shipping enclosure, size of the ventilation areas in transport containers, and temperature parameters for the shipment of animals. Though rodents such as laboratory rats and mice are excluded from the Animal Welfare Act at this time by definition, commercial carriers normally apply the transportation standards of this Act to all species.[17] Improvements in common carrier transportation practices have been instigated through efforts of the commercial production facilities which have recognized the futility of producing well-defined, disease-free animals and then subjecting them to the vagaries of transportation by common carriers. When feasible, and predominantly in metropolitan areas having large numbers of biomedical institutions, the larger commercial sources of laboratory rodents have developed corporate-owned and -managed carrier systems. These are normally surface vehicles which can mimic the controlled environment of the production laboratory and provide direct delivery of animals to the door of the research institution. Through the addition of commercial production sites and the expansion of corporate-owned surface transport systems, a growing proportion of all purpose-bred rats and mice used in research are delivered to facilities directly by the supplier. When common carriers must be used to provide all or part of the animal transportation requirements of an institution, coordination of shipping arrangements is necessary to minimize shipment stress. Ground transportation should be timed to minimize necessary holding periods at air terminals. When feasible, transport vehicles should be operated by the supplier and recipient institution at the respective points of departure and arrival, controlled to avoid environmental extremes, and manned by trained personnel. It is incumbent upon the receiving laboratory to be aware of all pertinent shipping information, including the waybill number, so that shipments can be traced in the event they are delayed or misdirected.

The design of shipping containers plays an important role in assuring that animals are comfortable, protected from trauma, subjected to a minimum amount of stress, and, in the case of microbiologically defined animals, protected from microbial contamination. They should be constructed of a material that is sturdy, appropriate for the species, and adequately ventilated. Molded plastic, fiberglass, or wood are usually used for the larger species such as dogs and cats because they are sturdy and lightweight. To prevent possible escape, screen-lined cardboard containers are most commonly used for shipment of laboratory rodents. They may be subdivided when it is necessary to ship animals of different sex, age, or strain, and the ventilation openings are covered with sheets of fiberglass filter material when barrier-sustained animals are being shipped. Metal or plastic isolators equipped with high-efficiency particulate air (HEPA) filters are used when germ-free or gnotobiotic animals are shipped.

Provision of food and water during transit times is an important consideration and may be provided to the smaller rodent species by the addition of moisture-laden fruit and vegetables such as apples or potatoes to shipping containers. Commercially available valved water pouch transit kits, (Transit Kit/Rodent, Cold Spring Products, Inc., Stanfordville, NY) are probably a more efficient and reliable method of ensuring a constant and sanitary water supply for rodents.[18] On trips of long duration, larger domestic species must be provided with water at intervals in accordance with requirements of the Animal Welfare Act.

Since animals arriving at a research facility represent a potential source of infection or, in the case of disease-free animals, may become exposed to diseases of the resident animal population, a separate sanitized area should be designated for animal receipt and appropriate procedures instituted to prevent the transference of infectious organisms in either direction. Shipping containers should be considered to be contaminated and not introduced into animal housing areas. Animals should be removed from their shipping containers as soon as possible after arrival, examined for signs of disease or injury, and placed in clean, sanitized cages provided with food and water. It may be desirable at this time to administer required prophylactic or therapeutic treatments. Elective, manipulative procedures which might induce additional stress should be avoided, however, until the animal has had sufficient time to adjust to its new environment.

When rodents or other animals have been procured as disease free and have been shipped in filtered containers, the condition of the containers should be observed and the shipment rejected if the filter or containers is torn, perforated, or wet. When removing disease-free animals from intact filtered containers, it is necessary to use aseptic techniques which prevent contaminating the animals with infectious organisms which may be on the carton or in the surrounding environment.

IV. QUARANTINE

Despite vast improvements over the past several years in the quality of commercially produced laboratory animals,[19] the introduction of newly acquired animals to a facility still poses a risk to resident experimental animals. This is particularly true if the resident animals are "disease free" or at least naive in their exposure to a specific pathogen harbored by the new arrival. Conversely, the incoming animal may represent the more microbiologically naive specimen and be at risk of acquiring infection from the resident animal population.

Optimally, newly procured laboratory animals, regardless of their presumed or documented health status, should be maintained in rooms separate from resident experimental animals for the duration of the experimental procedure with no crossover of personnel, equipment, supplies, or ventilation. As recommended in the NIH *Guide for the Care and Use of Laboratory Animals*,[20] (*Guide*) animals should be physically separated by species to reduce the anxiety which may be caused by interspecies conflict and prevent interspecies disease transmission. The *Guide* also recommends that animals of the same species but from different sources should be housed in different rooms, since animals from one source may harbor microbial agents not found in animals from another source. At a minimum, animals should be grouped based on their known exposure to microbial agents. When animals cannot be maintained in separate rooms by species and source for the duration of the study because of constraints on space and personnel or requirements of the experimental procedure, a period of quarantine aids in minimizing the risk associated with the introduction of new animals into existing colonies. Quarantine is considered essential in those situations where animals are to be used in studies which support applications for research or marketing permits for products regulated by the Food and Drug Administration. *The Good Laboratory Practice Regulations for Nonclinical Laboratory Studies*,[21] states that "All newly received animals from outside sources shall be placed in quarantine until their health status has been evaluated. This evaluation shall be in accordance with acceptable veterinary medical practice." The NIH *Guide*[19] also supports the need for a quarantine period and suggests that during this period of time some or all of the following should be achieved: "...diagnosis, control, prevention, and treatment of diseases including zoonoses; physiological and nutritional stabilization; and grooming, including bathing, dipping, and clipping, as required."

To be effective, quarantine programs should consist of several elements: a physical structure which affords adequate separation between newly acquired animals and resident populations; operational procedures which prevent cross-contamination; and surveillance for disease organisms or physiological abnormalities which might jeopardize the health of other animals or interfere with planned studies. Based on the experience and judgment of a knowledgeable professional responsible for animal health, these elements may be varied depending upon the reliability of the animal source and knowledge of their health surveillance program, the likelihood of exposure during transit, the animal species, and the intended use of the animal.

While various protective housing systems including quarantine cubicles, laminar-flow racks, ventilated cage racks, and filter-top cage systems have been advocated and used to quarantine laboratory rodents, such systems must rely on meticulous attention to animal care procedures if cross-contamination is to be prevented. The use of filter-top cages was first reported by Kraft[22] in 1958 to control the spread of a viral diarrhea in young mice. More recently, Sedlacek et al.[23] described a cage-level barrier system which has been successfully used to protect disease-free rodents from animals harboring various pathogens and housed in adjacent cages. These so-called "petri dish" cages have had wide acceptance in research facilities and have been successfully used to protect defined flora animals when coupled with the use of aseptic techniques and laminar-flow change stations.[24] The quarantine of larger random source species is often achieved in remote facilities owned and maintained by many larger public institutions for the initial separation of dogs, cats, non-human primates, and domestic farm species from experimental colonies.

V. IMMUNIZATION

Domestic, random source dogs, cats, and farm animals are normally vaccinated soon after arrival at a research facility with one or more of the vaccines which have proven clinically efficacious in protecting these species from the more common infectious diseases. A compendium of these vaccines commonly used in dogs and cats and recommended immunization procedures may be found in a text on current veterinary therapy.[25-27] Vaccines may not be wholly effective in increasing host resistance if the animal has been recently exposed to an infectious disease and is in the incubation period at the time of immunization or is in a severely debilitated condition. Many other factors affect the immune response of an animal to a particular vaccine and include the blocking effect of colostral antibodies or immune globulin, the route of immunization, and the age of the recipient.[28] It may not be necessary to vaccinate purpose-bred animals or animals obtained from reliable dealers upon arrival if recent immunizations can be documented. Depending on the vaccine, periodic revaccination is recommended at appropriate intervals for animals on long-term studies. Many wild species used in research are susceptible to some of the infectious diseases of domestic animals, particularly the viral diseases of the dog and cat, and should be vaccinated according to recommended procedures.[29] Live virus rabies vaccines developed for use in domestic animals are not approved for use in wild animal species because of the risk of inducing the disease. If it is deemed necessary to vaccinate confined wild animals, only inactivated rabies vaccine should be used.

The immunization of laboratory rodents and rabbits is not generally practiced, though development of a commercial vaccine for immunization against Sendai virus of mice has proven of value in controlling the clinical manifestations of the disease in the more highly susceptible inbred strains when other preventive measures fail or are impractical.[30] An IHD-T vaccinia strain has also been used to control outbreaks of ectromelia (mouse pox) in mice,[31] though because of the concern for the establishment of this disease in mouse colonies in the U.S., elimination of affected colonies is normally the recommended course of action. The morbidity and mortality associated with *Salmonella* infections of mice,[32] *Bordetella* infections of guinea pigs,[33] and measles infections of rhesus monkeys[34] have been reduced by experimental vaccines, though such vaccines have not found wide clinical applications, perhaps because of improved preventive methods for controlling these diseases. Work by several investigators,[35,36] however, continues on the development of a vaccine for pasteurellosis of rabbits, an ubiquitous disease of the domestic rabbit which seriously impedes the use of this species for many biomedical procedures.[37]

VI. NATURAL DISEASE RESISTANCE

It is known that many inbred strains of mice have an inherited resistance or susceptibility to specific infectious organisms such as *Salmonella typhimurium*, lymphocytic choriomeningitis virus, mouse hepatitis virus, as well as certain tumor viruses and helminths. Festing[38] provides a useful strain ranking for some of these diseases. As demonstrated by Parker et al.,[39] the 129/ReJ strain of mouse was 32,000-fold more sensitive to lethal infections with Sendai virus than the most resistant SJL/J strain. In some situations, such information can be used to the advantage of the research investigator during strain selection if the more resistant strain of animal also meets other experimental requirements. Increased interest in natural disease resistance of domestic animals has promoted research into the mechanisms which promote this intrinsic capacity of an animal to resist disease without prior exposure or immunization and several references to this work are cited in a review article by Templeton et al.[40] Through recent improvements in molecular biotechnology, genetic selection for disease resistance seems destined to play a more prominent role in controlling infectious diseases in all animal species.

VII. HEALTH SURVEILLANCE

Health surveillance of laboratory animals should take place during the initial quarantine period and continue, in some form, for the duration of the animal's stay in a research environment. Surveillance which involves only observation of animals for signs or symptoms of disease and does not involve procedures which would provide more comprehensive health information may be acceptable if there is recent documentation which would support confidence in the health of animals from a specific source. It should not be considered adequate if the animals will be housed in a manner that could jeopardize the health of large numbers of other animals in breeding or experimental colonies. It also provides no information of value in assessing the presence of latent or subclinical infections which might later serve to invalidate research data. Whenever possible, additional clinical and/or laboratory data should be used to substantiate clinical observation of animals. In the larger, domestic species, individual evaluations may include physical examinations, screening for endo- and ectoparasites, hematology, and serum biochemistries. Bacteriological testing is also used to screen for enteric pathogens which may be found in any species, but particularly non-human primates where they pose a threat to man as well as animals. Though such procedures may also be performed in rodents, emphasis in disease surveillance is normally directed towards the detection of serum antibodies and gross necropsies and tissue examinations to detect the presence of viral, bacterial, mycoplasmal, and parasitic organisms. Health screening of rodents should be accomplished randomly and on a sufficient number of animals to provide significant statistical evidence as to the health status of the entire group of animals.

When it is not possible to remove rodents on a periodic basis from an experimental colony for health screening because of their inherent value to ongoing studies, the so-called "sentinel" animal may be introduced into the colony. Such animals must have been previously determined to be free of all adventitious organisms and, following a suitable period of time in the experimental colony, removed and diagnostically evaluated to determine if their association with the experimental animals exposed them to pathogenic organisms. In general, detection procedures should include determination of serological titers for murine viral antibodies, mycoplasmal cultures, fecal examinations for endoparasites, cultures for enteric pathogens, and gross necropsies. References on the quarantine, conditioning and health surveillance of rodents,[12,41-45] rabbits,[37,44] dogs,[46,47] cats,[46,48] non-human primates,[49,50] and ungulates[51-53] should be reviewed for more specific information on these topics.

VIII. PERSONNEL

There are many elements involved in the effective prevention of disease in laboratory animals once they have been released from quarantine and placed on experiment. Perhaps more than any other single factor, personnel involved in animal care and use are capable of ensuring that animals remain in a state of health and well-being throughout the duration of the study. Conversely, they may serve as the weakest link in disease control through inattention to proper procedures established for the control of disease. Personnel may also serve as carriers of diseases to which naive animal populations are susceptible.[24,54]

Because of their close and daily contact with the laboratory animal, animal technicians can be instrumental in preventing the introduction of animal disease. When properly trained, such individuals are likely to be the first to observe abnormal signs or symptoms and alert the veterinary staff for prompt isolation and treatment or, if indicated, euthanasia. In some cases, such action may be sufficient to prevent spread to the remaining animals. Though the primary responsibility of animal technicians is the daily care of the animals, there is merit in involving such individuals in other manipulative aspects of the research study when feasible. Such participation alerts the animal technician to other aspects of the study and may encourage innovative contributions if problems arise. Assuming the animal technicians are properly selected and trained, their involvement in the study may also reduce stress to the animals caused through handling by personnel unfamiliar to the animal or inexperienced in proper and humane techniques of restraint. Because of the pivotal role of the animal technician in maintaining the health of the animal, and consequently the conduct of the experimental study, this position should not be relegated to untrained individuals. Most biomedical institutions using research animals have established in-house training programs for animal technicians or rely on training programs sponsored by local branches of the American Association for Laboratory Animal Science (AALAS), and individuals can attain certification at one of three proficiency levels. Many technicians working with laboratory animals also receive formal training in animal health technology through intensive 2-year training programs accredited by the American Veterinary Medical Association and established at universities and institutes throughout the U.S.

The role of the veterinarian in maintaining animal health is in part mandated by regulatory agencies, in particular, the U.S. Department of Agriculture (USDA) and the major agency responsible for funding biomedical research, the Public Health Service of the Department of Health and Human Services (PHS/DHHS). More importantly, as a professional qualified to recognize and treat diseases or injuries of animals, veterinarians have the ability to oversee the various aspects of animal care and use while contributing to the scientific aspects of the study. In general, institutional veterinarians responsible for laboratory animals must receive post-graduate training or experience in the specialty of laboratory animal science and medicine. The American College of Laboratory Animal Medicine (ACLAM) certifies veterinary specialists in this field following a combination of formal training, experience, and successful completion of board examinations on a range of topics dealing with laboratory animal medicine and science. Depending on the numbers of animals used in research and the nature of the research, large biomedical facilities may employ several full-time veterinarians, while smaller facilities may rely on part-time or consulting veterinarians. Regardless of the time commitment, the veterinarian must have the responsibility and authority for establishing and overseeing the programs for disease detection, prevention, treatment, and control. This necessitates the responsibility for establishing programs for animal health surveillance, environmental monitoring, quarantine, immunization programs, preventive treatments, therapy and general husbandry practices. In addition, the veterinarian must be kept informed on the nature and conduct of all protocols to be performed on animals under his or her charge and have the knowledge and authority to provide meaningful input into such procedures. The veterinarian is ultimately responsible for ensuring that animals are being observed by knowledgeable personnel on a daily basis and that animals requiring veterinary care are brought to his or her immediate attention. If disease or injury should occur to an animal on study, it is reasonable to expect that the veterinarian will consult with the research investigator responsible for the study regarding the appropriate treatment or disposition of the animal. Should there be disagreement over treatment or control measures, the veterinarian must have the authority and institutional support necessary to protect the health and well-being of the animal(s) and institute appropriate treatment or, if necessary, euthanasia.[56]

The best efforts of the professional staff responsible for animal health and well-being may be negated if research investigators and technicians do not have a general knowledge of the diseases that can affect their experimental animal, the potential impact of such disease on the outcome of their research, and control measures which can be taken to prevent such disease. There is little merit in implementation of disease control measures by animal care staff if research personnel cannot or will not comply with established precautions. For example, surveillance procedures will be of little avail if the research investigator introduces biologicals such as tumor transplants, hybridomas, and tissue transplants into an experimental rodent colony without prior mouse antibody production (MAP) testing,[57] to ensure that such tissue is not a vector for one or more pathogenic adventitious agents. The likelihood that these biologicals will carry at least one such virus is significant and has been reported to be between 25 and 86%.[24,58] While it is obviously in the best interest of the research investigator to cooperate with efforts to prevent and control animal disease, many investigators remain unaware of the implications of intercurrent infections in their experimental animals. It is the responsibility of all personnel involved in animal care and use to receive training and instruction that will serve to minimize animal distress as well as the numbers of animals used in research. While attainment of this goal can take many forms, a broader knowledge of disease prevention and control in a biomedical setting will aid these objectives and help to ensure validity of animal-related experimental data. Recent Public Health Service Policy,[59] and the Animal Welfare Act, PL-99-198,[60] now require such training.

IX. SANITATION

The goals of a comprehensive sanitation program in a laboratory animal facility are the establishment of an environment in which there are no disease-producing organisms and a significant reduction of all microbial contamination. Attainment of these goals is particularly difficult in a facility housing multiple species obtained from different sources which must be accessed by numerous investigators and technicians with varying research goals and objectives. Man is undoubtedly a serious hindrance to a reduction in microbial load of the environment, as he dispenses 4000 to 30,000 particles each minute[61] from his skin alone, and an untold number of organisms from the oropharynx and respiratory passages as he talks, coughs, and sneezes through the day. In case his endogenous flora is lacking one or more species from *Bergey's Manual,*[62] he may inadvertently correct this omission by conveying environmental pathogens to the animal on almost any inanimate object. There is little wonder, then, that man is often considered to be the weakest link in maintaining disease-free animals. Knowledgeable but frustrated scientists have gone to the extreme of suggesting robotic husbandry and other space-age technology to eliminate this contaminating influence from the environment of those animals which must be maintained free of all exogenous microorganisms.[63] In the practical realm, however, scientists must have access to their animals. The first step, therefore, in establishing a sanitation program is to determine the degree of control over microbial contamination that is required and feasible considering the species to be used, the facilities and personnel available, and the nature of the study. The general requirements for maintaining maximum, high, moderate, and minimal security barriers are outlined in the ILAR report, Long-Term Holding of Laboratory Rodents.[12] Since most random source animals, and many purpose-bred animals harbor a full complement of microbial flora, they are normally housed in so-called "clean" conventional facilities which permit ready access by research personnel. Well-defined purpose-bred rodents known to be free of infectious organisms are usually housed in various types of barriers where access may be restricted and sanitation programs are designed to exclude specific pathogenic microbes or, in the case of germ-free animals, all adventitious microorganisms. Animals with experimentally induced organisms, or those species such as non-human primates which have a great potential for harboring zoonotic infections,[50] must be housed in varying levels of containment based on their infectious potential to man and other animals. In these situations, the sanitation program must be designed to prevent the escape of infectious agents as well as monitor for breaks in control procedures.

The ability to sanitize an animal facility is influenced by many factors. The appropriate design, construction, and maintenance of facilities, cages, and equipment is critical to an effective sanitation program. Room surfaces should be smooth, impervious to moisture, and capable of withstanding scrubbing with detergents and disinfectants. The rooms should be free of cracks, unsealed penetrations, and imperfect junctures which might impede sanitation through harboring of microorganisms or vermin. Cages that are cracked or damaged by rust or corrosion similarly impede sanitation. Traffic flow of personnel and equipment as well as the exchange, distribution, and filtration of air all impact on the ability to control cross-contamination, and specialized animal isolation equipment may be necessary to protect the animal from infectious agents in the macroenvironment or contain infectious microorganisms used in experimental studies. A number of references may be consulted for further information on facility design.[20,43,64-66]

Regardless of the design and construction of facilities or the availability of specialized isolation or containment equipment, the control of microbial contamination is dependent upon the procedures established and the diligence used in their application. Procedures such as the pressure spraying of water or the disposal of litter within animal rooms aerosolizes a large number of microorganisms which may expose animals and personnel or contaminate recently sanitized equipment. Even the opening and closing of animal room doors disrupt pressure differentials and influence the transfer of contaminants.[67] Consequently, animal room doors should be kept closed at all times and access limited to those individuals with an essential need. The sequence in which animal rooms are entered is of importance, and personnel should enter the room housing the pathogen-naive animals first and progress to rooms housing animals of progressively less well-defined health status when access to more than one room is necessary. The risk of cross-contamination can also be minimized in these situations by using specialized caging and/or donning protective clothing, including separate footwear or shoe covers, prior to entering each room. Showering and donning of sterilized clothing, face masks, gloves, and shoe and hair covers are recommended in maximum- and high-security barriers.[12]

Many items of equipment within an animal facility may serve as fomites to transmit infectious organisms between rooms or cages. Cleaning equipment, such as wet mops,[68] are frequent culprits, and it is recommended that separate room-specific utensils be designated for each animal room or procedural area to prevent cross-contamination. The frequency of cleaning and sanitizing of rooms, cages, and ancillary equipment may vary with the species, the number of animals per enclosure, the caging system, and, in some instances, the experimental protocol. The NIH *Guide for the Care and Use of Laboratory Animals*[20] recommends that solid-bottom rodent cages and accessories be sanitized once or twice a week and wire bottom cages for any animal be sanitized at least every 2 weeks using high-temperature mechanical cage-washing equipment.

Heat is generally considered to be the most effective method of destroying pathogenic organisms, and moist heat is measurably more effective in destroying microorganisms than dry heat. It is the primary method used to sanitize or sterilize cages, racks, bottles, sipper tubes, and other ancillary equipment in animal facilities. Though most vegetative organisms are destroyed by a few minutes of exposure to temperatures of 130 to 150°F (54 to 65°C), rinse water temperatures for cage and bottle washers are recommended to be 180°F (82.2°C) to ensure destruction of these microbes.[20] Destruction of all living organisms, as required for equipment entering barrier animal facilities, requires direct contact of saturated steam at an elevated temperature and under high pressure to increase the boiling point of water. Though several time-temperature combinations can be used, autoclaves are generally operated at 250°F (121°C) for 15 min at a pressure of 15 psi to ensure sterilization. To minimize nutrient loss, variations in temperature and time may be necessary when autoclaving vitamin-fortified diets.[70] The efficacy of sanitation and sterilization procedures with this equipment must be monitored on a frequent basis using chemical and biological indicators. Small[71] provides a comprehensive review of factors to be considered in the proper operation and monitoring of cage-washing equipment and autoclaves.

Since heat cannot be efficiently applied to disinfect animal room surfaces, chemical disinfectants are commonly used for this purpose. Though numerous chemicals can be used to destroy disease-producing organisms, none are effective under all environmental conditions. Consequently, it has been recommended that complete reliance not be placed upon liquid disinfectants because their effectiveness can be altered by so many variables, including contact time, temperature, pH, disinfectant concentration, and the presence of organic materials or detergent residues.[72]

Most frequently, quaternary ammonium compounds, halogens, and phenolics are used in the disinfection of room surfaces of animal facilities. Quaternary ammonium compounds are relatively nontoxic and nonirritating to personnel and most animals. They are broadly bacteriocidal and destroy most vegetative forms of fungi. They are not effective against bacterial spores or the mycobacterium bacillus (tuberculosis), even at high concentration levels, and have only selective virucidal properties against lipophilic viruses. Their use in disinfecting areas subject to gross contamination from animal excrement may be questionable since they are susceptible to inactivation from organic material and they are not as effective against Gram-negative as they are against Gram-positive organisms.[73] Though the spectrum of biocidal activity varies widely with the specific phenolic compound, the phenolic class of disinfectants are not as adversely affected by organic soil or detergent residues and they generally have a broad spectrum of activity, including active destruction of vegetative bacteria and fungi, lipophilic viruses, and the tuberculosis organism.[74] Their major disadvantages revolve around their toxic effects, which may include skin irritation or depigmentation. Chlorine solutions (hypochlorous acid) also have a broad spectrum of biocidal activity and are virucidal for most hydrophilic type viruses. Disadvantages include the potential for toxicity, particularly if mixed with other chemicals, which cause the release of chlorine gas. Chlorine solutions are also readily inactivated by organic soils and hard water.[75] Scott found a dilution of 1 part of household bleach (sodium hypochlorite) to 32 parts of water to be one of the most effective virucidal disinfectants against a variety of feline viruses.[76] A newer chlorine oxide sterilant/disinfectant (Alcide-ABQ®, Alcide Corp., Norwalk, CT), is finding widespread use in research animal facilities because it is rapidly bacteriocidal, virucidal, fungicidal, and tuberculocidal, even in the presence of a significant (5% v/v) organic load.[77,78] Unlike some of the other agents mentioned it has been shown to be relatively nontoxic and nonirritating when properly mixed and diluted.[79]

Both chemical solutions and equipment used for sanitation or sterilization can serve to spread rather than control disease in an animal facility if improperly used or maintained. Incorrect dilution of a disinfectant, careless use of a mop, or inadequate maintenance of an autoclave or cage washer are just a few examples of practices which can invalidate an effective sanitation program and place animal health in jeopardy. To ensure that the sanitation program is accomplishing its purpose, it is advisable to undertake specific goal-oriented evaluations of

TABLE 3
Suggested Monitoring Protocol[a]

Specimen	Sample size	Frequency
Food, water, bedding (for Gram-negative organisms)	Not applicable	Weekly
Clean cages and water bottles (for Gram-negative organisms)	16	Weekly
Floor and work spaces	One swab from each animal room and several from corridors	Weekly
Water bottles (for *Pseudomonas aeruginosa*)	100[b]	Weekly
Cage fecal samples (for *salmonella* spp.)	100[b]	Weekly
Rodents — pathologic, microbiologic, and serologic examinations	25—30	Every 2 months

[a] Assumptions made in devising this plan: (1) The facility has 10,000—15,000 rodents; (2) the mean incidence of clean cage and water bottle is 100 per week, with standard deviation of 10; (3) on the average two cages per month (+1) are found to be contaminated with *P. aeruginosa* and *Salmonella* species.
[b] Sample size may be reduced if no positives are found in a period of 6 months.

Reprinted from Institute of Laboratory Animal Resources, National Research Council, *Laboratory Animal Management: Rodents*, National Academy of Sciences, Washington, DC, 1977.

certain sanitation and sterilization procedures. Though several classes of pathogens, including bacteria, viruses, protozoa, fungi, arthropods, and helminths, may be present on equipment and room surfaces, Weisbroth[80] recommends that bacteria be used as sentinels in a monitoring program, recognizing that some life forms, such as viral spheroids and coccidial oocysts may be more resistant to sanitation procedures than the bacterial agents. Using the methods and criteria outlined by this author, surfaces are to be considered to be adequately sanitized if free of Gram-negative rods and Gram-positive cocci. A sample plan for monitoring barrier rodent facilities as proposed by an ILAR committee provides recommendations as to sample size and frequency of specimen collection (Table 3), but is subject to modification based on animal populations, sanitation schedules, and the incidence of positive samples. Several authors[71,81-83] provide additional information on methods of monitoring and evaluating the efficiency of sanitation programs in laboratory animal facilities.

X. THERAPY

The emphasis in laboratory animal medicine is upon disease prevention, and it is recommended that the use of chemotherapeutic agents in experimental animals be limited for several reasons. Though antibiotics may be effective in producing clinical remission from some bacterial infections, the recovered animals may still harbor the organism in numbers sufficient to disseminate the disease to other laboratory animals. The use of any chemotherapeutic agents also alters, in some manner, the normal metabolism of the experimental animal, introducing an undesirable variable into the experimental study. As examples, animals with high blood levels of aminoglycoside antibiotics have altered cardiovascular dynamics,[84] while animals with high blood levels of chloramphenicol will have a prolongation of sleep time when sodium pentobarbital is administered.[85] Other drug interactions probably occur more frequently than recognized. For example, animals which have been administered coumarin derivatives as part of an experimental cardiovascular procedure should not receive concurrent treatment with a host of therapeutic agents including salicylates, chloramphenicol, griseofulvin, neomycin, and phenylbutazone, to name only a few agents, because of adverse side effects.[86] There are also species and strain variations in drug responses to the same chemotherapeutic agent, and several antibiotic toxicities have been reported in guinea pigs, hamsters, and mice when such agents are administered to these species orally or in high parenteral doses.[87-89]

When treatment of an experimental animal is necessary because of humane and/or experimental considerations, the research investigator must be informed of the nature of the intended therapy and, when known, the expected pharmacodynamics of the therapeutic agents. Animals suspected of harboring infectious agents should be isolated from other animals during the course of therapy until it can be demonstrated that they do not pose a threat to the health of other animals. Domestic species such as dogs, cats, and farm animals are normally treated on an individual basis using accepted practices as described in texts on current veterinary therapy.[90-92] Non-human primates, in many cases, may be individually treated with drugs used in human medicine and at dosage levels determined on a weight basis from those used in pediatric medicine. Though individual parenteral treatment of other laboratory animals such as rodents and rabbits may be indicated in some instances, these species are more frequently treated by the administration of drugs in the food and water. Harkness and Wagner[93] provide dose levels and regimens for several common problems of rabbits and rodents, and Flynn[94] describes several chemotherapeutic agents useful in controlling endo- and ectoparasites of small laboratory species.

XI. CONCLUSIONS

Basic knowledge concerning the nature of some prominent diseases of laboratory animals is still lacking, and much of the information regarding the significance of such diseases to experimental studies has been derived as a consequence of untoward experimental results. Despite these handicaps, the prevalence of many naturally occurring diseases, particularly those producing overt clinical signs or symptoms, has been dramatically reduced over the past 2 decades. Using experimental animals for vaccine development and testing new or improved vaccines has provided increased immunity against many infectious diseases of the common domestic species, whether used as pets or agricultural or laboratory animals. During the past decade, the use of gnotobiotic technology by many commercial animal suppliers has permitted large scale production of disease-free rodents. Though not always applied in a consistent fashion, procedures for quarantine, sanitation, husbandry, and protective housing, coupled with well-designed and -managed animal facilities have facilitated maintenance of disease-free animals in an environment where their pathogen-free status can be preserved. Diagnostic procedures for detection of laboratory animal disease have been improved and simplified, and general recommendations for environmental and animal health monitoring have been established. Diligent application of these recommendations and procedures by both the supplier and user have increased the confidence in the scientific validity of data obtained from studies using laboratory animals. The vital roles of the veterinarian, animal technician, and researcher in assuring laboratory animal health have been delineated, and the need for special training of this triad is recognized by policies of granting agencies as well as federal laws.

Despite the improvements in the prevention and control of animal disease, it does not seem likely that further advancements can continue at an acceptable rate without greater effort and, consequently, funding directed towards more basic research into the nature of specific diseases affecting laboratory animals. The more extensive use of aged and immunodeficient animals in biomedical research only serves to intensify this need because of the increased disease susceptibility associated with these groups of animals. It is not possible, or perhaps desirable, for all research endeavors employing laboratory animals to be conducted in a manner that will assure complete freedom from all disease organisms, especially those organisms which produce subclinical or latent disease. Consequently, more primary research effort should be directed towards determining the subtle but complex effects of these microbes on cellular kinetics and their subsequent biochemical responses. It seems likely that some studies can be conducted in the presence of such organisms without compromising scientific validity as long as the significance of concurrent infection can be interpreted. Humane considerations, however, must take precedence when an animal displays any evidence of clinical disease. Regardless of species, all laboratory animals must be afforded the benefits of current knowledge related to disease prevention and control. The assurance that animals are free of disease which would cause suffering or discomfort or jeopardize the outcome of an experimental study partially fulfills the ethical and scientific responsibilities incumbent upon the user of experimental animals.

REFERENCES

1. **Shadduck, J. A. and Pakes, S. P.,** Encephalitozoonosis (nosematosis) and toxoplasmosis, *Am. J. Pathol.*, 64, 657, 1971.
2. **Notkins, A. L.,** Enzymatic and immunologic alterations in mice infected with lactic dehydrogenase virus, *Am. J. Pathol.*, 64, 733, 1971.
3. **Hotchin, J.,** The contamination of laboratory animals with lymphocytic choriomeningitis virus, *Am. J. Pathol.*, 64, 747, 1971.
4. **Hall, W. C., Lubet, R. A., Henry, C. J., and Collins, M. J., Jr.,** Sendai virus-disease processes and research complications, in *Complications of Viral and Mycoplasmal Infections in Rodents to Toxicology Research and Testing*, Hamm, T.E., Jr., Ed., Hemisphere, Washington, 1986.
5. **Barthold, S. W.,** Research complications and state of knowledge of rodent coronaviruses, in *Complications of Viral and Mycoplasma Infections in Rodents to Toxicology Research and Testing,* Hamm, T.E., Jr., Ed., Hemisphere, Washington, 1986.
6. **Lindsey, J. R., Davidson, M. K., Schoeb, T. R., and Cassell, G. H.,** Murine mycoplasmal infections, in *Complications of Viral and Mycoplasmal Infections in Rodents to Toxicology Research and Testing,* Hamm, T.E., Jr., Ed., Hemisphere, Washington, 1986.
7. **Flynn, R. J.,** *Pseudomonas aeruginosa* infection and radiobiological research at the Argonne National Laboratory: effects, diagnosis, epizootiology and control, *Lab. Anim. Care,* 13, 25, 1963.
8. **Meshores, A.,** Hexamitiasis in laboratory mice, *Lab. Anim. Sci.,* 19, 33, 1969.
9. **Russell, W. M. S. and Burch, R. L.,** *The Principles of Humane Experimental Technique,* Methuen, London, 1959, chap. 6.
10. *Animals for Research — A Directory of Sources,* 10th ed., Greenhouse, D.D. and Cohen, A.L. Eds., Office of Publications, National Academy of Sciences, Washington, DC, 1979.
11. 1988 Buyer's Guide, *Lab Anim.,* 16, 1, 1988.
12. *Long-Term Holding of Laboratory Rodents,* Committee on Long-Term Holding of Laboratory Rodents, Institute of Laboratory Animal Resources (ILAR), National Academy Press, Washington, D.C., 1976.
13. **Fox, M. W.,** *Laboratory Animal Husbandry,* State University of New York Press, Albany, 1986, chap. 6.
14. *Animal Resources — A Research Resources Directory,* 6th ed., NIH Publication No. 86-1431, Bethesda, MD, 1985.
15. **Van Hoosier, G. L. and DiGiacomo, R.,** Barrier maintenance of rodents in multipurpose facilities, Paper 36th Annu. Meet. of the American Assoc. for Laboratory Animal Care, Baltimore, 1985.
16. **Anon.,** Public Law 89-544, August 24, 1966.
17. **Foster, J. C. and Myers, N. M.,** Regulatory considerations in the transportation of laboratory rodents, *Lab. Anim. Sci.,* 30, 312, 1980.
18. **Wiesbroth, S. H., Paganelli, R. G., and Salvia, M.,** Evaluation of a disposable water system during shipment of laboratory rodents, *Lab. Anim. Sci.,* 27, 186, 1977.
19. **Balk, M. W.,** Production of laboratory rodents free of disease, in *Complications of Viral and Mycoplasmal Infections in Rodents to Toxicology Research and Testing,* Hamm, T.E., Jr., Ed., Hemisphere, Washington, 1986.
20. *Guide for the Care and Use of Laboratory Animals,* U.S.D.H.E.W., NIH Publication No 86-23, 1985.
21. Good laboratory practice for nonclinical laboratory studies, Fed. Regist., 43, 209, 1978.
22. **Kraft, L. M.,** Observations on the control and natural history of epidemic diarrhea of infant mice (edim), *Yale J. Biol. Med.,* 31, 121, 1958.
23. **Sedlacek, R. P., Orcutt, R. P., Suit, H. D., and Rose, E. F.,** A flexible barrier at cage level for existing colonies: production and maintenance of a limited stable anaerobic flora in a closed inbred mouse colony, in *Recent Advances in Germfree Research,*. Sasaki, S., et al., Eds., Tokai University Press, Tokyo, 1981.

24. **Orcutt, R. P.,** How to conduct rodent research free of MHV, sendai virus, and other pathogens, *Lab. Anim.,* May/June, 31, 1987.
25. **Scott, F. W.,** Compendium of canine vaccines, 1985, in *Current Veterinary Therapy IX,* 9th ed., Kirk, R.W., Ed., W.B. Saunders, Philadelphia, 1986, 1292.
26. **Scott, F. W.,** Compendium of feline vaccines, 1985, in *Current Veterinary Therapy IX,* 9th ed., Kirk, R.W., Ed., W.B. Saunders, Philadelphia, 1986, 1292.
27. Compendium of animal rabies vaccines, 1986, Prepared by the National Association of State Public Health Veterinarians, Inc., in *Current Veterinary Therapy IX,* 9th ed., Kirk, R.W., Ed., W.B. Saunders, Philadelphia, 1986, 1288.
28. **Schnurrenberger, P. R., Sharman, R. S., and Wise, G. H.,** *Attacking Animal Disease,* Iowa State University Press, Ames, 1987, chap.10.
29. **Kirk, R. W.,** Immunization procedures, in *Current Veterinary Therapy IX,* Kirk, R.W., Ed., W.B. Saunders, Philadelphia, 1986, 1287.
30. **Eaton, G. J., Lerro, A., Custer, R. P., and Crane, A. R.,** Eradication of Sendai pneumonitis from a conventional mouse colony, *Lab. Anim. Sci.,* 32, 384, 1982.
31. **Briody, B. A.,** Response of mice to ectromelia and vaccinia viruses, *Bacteriol. Rev.,* 23, 61, 1959.
32. **Carnochan, F. G. and Cumming, G. N. W.,** Immunization against *Salmonella* infection in a breeding colony of mice, *J. Infect. Dis.,* 90, 242, 1952.
33. **Ganaway, J. R., Allen, A. M., and McPherson, G. W.,** Prevention of acute *Bordetella bronchiseptica* pneumonia in a guinea pig colony, *Lab. Anim. Care,* 15, 156, 1965.
34. **McLaughlin, R. M. and Sueur, R. E.,** Reduction of losses on newly imported *Macaca mulatta* by measles immunization, Paper 25th Annu. Sess. of the American Assoc. for Laboratory Animal Science, Cincinnati, 1974.
35. **Percy, D. H., Prescott, J. F., and Bhasin, J. L.,** *Pasteurella multocida* infection in the domestic rabbit: immunization with a streptomycin-dependent mutant, *Can. J. Comp. Med.,* 49, 277, 1985.
36. **Lu, Y. -S. and Pakes, S. P.,** Protection of rabbits against experimental pasteurellosis by a streptomycin-dependent *Pasteurella multocida* serotype 3: A type live mutant vaccine, *Infect. Immun.,* 34, 1018, 1981.
37. **Percy, D. H. and Bhasin, J. L.,** Control of Pasteurellosis in rabbits: an overview, *Lab Anim.,* 16, (3), 29, 1987.
38. **Festing, M. T. W.,** *Inbred Strains in Biomedical Research,* Oxford University Press, New York, 1979, 380.
39. **Parker, J. C., Whiteman, M. D., and Richter, C. B.,** Susceptibility of inbred and outbred mouse strains to Sendai virus and prevalence of infection in laboratory rodents, *Infect. Immun.,* 19, 123, 1978.
40. **Templeton, J. W., Smith, R., III, and Adams, L. G.,** Natural disease resistance in domestic animals, *J. Am. Vet. Med. Assoc.,* 192, 1306, 1988.
41. Laboratory animal management: rodents, ILAR News, 20, 13, 1977.
42. **Loew, F. M. and Fox, J. G.,** Animal health surveillance and health delivery systems, in *The Mouse in Biomedical Research,* Vol. 3, Foster, H. L., Small, J. D., and Fox, J. G., Eds., Academic Press, New York, 1983, chap.5.
43. **Lang, C. M.,** Design and management of research facilities for mice, in *The Mouse in Biomedical Research,* Vol. 3, Foster, H. L., Small, J. D., and Fox, J. G., Eds., Academic Press, New York, 1983, chap.3.
44. **Small, J. D.,** Rodent and lagomorph health surveillance-quality assurance, in *Laboratory Animal Medicine,* Fox, J. G., Cohen, B. J., and Loew, F. M., Eds., Academic Press, Orlando, 1984, chap.26.
45. *Manual of Microbiological Monitoring of Laboratory Animals,* Allen, A. M. and Nomura, T., Eds., U.S. Department of Health and Human Services, NIH Publication No. 86-2498, 1986.
46. **Ringler, D. H. and Peter, G. K.,** Dogs and cats as laboratory animals, in *Laboratory Animal Medicine,* Fox, J. G., Cohen, B. J., and Loew, F. M., Eds., Academic Press, Orlando, 1984, chap.9.
47. *Dogs-Standard and Guidelines for the Breeding, Care and Management of Laboratory Animals,* Subcommittee on Dog and Cat Standards, Institute of Laboratory Animal Resources (ILAR), National Academy Press, Washington, DC, 1973.
48. *Laboratory Animal Management — Cats,* Committee on Cats, Institute of Laboratory Animal Resources (ILAR), National Academy Press, Washington, DC, 1978.
49. Laboratory animal management: nonhuman primates, *ILAR News,* 23, P3, 1980.
50. *Biohazards and Zoonotic Problems of Primate Procurement, Quarantine and Research,* Publ. No (NIH) 76-890, U.S. Department of Health, Education and Welfare, Washington, DC, 1975.
51. **Brooks, D. L., Tillman, P. C., and Niemi, S. M.,** Ungulates as laboratory animals, in *Laboratory Animal Medicine,* Fox, J. G., Cohen, B. J., and Loew, F. M., Eds., Academic Press, Orlando, 1984, chap.10.
52. *Ruminants (Cattle, Sheep, and Goats): Guidelines for the Breeding, Care and Management of Laboratory Animals,* Committee on Standards, Institute of Laboratory Animal Resources (ILAR), National Academy of Sciences, Washington, DC, 1974.
53. *Swine: Standards and Guidelines for the Breeding, Care and Management of Laboratory Animals,* Subcommittee on Standards for Large (Domestic) Laboratory Animals, Institute of Laboratory Animal Resources, National Academy of Sciences, Washington, DC, 1971.
54. **Brede, H. D.,** The human factor-the weakest link, *Lab. Anim. Sci.,* 30, 451, 1980.

55. **The Animal Technician Certification Program,** Brochure available from the American Association of Laboratory Animal Science, 70 Timber Creek Dr., Suite 5, Cordova, TN 38018, 1988.
56. Report of the American College of Laboratory Animal Medicine on Adequate Veterinary Care, 1986.
57. **Smith, A. L.,** Serologic tests for detection of antibody to rodent viruses, in *Viral and Mycoplasmal Infections of Laboratory Rodents:Effects on Biomedical Research,* Bhatt, P. N., Jacoby, R. O., Morse, H. C., III, and New, A. E., Eds., Academic Press, Orlando, 1986, chap.35.
58. **Parker, J. C.,** The possibilities and limitations of virus control in laboratory animals, in *ICLAS Symp.,* Spiegel, A. et al., Eds., Gustav Fischer Verlag, Stuttgart, 1986, 161.
59. *Public Health Service Policy on Humane Care and Use of Laboratory Animals,* Office for Protection from Research Risks, National Institutes of Health, Bethesda, 1986.
60. Animal Welfare Act, Title 9, Subchapter A, Animal Welfare, PL 99-198, Animal and Plant Health Inspection Service, USDA, 1985.
61. **Atkinson, L. J. and Kohn, M. L.,** Asepsis and principles of sterile techniques, in *Berry and Kohn's Introduction to Operating Room Techniques,* 5th ed., Atkinson, L. J. and Kohn, M. L., Eds., McGraw-Hill, New York, 1978, chap.4.
62. *Bergey's Manual of Systematic Bacteriology,* Vol. 1, Kreig, N. R. and Holt, J. G., Eds., Williams & Wilkins, Baltimore, 1984; and Vol. 2, Sneath, P. H. A., Muir, N. S., Sharpe, M. E., and Holt, J. G., Eds., Williams & Wilkins, Baltimore, 1986.
63. **Wagner, J. E.,** Emerging problems in laboratory animal science and suggested direction for future research and debate, in *Viral and Mycoplasmal Infections of Laboratory Rodents: Effects on Biomedical Research,* Bhatt, P. N., Jacoby, R. O., Morse, H. C., III, and New, A. R., Eds., Academic Press, Orlando, 1986, chap. 38.
64. **Balk, M. W.,** Animal facility design criteria for a toxicologic testing laboratory, *Pharm. Technol.,* 4, 59, 1980.
65. **Henke, C. B.,** Design criteria for animal facilities, in *Symp. on Laboratory Animal Housing,* Institute of Laboratory Animal Resources, National Academy of Sciences, National Research Council, Washington, DC, 1978, 142.
66. **Hessler, J. R. and Moreland,** Design and management of animal facilities, in *Laboratory Animal Medicine,* Fox, J. G., Cohen, B. J., and Loew, F. M., Eds., Academic Press, Orlando, 1984, chap. 17.
67. **Keene, J. H. and Sansone, E. B.,** Airborne transfer of contaminants in ventilated spaces, *Lab. Anim. Sci.,* 34, 453, 1984.
68. **Westwood, J. C. N. and Mitchell, M. A.,** Hospital sanitation; the massive bacterial contamination of the wet mop, *Appl. Microbiol.,* 21, 663, 1971.
69. **Atkinson, L. J. and Kohn, M. L.,** Sterilization and disinfection, in *Berry and Kohn's Introduction to Operating Room Techniques,* 5th ed., Atkinson, L. J. and Kohn, M. L., Eds., McGraw-Hill, New York, 1978, chap. 5.
70. **Williams, F. P., Christie, R. J., Johnson, D. J., and Whitney, R. A.,** A new autoclave system for sterilizing vitamin-fortified commercial rodent diets with lower nutrient loss, *Lab. Anim. Sci.,* 18, 1968.
71. **Small, J. D.,** Environmental and equipment monitoring, in *The Mouse in Biomedical Research,* Vol. 3, Foster, H. L., Small, J. D., and Fox, J. G., Eds., Academic Press, Orlando, 1982, chap. 6.
72. *Microbiological Contamination Control: A State of the Art Report,* Prepared by the Biological Contamination Control Committee of the American Association for Contamination Control, 1, 1965.
73. **Petrocci, A. N.,** Surface-active agents: quaternary ammonium compounds, in *Disinfection, Sterilization and Preservation,* 3rd ed., Block, S. S., Ed., Lea and Febiger, Philadelphia, 1983, chap. 14.
74. **Prindle, R. F.,** Phenolic compound, in *Disinfection, Sterilization and Preservation,* 3rd ed., Block, S. S., Ed., Lea and Febiger, Philadelphia, 1983, chap.9.
75. **Dychdala, G. K.,** Chlorine and chlorine compounds, in *Disinfection, Sterilization and Preservation,* 3rd ed., Block, S. S., Ed., Lea and Febiger, Philadelphia, 1983, chap.7.
76. **Scott, F. W.,** Virucidal disinfectants and feline viruses, *Am. J. Vet. Res.,* 41, 410, 1980.
77. **Orcutt, R. P. et al.,** Alcide: an alternative sterilant to peracetic acid, in *Recent Advances in Germfree Research,* Sasaki, S., Ed., Tokai University Press, Tokyo, 1981.
78. Summary of Efficacy Data for Alcide Disinfectants and Sterilants, Alcide Corporation, Westport, CT, 1985.
79. Summary of Toxicology Results for Alcide Disinfectants and Sterilants, Alcide Corporation, Westport, CT. 1985.
80. **Weisbroth, S. H. and Weisbroth, S. P.,** A proposed microbial evaluation program, *Lab Anim.,* 11, 25, 1982.
81. **Shields, R. P., Schramm, B., and Braune, N. E.,** Evaluation of room cleaning procedures in a laboratory animal facility, *Lab Anim.,* 13, 253, 1979.
82. **Thibert, P.,** Control of microbial contamination in the use of laboratory rodents, *Lab. Anim. Sci.,* 30, 339, 1980.
83. **Wierup, M.,** Bacteriological examination of a modern animal house containing small laboratory animals, *Lab Anim.,* 13, 21, 1979.
84. **Adams, H. R.,** Cardiovascular depressant effects of the neomycin-streptomycin group of antibiotics, *Am. J. Vet. Res.,* 36, 103, 1975.
85. **Adams, H .R.,** Prolongation of barbiturate anesthesia by chloramphenicol in laboratory animals, *J. Am. Vet. Med. Assoc.,* 157, 1908, 1970.

86. **Koch-Weser, J. and Sellers, E. M.,** Drug interactions with coumarin anticoagulants, *N. Engl. J. Med.,* 285, 287, 1971.
87. **Eyssen, H., DeSomer, P., and Van Dijck, P.,** Further studies on antibiotic toxicity in guinea pigs, *Antibiot. Chemother.,* 7, 55, 1957.
88. **Kaipanem, W. J. and Faine, S.,** Toxicity of erythromycin, *Nature (London),* 174, 969, 1974.
89. **Schneierson, S. S. and Perlman, E.,** Toxicity of penicillin for the Syrian hamster, *Proc. Soc. Exp. Biol. Med.,* 91, 229, 1956.
90. *Current Veterinary Therapy IX — Small Animal Practice,* 9th ed., Kirk, R. W., Ed., W. B. Saunders, Philadelphia, 1986.
91. *Current Veterinary Therapy—Food Animal Practice 2,* Howard, J. L., Ed., W. B. Saunders, Philadelphia, 1986.
92. *Current Therapy in Theriogenology 2 — Diagnosis, Treatment and Prevention of Reproductive Diseases in Small and Large Animals,* Morrow, D. L., Ed., W. B. Saunders, Philadelphia, 1986.
93. **Harkness, J. E. and Wagner, J. E.,** *The Biology and Medicine of Rabbits and Rodents,* 2nd ed., Lea and Febiger, Philadelphia, 1983, chaps. 3 and 5.
94. **Flynn, R. J.,** *Parasites of Laboratory Animals,* 1st ed., Iowa State University Press, Ames, 1973.

Chapter 10

THE NEED FOR RESPONSIVE ENVIRONMENTS

Hal Markowitz and Scott Line

EDITOR'S PROEM

As soon as one has adopted anything like a moral *gestalt* on laboratory animals, one quickly realizes that the environments in which they are housed have been devised to serve human convenience, with little or no regard for the animals' nature or *telos,* as discussed in Chapter 2 by Rollin. That such a stance is no longer acceptable to society is plain from the 1985 amendment to the Animal Welfare Act, mandating exercise for dogs and enriched environments for primates. In this chapter, Drs. Markowitz and Line address the dual problem of designing and evaluating environments for laboratory animals, using the history of such innovations in zoos as a springboard, and forthrightly discussing the trade-offs and problems that such a commitment — which they see as a moral necessity — entails.

TABLE OF CONTENTS

I. INTRODUCTION

This chapter includes examples from both laboratory and zoo settings to illustrate that planning for animal environments has been based more upon human values, aesthetic and other, than on consideration of the opportunities that these environments provide for the residents. Some examples of attempts to produce environmental components which increase the power of the animals who must live in captivity will be provided. This power involves the ability of the animals to control some aspects of their living space. Most often this involves the opportunity to feed themselves whenever they wish rather than waiting for caretakers' scheduled feeding times; but the number of ways in which captive animals can be given greater control of environmental components is only limited by the imagination of designers and the availability of adequate budgets.

Varying theoretical positions with respect to the basis of species-typical behaviors, and pressures from inspectors demanding clean facilities, have led some planners of housing for experimental animals to believe that cosmetically attractive, easily sanitized environments are the essential ingredients for good animal quarters. The need for environments that are responsive to animals is ignored, in part because of the belief that each research subject brings with it a natural repertoire of behaviors. Others have taken the position that environments that look natural and have natural substrates will lead the animal to emit a full repertoire of natural behaviors. Neither of these positions is tenable. Attention to the contingencies which animals encounter in nature should be the focus in the development of quarters responsive to the animals that live in them. Husbandry and health care delivery can be facilitated with careful design of responsive environments, thus encouraging the adoption of "wellness" models of animal health. With increasing pressures from organized critics of animal research environments, the time is ripe for scientists to take the lead in demonstrating that they care about quality of life for animals in their projects.

One step toward this goal is to establish institutional policies for the implementation and assessment of environmental changes. Later in the chapter we outline the approach one of our institutions is taking to improve the psychological condition of the animals in our care. It involves not only expanding the behavioral opportunities for the animals and then carefully assessing the results, but also monitoring the animals' responses to routine management events and husbandry procedures.

II. SPECIES-TYPICAL BEHAVIOR AND MEDICAL MODELS OF HEALTH

If you want to use a prompt guaranteed to elicit evasive answers, ask any person who has extensively studied one of the more complex mammalian species to describe their "normal behaviors". Good scientists, unwilling to prematurely categorize behaviors, are at best willing to describe a few gross trends, species-typical signals, or general categories of occupation. Further, a careful analysis, such as those accomplished by Beach[1] or Lehrman,[5] shows that many of these generalized "instincts" or "typical behaviors" are more predictable on the basis of available resources, climatic conditions, and presence of other animals in the habitat than on the basis of species membership.

Yet, if one spends as little as an hour each in front of gorillas, chimpanzees, and orangutans during their active periods, even a fledgling zoo biologist is able to describe some salient differences in their typical behaviors. Thus, orangutans are often described as "engineers", interested in taking apart and reorganizing whatever complex materials they can get their hands on; chimpanzees as highly social, emotional, and, by some (e.g., Maple[7]), as highly intelligent. Gorillas are typically found in family groups led by one dominant male and are considerably less active than chimpanzees.

There are some substantial differences between field reports and the results of studies of captive gorillas with respect to the amount of interaction between adult males and their offspring, and this is also true for orangutans. Before we rush to the conclusion that the differences in reports such as these are a function of unnatural captive habitats, we must also consider that the limited data from the wild is subject to biases based upon presence of observers, situations in which unrestrained animals allow themselves to be observed, and provisioning of wild animals which accompanies some studies.

In spite of such difficulties in interpreting the results available concerning representative behaviors for various animals, progress in environmental engineering requires concentration on the identification of species-typical behaviors in research on both captive and wild animals. Rather than being trapped into seeing our occupation as identifying "abnormal" behaviors or "normal" conditions based upon outmoded dichotomous models of physical and psychological health, we should strive to provide quarters which engender opportunities for animals to exercise those behaviors for which they have become specialized in the course of their evolutionary histories.

Early band-aid efforts to improve the opportunities available in the environments of captive animals (e.g., Markowitz[9,10] and Schmidt and Markowitz[16]) and simple behavioral diagnostic procedures for severe anomalies[11] were shown to be effective. However, both humane considerations and the important matter of having fit experimental subjects can only be adequately approached by the development of environments rich enough that naturalistic contingencies provide the animals with a measure of wellness. This is the message which we offer in this chapter: let those of us who decide to try to improve captive environments for animals concentrate on the wealth of naturalistic behavioral opportunities which we can provide the inhabitants, rather than exclusively consulting on how to diagnose and cure their abnormalities.

Surely we need not make captive animals live through unnecessary discomforts by recapitulating the history of Western human medicine in excessive concentration on curing the ill and little attention to quality of life for those who are not substantially suffering. The solution does not lie in uncritically introducing some materials to make things appear natural. Neither zoos nor research facilities can be "natural" environments, nor, in truth, do their administrators wish to introduce the hazards of nature. This point has been made often in the past[9,10] and we will not bore the reader by recapitulating the irrefutable evidence. Instead, we will try to make the critical point — that most environment planners continue to erroneously look at behaviors as things that emerge from animals *in vacuo* — with some specific examples.

III. EXAMPLES FROM ZOOS

A. POLAR BEARS

Figure 1 illustrates a food delivery mechanism developed for use in an antiquated polar bear exhibit in Portland. Although the protocol used was exceedingly unnatural, it did some worthwhile things for bears in an even *more* highly unnatural environment.[11] The bears vocalized when they wished to order some food which consisted of catapulted fish and omnivore chow. Some of the food was delivered into their pond, and Figure 2 shows one of the bears who apparently enjoyed diving for the fish. Although this temporary procedure did provide some exercise and a small measure of control over their own schedules, it is hardly what any sane person would propose as an ideal captive circumstance for polar bears.

A much more adequate environment is seen in Figure 3, taken in the Point Defiance Zoo in Tacoma, WA. In our estimation, this is one of the most attractive polar bear exhibits ever established. The clever use of fences which disappear into the trees in the background, the unusually large area, and the extensive natural-appearing access to water are all ways to encourage visitors to think of these nomads of the north as animals that live in relatively boundless domains. The opportunity to observe these marine mammals swimming is certainly

FIGURE 1. Food delivery mechanism for polar bears used in the Portland Zoo.

FIGURE 2. Polar bear dives into pool to "capture" fish.

of great educational value. It is obscene that most zoo-goers traditionally delight in making terrestrial beggars of *Ursus maritimus*, animals that make their living in nature by capturing seals and fish.

This exhibit is beautiful, but it is only half excellent. Notice that in Figure 3 none of these aquatic mammals are swimming. Why? Because the exhibit is designed for natural *appearance*, but largely ignores the contingencies which lead to natural *behaviors*. In order to demonstrate the beauty of these bears swimming during a visit by the first author, one of the curators threw some fish into the water. In Figure 4 one of the bears has finally entered the water because, for the first time, there is a clear reason to do so.

The Zoo Director indicated his plans to have regular feeding times, when the public might have a chance to see these bears swimming to collect fish. While this is a step in the right direction, why should a zoo which has spent so much time and money developing an attractive *looking* exhibit not spend equal effort to make it an attractive *working* exhibit?

The answers to this question lie in long traditions of architectural lobbying *and* the fact that

FIGURE 3. Polar bear exhibit at Point Defiance Zoo.

people continue to expect that given a good looking, large environment, the animals' *natures* will produce a full repertoire of species-typical behaviors. The suggestion was made that it would be easy to design a protocol and equipment whereby occasionally, on a random naturalistic basis, fish would swim by. Responses to this suggestion were: "How much will it cost?" and "Can you promise that it will not require any maintenance?" Cost? A cinch: less than 5% of the total cost for the other *half* of the exhibit (the appearance). Maintenance free? Of course not, but requiring as little or less maintenance than the pumps and filters required for the polar bear aquatic areas.

Neither of these answers were attractive to the zoo administrators, who understood that façades, cages, and buildings were necessary, expensive, and in need of routine maintenance, but did not feel that providing behavioral opportunities for the captive animals could justify significant costs or maintenance time.

B. GREAT APES

Figure 5 is a photograph of a spacious, rather elegant gorilla exhibit in the San Francisco Zoo. When the animals were originally introduced to this domain, the adult male sat for days on the concrete at the entrance to the new open area; but he did ultimately join the others out in the grass, the aquatic areas (where he tried to reengineer the plumbing), and under the trees. The younger animals climbed the trees and seemed especially to enjoy those times when branches would break and drop them unexpectedly to the ground. Soon many of the branches were gone and the

FIGURE 4. Polar bear underwater at Point Defiance Zoo.

FIGURE 5. Gorilla grotto at San Francisco Zoological Gardens.

FIGURE 6. Gorillas at perimeter of grotto in San Francisco Zoological Gardens.

trees began to be stripped of some of their bark. In order to prevent the trees from dying, the zoo staff had the trunks surrounded with fencing material which allowed the animals to continue to climb, but largely prevented stripping the bark. Frequency of climbing was reduced, but some of the gorillas do still occasionally use the trees.

Recently, several zoo staff members have asked for ideas and encouraged research proposals to generate more active and extensive use of the exhibit. For example, they wish to know why gorillas spend much of their time in the perimeter moat, as seen in Figure 6. the answer seems very apparent: there is more for them to manipulate, more that is novel from day to day, more with which to impress each other in the litter near the perimeter than there is in the more static, albeit naturalistic-appearing, major exhibit area. In the wild, things change. Keepers do not bring measured diets each day, walls do not protect residents from potential harm, and older animals can demonstrate their power and skills in provisioning and defending the group. Seasonal changes and movement within habitats and between groups provide ever-changing stimulation and new contingencies to be learned. For obvious reasons, zoos are not really willing to introduce dangerous predators to exhibits with rare and endangered species. Neither are they anxious to have visitors observe exhibits in which animals regularly forage for their own food or are left to grow hungry. So, as has been noted,[8] the notion that zoos and wildlife parks are becoming increasingly natural is a fallacious one. What they are becoming, to their credit, is increasingly naturalistic in appearance and increasingly generous in space provided for captive animals. Most major parks are now spending millions of dollars in the development of these new habitats. Few of them are spending thousands of dollars in the at least equally important task of providing stimulating and dynamic environments for the resident species. In the course of natural selection these animals have been provided unique and sometimes amazing morphological characteristics which allow them to perform food-capturing feats which can only leave us in awe. Zoo visitors learn about the general appearance of wild habitats and see handsome creatures occasionally

FIGURE 7. Orangutan enclosure at San Francisco Zoological Gardens.

moving about in these exhibits, but only on those rare occasions when a bird happens to fly into an open exhibit, when an insect is unfortunate enough to move near a quick predator, or when a lunatic human climbs into one of the exhibits, do visitors typically see the excitement of the chase for which so many of their favorite zoo animals are specialized.

If zoos are to become the wonderful educational conservatories that they profess to wish to become, they cannot do so by continuing to pay only lip service to the behavioral needs of the animals. In the sense that natural contingencies have been largely eliminated, virtually all of those behaviors on which we tend to focus are abnormal (witness the average zoo visitor's delight in making faces or otherwise hazing primates to get responses from them, the crowd's delight in seeing an otherwise inactive bear or sea lion beg for food, etc.).

Stripped of the need to defend and provision themselves in the "natural" ways that their ancestors did, humans have evolved endless means of involving themselves in new activities to exhibit their talents and provide exercise and competition. Unless one thinks of human beings as having evolved in some special process, separate from other animals, it is most parsimonious to believe that other complex species would, given the chance to control more of their own destinies, opt for complexity, variety, and challenge rather than the behaviorally barren environments that we typically provide for them.

Figure 7 shows the traditional home of the orangutans in the San Francisco Zoo. Comparison with pictures of young orangs in the forest of the same zoo (Figure 8A and 8B) will help to introduce another important topic for consideration: how great a role should quality of life play in husbandry procedures and habitat development for valuable and increasingly rare captive species?

This work was part of a series of orangutan enrichment projects initiated by John Alcaraz and other keepers. The first project, which involved taking the young orangs on supervised outings in the woods, was discontinued partly because of discord among the staff with respect to

A

B

FIGURE 8. (A and B) Orangutans in trees within the grounds of the San Francisco Zoological Gardens.

volunteer participation, and partly because of the fear that the orangs were reaching an age at which they might inflict serious damage on handlers. The project illustrated here involved two 3-year-old animals, Oliver and Thelma. It served an important function in providing them opportunities to develop extensive behavioral repertoires and interactive skills. Introduction of these youngsters to the established group of orangutans seen in Figure 7 went very smoothly, and within the first day they were climbing on the wooden and steel structures and sharing brooms, garbage can hats, and other favorite items with the older animals.

The zoo director was lobbied to consider turning a substantial wooded area into a permanent home for an entire group of orangutans. While he was generally responsive to the idea, he had a number of well-founded reservations which may be the best brief way to emphasize the questions at hand. For example, suppose that as they grew larger one of the older animals followed the young ones into the less sturdy tree branches and took a disastrous fall. Could the zoo justify to its patrons and to welfare agencies having taken this risk with these increasingly rare and valuable apes? Suppose that one of the animals became ill, or was injured, and chose to climb high into the forest rather than allowing treatment to be easily administered. Could the zoo justify endangering the keepers and veterinary staff who would try to capture the orang under these precarious conditions? What about the danger to the ape of falling during tranquilization or other treatment? What about complaints from the public that they were unable to easily see the young animals in the large forested area?

All of the above questions illustrate the dilemma faced by progressive zoo administrators who would work to provide more substantially natural environments for the animals in their trust. There are no easy solutions to this problem of deciding whether the quality of life available for an orangutan group that can spend much of its time playing and exercising in the trees outweighs the inherent dangers of the situation. But there is an interesting way to consider this by analogy: with human beings, including our loved ones, we may make some judgements about protecting them from severe avoidable dangers; but should we protect them with the sterile, impoverished environments which we provide for animals that we claim to love, we would certainly be condemned by our peers. Can we really defend protecting these rare and sensitive animals to the extent that we reduce their quality of life to a mere shadow of what it might have been?

Obviously, this is a question that each reader might answer in a different fashion, and that cannot be resolved with simple presentations like this overview. But for us, even it it were the last pair of breeding orangutans on earth, we believe that it would be a selfish and cruel decision to place the preservation of the species before consideration of the quality of life for the individuals. After all, if we have been selfish enough to destroy the environments which provided their wild habitat, what are we "preserving" them for anyway?

IV. OPPORTUNITIES AND CONTINGENCIES

A. RIVER OTTERS

An exhibit in which Asian small-clawed river otters were given the opportunity to hunt for crickets has been described in detail elsewhere.[3] Although the appearance was somewhat naturalistic, the contingencies (which involved interrupting photo cells during hunting activities) were absolutely unnatural. But that does not mean that they were unhealthy or the result of some misbegotten conception of an animal's nature. They were the result of careful thought about what might be interesting and attractive modes of exercise for animals whose custodians have chosen to remove them from natural dangers and from opportunities to capture wild prey. They were the result of an admittedly empathetic approach in which the question was asked: "If we were this species, with this morphology, and with these typical behaviors in nature, what would help to return some of the excitement and opportunity that is missing in the captive environment?" The project was one in a long series of efforts to find ways for animals to exercise *when they wished to* in modes that were species appropriate. The history of that work shows that many of these efforts were attractive to animals[10] and, as a matter of good fortune, were also attractive to many visitors who were excited to see active, motivated animals.

B. SERVALS

Servals are vary fast hunters, able to flush game birds from the bush and capture them as they begin flight. Undoubtedly, there is some contingency in this behavior. If the cat were never to capture any of the game which it pursued, it would surely give up on this behavior and find some alternative means of nutrition, or die of starvation. Providing these felines with flying meatballs in captivity (Figure 9) was a successful "unnatural" means of motivating exciting capture behaviors. It involved contingencies. So does nature.

C. RHESUS MONKEYS

Our work in the laboratory has focused on rhesus monkeys, one of the most commonly kept laboratory species. In a series of experiments we have explored ways of providing singley caged adult rhesus monkeys with increased opportunities for contingent behavior. A variety of devices have been used readily by the monkeys and have been associated with increases in species-typical activity and decreases in aberrant behavior.

One of the most successful devices is an electromechanical apparatus that contains a radio and a food dispenser (Figure 10). Three contact detectors extending from the device approximately 4 cm into the cage allow the monkeys on/off control of the radio and activation of the food dispenser. The radio is preset to a "top 40" format radio station at a low volume. The food dispenser delivers a 45-mg banana-flavored food pellet into an attached trough immediately after the monkey touches the appropriate detector. All of the monkeys tested used the device throughout the time it was available (Figure 11), and self-abusive behaviors such as fur plucking and self-biting decreased significantly when the device was available. Stereotyped pacing declined, while nonstereotyped locomotion increased. Other behaviors, such as grooming, also changed in directions that resembled the activity patterns of free-ranging rhesus monkeys.

Another enrichment device that was effective in decreasing abnormal behavior is a plastic puzzle feeder (Figure 12). The device attached to the monkeys' home cage and allowed them to spend time gathering extra food, an activity similar in kind to the foraging wild monkeys perform daily. The device was loaded with monkey biscuits, which were also available by free choice during regular twice-daily feedings. It required little maintenance, and all the subjects learned to use it with no training. On average, they removed half the biscuits from the puzzle feeder daily, even though they could get the same food for no work at the regular feedings. There was a significant Spearman's correlation of 0.40 between the number of biscuits taken and the frequency of abnormal behavior displayed ($p < 0.001$). The term "abnormal" is used here to describe behaviors which are not typical of the species in the wild, but are typically produced in unresponsive, barren environments.

These two examples demonstrate that, like animals in zoos, monkeys housed in laboratories will readily make use of increased chances to exert control over their environment.

V. INSTITUTIONAL POLICY TOWARD EXPANDED BEHAVIORAL OPPORTUNITIES

The U.S. Department of Agriculture will soon be publishing regulations that will be used to implement the most recent amendment to the Animal Welfare Act. Part of the regulations will require institutions using non-human primates to take steps to ensure their "psychological well-being". At the California Primate Research Center (CPRC) we are addressing the issue of "psychological well-being" by adopting a general plan for assessing and improving the behavioral opportunities of all of the animals in our care. In the next part of this chapter we shall briefly outline this plan and give examples of how it is being implemented.

FIGURE 9. Serval captures flying meatball in Portland Zoo.

The first task was to reach a common understanding of what is meant by "psychological well-being". Given the difficulty in defining it as it applies to people, it is not surprising that there is a general reluctance among the scientific community to apply this term to non-human primates. Still, this is no reason to wring our hands in despair and give up. Rejecting the concept as unworkable and simply accepting the status quo is not an adequate response to the wishes of

FIGURE 10. Music-feeder device.

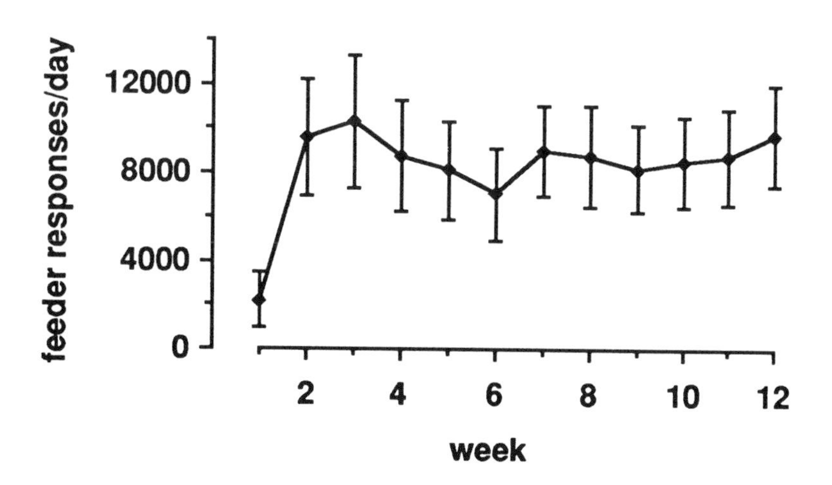

FIGURE 11. Use of music-feeder by ten adult female rhesus monkeys. This figure represents the mean number of responses per day to the feeder over a 12-week period.

FIGURE 12. Puzzle feeder.

Congress. Leaving aside the issue of changes necessitated by regulation, we also feel an ethical imperative to continually strive to improve conditions for our animals. The goal of laboratory animal science from its inception has been to continually improve animal care. Great strides have been made in the past several decades in improving the health and safety of laboratory animals. Attention to the behavioral opportunities afforded by the laboratory environment is a logical extension of that care.

An approach that could satisfy the intent of the legislation is to establish an operational definition of well-being by quantifying the monkeys' responses to environmental changes. A variety of behavioral and physiological measures such as the range and frequency of normal and abnormal behaviors displayed, rates of growth and reproduction, and the response to stressors in the environment have been suggested;[6,12] we are using various combinations of these to assess the effect of the behavioral enrichment program.

Most primate species are social, and yet many facilities keep primates in single cages. This is sometimes required by experimental protocol and has reduced the chance of trauma and disease transmission, but it severely limits the expression of social behavior. Placing compatible animals in pairs is an alternative to single caging that allows animals to perform many social behaviors.[15] If carefully done, it does not greatly increase the risk to the animals. At the CPRC we will begin by pairing adult male and female macaques with infants, since they are more readily accepted as social partners.

The goal for primates that still need to be housed alone will be to provide other forms of enrichment. As described above, this will include the use of devices that allow the animals to control delivery of food and music. Additional options will be the provision of cage toys and climbing structures, variations in the diet, and increased positive human interaction. These alternatives will also be provided to the animals housed in groups.

Increased contact between animals, and between animals and people, increases the health risks for both monkeys and personnel. As we stated before with respect to zoos, there are trade-offs that will have to be made in order to reduce the barrenness of many animal environments.

An integral part of the program is the formal evaluation of the effectiveness of various enrichment strategies. We feel it is vitally important to demonstrate through behavioral and physiologic testing that the changes that are made have a beneficial effect on the animals. As part of this effort, we are instituting a training program for the animal care staff and volunteers who will be helping evaluate the environmental changes. The next section of the chapter describes in more detail the methods we are using.

VI. METHODOLOGY FOR ASSESSING ANIMALS' RESPONSES TO ENRICHMENT

A wide variety of behavioral and physiologic variables can be monitored to assess responses to changes in laboratory environments. Hormonal levels in plasma or urine, heart rate and heart rate variability, and changes in immune status and reproductive function are just a few examples that could be measured. Changes in physiologic systems can provide additional objective means (besides behavioral measures) of determining the impact of environmental enrichment. They can be measured remotely in some cases, or quickly in others, thus avoiding any influence of the experimenter on the variable being measured.

The technique we are using for behavioral observations is based on a microcomputer/bar code system that we developed in collaboration with colleague Eric Carlson. A 24-kbyte portable computer is used to record frequency and duration of behavior in 51 categories. Each category is printed with a unique bar code on a plasticized sheet. A light pen reads the bar codes and enters the data into the portable computer, which can hold up to 15 h of observations. Data are automatically transferred from the portable computer to a personal computer, eliminating the need to transcribe them by hand. The result is rapid and accurate analysis of the raw data files.

We are using this system to construct a behavioral profile of rhesus monkeys housed in a typical laboratory environment. We are generating activity budgets that show how much time the monkeys spend sitting, standing, crouching, lying, moving, and pacing. We also record the amount of time grooming, sleeping, and looking at the observer. The frequency of other behaviors, including aggressive and submissive acts, vocalizations, cage manipulation, aberrant postures and stereotypic actions, eating, and drinking are also recorded. The result is a comprehensive picture of how the monkey spends its time in the laboratory.

The primary hormone we have been measuring is plasma cortisol, both basal levels and changes in response to mild, acute stressors. Interpretation of cortisol values can be problematic, but at least it serves as a short-term indicator of arousal in the subject. Plasma cortisol rises quickly in response to stressful events, including blood sampling, so it is important to get the sample within minutes of entering the animal room in order to accurately measure basal values. In practice this is easy to accomplish with monkeys, since they can readily be trained to present an arm for venipuncture and quickly adapt to the blood sampling procedure. We have also been recording the monkeys' cortisol responses to mildly stressful events, such as brief restraint and short confinement in a transfer cage, that routinely occur in the laboratory. This is done to see if provision of an enriched environment modulates the response to stressors.

The cardiovascular system also responds quickly to environmental changes, making it another useful variable to monitor. We have been recording heart rates with a biotelemetry system. A small transmitter is implanted subcutaneously in the animal and continually broadcasts a radio signal encoding the electrocardiogram. The system also gives information on the animal's body temperature and amount of activity. A receiver mounted on the cage decodes the signal and relays it to a computer in an adjacent room. In this way a large number of subjects can be simultaneously monitored, without disturbance. The telemetry equipment allows us to record responses to many more routine events, such as caretakers entering the room to feed or clean, and the like. Continuous recording also allows study of circadian patterns of heart rate and activity.

Other body systems, including the reproductive and immune systems, provide many variables that could be used to register the effect of environmental enrichment attempts. Menstrual cycle length, conception rate, and urinary levels of gonadal hormones are examples of reproductive function that can be recorded noninvasively. Preliminary results with lymphocyte mitogen stimulation at the CPRC have shown that simply moving monkeys from one room to another can suppress the mitogen response for 1 to 2 weeks. This test is sensitive, but is expensive and technically difficult, which may limit its utility. Measurement of other components of the immune system, such as levels of plasma globulins or numbers of different cell types in a complete blood count, may provide similar information.

Given the complexity inherent in biological systems, it is clear that no single behavioral or physiologic measure will adequately reflect an animal's response to changes in its environment. Quite often, in fact, one system may show no change to an event, while another is dramatically effected. For example, we found that in rhesus monkeys plasma cortisol and heart rate both increase in response to brief restraint. However, if the procedure is repeated several times, the cortisol response rapidly diminishes while the heart rate response continues to occur. This is the reason we record combinations of physiologic and behavioral variables. This approach is more likely to provide a comprehensive assessment of the monkeys' responses, and when multiple variables are affected in the same way, stronger conclusions about the impact of the environmental change can be made.

VII. COMPARISONS BETWEEN ZOOS AND RESEARCH FACILITIES

The design of zoo exhibits does not require choosing between natural appearances and behavioral opportunities. Quick[14] has argued for "An integrative approach to environmental engineering in zoos." Quick has echoed our sentiment that the presence of natural materials may not suffice in generating natural behaviors. But zoos do have an obligation to teach people appropriate things about the natural habitats of animals and to replace traditional concrete substrates with ones more comfortable for the animals. The efforts of zoo directors who have committed themselves to exhibiting fewer animals in more ample naturalistic enclosures and to moving away from the traditional "postage stamp" or "trophy" collections are to be applauded. For some animals, such as certain ruminants, allowing ample natural grazing area may provide much of what is necessary for natural behavior.

However, for many of the most popular zoo animals, more ideal exhibits require intensive cooperative efforts between zoo biologists and exhibit planners to insure that behavioral opportunities will accompany more adequate exhibit appearances. These efforts should result in animal habitats which look naturalistic and lead to naturalistic behaviors. With modern technology, wonderful looking artificial rock facades may be used to enhance the physical appearance of exhibits where it is often impossible to use real rock of the same dimension. In an analogous fashion, we may not be able to divert a river stocked with fish through every exhibit to provide prey for fish-eating mammals or birds, but we can provide attractive opportunities for fish capture through use of equipment which releases fish on a periodic basis into aquatic areas of exhibits. Both artificial rocks and artificial fish delivery systems are unnatural, but both can provide more appropriate educational opportunities for zoo visitors than traditional barren exhibits.

What benefits, in addition to the knowledge that they have done something for the animals, can zoos accrue from behavioral enrichment work? Obviously, as Chasan pointed out,[2] the first bonus is that zoo visitors will learn much more about the animals' abilities and spend more time watching the residents of cages in which behavioral opportunities lead to complex activity. Carefully engineered behavioral enrichment designs can be used to enhance husbandry procedures, for example, by ensuring that animals regularly come to places where they may be weighed or routinely examined. Well-conceived protocols may also largely eliminate the need for excessive capture procedures in which keepers sometimes chase animals around their environments with brooms, nets, and hoses. As mentioned earlier, diagnosis of illness in these animals incapable of symptomatic report is enhanced when they are active and when some of the ways they make a living are predictable.

Another factor which should help spur action and encourage the establishment of budgets for behavioral enrichment is the increasing effectiveness of animal welfare groups in lobbying against traditional methods of confinement. Sometimes, as we have indicated in the past,[10] these active protestors miss the point and may indict zoos on faulty bases, such as foreshortening the lives of animals. Animals live longer on the average in zoos than in the wild. Sometimes there has also been the misconception that all pacing is abnormal or a sign of boredom, when a more careful analysis may show that this behavior may be typical of the same species in the wild. But, there is one area of criticism from those who condemn zoos that cannot be turned aside so easily. This is the argument about quality of life, typically made by analogies to prisons and other human "corrective" institutions. While some of these complaints may lose credence because of their overzealous presentation, it is certainly true that there are some decided similarities between prison protocol and zoo protocol. In both places there is an overwhelming emphasis on having the residents meet schedules of the staff. Typically, both kinds of institutions have been stripped of almost all opportunities for individuals to gain any substantial control over their own environments. Independent, self-assertive behaviors are frowned upon.

In outmoded penal institutions, based on retributive models of justice, these procedures are justified by some people as appropriate punishments, or deterrents to others who might commit crimes, even though the data clearly contradict the latter premise. In zoos, the argument is that our job is to preserve these precious animals; to protect them from dangers such as predators, or falls from trees that sometimes take their lives in the wild; to protect those young that might be culled by natural selection; and even to protect our charges from the wrath of those who might condemn them for natural behaviors such as eating other live creatures.

All of us have become increasingly aware of the dangers of overprotection for those humans that we love. The same logic applies to other animals for whom we care. To deny them all that makes life vital, in the name of conservation and protection, is to preserve mere vestiges of the animals. There should be a substantial difference between taxidermy and husbandry!

Behavior counts. This clear fact should be a powerful persuader for those readers who would help improve captive animal situations. Zoo directors will be able to highlight your efforts in showing critics that they do recognize the need for active opportunities for those creatures in their care. Ultimately, the very survival of zoos may be at stake.

Racing down to your local zoo and asking for truly adequate budgets to begin respectable work in the development of behavioral opportunities for the resident collection is not likely to be very rewarding. Zoos are *prepared* to pay fortunes to planners and architects who provide them attractive façades. It is their heritage! Suggesting that a small fraction of those costs be designated for attention to naturalistic behavioral opportunities is not a tradition. Yet it is clearly apparent that an animal deprived of opportunities to control some of its own critical contingencies is living in an unhealthy environment.

It is time for us to stop accepting the absence of physical lesions and the presence of offspring as the ultimate criteria for adequate captive environments. It is time to champion the quality of life for those captive creatures which many of us claim to love. It is time for experts on behavior who choose to work in zoos to stop being coopted by the establishment. It is time to recognize that our greatest contributions may be made by designing *behaviorally* adequate environments and daily protocols, rather than passively assessing the effects of the highly inadequate domains which we continue to build for most species.

Those readers who are primarily interested in assessing the adequacy of environments for research animals are doubtless wondering what all of this zoo business has to do with their charges. Although our research facilities are increasingly visited by those interested in the humane care of animals, there is no great demand to make research quarters look zoogeographically appropriate. However, with new regulations asking that we address the "psychological well-being" of at least some research animals, there seems to be increasing recognition that the behavioral opportunities available to animals are of consequence. There is also increasing awareness that living in a responsive environment may be essential to the maintenance of research subjects that are adequate representatives of their species for most research purposes.[10,13]

ACKNOWLEDGMENTS

This research was supported in part by NIH grant RR00169. We thank Dr. Dave Hess, Oregon Regional Primate Research Center, for performing the cortisol assays, Kathy Morgan, Sharon Strong, Mark Nakazano, and Bill Wolden for assistance in data collection, and the CPRC animal care staff for their cooperation and assistance.

REFERENCES

1. **Beach, F. A.,** The descent of instinct, *Psychol. Rev.,* 62, 401, 1955.
2. **Chasen, D. Z.,** In this zoo visitors learn, though no more than animals, *Smithsonian,* 5(4), 22, 1974.
3. **Foster-Turley, P. and Markowitz, H.,** A captive behavioral enrichment study with Asian small-clawed river otters (*Aonyx cinerea*), *Zoo Biol.,* 1, 29, 1982.
4. **Hancocks, D.,** Social biology, bioclimatic zones and the comprehensive master plan for the Woodland Park Zoological Gardens, in *Applied Behavioral Research,* Crackett, C. and Hutchins, M., Eds., Pika Press, Seattle, 1978, 1.
5. **Lehrman, D. S.,** Problems raised by instinct theories, *Q. Rev. Biol.,* 28, 337, 1953.
6. **Line, S. W.,** Environmental enrichment for laboratory primates, *J. Am. Vet. Med. Assoc.,* 190, 854, 1987.
7. **Maple, T. L.,** Great apes in captivity: the good, the bad, and the ugly, in *Captivity and Behavior,* Erwin, J., Maple, T.L., and Mitchell, G., Eds., Van Nostrand Reinhold, New York, 1979.
8. **Markowitz, H.,** On natural zoos and unicorns. Keynote address, Western AAZPA Conf., Seattle, 1977.
9. **Markowitz, H.,** Engineering environments for behavioral opportunities in the zoo, *Behav. Anal.,* 1(1), 34, 1978.
10. **Markowitz, H.,** *Behavioral Enrichment in the Zoo,* Van Nostrand Reinhold, New York, 1982.
11. **Markowitz, H., Schmidt, M. J., and Moody, A.,** Behavioral engineering and animal health in the zoo, *Int. Zoo Yearb.,* 18, 190, 1978.
12. **Novak, M. A. and Suomi, S. J.,** Psychological well-being of primates in captivity, *Am. Psychol.,* 43, 765, 1988.
13. **Preilowski, B., Reger, M., and Engele, H.,** Combining scientific experimentation with conventional housing, a pilot study with rhesus monkeys, *Am. J. Primatol.,* 14, 223, 1988.
14. **Quick, D. L. F.,** An integrative approach to environmental engineering in zoos, *Zoo Biol.,* 3, 65, 1984.
15. **Reinhardt, V., Houser, D., Eisele, S., Cowley, D., and Vertein, R.,** Behavioral responses of unrelated rhesus monkey females paired for the purpose of environmental enrichment, *Am. J. Primatol.,* 14, 135, 1988.
16. **Schmidt, M. J. and Markowitz, H.,** Behavioral engineering as an aid in the maintenance of healthy zoo animals, *J. Am. Vet. Med. Assoc.,* 171(9), 966, 1977.

Part VI
Stress, Pain, and Suffering

Chapter 11

STRESS AND COGNITION IN LABORATORY ANIMALS

Seymour Levine

EDITOR'S PROEM

For much of the 20th century, stress in animals was seen as a mechanistic, nonspecific response to a variety of noxious stimuli. This purely physiological approach to stress has increasingly come to be seen as inadequate. In this chapter, Dr. Levine details the experimental evidence for the claim that the phenomenon of stress cannot be adequately discussed apart from the animals' cognitive states. It has been shown that stress responses are under psychological control, that situations of, for example, frustration, uncertainty, and novelty can produce different physiological effects depending on the animal's psychological state. As Levine shows, this new understanding of stress has major implications for the proper treatment of animals. In particular, he emphasizes the extent to which this cognitive approach to stress needs to be considered in giving meaning to new regulations mandating concern for the psychological well-being of primates.

TABLE OF CONTENTS

I. INTRODUCTION

The issues concerning the moral justification of the use of animals for research has resulted in an impassioned, often accusatory, and somewhat irrational emotional debate. One of the extreme positions would legislate all animal research out of existence. Justifications for this view stem primarily from moral and sometimes religious arguments. The other extreme would place no restrictions on the endeavors of the scientists based upon the assumption that pursuits aimed at improving the human condition supersede all other considerations. Thus, all decisions concerning the use of animals should be made by scientists whose wisdom and judgments are beyond reproach. Clearly, acceptance of either of these extreme positions would have profound consequences for mankind. There are so many examples of the importance of animal models in solving problems of human pathology that the loss of these tools would impair, if not destroy, biomedical research. Computer modeling or the exclusive use of *in vitro* techniques cannot replace the use of a living organism, an exquisitely beautiful and complex system which consists of an integrated series of interactions and communication between the numerous components of this system.

Over the past decades we have seen an erosion of traditionally held concepts that imply independence of biological systems. For example, it has become apparent that the nervous, endocrine, and immune systems act in concert to regulate basic biological phenomena. Without the knowledge of all the required information, computer models at best provide hypotheses that require testing in the living organism. *In vitro* systems have the power to isolate various components of the living systems and thus to examine specific activities, but it is not possible to predict that the *in vivo* action of these processes will be concordant with the integrated action of biological processes. Since it does not appear likely that we can readily dispense with animals in biological research, must we thus accept the opposite extreme?

Clearly, this is also not the solution. Science is a human endeavor and, thus, is also victim to the numerous frailties that plague humans. All scientists are not wise or capable of good judgment. Numerous abuses, well documented, have occurred. Falsification of information, although hopefully not common, has also been well documented. Therefore, we must accept the fact that regulations must be imposed upon the scientific community regarding the care and use of animals in research. However, care must be exerted not to throw the baby out with the bath water. Some of the current regulations are so vague and undefined that they already threaten to stifle research, a goal which is clearly the mandate of the extreme animal rights activists. One such example is the regulation that requires that the psychological well-being of primates in the laboratory be maintained. This requirement is a direct result of the increased pressure for legislation and regulation intended to protect animals against the infliction of unwarranted pain and distress and to establish guidelines that assure the well-being of laboratory animals. The problem in developing such guidelines is to establish and define what constitutes well-being for animals. In the absence of such a definition, any legislation regarding the care and use of laboratory animals will be based more on emotional responses rather than objective information.

II. STRESS AND COGNITION

There appear to be several underlying assumptions that have generated the most current recommendations concerning the well-being of laboratory animals. The first assumption is based upon the argument that stress is detrimental to living organisms and, therefore, as much as possible, laboratory animals should be maintained in a stress-free environment. However, many of the physiological responses that are elicited by stress are part of the basic adaptive mechanisms of animals.[1] A second assumption is that stress is predominantly a result of physical insult and trauma. Among the numerous examples cited are pain, surgical trauma, and exposure to toxins. Although these stimuli clearly elicit many of the physiological and behavioral manifestations of stress, there are numerous stimuli which elicit identical responses which apparently have no obvious noxious properties.[2] Further, even physical insults, such as hemorrhage or exposure to heat, may elicit stress-related physiological responses as a consequence of inherent psychological factors.[3,4] Perhaps the most difficult aspect of the problem of well-intentioned guidelines is the assumption that there exists a universally accepted definition of stress.

However, despite numerous and valiant attempts to define stress, a clear and universally accepted definition still remains, to say the least, elusive. It is important to note, however, that historically the concept of stress has always been associated with changes in the endocrine system. Initially, these changes were specifically related to either increased secretion of catecholamines or activation of the pituitary-adrenal (P-A) system. It has now been clearly demonstrated that many endocrine changes occur following those environmental events which are typically called stressful. However, the focus of a great deal of research related to stress physiology is still on the P-A system. It is, of course, impossible to discuss any facet of the P-A system and its relationship to environmental factors without acknowledging the landmark contributions of Hans Selye. His formulation of the general adaptation syndrome (GAS) highlighted the importance of this hormone system and the diverse physiological effects of glucocorticoids following adverse stimulation.[5]

Selye's emphasis on the role of the glucocorticoids in biological adaptation was indeed prophetic. With the advent of modern molecular biology there has been increasing evidence of the fundamental role that the adrenal glucocorticoids play in a wide variety of physiological processes. Glucocorticoids have an effect on many aspects of development, on metabolism, on the expression of genes in essentially all tissues of the body, on basic immune processes, and there is increasing evidence that they have important influences on many functions of the central nervous system, including learning and memory.[1,6] In view of the essential function that the glucocorticoids have in the maintenance of numerous physiological functions, it is not surprising that the regulation of the secretion of glucocorticoids has been studied extensively. The regulatory mechanisms which govern the secretory rates of glucocorticoids have been examined at every level, including the adrenal, pituitary, hypothalamus, midbrain, and limbic system.

In Selye's original theory, which elaborated the elements of the GAS, one of the major assumptions was that the stimuli eliciting the stress response were nonspecific. This hypothesis was based largely on the results of early research which demonstrated that a variety of physical and chemical challenges seemed to elicit the same pattern of endocrine responses. Mason,[3,4] in his critique of Selye's theory, pointed out that there were important psychological variables embedded in many of the experimental situations in which P-A activity was observed to increase. Moreover, in some cases the psychological reaction appeared to be the primary stimulus for initiating the cascade of neuroendocrine responses observed following stressful stimulation. Mason cited several experiments in which seemingly equivalent physical stimuli had different effects on the P-A system, depending upon temporal aspects of the challenge. He hypothesized that the psychological aspects of the situation, related to novelty, had to be considered and controlled for by altering the rate of presentation. One of the best examples of graded adrenocortical responses, which are dependent upon the rate of presentation, was reported by Gann.[7] In this experiment, dogs were first hemorrhaged 10 ml/kg at the rate of 6.6 ml/kg/min. This rate of hemorrhage actively stimulated the adrenal cortex of the dog. In contrast, if the same volume of blood loss was achieved at a slower rate of hemorrhage (i.e., 3.3 ml/kg/min), the P-A system would not activate. That rapid hemorrhaging induced adrenocortical activity, while a slow rate of hemorrhaging did not, indicates that the rate of stimulus change is one important parameter for the induction of P-A activity. The fact that dexamethasone blocked the P-A response at the high rate of hemorrhaging indicated that the neuroendocrine system was involved and the effect was not mediated at the adrenal level. These data along with the studies presented by Mason indicate that there are many appraisal processes located in the CNS which determine whether or not the P-A system will respond to a specific set of stimuli.

In this chapter we will attempt to describe the specific psychological variables which are involved in the regulation of the P-A activity. A psychobiological approach to understanding the participation of psychological factors in the regulation of the P-A system cannot escape making reference to cognitive processes.[8,9] Berlyne's arousal theory provides a framework for the description of the processes by which stimulators of arousal operate. Novelty, uncertainty, and conflict are considered arousing. These have been labeled by Berlyne as collative factors, because in order to evaluate them it is necessary to compare similarities and differences between stimulus elements (novelty) or between stimulus-evoked expectations (uncertainty). Hennessy and Levine[2] have assembled evidence in support of the hypothesis that the responses of the P-A system are a reflection of changes in the level of the hypothetical construct of arousal. The approach that was followed was to demonstrate that the psychological arousal and the P-A response were both subject to the same parametric relationships between stimulus and training variables. It thus follows that the basic cognitive process involved in regulation of the P-A system is one of comparison. The cognitive processes of comparison can best be understood by considering the concept of uncertainty.

Neuroendocrine activation in response to uncertainty is best explained by Sokolov's model[10] accounting for the general process of habituation. The pattern of habituation has been well described. A subject is presented with an unexpected stimulus and initially shows an orienting reaction. Physiological components of the orienting reaction include: general activation of the brain, decreased blood flow into the extremities, changes in electrical resistance of the skin, and increases in adrenal hormones. If the stimulus is frequently repeated, most of these reactions gradually diminish and eventually disappear, and the subject is said to be habituated. It does appear, however, that some physiological responses may habituate more slowly than the overt behavioral reactions.[2] Sokolov's model, in essence, is based on a matching system in which new stimuli or situations are compared with a representation in the central nervous system of prior events. This matching process results in the development of expectancies whereby the organism is either habituated or gives an alerting reaction to a mismatch.[11] If the environment does not contain any new stimuli or contingencies, the habituated organism no longer responds with the physiological responses related to the alerting reaction. Activation of the P-A system to novel stimuli can also be accounted for by invoking the powerful explanatory principles of the Sokolov model.

Exposure of an animal to novelty is one of the most potent experimental conditions leading to an increase in adrenal activity.[2] Novelty can be classified as a collative variable, since the recognition of any stimulus situation as being novel requires a comparison between present stimulus events and those experienced in the past. Studies on animals indicate an important characteristic of the cognitive process which results in P-A activation, that is, the ability of the animal to discriminate familiar vs. unfamiliar stimulus elements. In a series of experiments on rats and mice it was demonstrated that if novelty was varied along a continuum with increasing changes in the stimulus elements, there was a graded adrenocortical response according to the degree to which the environment represented a discrepancy from the normal living environment of the organism.[12,13] Thus, minor changes, such as placing the animal in a different cage, but one identical with its home cage, resulted in a moderate elevation of plasma corticosterone, but the response was significantly less than when the animal was placed in a totally novel cage that was distinctly different from its familiar home cage. This capacity for graded elevations of P-A activity clearly demonstrates the remarkable capacity of the central nervous system to regulate the output of the adrenal glands.

Novelty, according to the theory presented by Sokolov, should be one of the most potent variables that elicits increases in P-A activity. Insofar as an organism has no familiarity with a novel environment, that environment should initiate a degree of uncertainty that maximally elicits increases in neuroendocrine activity. Although novelty can be subsumed under the general heading of uncertainty, not all conditions which create uncertainty are novel. Uncertainty can also be evoked by insufficient information concerning the nature of upcoming events. Uncertainty can be seen to vary along the continuum from highly certain predictable events to highly uncertain unpredictable events. The presentation of a novel stimulus is likely to lead to an increase in uncertainty, by definition, because there is little information the organism can use to predict forthcoming events. Uncertainty can also be defined in terms of contingencies between environmental events. Experimentally, the dimension of uncertainty can be controlled by limiting the amount of information available to the organism to predict the occurrence of a specific event. Thus, one would hypothesize that if an organism is given information about the occurrence of either an appetitive or an aversive stimulus, such predictability should lead to a reduction in the P-A response. Further, situations in which there is an absence of predictability should lead to a marked increase.

There are many experiments that illustrate the value of predictability in modifying the P-A response to a variety of stimuli.[14] One illustration of the effects of reducing uncertainty by providing predictability can be seen in a study by Dess-Beech and colleagues.[15] Dogs were subjected to a series of electric shocks which were either predictable or unpredictable. The predictable condition involved presenting the animal with a tone prior to the onset of shock. In the unpredictable condition, no tone was presented. The adrenocortical response observed on subsequent testing of these animals clearly indicated the importance of reducing uncertainty by predictability. Animals that did not have the signal preceding the shock showed an adrenocortical response which was two to three times that observed in animals which had previous predictable shock experiences. It should be noted that the procedures used in this experiment are typical of those utilized in experiments examining learned helplessness.[16] Learned helplessness refers to the proactive effects resulting from prolonged exposure to unpredictable and uncontrollable stimuli of an aversive nature. It has been observed that organisms exposed to this type of an experimental regime show long-term deficits in their ability to perform appropriately under subsequent testing conditions. Further, these animals show a much greater increase in adrenocortical response when exposed to novelty[17] than do control animals. Thus, an organism exposed to an uncontrollable and unpredictable aversive stimuli not only shows increases in adrenocortical activity while exposed to these conditions, but there are also long-term consequences observed in other test conditions.

There is yet another series of experiments related to the issue of uncertainty. These experiments do not utilize aversive stimuli usually typical of stress research, but are more directly related to a psychological process commonly described as frustration. Frustration can be evoked when the organism fails to achieve a desired goal following a history of successful fulfillment of these goals. Experimentally, the operations utilized to produce frustration involve either preventing an animal from making the appropriate response to achieve a desired object, or not reinforcing the animal for a response that has had a prior history of reinforcement. In a broader sense, frustration involves the failure of an animal to fulfill expectancies developed through previous experiences and, thus, can be subsumed under the larger heading of uncertainty. For example, rats trained to press a lever for water, in which each lever-press delivered a small amount of water, showed an elevation of plasma corticosterone when the water was no longer available following the lever-press response.[18,19] Elevations of plasma corticosterone have been shown to be a robust and reliable phenomenon occurring under many experimental conditions in which reinforcement contingencies are altered. Thus, not only is an elevation of plasma corticosterone observed when reinforcement is eliminated, but also if the animal receives less reinforcement than it has previously become accustomed.

A similar phenomenon can be observed when using aversive stimuli. If an animal has learned to make an appropriate avoidance response that eliminates the occurrence of an electric shock, and this animal is then prevented from making the response, an increase in P-A activity occurs even when no electric shock is delivered.[20] These experiments have led to the hypothesis that one of the primary conditions for activation of neuroendocrine responses is a change in expectancies concerning well-established behaviors. In the case of the appetitive learning situation when reinforcement is eliminated, as well as in avoidance conditions, activation of the P-A system is a common occurrence following disruption of previously predictable behavior and outcomes.

The consequences of inappropriate adrenal secretion are evident from the effects of both hyper- and hypoadrenal output. Organisms deprived of adrenocorticoids are clearly in jeopardy and unable to deal effectively with even minor stresses that are of little consequence to the intact organism such as water or salt restriction. Conversely, chronic excessive secretion of glucocorticoids is also maladaptive. Prolonged elevations of these hormones can have a high biological cost leading to a number of immunological changes, as well as effects on digestive and cardiovascular physiology.[1,21] It would follow, therefore, that there must have evolved a set of mechanisms available to the organism whereby it can regulate and modulate excessive output of glucocorticoids. We believe that many of these mechanisms are predominantly psychological. Perhaps the most important single psychological factor involved in modulating hormonal responses to aversive stimuli is the dimension of control. Control can best be defined as the capacity to make active responses during the presence of an aversive stimulus. These responses are frequently effective in allowing the animal to avoid or escape from the stimulus, but they may also function by providing the animal with the opportunity to change from one set of stimulus conditions to another, rather than to escape the aversive stimulus entirely. Control can reduce an organism's physiological response to noxious stimuli. It has been observed that rats able to press a lever to terminate shock show less severe physiological disturbances (e.g., weight loss and gastric lesions) than yoked controls which cannot respond, even though both animals receive identical amounts of shock.[22] Similarly, animals able to escape from shock show a reduction in plasma corticosterone following repeated exposures of shock.[23] The effects of control have been demonstrated in primates in an experiment by Hanson et al.[24] These investigators studied rhesus monkeys that were exposed to the noxious stimulus of loud noise. One group of monkeys was permitted to control the duration of the noxious stimulus by making a lever-press response to terminate the noise. A comparable group of monkeys was given the identical amount of noise, but was not permitted to regulate the duration. The animals that were allowed to control the stimulus showed plasma cortisol levels like those of undisturbed subjects, whereas their yoked counterparts showed extremely high levels of plasma cortisol.

The effect of control over stimuli on adrenal responses was also demonstrated very clearly in an experiment using dogs.[15] These animals were subjected to a standard procedure used to produce learned helplessness. They were placed in a hammock and given uncontrollable and unpredictable threshold level shocks. Other dogs were allowed to control the shock and terminate it by making a panel-press response with their heads. We have previously discussed the role of predictability in this experiment. The results further indicated that controlability also affected the magnitude of the cortisol response to the shock. Having neither control nor predictability elicited the maximum cortisol response, while having both minimized the impact of the shock. In addition, the capacity to control the stimuli, even in the absence of predictability, resulted in reducing the cortisol response to shock.

Although it is clear that control is a major factor involved in modulating endocrine responses induced by stress, there is yet another factor which is also involved in this process: feedback. Feedback refers to stimuli or information occurring after a behavioral response has been made in reaction to an event. These stimuli may be used to convey information to the responding organism, indicating that it has made the correct response to a noxious event or that the aversive event is terminated for at least some interval of time. According to Weiss,[25-27] the amount of stress an animal actually experiences when exposed to noxious stimuli depends upon the number of coping attempts the animal makes (control) and the amount of relevant information the coping response produces (feedback). As the required number of coping responses increases and/or the amount of relevant feedback decreases, the amount of stress experienced increases. He has also argued that high operant responders — as an individual trait characteristic — will show more stress pathophysiology than low responders. In an extensive series of studies, Weiss demonstrated that if two groups of rats were subjected to the same amount of electric shock, the severity of gastric ulceration was reduced if the animal could respond — avoid and escape — and if the situation had some feedback information, i.e., a signal following the termination of shock. Although feedback information usually occurs in the context of control, namely, information about the efficacy of a response, it has been reported that feedback information per se, even in the absence of control, can reduce the P-A response to noxious stimuli. Hennessy et al.[28] reported that the presence of a signal following the delivery of shock resulted in a reduced adrenocortical response even in the absence of control. In contrast, the P-A response of animals given a random signal was not significantly different from those animals that had no signal at all.

When one closely examines the three factors — control, predictability, and feedback, they all have a common element which can be viewed within the framework of a determinant of P-A activation proposed earlier, namely, uncertainty. Each of these factors, either acting alone or more probably in concert, has the capacity to reduce uncertainty. Control provides the organism with the capacity to eliminate or at least to regulate the duration of the aversive stimuli. Thus, the uncertainty involved in an unpredictable and uncontrollable situation is reduced. Predictability, by definition, serves to reduce uncertainty. Feedback can also be viewed in terms of reducing uncertainty, since feedback provides information to the organism about the efficacy and success of the response being emitted. We can therefore speculate that any cognitive or behavioral process that reduces uncertainty can result in a reduction or elimination of the endocrine response to a stressful situation.

There has been a great deal of speculation concerning the importance of social relationships in determining an individual's response to stress.[29,30] The effects of social variables on P-A activity have been studied in an extensive series of studies using non-human primates. These studies have shown that at almost every stage in development social variables can modulate the level of endocrine activity, both in terms of increased hormone release and/or the inhibition of endocrine responses to environmental perturbations. Disruption of social relationships is a potent psychogenic stressor leading to increased P-A activity in non-human primates. Conversely, the presence of social partners can reduce, and at times completely eliminate, the normal P-A response which would occur when social companions were not available. The neuroendocrine response to the disturbance of social relationships emerges early in life. This is best exemplified by the mother-infant relationship, which for the infant is the basis of its first experience with any kind of social interaction. Beginning with the work of Harry Harlow,[31] numerous studies have shown that the mother serves not only to provide sustenance, but also to function as a source of emotional security.

However, for the infant it has been demonstrated that there are relevant social variables that can modify the response to separation. In several experiments it has been demonstrated that the activity of the P-A system can be significantly reduced if the infant while separated from the mother remains with its familiar social partners. This is not due to substitute maternal care, since the remaining females are actively caring for their own infants. It is critical for the infant that the social partners be *familiar*. The experimental evidence indicates that for the separated infants the most profoundly disturbing set of conditions is to be removed from the mother and placed in an unfamiliar social group. For infants separated under these circumstances the endocrine response to separation was even further augmented. In view of our previous discussion on cognitive factors responsible for eliciting the stress response, all of the elements required to produce uncertainty are present in this experimental paradigm. Thus, the infant is confronted with loss of control (no mother) and novelty, both from a strange environment and from strange conspecifics. Thus, it is not surprising that these infants would show close to a maximal stress response.

The early studies on separation in primates interpreted the infant's response to separation primarily in behavioral and emotional terms.[32-34] Based on the theoretical formulations of Bowlby,[35] the infant's agitated activity and vocalizations were described as indicative of a "protest" response. As the infant's agitated behavior subsided, it was described as entering a stage of "despair". The emphasis on the emotional basis of the reaction can be thought of as a ballistic model, in that the infant's behavior follows a fixed course following maternal loss.[36] From this perspective, the infant is viewed as a passive respondent. However, the infant's behavior following maternal loss can also be described in functional and cognitive terms as active "coping" attempts to resolve the aversive aspects of the situation. Thus, the infant's calls are not simply a reflection of affective disturbance, but actually represent a concerted effort to reestablish contact with the mother. The agitated activity would normally manifest itself as active searching behavior to locate the mother, and the high-intensity calls would elicit retrieval efforts by the mother. In this view, the infant is actively attempting to exert control over the situation. By resuming contact with the mother, the infant is attempting to use one of the most important behavioral responses for lowering its arousal levels (i.e., interaction with its mother). Thus, control over the reunion process is extremely important for the infant, and one of the aversive aspects of maternal separation appears to be loss of this behavioral control. We have argued previously that one of the most potent psychological variables modulating the P-A response to stress is control; conversely, the loss of control, with the resulting increase in uncertainty, would activate the P-A axis.

Probably as an outgrowth of the primate infant's reliance on its mother and other conspecifics, social factors also significantly influence the P-A response to stressful stimuli in adult monkeys. In the following studies, it has been demonstrated that the presence of a familiar social group can ameliorate the neuroendocrine response to aversive stimuli that normally evoke a marked adrenal response in individuals housed alone. Non-human primates typically show strong behavioral reactions indicative of fear when exposed to a snake or a snake-like object, and these reactions do not readily extinguish following repeated exposure.[37] In order to determine whether a social group can serve as an effective modulator of physical responses to this type of fear stimulus, squirrel monkeys were exposed to a live boa constrictor which was presented in a plexiglas container above their cage.[38] Monkeys living in four social groups, each consisting of two males and four females, were exposed to a snake for 30 min while in the group, and also after being removed from the group and placed in an individual cage. To control for the effects of general disturbance and handling, an empty stimulus box was placed on top of the group and individual cages on different test days. All of the monkeys showed increased vigilance, agitated activity, and total avoidance of the snake in both the group and individual conditions. However, while the behavioral response was consistent in both housing conditions, the snake did not elicit an adrenal response when the monkeys were tested as a group. This study, therefore, indicates that a familiar social group could prevent a physiological response from occurring in a potentially threatening situation. In further experiments, we have tried many other stimuli, including a loud, mobile robot with flashing lights, and have not observed an adrenal response in group-housed subjects tested in their home cage.

One interesting dimension of this social buffering is that it appeared to be dependent upon the number of available partners. In a subsequent study, we evaluated the response of adult males tested either alone or as a pair.[39] The results indicated that a single partner was not sufficient to ameliorate the adrenal response to snake presentation and that, in some cases, the reaction was even aggravated. In an extension of these studies, we have carried out a classical conditioning study which more clearly delineates the effect of two or more partners in ameliorating stress responses.[40] In this study, the stressor and behavioral reactions were also more clearly delineated by associating a previously neutral light stimulus with an aversive event (electric shock). Thus, animals were individually presented in a test chamber with multiple pairings of the light (conditioned stimulus = CS) and the shock (unconditioned stimulus = UCS). All subjects were later tested in the home cage and their adrenal responses monitored when they were housed individually, as pairs, or in a group of six animals. When the monkeys were housed either alone or in pairs, the presentation of the flashing light led to a significant elevation in plasma cortisol levels. In contrast, when the light was presented to the animals in a social group with six subjects, there was no elevation in plasma cortisol levels, even though increased behavioral activity was observed.

While there were several possible interpretations for the initial experiment on responses to snakes, including that the group test provided more behavioral options (i.e., avoidance) and control, the conditioning experiment unequivocally demonstrated that social factors can modulate the P-A response to fear-eliciting stimuli. These data concur with the view held by ethologists that one of the most important adaptations developed by higher primates is a sustained mother-infant relationship and group living. As Hans Kummer has stated, "...primates seem to have only one unusual asset in coping with their environments: a type of society which, through constant association of young and old and through a long life duration, exploits their large brains to produce adults of great experience. One may, therefore, expect to find some specific primate adaptations in the way primates do things as groups."[41]

As we have indicated earlier, we have hypothesized that one of the primary psychological stimuli for eliciting P-A activation is uncertainty. How then is it possible to fit the data on social support within the general context of the proposition that uncertainty leads to an activation of the P-A response, and that the reduction of uncertainty diminishes or eliminates this response? Cobb[42] discusses social support as a moderator of life stress and sees it as providing information that falls into three major categories. The first leads the individual to believe that he is loved or cared for. The second leads the individual to have higher self-esteem as a public expression of approval. The third is a perception of social congruity derived from a shared network of information and mutual obligation within which each member participates and a common knowledge is shared and accepted by all. It is possible to speculate that all three types of information derived from social support serve to minimize an individual's level of uncertainty about aversive situations. Although Cobb's propositions are expressed in humanistic terms, we can propose that the availability of stable and familiar social relationships also provides a set of predictable outcomes, due in part to the long history of previous interactions and experience. Thus, the predictability of the social interactions in a group of familiar conspecifics would tend to reduce uncertainty. This hypothesis would lead to the prediction that an unfamiliar social group would provide none of these beneficial features and, in fact, should evoke a state of high uncertainty leading to an elevation of P-A activity. In several experiments we have shown that there was a marked elevation of plasma cortisol when animals are placed in a new social group, and that this elevation continued for several months.[43,44]

III. CONCLUSION

Recently we have seen a great deal of discussion regarding certain amendments to the Animal Welfare Act of 1985. These concern exercise for dogs and the psychological well-being of primates. A series of recommendations have been put forth to implement the primate portion of the amendment.[45]

However, in view of our previous discussion in this paper several of these proposals appear questionable, to say the least. The draft states that primates are "social beings in nature and require contact with other non-human primates for their psychological well-being." What is clear from the evidence presented is that social contact may be beneficial only under certain conditions and that the imposition of social contact may produce detrimental effects both behaviorally and physiologically if social contact is forced among strange conspecifics. The benefits of social contact are derived from the stability of the social relationship. Continual changes in social interactions, which are more likely to occur under most laboratory conditions, maintain a heightened level of arousal including chronically elevated cortisol levels, most likely compromise the immune system, and often lead to wounding and death due to the increased levels of aggression. The proposal goes on to state that "those that must be isolated from other animal contact must be accorded positive physical contact or other interaction with a human for at least one hour a day." Certainly, one hour of contact with an adult chimpanzee would hardly be positive for the human, let alone for the animal. For most primates, human handling rarely represents a positive condition. Simply handling most primates is clearly stressful.

The draft document also proposes that "the method of feeding non-human primates must be varied daily in order to promote their psychological well-being." Varied feed is presumed to mimic foraging as it occurs in nature. Recently, there have been several reports which have shown that the introduction of foraging procedures have detrimental effects on macaque social relationships and markedly disrupt mother-infant relations.[46] We have evidence that varying feeding demands (foraging) produces elevated levels of cortisol similar to those observed using more traditionally defined stress stimuli. These data fit well within our theoretical position that uncertainty is a primary factor in activating stress-related hormones.

It should be apparent that before these or any set of regulations are imposed on the scientific community, a real awareness of the psychological factors involved in the stress response must be maintained. Specific attention must be paid to those cognitive factors which appear to play a fundamental role in regulating not only endocrine systems, but many other basic physiological processes as well. What has become increasingly apparent is that organisms live in concert with their environment and that many external factors have profound consequences on endogenous processes.

ACKNOWLEDGMENTS

This research was supported by a grant from the National Institute of Mental Health (NIMH) HD-02881 and by a Research Scientist Award MH-19963 to Seymour Levine.

REFERENCES

1. **Munck, A., Guyre, P. M., and Holbrook, N. J.,** Physiological functions of glucocorticoids in stress and their relation to pharmacological actions, *Endocrinol. Rev.*, 5, 25, 1984.
2. **Hennessy, J. W. and Levine, S.,** Stress arousal and the pituitary-adrenal system: a psychoendocrine model, in *Progress in Psychobiology and Physiological Psychology*, Sprague, J. M. and Epstein, A. N., Eds., Academic Press, New York, 1979.
3. **Mason, J. W.,** A historical view of the stress field, *J. Hum. Stress*, 1, 6, 1975a.
4. **Mason, J. W.,** A historical view of the stress field, *J. Hum. Stress*, 1, 22, 1975b.
5. **Selye, H.,** Stress, *Acta*, Montreal, 1950.
6. **De Kloet, E. R. and Reul, J. M. H. M.,** Feedback action and tonic influence of corticosteroids on brain function: a concept arising from heterogeneity of brain receptor system, *Psychoneuroendocrinology*, 12, 83, 1987.
7. **Gann, D. C.,** Parameters of the stimulus initiating the adrenocortical response to hemorrhage, *Ann. N.Y. Acad. Sci.*, 156, 740, 1969.
8. **Berlyne, D. E.,** *Conflict, Arousal and Curiosity*, McGraw-Hill, New York, 1960.
9. **Berlyne, D. E.,** Arousal and reinforcement, in *Nebraska Symp. on Motivation*, Levine, D., Ed., University of Nebraska Press, Lincoln, 1967.
10. **Sokolov, E. N.,** Neuronal models and the orienting reflex, in *The Central Nervous System and Behavior*, Brazier, M. A. B., Ed., Josiah Macy, Jr. Foundation, New York, 1960.
11. **Pribram, K. H. and Melges, F. T.,** Psychophysiological basis of emotion, in *Handbook of Clinical Neurology*, Vinken, P. J. and Bruyn, G. W., Eds., North-Holland, Amsterdam, 1969.

12. **Hennessy, M. B., Heybach, J. P., Vernikos, J., and Levine, S.,** Plasma corticosterone concentrations sensitively reflect levels of stimulus intensity in the rat, *Physiol. Behav.*, 22, 821, 1979.
13. **Hennessy, M. B. and Levine, S.,** Effects of various habituation procedures on pituitary-adrenal responsiveness in the mouse, *Physiol. Behav.*, 18(5), 799, 1977.
14. **Weinberg, J. and Levine, S.,** Psychobiology of coping in animals: the effects of predictability, in *Coping and Health,* Levine, S. and Ursin, H., Eds., Plenum Press, New York, 1980.
15. **Dess-Beech, N. K., Linwick, D., Patterson, J., and Overmier, J. B.,** Immediate and proactive effects of controllability and predictability on plasma cortisol responses to shocks in dogs, *Behav. Neurosci.*, 97, 1005, 1983.
16. **Seligman, M. E. P.,** *Learned Helplessness: On Depression, Development and Death,* W.H. Freeman, San Francisco, 1975.
17. **Levine, S., Madden, J., IV, Conner, R. L., Moskal, J. R., and Anderson, D. C.,** Physiological and behavioral effects of prior aversive stimulation, (preshock) in the rat, *Physiol. Behav.*, 10, 467, 1973.
18. **Coover, G. D.,** The rat's reward environment and pituitary-adrenal activity, in *Biological and Psychological Basis of Psychosomatic Disease,* Ursin, H. and Murison, R., Eds., Pergamon Press, Oxford, 1983.
19. **Levine, S., Goldman, L., and Coover, G. D.,** Expectancy and the pituitary-adrenal system, in *Physiology, Emotion and Psychosomatic Illness* (Ciba Found. Symp. 8), Porter, R. and Knight, J., Eds., Elsevier, Amsterdam, 1972, 281.
20. **Coover, C. D., Ursin, H., and Levine, S.,** Plasma corticosterone levels during active avoidance learning in rats, *J. Comp. Physiol. Psychol.*, 82, 170, 1973.
21. **Coe, C. L., Rosenberg, L. T., Fischer, M., and Levine, S.,** Psychological factors capable of preventing the inhibition of antibody responses in separated infant monkeys, *Child Dev.*, 58, 1420, 1987.
22. **Weiss, J. M.,** Behavioral and psychological influences on gastro-intestinal pathology, in *Handbook of Behavioral Medicine,* Gentry, W. D., Ed., Guilford Press, New York, 1984.
23. **Davis, H., Porter, J. W., Livingstone, J., Herrmann, T., MacFadden, L., and Levine, S.,** Pituitary-adrenal activity and lever press shock escape behavior, *Physiol. Psychol.*, 5, 280, 1977.
24. **Hanson, J. D., Larson, M. E., and Snowdon, C. T.,** The effects of control over high intensity noise on plasma cortisol levels in rhesus monkeys, *Behav. Biol.*, 16, 333, 1976.
25. **Weiss, J. M.,** Effects of coping behavior in different warning signal conditions on stress pathology in rats, *J. Comp. Physiol. Psychol.*, 77, 1, 1971a.
26. **Weiss, J. M.,** Effects of punishing the coping response, (conflict) on stress pathology in rats, *J. Comp. Physiol. Psychol.*, 77, 14, 1971b.
27. **Weiss, J. M.,** Effects of coping behavior with and without a feedback signal on stress pathology in rats, *J. Comp. Physiol. Psychol.*, 77, 22, 1971c.
28. **Hennessy, J. W., King, M. G., McClure, T. A., and Levine, S.,** Uncertainty, as defined by the contingency between environmental events, and the adrenocortical response of the rat to electric shock, *J. Comp. Physiol. Psychol.*, 91, 1447, 1977.
29. **Hamburg, D. A. and Adams, J. E.,** A perspective on coping behavior: seeking and utilizing information in major transitions, *Arch. Gen. Psychiatry,* 17, 277, 1967.
30. **Cohen, S. and Wills, T. A.,** Stress, social support and the buffering hypothesis, *Psychol. Bull.*, 98(2), 310, 1985.
31. **Harlow, H. F. and Harlow, M. K.,** Effects of various mother-infant relationships on rhesus monkey behaviors, in *Determinants of Infant Behaviour,* Foss, B. M., Ed., Methuen, London, 1969.
32. **Rosenblum, L. A. and Kaufman, I. C.,** Laboratory observations of early mother-infant relations in pigtail and bonnet macaques, in *Social Communication among Primates,* Altmann, S. A., Ed., University of Chicago Press, Chicago, 1967, 33.
33. **Hinde, R. A. and Spencer-Booth, Y.,** Effects on brief separations from mothers on rhesus monkeys, *Science,* 173, 111, 1971.
34. **Mineka, S. and Suomi, S. J.,** Social separation in monkeys, *Psychol. Bull.*, 85, 1376, 1978.
35. **Bowlby, J.,** *Attachment and Loss, Vol. 2,* Basic Books, New York, 1973.
36. **Rosenblum, L. A. and Plimpton, E. H.,** Adaptation to separation: the infant's effort to cope with an altered environment, in *The Uncommon Child: Genesis of Behavior,* Lewis, M. and Rosenblum, L. A., Eds., Plenum Press, New York, 1981, 225.
37. **Mineka, S., Keir, R., and Price, V.,** Fear of snakes in wild- and lab-reared rhesus monkeys, *Anim. Learn. Behav.*, 8, 653, 1980.
38. **Vogt, J. L., Coe, C. L., and Levine, S.,** Behavioral and adrenocorticoid responsiveness of squirrel monkeys to a live snake: is flight necessarily stressful?, *Behav. Neural Biol.*, 32, 391, 1981.
39. **Coe, C. L., Franklin, D., Smith, E. R., and Levine, S.,** Hormonal responses accompanying fear and agitation in the squirrel monkey, *Physiol. Behav.*, 29, 1051, 1982.
40. **Stanton, M. E., Patterson, J. M., and Levine, S.,** Social influences on conditioned cortisol secretion in the squirrel monkey, *Psychoneuroendocrinology,* 10, 125, 1985.

41. **Kummer, H.,** *Primate Societies: Group Techniques of Ecological Adaptation,* Aldine-Atherton, Chicago, 1971, 37.
42. **Cobb, S.,** Social support as a moderator of life stress, (Presidential address), *Psychosom. Med.,* 38, 300, 1976.
43. **Gonzalez, C. A., Hennessy, M. B., and Levine, S.,** Subspecies differences in hormonal and behavioral responses after group formation in squirrel monkeys, *Am. J. Primatol.,* 1, 439, 1981.
44. **Coe, C. L., Smith, E. R., Mendoza, S. P., and Levine, S.,** Varying influence of social status on hormone levels in male squirrel monkeys, in *Hormones, Drugs and Social Behavior in Primates,* Steklis, H. D. and Kling, A. S., Eds., Spectrum, New York, 1983.
45. **Holden, C.,** Billion dollar price tag for new animal rules, *Science,* 242, 662, 1988.
46. **Rosenblum, L. A. and Sunderland, G.,** Feeding ecology and mother-infant relations, in *Parenting: Its Causes and Consequences,* Hoffman, L. W., Gandelman, R., and Schiffman, H. R., Eds., Erlbaum, Hillsdale, New Jersey, 1982, 75.

Chapter 12

THE NATURE OF PAIN IN ANIMALS

Ralph L. Kitchell and Michael J. Guinan

EDITOR'S PROEM

As moral concern for animals has developed in society, it has naturally focused on the most obvious morally problematic dimensions of animal treatment — pain experienced by animals as a result of human manipulations. This social demand for minimization of animal pain has in turn ramified in catalyzing scientific attention to animal pain in itself rather than as a "model" of human pain, a subject virtually ignored in 20th century science until the last few years. As a result, more literature on animal pain and its assessment and alleviation has appeared during the past few years than in the previous century. In his paper, Professor Kitchell, a pioneer in exploring animal pain, outlines the neurophysiological substratum underlying the animals' experience of physical pain, a substratum not significantly different from the same machinery extant in humans.

TABLE OF CONTENTS

I. INTRODUCTION

Most informed scientists would agree with Kelly[1] that pain is a protective experience that humans share with almost all animals. Vyklicky[2] states, "It is generally accepted that animals, at least mammals, suffer pain similar to human pain." Some individuals, however, believe that animals give only mechanistic (reflexic) responses to damaging or potentially damaging stimuli (noxious stimuli) and believe that we have no knowledge about pain perception in animals. The purpose of this chapter is to present various lines of evidence which support the premise that pain is perceived by animals in the same manner as by people.

This discussion will be limited to the pain which results when a part of the body is damaged, or threatened with damage. This discussion will not include the use of the term pain to describe perceptions caused by such experiences as when one loses a loved one, is severely distressed, or is verbally punished.

Pain, produced by noxious stimuli, results from the excitation of special receptors called "nociceptors". The reception, conduction, and central nervous system processing of nerve signals generated by the stimulation of nociceptors is called "nociception". The neural structures involved in these events are the nociceptive or nocifensive system. Pain is a perception, the subjective sensation arising as a consequence of activation of the nociceptive system. It is a common error to use the terms "pain stimuli", "pain receptors", "pain nerve fibers", "pain reflexes", "pain tracts", and "pain centers" instead of nociception. The error is that all of these can be active without the sensation of pain being felt. It is now accepted that pain, like all the other conscious sensations — hearing, seeing, touch — is impossible without the participation of the cerebral cortex.[3]

Contrary to the commonly held concept, the intensity of the pain felt in response to a given noxious stimulation is seldom precisely the same between two individuals. The nociceptive system has been shown to be the most strongly modulated (altered by inherent inhibitory and excitatory mechanisms) of all the sensory system of the body. Evidence supporting this statement will be presented later in this chapter.

Pain can be categorized as being either somatic pain (arising from the body wall) or visceral pain (arising from the viscera). If somatic pain arises from stimulation of the skin, it is referred to as superficial pain, as opposed to deep pain, arising from the deeper tissues of the body wall such as muscles, joints, and bones. Superficial pain has two qualities, fast (initial, sharp) or slow (second, dull, or burning pain).

Pain can also be divided into acute or chronic pain on the basis of the duration of the pain. Acute pain is of short duration. If the pain persists for over 6 months, it is considered chronic pain. Little is known about chronic pain in animals; thus, this discussion will deal with acute pain.

Pain has three major psychological dimensions: sensory-discriminative, affective-motivational, and cognitive-evaluative,[4] which are subserved by physiologically specialized systems.[5] The sensory-discriminative dimension of pain provides information about the onset of the stimulation, the location of the stimulus, the intensity of the stimulation, the type of stimulation (mechanical, thermal, chemical, or combinations), and the duration of the stimulation. The affective-motivational dimension of pain disturbs the feeling of well-being of the individual; in other words, the pain really hurts and suffering may occur. This dimension of pain is closely linked to the autonomic nervous system of the body; thus, respiratory, cardiovascular, and gastrointestinal responses often are associated with this dimension of pain, although all of these can occur reflexly. The cognitive-evaluation dimension of pain encompasses the effects of prior experiences, social and cultural values, anxiety, attention, and conditioning. These activities are subserved largely by the cerebral cortex, although the reticular formation responsible for cortical activation is critical to cortical functioning. The classic example of cognitive-evaluation dimension of pain in animals has been Pavlov's conditioning of dogs by repeatedly giving food to the dogs after a particular area of skin is shocked, burned, or cut.[6] The dogs soon responded to these stimuli as signals for food and would salivate, without showing any signs of pain, yet would howl when the stimuli were applied to other sites on the body. All three dimensions of pain most often occur together. However, as we shall see later, it is possible for them to be separated from each other surgically and by the use of certain drugs.

There are four useful terms relating to nociception. These are (1) nociceptor threshold: the strength that the noxious stimulus must have in order to cause a nociceptor to generate a nerve impulse. (2) Pain detection threshold: that point at which an individual, when a noxious stimulus is being applied, just begins to feel pain in an ascending trial or at which the pain disappears in a descending trial. This threshold is relatively constant among individuals and species. In most instances the pain detection threshold is higher than the nociceptor threshold.[7] (3) Pain tolerance threshold: that point which is the upper limit of pain that an individual will accept. This threshold varies considerably among individuals, both humans and animals. It is influenced greatly by the prior experience of an individual (cultural and social backgrounds), by the environment, by stress, and by drugs. In Pavlov's experiments (mentioned above), the conditioning reward (food) increased the dogs' pain tolerance threshold. Similarly, athletes are rewarded for tolerating more pain by fame and financial rewards. (4) Pain tolerance range: the arithmetical difference in strength between the pain detection threshold and the pain tolerance threshold. Individuals with a low pain tolerance threshold will have a short tolerance range.

The conclusion that both animals and humans experience pain is based upon an almost overwhelming amount of scientific evidence that nociceptive mechanisms exist in animals which are exactly like or very similar to those found in humans. This evidence has been, and continues to be, used as the rationale for using animals in pain research aimed at finding means of relieving pain in humans.

Historically, Lewis[8] described pain as a subjective phenomenon "known to man only by experience and described by illustration". Poggio and Mountcastle[9] stated, "There is no 'a priori' reason to suppose that in the evolution the perception of pain appears as a wholly new sensory phenomenon in man." Lord Brain[10] stated, "I see no reason for conceding mind to my fellow man and denying it to animals." The senior author stated in 1961[11] that knowledge about pain perception in animals involved a distinction between knowledge which an individual acquires by *acquaintance* (our own perceptions) and knowledge acquired by *description* (knowledge of pain in a fellow human being or an animal). Modern theory of knowledge suggests that we do not need to know a state or a substance directly through use of our own senses in order to establish that a state or substance possess a character of being a real event. Science is primarily knowledge obtained by description. Much is made of the fact that animals can't speak in a language that we can understand. This is also true of babies and of mute persons. Little doubt exists that infants who cannot talk perceive pain.[12] Speech has been referred to as nothing other than a form of behavior.[13,14] In instances where a human being can't speak, knowledge about whether or not they are in pain must be *inferred,* by analogy, from their behavior. The similar situation exists regarding the existence of pain in animals. The existence of pain must be *inferred,* by analogy, from their behavior and by drawing analogies with what occurs in human beings.[11] If a stimulus evokes painful sensations in humans, is noxious or potentially noxious to tissues, and provokes escape and "emotional" responses in an animal, that stimulus must be considered a painful stimulus to that animal.[15]

We will present analogies between the neural mechanisms involved in pain perception in humans and animals under five categories: (1) peripheral aspects of nociception, (2) nociceptive ascending pathways, (3) supraspinal mechanisms associated with nociception, (4) factors which modulate pain, and (5) nociceptive reflexes, reactions, and responses.

II. PERIPHERAL ASPECTS OF NOCICEPTION

In nonpathological conditions, a sensory experience requires the activation of neurons in the periphery which carry this information to the central nervous system. These neurons constitute the first of several relays to the brain and are called receptors or primary afferents. They have receptive endings in the periphery for the transduction of mechanical, thermal, or chemical stimuli into electrical impulses. These receptive endings may be very specific to only one type of stimuli (e.g., hair movement, vibration, cooling, etc.). Early notions of pain reception suggested that pain was due to the excessive stimulation of receptors responsive to nonnoxious stimuli. This idea was abandoned when it was shown that there were somatic primary afferents which responded exclusively to noxious stimuli.[16] Since then, such nociceptors have been found in muscles, joints, and viscera, as well as the skin.

Skin or cutaneous nociceptors are the best studied to date. They can be grouped into one of several classifications based on the types of stimuli they respond to. These three populations of nociceptors have been found in both humans and animals.

Mechanonociceptors — These are activated by intense mechanical stimuli. They have conduction velocities of small myelinated fibers (5 to 35 m/s) and so are often called Aδ nociceptors. While not initially responsive to heat, they may become sensitized to repeated or prolonged applications of noxious heat to their receptive fields.[17]

Polymodal nociceptors — These respond to noxious mechanical stimuli, noxious or near noxious heat (>40° C), and chemical irritants. They are mostly slowly conducting unmyelinated C fibers (0.5 to 1.5 m/s) with small receptive fields. C polymodal nociceptors are the predominant type in nonhairy skin.[18]

Mechanothermal nociceptors — These are activated by noxious mechanical or thermal stimuli. They are Aδ fibers and probably function in reflex responses to noxious heat as impulses in C fibers conduct too slowly to be involved in this reflex.[19]

The mechanism of activation of nociceptors is unclear. The rapidity of response of some primary afferents to a stimulus argues for some type of direct mechanical transduction mechanism.[20] While pain receptors are most often described as free or bare, mechanonociceptors do seem to have a specialized kerotinocyte/Schwann cell covering over their axonal endings in the epidermis which may contribute to their lack of response to innocuous stimuli.[21] The sensitivity of polymodal nociceptors to several types of stimuli suggests some sort of chemical intermediate for activation. Several chemical agents (H^+, K^+, histamine, bradykinin, prostaglandins) are released from damaged tissues. These chemicals may mediate neuronal responses to noxious stimuli as well as the local vascular response. In addition, some chemicals can sensitize nociceptors to respond to innocuous stimuli. This sensitization may result in important adaptive behavioral responses (e.g., disuse) to aid healing.

As might be expected, stimuli which activate C or Aδ fibers produces the sensation of pain in humans and pain reactions in animals. Under experimental conditions, the activation of a single fiber in humans can result in the perception of pain,[22] but the activity of several neurons is probably necessary to produce pain by natural stimulation.[23] Nociceptors in both humans and animals encode the intensity of noxious stimuli by increasing their impulse discharge frequency. This discharge frequency is well correlated with the magnitude of the pain sensation in humans.[24] Animals also appear to be able to perceive differences in the magnitude of noxious stimuli.[25] They will react faster or more vigorously on a learned escape task to terminate a moderately noxious stimuli than a mildly noxious stimuli.[26]

Except for the head, nociceptive primary afferents relay their impulses to the dorsal horn of the grey matter of the spinal cord via the dorsal roots (Figure 1). After the primary afferents enter the spinal cord, their fibers split and may ascend or descend in the tract of Lissauer for several segments before synapsing in the dorsal horn. The dorsal horn has been subdivided into ten

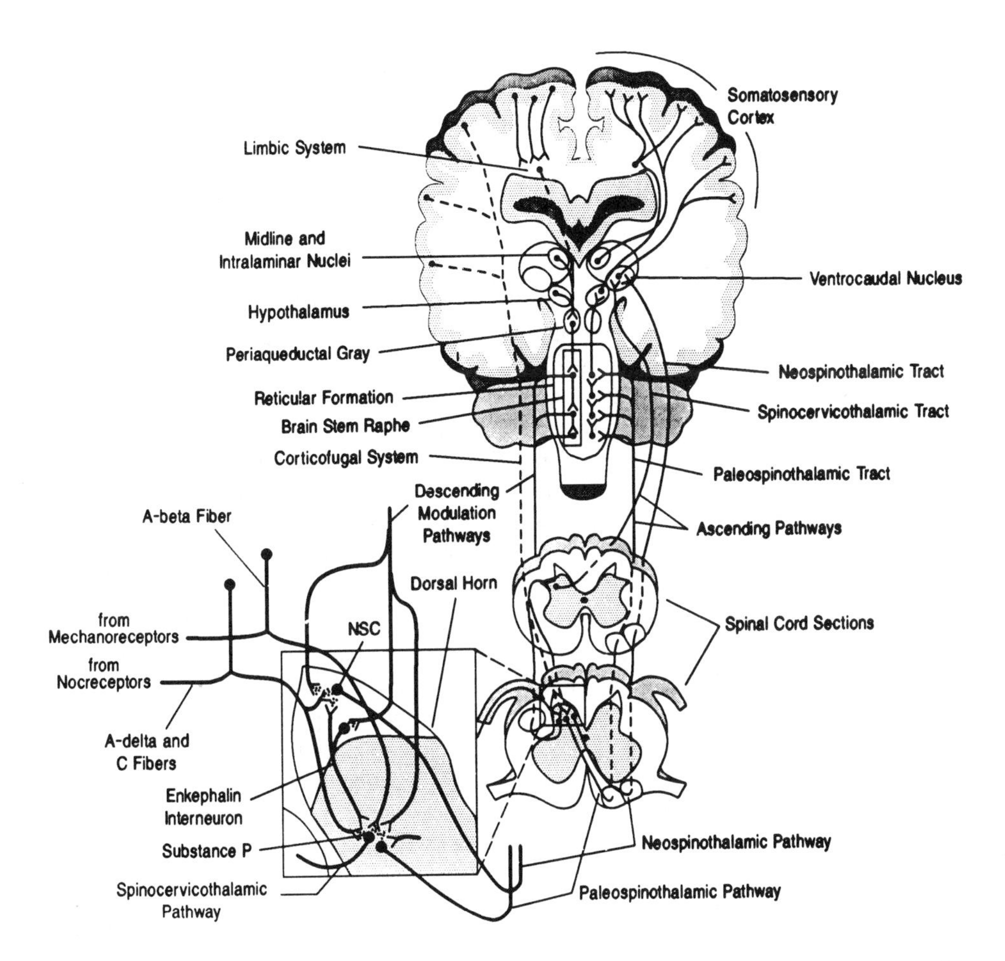

FIGURE 1. Nociceptive pathways in the cat. Nociceptive Aδ and C primary afferent fibers carry ascending nociceptive impulses from the periphery to the dorsal horn of the spinal cord (insert). Small interneurons containing enkephalins are acted upon by descending pathways and larger afferents (Aδ) fibers to modify (suppress) the nociceptive signals. Impulses ascend to the ventrocaudal nucleus of the thalamus via contralateral neospinothalamic and ipsilateral spinocervicothalamic pathways. The contralateral paleospinothalamic pathway ascends to the reticular formation and the midline and intralaminar nuclei of the thalamus. Descending pathways carry impulses from higher brain centers through the periaqueductal gray and the brain stem raphe back to modulate the nociceptive activity in the dorsal horn using serotonergic as well as the enkephalinergic mechanisms. All descending pathways come down both sides of the spinal cord. They are shown only on one side for simplicity. A number of other pathways are not shown.

layers or laminae based on cell types and connections.[27] The synaptic terminals of nociceptors are predominantly located in the superficial layers (lamina I and the outer part of II). A significant number of Aδ nociceptive fibers also go to lamina V.[28] Laminae I, IV, V, VII, and VIII, contain cell bodies of neurons which project to the brainstem and thalamus (Figure 1).

The particular types of synaptic arrangements have not been determined for C fibers due to their small size, but synaptic contacts have been studied for nociceptive Aδ fibers. In the superficial layers, these fibers make mostly axo-dendritic synapses. Their terminals may also be postsynaptic to other axons or dendrites.[29] These latter two synapses are likely to be involved in modulation of ascending input by peripheral and central mechanisms (see Section V).

There is much evidence that substance P is the neurotransmitter released from nociceptive primary afferents.[19,20,30] Substance P is excitatory to nociceptive spinal neurons,[31] and depletion of substance P can cause hypalgesia.[32] However, conclusive evidence that substance P is specific to nociceptive afferents is lacking. Many other potential transmitters involved in nociception (somatostatin, vasoactive inhibitory peptide, cholecystokinin, glutamate, aspartate, bombesin) are found in primary afferents and in some cases are colocalized with substance P.[33] As yet, no distinct correlation of sensory receptor types with specific transmitters is apparent.

III. NOCICEPTIVE ASCENDING PATHWAYS

An ascending pathway, or tract, is a collection (bundle) of nerve fibers which have their cell bodies in the gray matter of the spinal cord (or of the brain stem) and terminate by synapsing on cells of the brain, usually a part of the reticular formation or of the thalamus. All nerve fibers in a named pathway do not serve the same sensory function (i.e., touch, temperature, or nociception). A pathway can be considered a nociceptive pathway as well as a mechanoreceptor pathway if it has ascending fibers in it serving each function.

The majority of the cell bodies giving rise to fibers which form ascending pathways are found in the dorsal horn of the spinal cord where they form columns (sheets) of cells called laminae. The neurons forming ascending pathways are frequently referred to as relay cells because they are acted upon by terminals from primary afferent neurons and, if excited, "relay" the activity to other parts of the nervous system. If the primary afferent synapsing with a relay neuron is a nociceptive primary afferent, the relay cell and the pathway it forms are referred to as belonging to a nociceptive ascending pathway. In all animals studied, two classes of relay neurons are found in the dorsal horn which are involved in nociception. One set responds only to stimuli which are noxious. These cells are referred to as "nociceptive specific" (NS) neurons. The second set, called "wide dynamic range" (WDR) neurons (nonspecific nociceptive or multireceptive neurons) respond slightly to innocuous (light) mechanical stimuli, but give a much stronger response when a noxious stimulus is applied in the periphery.

Nociceptive specific cells — The nociceptive specific neurons are found in both the superficial layer (marginal layer or lamina I) and the deeper layers of the dorsal horn. They usually have discrete receptive fields in the periphery, one to several square centimeters in size in the monkey. The cells are of two functional types, those which respond only to noxious mechanical stimuli, and the second group which responds to both noxious mechanical and to noxious heat stimuli.

Wide dynamic range cells — These cells are seldom found in lamina I, but are mostly located in the deeper layers of the dorsal horn (laminae IV to VI) and in laminae VII and VIII of the ventral horn in the lumbosacral enlargement of the spinal cord. The receptive fields of these neurons vary considerably in size. In some instances they may be confined to a single digit, in others, to a major portion of the leg. The receptive fields usually have a central zone where both innocuous and noxious stimuli excite the cell and a peripheral zone where only noxious stimuli can excite the cell. These cells respond to a variety of energies. Some respond to mechanical stimuli at the innocuous level, but not to warm stimuli, yet respond only to noxious heat, not to noxious mechanical stimulation. In the peripheral zone of the receptive fields, these cells act like specific heat nociceptors. There are some serious questions being raised about the WDR cells because in the unanesthetized animal, they do not respond to noxious stimuli and respond to noxious stimuli only when anesthesia is administered.[34] This suggests that the WDR cells may be tonically inhibited in the awake animal and that this inhibition is suppressed by the barbiturate anesthesia. None of these cells have been recorded from in humans, thus, their functioning in humans must be inferred by drawing analogies based upon their anatomical similarities.

Inhibition of nociceptive relay cells — Both types of cells are inhibited (modulated) by local activity generated by adjacent primary afferent fibers as well as descending inhibition from the brain. Failure to understand the fact that these cells are under both tonic (constant) and phasic (inconstant, irregular) inhibition often leads to misconceptions about the functioning of the nervous system. It is a common misconception that impulses generated by a noxious stimulus are automatically relayed along various parts of the nervous system to the brain where perception will always follow the application of a noxious stimulus. What actually happens is that the activity conveyed from the periphery via primary afferent fibers often does not result in the production of activity in the ascending nociceptive pathway fibers, thus, the original noxious stimulus is not perceived. Local inhibitory circuits drastically alter the excitability of these pathway neurons, reducing their excitability. Thus, little or no activity is relayed to the brain where no pain, or only mild pain is perceived. Similarly, descending inhibition from the brain (pain suppression) may also result in the relay neuron not producing nerve impulses. This will be discussed at length in the pain modulation section of this chapter.

Nociceptive information is transmitted from the spinal cord to the brain by multiple ascending systems.[35] The systems include a "lateral" group, consisting of (1) the neospinothalamic tract, which ascends directly from the spinal cord to the thalamus; (2) the spinocervical tract, which relays to the thalamus through the lateral cervical nucleus; and (3) the dorsal column-postsynaptic tract, which relays to the thalamus through the dorsal column nuclei. The "medial" group consists of (1) the paleospinothalamic tract, which projects to the midline-intralaminar thalamic regions; (2) the spinoreticular tract, which sends fibers throughout the reticular formation; (3) the spinomesencephalic tract, which ends in the mesencephalon; and (4) the propriospinal system, which ascends through the spinal cord using a diffuse, polysynaptic network of fibers and terminates in the reticular formation.

Although there are major differences in the anatomy and physiology of these ascending systems among animals and humans, the general similarities among the species far outweigh the specific species dissimilarities. In humans and sub-human primates, the principal ascending nociceptive pathway appears to be the neospinothalamic pathway, which is a spinal cord pathway which ascends the spinal cord on the opposite side of the body from that where the nociceptors and primary afferent fibers are located (Figure 2). The neospinothalamic pathway consists of both specific and wide dynamic range neurons. Located medial (deep) to the neospinothalamic pathway is the much smaller paleospinothalamic pathway (Figure 2). This pathway consists primarily of wide range dynamic cells. The spinoreticular pathways are located posterior and deep to the spinothalamic pathways. This pathway also is largely made up of wide dynamic range type cells. Anatomical evidence supports the presence of a spinomesencephalic, a dorsal column-postsynaptic pathway, and a multi-neuronal pathway called the propriospinal pathway in humans as well as in animals. Until recently, neurosurgeons would incise both spinothalamic and the spinoreticular pathways on one side of the spinal cord in order to relieve intractable pain from the contralateral side of the body (tractotomy). This often gave complete relief of pain for a year or more, but the pain often returned with an even more disagreeable component.[36] The return of sensation after a period of 6 months or more has been attributed to the propriospinal and dorsal column-postsynaptic pathways taking over the functions of the spinothalamic and reticulospinal pathways.

In most non-primate animals, the nociceptive pathways are much more numerous, more diffuse, and often ascend the spinal cord bilaterally (Figure 1). This has been definitively shown in the cat where, in the cervical region, the neospinothalamic pathway has a bilateral origin.[37] In the cat and swine, perception of pain arising from noxious stimulation of either side of the body is not lost by severing one side of the spinal cord as it is in primates.[38,39] This may be due to the relatively larger propriospinal pathway in those species.

The lumbosacral portions of the neospinothalamic pathway arise from cells in quite a different location in the cat than in the rat and the monkey. There are far fewer spinothalamic cells located in the neck of the dorsal horn of the cat than there are in the other two species. The majority of spinothalamic cells in the lumbosacral enlargement region of the cat are located in the ventral horn in laminae VII and VIII[40,41] whereas in the other two species the cells are located

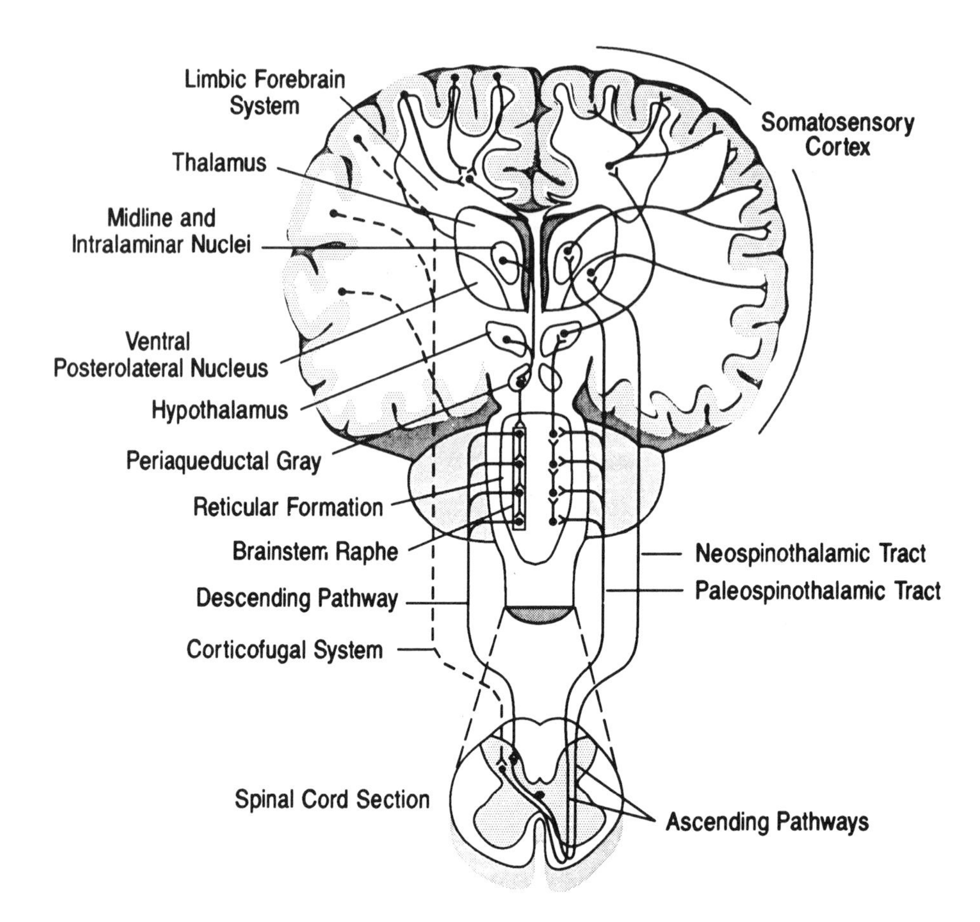

FIGURE 2. Ascending pathways in humans (compare with Figure 1). Note that the ascending pathways are contralateral in humans as compared to some ascending ipsilaterally and some contralaterally in the cat.

in principally in the laminae I through V although there a few cells located in laminae VII and VIII.[40,42] Most of the cells from the lumbosacral enlargement of the rat and the monkey project to the contralateral thalamus, whereas in the cat a higher percentage of these cells project to the ipsilateral thalamus.[42]

The sub-primates have a large spinocervicothalamic pathway which ascends the spinal cord ipsilaterally and crosses to the opposite side in, or cranial to, the first cervical segment (Figure 1). This pathway consists largely of low threshold mechanoreceptive cells (relay for hair follicle receptor activity). However, it has been shown to include some cells that respond to noxious stimulation.[43,44]

The dorsal column-postsynaptic pathway has been extensively investigated in the rat, cat, and the monkey. This pathway consists largely of cells which respond to low-threshold mechanical stimulation, although a number of wide range dynamic and a few nociceptive specific cells have been found.[44]

The terminations of the spinothalamic tract are quite different in the rat and the monkey than they are in the dog and the cat. In the rat and the monkey, substantial numbers of spinothalamic tract fibers end in the ventral posterior lateral nucleus of the thalamus (VPL).[46,47] In the cat, very few fibers end in the central part of the VPL.[47,48] The majority of the spinothalamic fibers in these species end in the "shell" region which is located dorsally between the VPL and the ventral lateral nucleus, and ventrally between the VPL and both the ventral posterior inferior and the posterior nuclei. The possible significance of the location of these endings will be addressed during the discussion of supraspinal nociceptive mechanisms.

Melzack and Casey[5] have suggested that the lateral ascending systems, consisting of the neospinothalamic pathway in humans and the neospinothalamic, dorsal column-postsynaptic, and the spinocervicothalamic pathways in animals, largely transmit information about the sensory-discriminative dimensions of pain (spatial, temporal, and intensity). The more medial systems, consisting of the paleospinothalamic, spinoreticulothalamic, spinomesencephalic, and the propriospinal pathways convey information about the motivational-affective dimensions of pain. These latter pathways are not organized to carry discrete spatial and temporal information. Their target cells in the brain usually have wide receptive fields, sometimes covering half of the body surface. The terminals from other sensory systems, such as somatosensory, visual, and auditory, often terminate on these same target cells. The lateral ascending systems have most of their terminations in the ventrobasal nucleus of the thalamus (Figures 1 and 2) which, in turn, relays the activity to the somatosensory part of the cerebral cortex. Lesions in the lateral ascending systems tend to interfere with the individual's ability to recognize the kind of energy being applied and to accurately determine which area of the body is being stimulated without appreciably affecting the aversive and emotional aspects of pain. The medial ascending systems have most of their terminations in the reticular formation and midline or intralaminar (nonspecific) nuclei of the thalamus (Figures 1 and 2). Lesions in these medial ascending pathways generally result in a human being stating that he/she still perceives the pain, they can localize it, and they can tell what kind of a stimulus gave rise to the perception, but that it is not a stimulus which is intolerable. In an animal, a lesion in the medial ascending pathways generally results in an animal that shows much less vigorous aversive responses than does a normal animal.[49-51]

The lateral ascending systems are not as effective as the medial ascending systems in mediating reflexes or in altering generalized brain functions (e.g., alertness or arousal).[52] The terminations of the medial systems in the reticular formation and the midline and intralaminar thalamic nuclei enable these systems to establish widespread connections with the hypothalamus and other parts of the limbic system. The hypothalamic-limbic system is thought to be the principal neuronal substrate for emotional states. Dennis and Melzack[53] have noted that from a phylogenic basis, the relative size of the medial ascending pathway systems have remained quite constant relative to brain size, whereas the lateral ascending pathways are more highly developed in primates. They suggest that the lateral ascending nociceptive systems may be tuned preferentially to the onset of sudden changes in noxious stimulation (phasic pain), i.e., related to the expression of the life-threatening potential of the noxious stimulus. They propose that the medial systems are better tuned to the persistent or tonic signalling of actual tissue damage. The activation of the lateral ascending systems could be instrumental in escape responses leading to the animal fleeing and eliciting behavioral responses (including vocalization) to warn other animals of that species from the threat of further damage. They suggest that the medial system could be instrumental in species specific behavior directed toward minimizing further tissue damage and to gaining assistance in healing, i.e., being fed, groomed, and otherwise cared for.

The differences in numbers of ascending fibers in the lateral system of humans compared to animals suggest that non-primate animals may not be able to receive as much information about the sensory-discriminatory dimensions of a stimulus as humans can, i.e., they may have a less refined ability to localize the stimulus, determine the energy source of the stimulus, etc. In contrast, non-primate animals have medial ascending pathways which are comparable, if not larger, than primates. This suggests equal or even greater ability to assess the affective-motivational dimensions of the stimulus, i.e., autonomic responses, unpleasant qualities of the stimulus, life-threatening consequences of the tissue damage. Precise information about the meaning of these anatomical and physiological differences are not available; thus, these observations remain in the realm of speculation.

Nociceptive pathways from the head region — The trigeminal nerve conveys the majority of the primary afferents from the skin of the head region, the nasal and oral cavities to the brainstem. Nociceptive fibers from the concave surface of the ear enter the brainstem via the facial nerve. Similar fibers from the external acoustic meatus enter the brainstem as a part of the vagus nerve. Within the brainstem, all nociceptive fibers enter the spinal, or descending, tract of the trigeminal nerve. The nociceptive fibers of this tract synapse in the caudal nucleus of the trigeminal nerve located in the caudal part of the medulla oblongata and the first two cervical spinal cord segments. These fibers synapse with nociceptive specific cells located in a layer that corresponds with the marginal layer (lamina I) or with dynamic wide-range cells in layers which correspond with laminae V and VI of the spinal cord.[54-58] The ascending pathways to the thalamus and other regions of the brain run in ascending pathways commonly referred to as specific and nonspecific trigeminothalamic pathways. The specific trigeminothalamic pathway conforms, in general, to the lateral system of ascending pathways, and the nonspecific pathway conforms, in general, to the medial system of ascending pathways.[58]

IV. SUPRASPINAL STRUCTURES INVOLVED IN NOCICEPTION

The involvement of a supraspinal structure in nociception should be based on four general observations: (1) the structure receives fibers from ascending nociceptive pathways; (2) the structure contains cells which respond during the application of a noxious stimulus in the periphery; (3) stimulation of the structure produces pain in humans and aversive responses in animals; and (4) destruction of the structure alters pain perception in humans and, in animals, alters aversive behavior in response to the application of a noxious stimulus.

Four major regions of the brain are involved in nociception: (1) the medulla oblongata; (2) the mesencephalon; (3) the diencephalon; and (4) the cerebral cortex. The reticular formation constitutes part of the first three regions. Functionally the reticular formation can be considered, as far as nociception is concerned, as consisting of three important subdivisions: (1) the ascending reticular activating system (ARAS), a network of long fibers which interrelate with each other to project rostrally into wide areas of the cerebral cortex to control the excitability of the cerebral cortex and play a major role in consciousness; (2) a series of descending systems which project down to the spinal cord to modulate (or suppress) the relaying of nociceptive information to the brain; and (3) specific nuclei which send projection fibers to other parts of the brain which have rather specific effects upon nociceptive behavior.

The medulla oblongata — Electrophysiological studies have shown that many cells in the medullary reticular formation respond preferentially or exclusively to noxious stimuli.[59,60] The receptive fields of these cells are large and often are located bilaterally, probably due to the bilateral conduction of the spinoreticular and propriospinal pathways. Cells contribute to the ARAS by projecting to the cerebral cortex through the medial and intralaminar thalamic nuclei. The medullary reticular formation contains a number of nuclei which have been implicated in stimulation produced analgesia (see pain modulation section of this chapter).

Mesencephalon — The mesencephalic reticular formation and the periaqueductal gray (PAG) receive nociceptive inputs from the medullary part of the reticular formation and the spinomesencephalic tracts. The spinal input to the PAG appears to be, in part, through collaterals from spinothalamic tract fibers destined for the VPL.[61] The cells of the mesencephalon have large receptive fields that may be bilateral.[62,63] These cells contribute to the ARAS. Anatomical and electrophysiological studies indicate that the cells of the PAG project rostrally to synapse in midline and intralaminar nuclei of the thalamus;[64-67] as well as caudally to the nuclei of the medulla. Electrical stimulation of the mesencephalic PAG can produce a profound analgesia. Electrical stimulation of deeper parts of the mesencephalon produces strong aversive responses in both humans[68] and animals.[69-72] Lesions in the mesencephalic region have not been associated with significant changes in pain detection thresholds.[72]

Diencephalon — The thalamus and the hypothalamus are both involved in nociception. The thalamus serves as the relay for most of the activity entering the cerebral cortex from the various sensory systems of the body, other than olfaction. The thalamus consists of a large number of highly complex nuclei. For the purposes of this presentation, the thalamic nuclei associated with nociception can be divided into a ventral group (the ventral lateral nucleus [VL] and the ventrobasal nuclei, which consists of the ventral posterolateral nucleus [VPL] and the ventral posteromedial nucleus [VPM]); a medial posterior thalamic complex (PO_m); an intralaminar group (especially the parafascicular and the central lateral nuclei); a midline nuclear group; and a submedius nucleus.

Dennis and Melzack[35] emphasize that the lateral ascending spinal cord nociceptive pathways, which are associated with the sensory-discriminative dimension of pain, terminate predominately in the more laterally located nuclei of the thalamus (VPL, VPM, VL, and the PO_m). The medial ascending pathways terminate in more medial structures such as the intralaminar and midline nuclei of the reticular formation.

The VPL relays information from the body to the somatosensory cerebral cortical areas (SI and SII), and the VPM relays information from the head to these same areas. These are the areas where specific mechanoreceptor system terminates. Selective lesions in the VPL, the VPM, or in the SI area will result in the loss of the ability to perceive mechanical stimuli applied to the specific regions in the periphery. The VPL has been extensively studied regarding the projections of nociceptive pathways to this nucleus (for a review, see References 74 to 76). The results will only be summarized here. Cells responding to the application of nociceptive stimuli have been recorded from the VPL of all species studied. The cells were few in number compared to the number of low-threshold mechanical cells found. Both nociceptive specific and wide dynamic range cells have been found. Most of the receptive fields were small and located on the contralateral side of the body. The receptive fields bore a somatotopic relationship to the location of the cells in the VPL. Major differences were found between the location of the cells in the monkey and the rat as compared to the cat. The majority of the cells in the cat were found in the "shell" region of the VPL[77,78] similar in location to the terminations of the neospinothalamic tract.[47,48,79] A question about the nociceptive specific cells found in the diencephalon and mesencephalon has been raised by Brinkhus and Zimmermann[80] who could not find any nociceptive specific cells in awake cat undergoing behavioral experiments. They suggest that the nociceptive specific cells are present only if anesthesia is present.

Other nociceptive nuclei in the thalamus are the medial part of the PO_m, the CL, and the nucleus submedius. The cells of these nuclei, in general, respond to the application of strong noxious stimuli to large, bilateral receptive fields. These findings are consistent with the fact that only the nociceptive tracts belonging to the medial system (affective-motivational dimension of pain) terminate in these nuclei.

The cerebral cortex — Historically, it has been only recently that the cerebral cortex has been definitely implicated as playing a major role in pain perception. Early neurologists proposed that pain, thermal sense, and gross touch were sensed in the thalamus rather than the cerebral cortex.[81-83] Recent work demonstrates that pain has multiple representations in the cerebral cortex rather than none.[76] Willis[76] sites several lines of evidence that the cerebral cortex has an important role in the processing of pain sensation: (1) stimulation of the exposed cerebral cortex in humans sometimes produces pain; (2) lesions of the postcentral gyrus may reduce pain; (3) pain is sometimes experienced in epileptic auras; and (4) lesions of the cortex can produce a syndrome resembling "thalamic" pain.

Several investigators have been able to record from cerebral cortical neurons which respond to noxious stimuli. Lamour et al.[84] identified neurons in the SI area of the rat cerebral cortex which responded to nociceptive stimuli. Kenshalo and Isnsee[85] reported similar findings in the monkey. Robinson and Burton[85] found cells in the SII area of the cerebral cortex of the monkey, while Burton et al.[87] found similar cells in the SII cortex of the cat. The fronto-orbital cortex has been shown to contain neurons responsive to noxious cutaneous stimuli.[88,89]

The limbic system plays and important role in pain.[5] The mesencephalic PAG is a part of the midbrain area that projects to the medial thalamus and the hypothalamus,[90] which, in turn, project to the limbic system. Many of these areas interact with the frontal cortex. Electrical stimulation of parts of the limbic system, such as the hippocampus and the amygdala, evokes escape or other attempts to stop the stimulation.[69] After ablation of the amygdala and the overlying cerebral cortex, cats show marked changes in affective behavior, including decreased responses to nociceptive stimulation.[91] Lessening of the cingulate gyrus will drastically alter the motivational aspects of pain in the chronically ill human patient.[92,93] The evidence indicates that limbic structures, although they play a role in many other functions, provide a neural basis for the aversive drive and affect that comprise the motivational dimensions of pain.[4]

V. FACTORS WHICH MODULATE PAIN

In general, a given noxious stimulus results in a fairly predictable pain experience or pain response. Normally, a more intense noxious stimulus causes increased activation of nociceptive afferents and a greater appreciation of pain, both in terms of sensory-discriminative and affective-motivational dimensions (e.g., a pinch hurts less than smashing a finger with a hammer). However, nociception is subject to a high degree of modulation in both animals and humans. These modulating factors include lesions of the central nervous system, the action of drugs, effects of focal brain stimulation, as well as more natural factors such as anxiety, attention, prior experiences, and the co-occurrence of other noxious and nonnoxious stimuli. Of special interest are intrinsic analgesic systems which are common to both animals and humans.

In addition to the complex system of ascending pathways necessary for pain perception, there are both local segmental systems and descending supraspinal systems which inhibit pain signals. A functional model of these inhibitory systems is known as the "gate control" theory of pain.[94] In this model, the output of transmission (T) cells which relay nociceptive signals to the brain is inhibited by cells in the substantia gelatinosa (SG; lamina II) of the dorsal horn. The SG is under the influence of both other afferent input and supraspinal controls. Thus, the SG acts as a gate to either allow or block the transmission of pain information to the brain.

The segmental control depends on the relative amount of large-fiber (innocuous) afferent activity vs. small-fiber (noxious) afferent activity. Large-fiber activity inhibits the T cells via presynaptic mechanisms in the SG. Small-fiber activity excites the T cells directly and reduces the inhibition from the SG. While the functional aspects of the theory have been supported, much debate remains regarding the specific anatomical substrates. Regardless of the precise circuitry involved, opiates seem critical to the functioning of the gate and GABA has been implicated in presynaptic mechanisms. Such segmental control of pain transmission may underlie the effectiveness of transcutaneous electrical nerve stimulation and acupuncture analgesia in both animals and humans.

Analgesic mechanisms involving supraspinal structures have been under intense study since the demonstration that profound surgical analgesia could be produced in rats by electrical stimulation of the midbrain PAG.[95] Such stimulation-produced analgesia (SPA) can be elicited from stimulation of a variety of other brain areas including the ventromedial medulla, midbrain tegmentum, reticular formation, periventricular gray, lateral hypothalamus, and septal area.[96] In many cases the stimulation effects are thought to be specific for pain inhibition as they block behavioral responses to noxious stimuli without seeming to interfere with motor function, motivation, or perception of nonnoxious stimuli.[96] This conclusion is supported by the observation that animals will press a bar to get brain stimulation in response to a noxious stimuli.[97] SPA is also effective in humans suffering from chronic pain.[98]

The mechanism of SPA likely involves inhibition of terminals of nociceptive primary afferent terminals and/or spinal cord relay cells (T cells) responses to nociceptive inputs. The response of a T cell to a noxious heat or pinch can be greatly reduced or even completely suppressed by stimulation of any one of the previously mentioned SPA sites. This descending inhibition appears to be more effective for noxious than nonnoxious inputs. There is evidence for both postsynaptic[99] and presynaptic[100] mechanisms of inhibition. The anatomical pathway for descending inhibition involves fibers in the spinal cord dorsolateral funiculus (DLF). These fibers originate in the ventral portion of the rostral medulla and have terminations in the dorsal horn.

These medullo-spinal neurons function as the final common pathway for more rostral structures involved in descending inhibition.[101] Transection of the DLF results in loss of SPA and descending inhibition.[102,103] However, descending inhibition is not a completely satisfactory explanation of SPA. SPA and descending inhibition differ in that (1) inhibition usually only lasts for the duration of the stimulation, while SPA may outlast the stimulation for minutes or hours; (2) stimulation can cause inhibition without analgesia.[104] Perhaps effective SPA sites have additional effects at more rostral levels.

Identifying the neurotransmitters involved in SPA has been a confusing process and is far from complete. Multiple sites of action of a single transmitter, inconsistent effects at a given site, lack of selective antagonists, and colocalization of transmitters in single neurons have confounded attempts to elucidate specific actions of various transmitters. Different structures are likely to use different pharmacological agents. Norepinephrine, substance P, acetylcholine, neurotensin, as well as others have been implicated in analgesic mechanisms, but their functional importance and synaptic relationships are not yet clear. Nevertheless, there is good evidence that serotonin (5-HT) and some endogenous opiates (enkephalin and dynorphan) play important roles.[101]

Serotonin is found in high concentrations in the nucleus raphe magnus in the rostral medulla, and its application at lamina I and II in the spinal cord inhibits nociceptive cells. This inhibition may be direct postsynaptic inhibition or indirect via an inhibitory opiate containing neuron.[101] In addition, 5-HT may be released at medullary sites where it can excite medullo-spinal cells.

The endogenous opiates are clearly involved in some but not all pain control systems. Opiate receptors are found at midbrain, medullary, and spinal sites.[20] Injection of morphine at any one of these levels can cause analgesia; spinal application of opiates is now a common practice in humans. In animals, both SPA and descending inhibition from some brain sites can be blocked by the opiate antagonist naloxone. This antagonism of SPA has also been observed in humans.[105]

How endogenous analgesic systems function in the normal animal is an important question. The supraspinal nociceptive modulating systems appear to work tonically to some degree[106] and may be activated phasically in response to nociceptive input.[107] Ascending nociceptive tracts send collaterals to the PAG and rostral ventromedial medulla. Such interactions between ascending nociceptive systems and areas known to be involved in descending inhibitory systems may explain the phenomenon known as diffuse noxious inhibitory controls (DNIC). DNIC refers to the ability of noxious input from one body area to inhibit spinal neuronal responses to a coincident noxious input from a second body area.[107] DNIC is dependent on a supraspinal loop and has recently been shown to work on human perception of pain intensity.[108]

Intrinsic pain modulation systems may also be activated in animals by a variety of environmental stimuli, some of which are nonnoxious (e.g., restraint, cold water swims, hypoglycemia, hypertension, footshock). This phenomenon has been labeled stress-induced analgesia or environmentally induced analgesia (EIA).[109] That there are multiple pharmacologically distinct pain modulation systems at work is illustrated by the fact that EIA may be opiate mediated or not depending on the particular type of stimulus used. Another important factor seems to be the control ability of the stimulus. Animals given identical amounts of footshock only become analgesic via opiate mechanisms if they learn that they have no control over the stimuli.[110] This certainly implicates the importance of higher brain centers in the processing of pain information in animals. Anecdotal report of humans becoming analgesic in battle[111] or sporting events are suggestive of mechanisms similar to EIA at work.

VI. NOCICEPTIVE REFLEXES, REACTIONS, AND RESPONSES

In understanding pain, particularly animal pain, where pain assessment must be made by observing or measuring behavior, it is of major importance to understand the interrelationships among reflexes, reactions, and responses. This is of importance because reflexes and reactions can occur in both humans and animals with or without the individual perceiving the originating stimulus as being painful. These reflexes or reactions are referred to as being **perception linked or non-perception linked.**

A reflex is an involuntary, purposeful, and orderly response to a stimulus. The anatomical basis for the reflex is the reflex arc consisting of (1) a receptor, (2) a primary afferent nerve fiber associated with the receptor, (3) a region of integration in the spinal cord or the brain stem (synapses), (4) a lower motor neuron leading to (5) an effector organ such a skeletal muscle (somatic reflexes), smooth muscles, or glands (visceral reflexes).

A reaction, in neurological terminology and as it will be used in this chapter, consists of a combination of reflexes designed to produce widespread movement in relation to the application of a stimulus. Reactions will be considered mass reflexes not under voluntary control, i.e., that do not involve the cerebral cortex.

A response, in neurological terminology and as it will be used in this chapter, consists of willful movement of the body or parts of the body. A response cannot be performed without involvement of the somatosensory parts of the cerebral cortex. A decerebrate animal can give a reaction, but not a response.

Reflexes and reactions **may or may not be perception linked**, whereas responses indicate that the somatosensory parts of the cerebral cortex are functioning; thus, the individual is **probably capable of perceiving the stimulus**.

Responses are of major importance to the veterinary clinical neurologist in determining whether or not the animal is capable of perceiving the stimulus. It is important to realize that an animal having a motor paralysis (such as paralyzed with a curariform drug), but has no sensory loss, is incapable of giving any somatic reflexes, somatic reactions, or responses. Autonomic reflexes and reactions, such as pupillary light reflexes, pupillary dilatation, and cardiovascular reactions to noxious stimuli, can still occur in curarized animals.

A common misconception is that reflexes and reactions are always of the same magnitude if the stimulus strength is constant. This is simply not true. Excitatory and inhibitory synaptic activity converge on the lower motor neuron, and depending upon the integrated result of this activity, the magnitude of the reflex or the reaction will vary. Classic examples of this are seen in reflexes distal to the transected area of the spinal cord. After spinal shock has subsided, the reflexes are exaggerated.

Nociceptive reflexes and reactions are classified as protective reflexes or reactions. Some reflexes are innate or unconditioned, being based on fixed neural connections between receptors and effectors (skeletal and smooth muscles and glands). Other reflexes or reactions are acquired, i.e., learned or conditioned in relation to a sensory input.

Nociceptive responses can also be innate or unconditioned, being based on fixed inherent patterns of behavior which are often unique to a given species, i.e., species-specific behavior. Other nociceptive responses are acquired, i.e., learned or conditioned in relation to a given sensory input. Both innate and conditioned responses will be lost if the animal is decerebrated.

The most common protective reflex is the flexion or nociceptive withdrawal reflex. This involves the involuntary withdrawal of a limb from the application of a noxious stimulus. The designation of this reflex as the flexion reflex is not as exact as calling it a nociceptive withdrawal reflex because flexion of the limb reflexly occurs as a part of other neural activity, such as the swing phase of the step cycle.

Other examples of common nociceptive reflexes are the tooth pulp-jaw opening reflex (TP-JOR), the skin twitch reflex (STR), the corneal reflex, and the facial twitch reflexes. These reflexes are often used to test analgesic agents where they are interpreted to be perception-linked. When using these procedures, dissociation of the reflex from the animal's response (indicating perception of the stimulus as painful) is a major problem.[2] Another major problem in using these reflexes in analgesic testing is the fact that the animals become conditioned to the application of the test stimulus, particularly if it is a heat stimulus, and respond before the stimulus becomes noxious.[112]

The application of a noxious stimulus to an animal, if perceived, should elicit an aversive response by the animal. These responses have been described by Vierck and Cooper[112] as (1) reflexive withdrawal and postural adjustments to accommodate the escape response; (2) conscious, energetic escape from the pain source (e.g., running, jumping); (3) prolonged protective activity (e.g., fighting, fleeing, or submitting); and (4) retreat and withdrawal to recuperate, including inhibition of activity. A difficulty in interpretation of these responses has been pointed out by the work of Woolf,[113] who demonstrated that decerebrate rats, following the application of a noxious stimulus, show reflexive withdrawal, postural adjustments, and running and jumping to escape the stimulus. However, the rats would not show prolonged protective activity. The removal of the noxious stimulus was followed by the rats returning to passive or grooming states as if nothing had occurred. The rats were blind, thus they were not able to be tested for such cortically dependent responses such as the menace response and visual placing responses. Decerebrate animals cannot give tactile placing reactions, e.g., place their feet on a table surface if the limbs touch the edge of the table, or conscious placing responses, e.g., straighten a paw that had been placed so that the animal was resting its weight on the dorsal surface of the paw. These tests are standard tests used by veterinary clinical neurologists to examine cerebral cortical function.

Classical conditioning procedures are very useful in determining the ability of the cerebral cortex to function. These procedures involve the sequential pairing of a nonnoxious conditioning stimulus (CS), such as an auditory click with a noxious stimulus (unconditioned stimulus, UCS). At first, before the animal is conditioned, the animal responds by withdrawing the stimulated part (unconditioned response, UCR). After a series of presentations the animal will withdraw the limb before the UCS is presented. This withdrawal is referred to as the conditioned response (CR). This relationship can be illustrated as follows on a time base:

$$\text{Conditioning trials CS} \rightarrow \text{UCS} \rightarrow \text{UCR}$$

$$\text{Conditioned animal CS} \rightarrow \text{CR}$$

If the animal loses its capacity for cerebral functioning, the CR will be lost. A conditioned response then is a very good way of determining cerebral cortical functioning in animals.[11,15,114] Another technique for measuring the capability of the cerebral cortex to function is to average evoked auditory, visual, and somatosensory inputs to the cerebral cortex. The disadvantage of these procedures are that evoked potentials can be averaged in animals which are unconscious. All these techniques do is to test the integrity of the ascending pathways to conduct nerve impulses to the cerebral cortex.

A major problem in assessing pain perception in animals is attempting to quantitate the intensity of the stimulus (or the amount of pain felt by the animals, i.e., how much does it hurt). Vierck and Cooper[115,116] conducted a carefully controlled study in monkeys and humans. They used a carefully quantitated series of electrical shock intensities which were psychophysically rated by their human subjects. They found that the strength (force) of the reflexive response of the monkeys was a good indicator that the pain detection thresholds had been surpassed. However, the reflexive response was a poor indicator of pain for several reasons: (1) the presence of a reflex response did not indicate pain because the monkeys gave reflex responses to stimulus intensities as low as $^1/_5$ that of the pain detection threshold as reported by their human subjects;

(2) the reflex magnitude did not increase throughout the range of suprathreshold intensities; moderate and strong stimuli did not elicit any change in the strength or the magnitude of the reflex response. They also recorded the monkeys' vocal responses during these experiments. They could not relate the intensity or the pattern of the monkeys' vocalizations to the presentation of noxious stimuli of differing strengths.[116] Stating their findings in another way, only the strength, not just the presence or absence, of a reflex response was a reliable indicator that the pain detection threshold had been exceeded and that the animal was perceiving pain. The speed and magnitude of the reflex response did not indicate the intensity of the pain which the animals felt. The vocalizations emitted by the monkeys also did not indicate whether or not the stimulus was or was not painful nor the intensity of the stimulation. These findings suggest that subjective assessments of the amount of pain that an animal is feeling based on the reflex responses and the vocalizations the animals are emitting probably are not very good indicators of the amount of pain the animals are perceiving, i.e., it is hard to judge actually how much pain an animal is perceiving by just looking at an animal. Their study did indicate that the force and speed with which the monkeys pulled a lever to avoid the receiving of a noxious stimulus (a conditioned response) was a good indicator of the intensity of the pain being felt.

Another procedure used to test an animal's evaluation of the intensity of the stimulus is the use of operant nonreflexive avoidance behavior (pulling a lever to terminate an electrical stimulus to another part of the body).[26] The speed and vigor with which an animal pulled the lever correlated directly with the intensity of the noxious stimulus.

The pain detection thresholds were determined in cats by applying a noxious stimulus and observing the interruption of eating (or the exploration for food), hindlimb movements, and vocalization.[117] The lowest pain detection thresholds were for interruption of eating, next for limb movement, and last for vocalization.

VII. SUMMARY AND CONCLUSIONS

Pain is a perception dependent upon the cerebral cortex being in a desynchronized state before it can occur. In order for a noxious stimulus to lead to the perception of pain, the stimulus must be of sufficient intensity to cross the pain detection threshold. Evidence is available which indicates that an animal can detect pain at the same stimulus intensity level that humans detect pain. In animals, the intensity of the pain above the pain detection level is very difficult to assess from observations of the behavior of the animal, particularly the animal's reflex reactions to the stimulus. Animals show considerable variation in their tolerance to a particular stimulus depending upon the species, the breed, their previous experience with similar stimuli, their anxiety state, their environment, and many other factors. In the assessment of pain in animals, the intensity of a particular stimulus should be correlated with an adult human's assessment of a similar stimulus, not unlike what one would do in assessing a child's reactions to a noxious stimulus. Stoic animals may give few indications that they are detecting a stimulus as pain; however, above the pain detection level, they may be suppressing not only their responses to the pain, but also the amount of pain being felt through endogenous pain suppression mechanisms. The pain suppression capability varies considerably among humans because of inherent as well as learned suppression. Evidence is available that animals have inherent as well as learned pain suppression.

This chapter has illustrated the high degree of similarity between humans and other animals in terms of (1) the physiological substrates of nociception, (2) the behavioral responses to noxious stimulation, and (3) the capacity to modulate nociceptive information and/or responses. Nevertheless, it can be argued that we still know nothing about the subjective feelings of an animal in pain. The existence of analogous structures in humans and animals which mediate the affective aspects of pain does not prove that animals have the same emotional response to pain that humans do; but, their existence strengthens the possibility. One could also argue that in animals these structures simply mediate operantly conditioned avoidance behaviors through negative reinforcement. Yet one has to ask what the nature of such powerful negative reinforcement is if it is not unpleasant or hurtful?[118] The majority of scientific evidence leads us to conclude that stimuli which both activate nociceptors and produce aversive responses should be regarded as painful.

REFERENCES

1. **Kandell, E. R. and Schwartz, J. H.,** Sensory systems of the brain: sensation and perception. Central representation of pain and analgesia, in *Principles of Neuroscience, Part V,* Elsevier, Amsterdam, 1985, 331.
2. **Vyklicky, L.,** Methods of testing pain mechanisms in animals, in *Textbook of Pain,* Wall, P. D. and Melzack, **R.,** Eds., Churchill Livingstone, Edinburgh, 1984, 178.
3. **Schmidt, R. F.,** Nociception and pain, in *Fundamentals of Sensory Physiology,* Schmidt, R. F., Ed., Springer-Verlag, Berlin, 1986, 117.
4. **Melzack, R.,** Neurophysiological foundations of pain, in *The Psychology of Pain,* Sternbach, R. A., Ed., Raven Press, New York, 1986, 1.
5. **Melzack, R. and Casey, K. L.,** Sensory, motivational and central control determinants of pain, in *The Skin Senses,* Kenshalo, D., Ed., Charles C Thomas, Springfield, IL, 1968, 423.
6. **Pavlov, I. P.,** *Lectures on Conditioned Reflexes,* International Publishers, New York, 1928.
7. **Zimmermann, M.,** Neurobiological concepts of pain, its assessment and therapy, in *Neurophysiological Correlates of Pain,* Bromm, B., Ed., Elsevier, Amsterdam, 1984, 15.
8. **Lewis, T.,** *Pain,* Macmillan, New York, 1942.
9. **Poggio, C. F. and Mountcastle, V. B.,** A study of the functional contribution of the lemniscal and spinothalamic systems to somatic sensibility, *Bull. Johns Hopkins Hosp.,* 108, 266, 1960.
10. **Brain, R.,** Presidential address, in *The Assessment of Pain in Man and Animals,* Keele, C. A. and Smith, R., Eds., Churchill Livingstone, Edinburgh, 1962, 3.
11. **Kitchell, R. L., Naitoh, Y., Breazile, J. E., and Lagerwerff, J. M.,** Methodological considerations for assessment of pain perception in animals, in *The Assessment of Pain in Man and Animals,* Keele, C. A. and Smith, R., Eds., Churchill Livingstone, Edinburgh, 1962, 244.
12. **Owens, M. E.,** Pain in infancy: conceptual and methodological issues, *Pain,* 20, 213, 1984.
13. **Sternbach, R. A.,** *Pain: A Psychophysiological Analysis,* Academic Press, Edinborough, 1968, 1.
14. **Dudel, J.,** General sensory physiology, psychophysics, in *Fundamentals of Sensory Physiology,* Schmidt, R. F., Ed., Springer-Verlag, Berlin, 1986, 1.
15. **Breazile, J. E., Kitchell, R. L., and Naitoh, Y.,** Neural basis of pain in animals, in *Proc. 15th Research Conf. Chicago, 1963,* American Meat Institute Foundation, Chicago, 1963, 53.
16. **Zotterman, Y.,** Studies in peripheral nervous mechanisms of pain, *Acta Med. Scand.,* 80, 185, 1933.
17. **Burgess, P. R. and Perl, E. R.,** Myelinated afferent fibers responding specifically to noxious stimulation of the skin, *J. Physiol.,* 190, 542, 1967.
18. **Willis, W. D.,** *The Pain System: The Neural Basis of Nociceptive Transmission in the Mammalian Nervous System,* Gildenberg, Ph.L., Ed., S. Karger, Basel, 1985.
19. **Dubner, R. and Bennett, G. J.,** Spinal and trigeminal mechanisms of nociception, *Annu. Rev. Neurosci.,* 6, 381, 1983.
20. **Besson, J. M. and Chaouch, A.,** Peripheral and spinal mechanisms of nociception, *Physiol. Rev.,* 67, 67, 1987.
21. **Kruger, L. and Rodin, B. E.,** Peripheral mechanisms involved in pain, in *Animal Pain. Perception and Alleviation,* Kitchell, R. L. and Erickson, H. H., Eds., American Physiological Society, Bethesda, MD, 1983, 1.
22. **Torebjörk, H. E. and Ochoa, J. L.,** Specific sensation evoked by activity in single identified sensory units in man, *Acta Physiol. Scand.,* 110, 445, 1980.
23. **Adriaensen, H., Gybels, J., Handwerker, H. O., and Van Hees, J.,** Response properties of thin myelinated (A-δ) fibers in human skin nerves, *J. Neurophysiol.,* 49, 111, 1983.
24. **Torebjörk, H. E. and Hallin, R. G.,** Identification of afferent C units in intact human skin nerves, *Brain Res.,* 67, 387, 1974.

25. **Dubner, R., Price, D. D., Beitel, R. E., and Hu, J. W.,** Peripheral neural correlates of behavior in monkey and human related to sensory-discriminative aspects of pain, in *Pain in the Trigeminal Region,* Anderson, D. J. and Mathews, B., Eds., Elsevier, Amsterdam, 1977, 57.
26. **Vierck, C. J., Cooper, B. Y., and Cohen, R. H.,** Human and nonhuman primate reactions to painful electrocutaneous stimuli and to morphine, in *Animal Pain. Perception and Alleviation,* Kitchell, R. L., Erickson, H. H., Carstens, E., and Davis, L. E., Eds., American Physiological Society, Bethesda, MD, 1983, 117.
27. **Rexed, B.,** The cytoarchitectonic organization of the spinal cord in the cat, *J. Comp. Neurol.,* 96, 415, 1952.
28. **Light, A. R. and Perl, E. R.,** Spinal termination of functionally identified primary afferent neurons with slowly conducting myelinated fibers, *J. Comp. Neurol.,* 186, 133, 1979.
29. **Knyihar-Csillik, E., Csillik, B., and Rakic, P.,** Ultrastructure of normal and degenerating glomerular terminals of dorsal root axons in the substantia gelatinosa of the rhesus monkey, *J. Comp. Neurol.,* 210, 357, 1982.
30. **Hökfelt, T., Kellerth, J. O., Nilsson, G., and Pernow, B.,** Experimental immunohistochemical studies on the localization and distribution of substance P in cat primary sensory neurons, *Brain Res.,* 100, 235, 1975.
31. **Henry, J. L.,** Effects of substance P on functionally identified units in the cat spinal cord, *Brain Res.,* 114, 493, 1976.
32. **Yaksh, T. L., Farb, D. H., Leeman, S. E., and Jessell, T. M.,** Intrathecal capsaicin depletes substance P in the rat spinal cord and produces prolonged thermal analgesia, *Science,* 206, 481, 1979.
33. **Leah, J. D., Cameron, A. A., and Snow, P. J.,** Neuropeptides in physiologically identified mammalian sensory neurons, *Neurosci. Lett.,* 56, 257, 1985.
34. **Collins, J. G. and Ren, K.,** WDR profiles of spinal dorsal horn neurons may be unmasked by barbiturate anesthesia, *Pain,* 28, 369, 1987.
35. **Dennis, S. G. and Melzack, R.,** Pain-signalling systems in the dorsal and ventral spinal cord, *Pain,* 4, 97, 1977.
36. **White, J. C. and Sweet, W. H.,** *Pain and the Neurosurgeon: a Forty-Year Experience,* Charles C Thomas, Springfield, IL, 1969, 850.
37. **Carstens, E. and Trevino, D. L.,** Anatomical and physiological properties of ipsilaterally projecting spinothalamic neurons in the second cervical segment of the cat's spinal cord, *J. Comp. Neurol.,* 182, 167, 1978.
38. **Kennard, M. A.,** The course of ascending fibers in the spinal cord of the cat essential to the recognition of painful stimuli, *J. Comp. Neurol.,* 199, 511, 1954.
39. **Breazile, J. E. and Kitchell, R. L.,** Ventrolateral spinal cord afferents to the brain stem in the domestic pig, *J. Comp. Neurol.,* 133, 363, 1968.
40. **Trevino, D. L., Coulter, J. D., and Willis, W. D., Jr.,** Location of cells of origin of the spinothalamic tract in lumbosacral enlargement of the monkey, *J. Neurophysiol.,* 36, 750, 1973.
41. **Carstens, E. and Trevino, D. L.,** Laminar origins of spinothalamic projections in the cat as determined by the retrograde transport of horseradish peroxidase, *J. Comp. Neurol.,* 182, 151, 1978.
42. **Trevino, D. L. and Carstens, E.,** Confirmation of the location of spinothalamic neurons in the cat and the monkey by the retrograde transport of horseradish peroxidase, *Brain Res.,* 98, 177, 1975.
43. **Brown, A. G. and Franz, D. N.,** Responses of spinocervical tract neurones to natural stimulation of identified cutaneous receptors, *Exp. Brain Res.,* 7, 231, 1969.
44. **Cervero, F., Iggo, A., and Molony, V.,** Responses of spinocervical tract neurones to noxious stimulation of the skin, *J. Physiol.,* 267, 537, 1977.
45. **Angaut-Petit, D.,** The dorsal column system. II. Functional properties and bulbar relay of the postsynaptic fibres of the cat's fasciculus gracilis, *Exp. Brain Res.,* 22, 471, 1975.
46. **Boivie, J.,** An anatomical reinvestigation of the termination of the spinothalamic tract in the monkey, *J. Comp. Neurol.,* 186, 343, 1979.
47. **Berkley, K.J.,** Spatial relationships between the terminations of somatic sensory and motor pathways in the rostral brainstem of cats and monkeys, *J. Comp. Neurol.,* 193, 283, 1980.
48. **Boivie, J.,** The termination of the spinothalamic tract in the cat. An experimental study with silver impregnation methods, *Exp. Brain Res.,* 12, 331, 1971.
49. **Mitchell, C. L. and Kaebler, W. W.,** Effect of medial thalamic lesions on responses elicited by tooth pulp stimulation, *Am. J. Physiol.,* 210, 263, 1966.
50. **Mitchell, C. L. and Kaebler, W. W.,** Unilateral vs bilateral lesions and reactivity to noxious stimuli, *Arch. Neurol.,* 17, 653, 1967.
51. **Kaebler, W. W., Mitchell, C. L., Yarmat, A. J., Affifi, A. K., and Lorens, S. A.,** Centrum medianum-parafasciculus lesions and reactivity to noxious and non-noxious stimulus, *Exp. Neurol.,* 46, 282, 1975.
52. **Cassem, N. H.,** Current topics in medicine. II. Pain, in *Scientific American Medicine, 11/1983,* Rubenstein, E. and Federman, D. D., Eds., Scientific American, New York, 1987, 1.
53. **Dennis, S. G. and Melzack, R.,** Perspectives on phylogenetic evolution of pain expression, in *Animal Pain: Perception and Alleviation,* Kitchell, R. L., Erickson, H. H., Carstens, E., and Davis, L. E., Eds., American Physiological Society, Bethesda, MD, 1983, 151.
54. **Gordon, G., Landgren, S., and Seed, W.,** The functional characteristics of single cells in the caudal part of the spinal nucleus of the trigeminal nerve of the cat, *J. Physiol. (London),* 158, 544, 1961.

55. **Mosso, J. A. and Kruger, L.,** Spinal trigeminal neurons excited by noxious and thermal stimuli, *Brain Res.,* 38, 206, 1972.
56. **Kruger, L. and Mosso, J. A.,** An evaluation of the duality in the trigeminal afferent system, *Adv. Neurol.,* 4, 73, 1974.
57. **Price, D. D., Dubner, R., and Hu, J. W.,** Trigeminothalamic neurons in nucleus caudalis responsive to tactile, thermal and nociceptive stimulation of monkey's face, *J. Neurophysiol.,* 39, 936, 1976.
58. **Hu, J. W., Dostrovsky, J. O., and Sessle, B. J.,** Functional properties of neurons in cat subnucleus caudalis (medullary dorsal horn). I. Responses to oral-facial noxious and nonnoxious stimuli and projections to the thalamus, *J. Neurophysiol.,* 45, 173, 1981.
59. **Bowsher, D.,** Role of the reticular formation in responses to noxious stimuli, *Pain,* 2, 361, 1976.
60. **Casey, K. L.,** The reticular formation and pain: towards a unifying concept, in *Pain Research Publications: Association for Research in Nervous and Mental Disease.,* Bonica, J.J., Ed., Raven Press, New York, 1980, 63.
61. **Price, D. D., Hayes, R. L., Ruda, M. A., and Dubner, R.,** Spatial and temporal transformations of input to spinothalamic neurons and their relation to somatic sensation, *J. Neurophysiol.,* 41, 933, 1978.
62. **Barnes, K. L.,** A quantitative investigation of somatosensory coding in single cells of the cat mesencephalic reticular formation, *Exp. Neurol.,* 50, 180, 1976.
63. **Eickhoff, R., Handwerker, H. O., McQueen, D. S., and Schick, E.,** Noxious and tactile input to medial structures of the midbrain and the pons in the rat, *Pain,* 5, 99, 1978.
64. **Hamilton, B. L. and Skultety, M.,** Efferent connections of the periaqueductual gray matter in the cat, *J. Comp. Neurol.,* 139, 105, 1970.
65. **Robertson, R. T., Lynch, G. S., and Thompson, R. F.,** Diencephalic distributions of ascending reticular systems, *Brain Res.,* 55, 309, 1973.
66. **Bowsher, D.,** Diencephalic projections from the midbrain reticular formation, *Brain Res.,* 95, 211, 1975.
67. **Mancia, M. G., Broggi, G., and Margnelli, M.,** Brain stem reticular effects on intralaminar thalamic neurons in the cat, *Brain Res.,* 25, 638, 1971.
68. **Nashold, B. S., Jr., Wilson, W. P., and Slaughter, D. G.,** Sensations evoked by stimulation of the midbrain of man, *J. Neurosurg.,* 30, 14, 1969.
69. **Delgado, J. M., Rosvold, H. E., and Looney, E.,** Evoked conditoning fear by electrical stimulation of subcortical structures in the monkey brain, *J. Comp. Psychol.,* 49, 373, 1956.
70. **Kiser, R. S., Lebovitz, R. M., and German, D. C.,** Anatomic and pharmacologic differences between two types of aversive midbrain stimulation, *Brain Res.,* 155, 331, 1978.
71. **Skultety, F. M.,** Stimulation of the periaqueductal gray and hypothalamus, *Arch. Neurol.,* 8, 608, 1963.
72. **Spiegel, E. A., Keiltzkin, A. M., and Szekely, E. G.,** Pain reactions upon stimulation of the tectum mesencephali, *J. Neuropathol. Exp. Neurol.,* 13, 212, 1954.
73. **Yaksh, T. L. and Hammond, D. L.,** Peripheral and central substrates involved in the rostrad transmission of nociceptive information, *Pain,* 13, 1, 1982.
74. **Willis, W. D.,** Thalamocortical mechanisms of pain, *Adv. Pain Res. Ther.,* 9, 245, 1985.
75. **Willis, W. D., Jr.,** The origin and destination of pathways involved in pain transmission, in *Textbook of Pain,* Wall, P. D. and Melzack, R., Eds., Churchill Livingstone, Edinburgh, 1984, 88.
76. **Willis, W. D., Jr.,** Nociceptive transmission to the thalamus and the cerebral cortex, in *The Pain System. The Neural Basis of Nociceptive Transmission in the Mammalian Nervous System,* Willis, W. D., Jr., Ed., S. Karger, Basel, 1985, chap. 6.
77. **Honda, C. N., Mense, S., and Perl, E. R.,** Neurons in ventrobasal region of cat thalamus selectively responsive to noxious mechanical stimulation, *J. Neurophysiol.,* 49, 662, 1983.
78. **Kniffi, K. D. and Mizumura, K.,** Responses of neurons in the VPL and VL-VPL region of the cat to algesic stimulation of muscle and tendon, *J. Neurophysiol.,* 49, 649, 1983.
79. **Jones, E. G. and Burton, H.,** Cytoarchitecture and somatic sensory connectivity of thalamic nuclei other than the ventrobasal complex in the cat, *J. Comp. Neurol.,* 154, 395, 1974.
80. **Brinkhus, H. B. and Zimmermann, M.,** Posterior thalamic neuronal activity during conditioned behavior of cats, *Neurosci. Lett. Suppl.,* 14, S43, 1983.
81. **Head, H. and Holmes, G.,** Sensory disturbances from head lesions, *Brain,* 34, 102, 1911.
82. **Head, H.,** *Studies in Neurology,* Oxford University Press, London, 1920.
83. **Penfield, W. and Jasper, H.,** *Epilepsy and the Functional Anatomy of the Human Brain,* Little, Brown, Boston, 1954.
84. **Lamour, Y., Guilbaud, G., and Willer, J. C.,** Altered properties and laminar distribution of neuronal responses to peripheral stimulation in the SmI cortex of the arthritic rat, *Brain Res.,* 273, 183, 1983.
85. **Kenshalo, D. R., Jr. and Isnsee, O.,** Response of primate SI cortical neurons to noxious stimuli, *J. Neurophysiol.,* 50, 1479, 1983.
86. **Robinson, C. J. and Burton, H.,** Somatic submodality distribution within the second somatosensory (SII), 7b, retroinsular, postauditory, and granular insular cortical areas of *M. fascicularis, J. Comp. Neurol.,* 192, 93, 1980.

87. **Robinson, C. J. and Burton, H.,** Somatic submodality distribution within the second somatosensory (SII), 7b, retroinsular, postauditory, and granular insular cortical areas of *M. fascicularis*, *J. Comp. Neurol.,* 192, 93, 1983.
88. **Korn, H., Wendt, R., and Albe-Fessard, D.,** Somatic projections to the orbital cortex of the cat, *EEG Clin. Neurophysiol.,* 21, 209, 1966.
89. **Korn, H.,** Splanchnic projections of the cerebral cortex of the cat, *Brain Res.,* 16, 23, 1969.
90. **Nauta, W. J. H.,** Hippocampal projections and related neural pathways to the midbrain in the cat, *Brain,* 81, 319, 1958.
91. **Schriener, L. and Kling, A.,** Behavioral changes following rhinencephalic injury in the cat, *J. Neurophysiol.,* 15, 643, 1953.
92. **Folz, E. L. and White, L. E.,** Pain "relief" by frontal cingulumotomy, *J. Neurosurg.,* 19, 89, 1962.
93. **Corkin, S., Twitchell, T. E., and Sullivan, E. V.,** Safety and efficacy of cingulotomy for pain and psychiatric disorders, in *Modern Concepts in Psychiatric Surgery,* Hitchcock, E. R., Ballentine, H. T., Jr., and Myerson, B. A., Eds., Elsevier, New York, 1979, 253.
94. **Melzack, R. and Wall, P. D.,** Pain mechanisms: a new theory, *Science,* 150, 971, 1965.
95. **Reynolds, D. G.,** Surgery in the rat during electrical analgesia induced by focal brain stimulation, *Science,* 164, 444, 1969.
96. **Mayer, D. J. and Price, D. D.,** Central nervous system mechanisms of analgesia, *Pain,* 2, 379, 1976.
97. **Mayer, D. J., Wolfle, T. L., Akil, H., Carder, B., and Liebeskind, J. C.,** Analgesia from electrical stimulation in the brainstem of the rat, *Science,* 174, 1351, 1971.
98. **Richardson, D. E. and Akil, H.,** Pain reduction by electrical brain stimulation in man, *J. Neurosurg.,* 47, 178, 1977.
99. **Shah, Y. and Dostrovsky, J. O.,** Postsynaptic inhibition of cat medullary dorsal horn neurons by stimulation of nucleus raphe magnus and other brain stem sites, *Exp. Neurol.,* 77, 419, 1982.
100. **Dostrovsky, J. O., Sessle, B. J., and Hu, J. W.,** Presynaptic excitability changes produced in brain stem endings of tooth pulp afferents by raphe and other central and peripheral influences, *Brain Res.,* 218, 141, 1981.
101. **Basbaum, A. I. and Fields, H. L.,** Endogenous pain control systems: brainstem spinal pathways and endorphin circuitry, *Annu. Rev. Neurosci.,* 7, 309, 1984.
102. **Basbaum, A. I., Marley, N., O'Keefe, J., and Clanton, C. H.,** Reversal of morphine and stimulation produced analgesia by subtotal lesion of the spinal cord, *Pain,* 3, 43, 1977.
103. **Fields, H. L., Basbaum, A. I., Clanton, C. H., and Anderson, S. D.,** Nucleus raphe magnus inhibition of spinal cord dorsal horn neurons, *Brain Res.,* 126, 441, 1977.
104. **Culhane, E. S. and Carstens, E.,** Medial hypothalamic stimulation supresses nociceptive spinal dorsal horn neurons but not the tail flick reflex in the rat, *Brain Res.,* 438, 137, 1988.
105. **Hosobuchi, Y., Adams, J. E., and Linchitz, R.,** Pain relief by electrical stimulation of the central gray matter in humans and its reversal by naloxone, *Science,* 197, 183, 1977.
106. **Dickhaus, H., Pauser, G., and Zimmermann, M.,** Tonic descending inhibition affects intensity coding of nociceptive responses of spinal dorsal horn neurons in the cat, *Pain,* 23, 145, 1985.
107. **Le Bars, D., Dickenson, A. H., and Besson, J. M.,** Diffuse noxious inhibitory controls (DNIC). I. Effects on dorsal horn convergent neurons in the rat, *Pain,* 6, 283, 1979.
108. **Talbot, J. D., Duncan, G. H., Bushnell, M. C., and Boyer, M.,** Diffuse noxious inhibitory controls (DNIC): psychophysical evidence in man for intersegmental suppression of noxious heat perception by cold pressor pain, *Pain,* 30, 221, 1987.
109. **Watkins, L. R. and Mayer, D. J.,** Multiple endogenous opiate and non-opiate analgesia systems: evidence for their existence and clinical implications, in *Annals of the New York Academy of Sciences, First Int. Conf. on Stress Induced Analgesia,* Vol. 467, Kelly, D. D., Ed., NYAS Press, New York, 1986, 273.
110. **Maier, S. F., Drugan, R. C., and Grau, J. W.,** Controllability, coping behavior, and stress induced analgesia in the rat, *Pain,* 12, 47, 1982.
111. **Beecher, H. K.,** *Measurement of Subjective Responses,* Oxford University Press, New York, 1959.
112. **Vierck, C. J. and Cooper, B. Y.,** Guidelines for assessing pain reactions and pain modulation in laboratory animal subjects, in *Advances in Pain Research and Therapy.* Vol. 6, Kruger, L. and Liebeskind, J. C., Eds., Raven Press, New York, 1984, 305.
113. **Woolf, C. J.,** Long term alterations of the excitability of the flexion reflex produced by peripheral tissue injury in the chronic decerebrate rat, *Pain,* 18, 325, 1984.
114. **Kitchell, R. L., Stromberg, M. W., Lagerwerff, J. M., and Arnold, J. P.,** Basis for evaluation of pain in animals, in *Proc. 12th Research Conf. American Meat Inst.,* American Meat Institute Foundation, Chicago, 1960, 25.
115. **Cooper, B. Y. and Vierck, C. J.,** Measurement of pain and morphine hypalgesia in monkeys, *Pain,* 26, 361, 1986.
116. **Cooper, B. Y. and Vierck, C. J.,** Vocalizations as measures of pain in monkeys, *Pain,* 26, 393, 1986.
117. **Casey, K. L., and Morrow, T. J.,** Nocifensive responses to cutaneous thermal stimuli in the cat: stimulus-response profiles, latencies, and afferent activity, *J. Neurophysiol.,* 50, 1497, 1983.
118. **Rollin, B. E.,** Ideology, ethics, and history: a reply to Feyerabend, Rachlin, and Leahey, *New Ideas Psychol.,* 4, 165, 1986.

Chapter 13

CLINICAL ASSESSMENT OF PAIN IN LABORATORY ANIMALS

Eugene M. Wright, Jr.and Judith F. Woodson

EDITOR'S PROEM

At the core of new laws, regulations, and policies pertaining to laboratory animal welfare, and at the core of the social moral concern prompting these regulations, is the demand that pain be minimized and controlled in experimental animals. In order to accomplish this, those who work with laboratory animals should be familiar with what conditions engender pain in animals and with the ways in which different species of animals provide evidence of being in pain or other forms of discomfort. In this paper, Dr. Wright summarizes some of the major signs of pain and suffering in laboratory animals. At the same time, he stresses the need for baseline familiarity with one's animals, under normal conditions, and the likelihood of individual differences playing a role in response to pain among animals. For this reason, Wright, like many other authorities on animal pain, emphasizes the important role of animal caretakers in recognizing deviations from normalcy.

TABLE OF CONTENTS

I. INTRODUCTION

Regulation of animal pain, distress, stress, and suffering is not new to the veterinarian, scientist, or society. As early as 1876, the interest in animal welfare led to the passage of the Cruelty to Animals Act in the U.K. The act was intended to "permit the infliction of pain on animals for scientific purposes, but with safeguard."[1] Since that time, many definitions of pain have evolved, and the question of pain has been recognized as one of the most challenging problems in medicine and biology.[2]

While the overall objective of pain assessment is alleviation, one must remember that pain also serves as a beneficial diagnostic aid. At a recent symposium on pain, Dr. Ralph Kitchell very aptly referred to this as the "positive" side of pain. Recognition of disease processes and injuries can be identified more easily when signs of pain are present. After such identification has been made and an appropriate diagnosis reached, it must be remembered that painful injuries may inhibit an animal from moving, thereby preventing further damage. Only after the above assessment has been made, should a regimen of pain relief be initiated. The issues at hand must not be confused, however; we are not suggesting that animals be left to suffer for the lack of an adequate diagnosis. Pain and suffering may actually constitute a situation in which, if in doubt, one should go ahead and treat.[3]

Assessment of pain has been complicated by the discovery that the brain and spinal cord have neural mechanisms that, when activated, can result in the release of endogenous opiate-like substances (enkephalins and endorphins). The nervous system also possesses descending nonopiate systems which modulate the transmission of nerve impulses that give rise to pain.[4] Assessment of pain and suffering in animals which (1) have sustained injury to the nervous system, (2) have been administered anesthesia, or (3) have received drugs which alter the central nervous system may be very difficult. In order for pain perception to occur, a functional cerebral cortex must be present. Once a noxious (harmful) stimulus comes in contact with the nociceptors, nerve impulses are generated in the nociceptor fibers. These impulses must reach the cerebral cortex before the stimulus actually can be interpreted. A clinical assessment of cerebrocortical function then can be made by the use of electroencephalographic (EEG) patterns. Knowledge of the functions and processes associated with nociceptor fibers, pain pathways, and their role in pain perception and tolerance is important in the assessment of pain.

Interpretation of the responses elicited to noxious forms of stimuli also is important. One must determine the difference between a reflex, which does not depend on a functional cerebral cortex, and a reaction, which does require a functional cerebral cortex.[4] Dr. C. G. Lineberry has referred to three criteria for determining whether a stimulus elicits a pain response, or if what is shown is simply a stereotypic movement of the body (a reflex).[5] First, the stimulus must be perceived as pain if applied to a normal human; second, the stimulus must produce tissue damage; and third, the animal must show an aversive response to the stimulus.

Pain assessment has to be accomplished in two different settings: (1) experimental situations, where sophisticated equipment may be used to measure intensity levels and the efficiency of methods for pain relief, and (2) the clinical situations where one must assess quickly the existence and intensity of pain without the use of sophisticated equipment. At times, the evaluation may be done by someone without prior training or knowledge of the animal involved.[6]

In order to "assess" pain, it is necessary to be familiar not only with the many terms used throughout the literature, but also to be able to interpret the clinical, psychological, and biochemical signs that indicate well-being of laboratory animals. Even the most apparent indicators of pain, such as injuries and severe body damage, suggest that pain is likely to be present, but cannot be considered as complete evidence that an animal actually feels pain.[6]

Along with the increased awareness of animal welfare in society and in the scientific

TABLE 1
Types, Characteristics, and Duration of Pain

Postoperative	Discomfort realized immediately following surgery and lasting several days
Psychogenic	Undesirable physical or mental state resulting from stress, anxiety, or fear
Superficial or cutaneous	Skin, fascia, horn, hoof, nail
Deep	Muscles, ligaments, joints, bones
Visceral	Intestinal, mesenteric attachments, abdominal, thoracic
Referred	Amputation, renal disease, cardiac disease

community, many definitions for various parameters of uncomfortable sensations in animals have evolved. Pain assessment in animals requires that one must be able to identify these sensations and relate a level or degree of discomfort elicited by them. Some of the basic notions of pain and/or discomfort and their working definitions include the following.

II. DEFINITIONS

Pain — An unpleasant sensory and emotional experience associated with actual or potential tissue damage or described in terms of such damage;[7] pain in animals is an aversive sensory experience that elicits protective motor actions, results in learned avoidance, and may modify species specific traits of behavior, including social behavior;[8] an awareness of acute or chronic discomfort occurring in varying degrees of severity resulting from injury, disease, or emotional distress as evidenced by biological or behavioral changes or both.[3]

The definitions go on and on, but the end point is that pain is an unpleasant phenomenon which elicits a variety of symptoms in varying degrees. It is obviously a departure from the normal (as we know it) condition and well-being of animals and/or humans. For proper assessment and treatment, pain must be broken down into categories by degree of intensity and duration. It may be helpful also to classify pain by anatomical/physiological type. This mode of classification groups similar types of pain together, regardless of the cause.

Acute pain — Results of a traumatic, surgical, or infectious event that is abrupt in onset and relatively short in duration. It is generally alleviated by analgesics.[3]

Chronic pain — Results from a long-standing physical disorder or emotional distress that is usually slow in onset and has a long duration. It is seldom alleviated by analgesics, but frequently responds to tranquilizers combined with environmental manipulation and behavioral conditioning.[3]

Table 1 illustrates types, characteristics, and duration of pain.[3]

Distress — Distress is described as an undesirable physical or mental state resulting from pain, anxiety, or fear.[8] Unlike pain, which usually has definite cause, distress is more aptly described as an unpleasant emotion which people (or animals in this case) would normally prefer to avoid.[9] Distress can be provoked by other non-tissue-damaging external stimuli such as denial of the fulfillment of an animal's natural instincts as in maternal deprivation, social contacts, and perhaps also appropriate to this category, food and water deprivation.

Discomfort — The *American Heritage Dictionary* describes discomfort as mental or bodily distress; something that disturbs one's comfort.[10] The *Reader's Digest Great Encyclopedic Dictionary* suggests it means to make uneasy, distress.[11] In a recent talk by Dr. Joseph Spinelli, it was suggested that discomfort is a state of being in which one feels poorly, even in the absence of physical or emotional pain. In humans such cases could be characterized by hangovers or motion sickness.[12]

Stress — Defined as a condition of tension or anxiety predictable or readily explicable from environmental causes, whether distinct from or including physical causes.[1] Stress has been referred to as a series of physiological changes such as release of hormones, which go on within the body when an animal is subjected to injury, extremes of temperatures, and so on.[13] Definitions of stress do not necessarily suggest or indicate a relation to physiological symptoms. So broadly is stress used that it has been referred to as "any aversive stimulus".[14] Stress is very often given credit for being a contributing factor in otherwise "undefinable" reactions. A good example of this occurs when hens that won't lay, or have a reduction in production.[15] Dr. G. P. Moberg very nicely summarizes stress as "a syndrome with no discrete etiology, no consistent biological response, nor even a single effect on the individual."[16]

Suffering — Suffering could well be defined as the end product of prolonged exposure to any or all of the aforementioned conditions (pain, distress, discomfort, stress). It could be the result of procedures which are more than momentary in duration, which are of more than minor intensity, and which cannot be easily alleviated. Suffering is an emotional state. It therefore requires a state of consciousness and a functional cerebral cortex.[12] Suffering can refer to a wide range of intense and unpleasant subjective states of people or other animals, such as fear and frustration.[13] It generally could be stated that suffering reflects a level of pain which is relatively constant for an extended period of time; however, one must remember also that not all suffering is a result of a "physical" stimulus. As with distress, emotional trauma, e.g., deprivation of maternal activities, also can produce a degree of suffering in animals.[17]

Most animal clinicians will have a working knowledge of the definitions previously mentioned. Before pain can be properly assessed, however, one should be familiar with the following definitions, many of which have been extracted from publications by Spinelli[18,19] and further defined by the *Dorland Medical Dictionary*:[20]

Noxious — Hurtful.
Noxious stimulus — Stimulus that threatens or produces damage to tissue, such as mechanical, thermal, or chemical energies applied to the body, or strong electrical stimulation.
Pain detection threshold — The minimal amount of stimulus intensity perceived as being painful 50% of the time. This may vary between animals.
Pain tolerance threshold — In humans, this is defined as the highest intensity of a series of stimuli that the subject will permit the experimentor to deliver. When tolerance levels of pain are approached, it causes intense emotional reactions. Human clinical literature reports that high levels of pain invoke suffering. It is probably suffering more than pain, as such, that we should prevent in laboratory animals.
Noci — A Latin term meaning to injure.
Receptor — A sensory nerve terminal that responds to stimuli of various kinds.
Nociceptor — A receptor which is stimulated by injury (noxious stimulus). A nociceptor may be unimodal or polymodal.
Nociceptor threshold — The lowest amount of noxious stimuli required to generate nerve impulses in fibers of the nociceptor. The nociceptive threshold is always below the pain detection threshold.
Nocifensor — System of nerves in the skin and mucus membranes which are concerned with local defense against injury.
Nociceptive reflex — The nociceptive reflex (nocifensive withdrawal) is a sterotyped withdrawal or movement of a part of the body which may or may not be perception linked.

III. ASSESSMENT OF PAIN AND SUFFERING

As can be seen from the definitions already stated, when veterinary clinicians, technicians, medical doctors, or scientists have to assess pain, they must be aware of more than just vocabulary words. The many factors that affect the evaluation or assessment is emphasized in

TABLE 2A
Signs of Acute Pain

Guarding (of affected area)	Protect or move away
Crying or vocalizing	On movement or palpation
Mutilation	Excessive licking, biting, scratching
Restlessness	Pacing, lying down, getting up
Sweating	Horses with colic, stressed mice, rats
Recumbency (especially notable in large animals)	Lying down for a long period
Heavy breathing	Rats, mice chatter, pneumonic

TABLE 2B
Assessment of Acute Pain

Protective motor reflexes
Vegetative reflexes
Vocalizations
Defense reactions including freezing
Aggressive reactions
Avoidance learning

TABLE 3A
Signs of Chronic Pain or Illness

Limping or carrying limb
Licking area of body
Reluctance to move
Loss of appetite
Change in personality
Dysuria (painful urination)
Bowel lassitude
Animals not up 24 h postsurgery

TABLE 3B
Assessment of Chronic Pain

Guarding behavior in movement and posture
Avoidance of pain aggravating influences
Seeking of pain-relieving factors and environments
Self-care of a painful region
Self-mutilation
Changes in sleeping behavior
Changes in feeding behavior (e.g., decrease of food intake)
Changes in explorative behavior
Physiognomic expressions
Decrease in productivity (farm animals)
Modifications of social behavior

medical literature all over the world. It is our conclusion that the assessment list starts with pain and ends with suffering. How people categorize what falls in between will depend on their scientific knowledge and experience of and familiarity with the many factors which may affect the species involved. Criteria used in pain evaluation are applied differently by different people.[3] As can be seen in Tables 2A,[8,21] 2B,[3] 3A[8,21] and 3B[3] individual interpretations of various signs and symptoms of pain responses cause an overlap in classification and/or categorization.

TABLE 4
Species-Specific Behavioral Signs of Pain

Species	Vocalizing	Posture	Locomotion	Temperament
Dog	Whimpers, howls, growls,	Cowers, crouches; recumbent	Reluctant to move; awkard, shuffles	Varies from chronic to acute; can be subdued or vicious; quiet or restless
Cat	Generally silent; may growl or hiss	Stiff, hunched in sternal recumbency; limbs tucked under body	Reluctant to move limb, carry limb	Reclusive
Primate	Screams, grunts, moans	Head forward, arms across body; huddled and crouching	Favors area in pain	Docile to aggressive
Mice, rats, hamsters	Squeaks, squeals	Doormouse posture; rounded back; head tilted; back rigid	Ataxia; running in circles	Docile or aggressive depending on severity of pain, eats neonates
Rabbits	Piercing squeal on acute pain	Hunched; faces back of cage	Inactive; drags hind legs	Apprehensive, dull, sometimes aggressive depending on severity of pain; eats neonates
Guinea pig	Urgent repetitive squeals	Hunched	Drags hind legs	Docile, quiet, terrified, agitated
Horses	Grunting, nicker	Rigid; head lowered	Reluctant to move; walk in circles "up and down" movement	Restless, depressed
Chickens	Gasping	Stand on one foot, hunched, huddled	None	Lethargic, allows handling
Cows, calves, goats	Grunting; grinding teeth	Rigid; head lowered; back humped	Limp; reluctant to move the painful area	Dull, depressed; act violent when handled
Sheep	Grunting; teeth grinding	Rigid; head down	Limp; reluctant to move the painful area	Disinterested in surroundings; dull, depressed
Pigs	From excessive squealing to no sound at all	All four feet close together under body	Unwilling to move; unable to stand	From passive to aggressive depending on severity of pain
Birds	Chirping	Huddled, hunched	From excessive movement to tonic immobility depending on severity of pain	Inactive; drooping, miserable appearance

While interpretation and explanations vary throughout the literature, the end concepts for assessment are basically the same. Table 4 is a guide to some of the most frequently recognized signs and symptoms as presented throughout the literature, observed by veterinary clinicians, medical doctors, scientists, and lay people.[3,9] Also included are the results of a survey of observations by the animal care technicians at the University of Virginia Medical Center.

Signs or symptoms of pain, distress, stress, discomfort, and suffering do not always present as species-specific behavior. Similar symptoms may be seen in a variety of animals for many other contributing reasons. Many physiological (i.e., respirations, weight, etc.) and biochemical

TABLE 5
Clinical and Physiological Signs of Pain in Laboratory Animals[a]

Species	Weight	Temperature	Heart rate	Respiration	Other
Dogs	Decreased due to inappetence, dehydration	b	Tachycardic in acute; bradycardic in chronic	Increased with panting in acute; decreased and labored in chronic	Increase in specific gravity and decrease in volume of urine; pupils dilated; eyes glazed
Cats	Decreased due to inappetence, dehydration	b	Tachycardic in acute; bradycardic in chronic	Increased, panting in acute; labored, decreased in chronic	Circumanal gland discharge; 3rd eyelid may protrude
Primates	Food and water usually refused, dehydration	b	Increased	Increased	General appearance of misery; sad facial expression; glassy eyes; lack of grooming
Mice, rats	Inappetence causes weight loss, dehydration	b	Increased	Congested	Change of normal group behavior; eats neonates; excessive licking and scratching
Rabbits	Inappetence often prolonged, dehydration	b	Increased	Increased	
Guinea pigs	Dehydration	b	Increased	Increased	Congested
Horses	Dehydration	b	Increased	Increased with flared nostrils	Interrupted feeding with food held in mouth uneaten; pupils dilated, eyes glassy; limbs held in unusual positions; reluctance to move or be handled
Chicken	Dehydration	b	Increased	Increased	Allow handling
Cows, calves	Decrease in weight due to inappetence, dehydration	b	Increased	Rapid, shallow in severe pain	Grinding teeth; lack of grooming; acts violent when handled
Sheep	Decrease in weight due to inappetence, dehydration	b	Increased	Rapid, shallow	Grinding of teeth and grunting
Pigs	Decreased; will still approach food, dehydration	b	Increased	Increased	Allow handling; hide in bedding
Birds	Decreased, dehydration	b	Increased	Increased	

[a] The physiological signs can vary greatly with intensity of pain.
[b] Temperature variation results from the cause of pain, rather than pain itself.

(urinary and blood serum parameters) indices are indicative of pain or distress. In order to properly diagnose injuries and disease, clinical veterinarians must look for the not-so-obvious signs and symptoms of pain and distress. The ability to distinguish where pain originates or to what degree it is felt is not always an easy task. Table 5 represents some general clinical and physiological signs observed by clinicians as noted throughout the literature.[2,3,6,8,9,22] While

TABLE 6
Postoperative Pain Evaluation

Type of surgery	Indications of pain response	Expected pain level	Duration
Head, ear, throat, dental	Rubbing, shaking, self-mutilation, depression, and reluctance to swallow, eat, drink, or move	Moderate to high	Generally intermittent
Ophthalmologic	Reluctance to move, scratching, rubbing	High	Intermittent to continual
Orthopedic	Abnormal posture, abnormal gait, reluctance to move, guarding, licking, self-mutilation	Moderate	Intermittent
Abdominal	Guarding, abnormal posture, vomiting, anorexia	Mild to moderate	Intermittent
Cardiovascular	Depression, reluctance to move, anxiety	Mild to moderate	Continual
Thoracic	Changes in respiratory rate and pattern, anxiety, reluctance to move	Mild to moderate	Continual
Perirectal	Scooting, biting, licking, self-mutilation	Moderate to high	Intermittent to continual

temperature variation is a pertinent physiological index, it normally results from a disease or traumatic process rather than from "pain" itself.

Postsurgical pain is probably the least difficult to interpret. One knows the general parameters expected from a procedure and can devise a pain relief regimen accordingly. If symptoms exceed the expected level of pain or discomfort for a given procedure, the clinician needs to reevaluate the surgery and surgical site and look for other complications. Table 6[8] lists a variety of surgical procedures and corresponding pain responses. Since variation in responses should be expected from species to species, as well as individuals within a species, this table may be used best as a general guideline.

Being able to categorize pain and compound lists of various signs and symptoms is only the beginning in pain assessment. Pain-like symptoms in animals also can be elicited from a wide variety of external factors.[13,23] Many of these symptoms, while seemingly abnormal, may not mean necessarily that the animal is in pain or is suffering. Abnormal responses may be provoked by drugs, fear, hypnosis, acupuncture, frustration, confinement, loss of social companions, and food, water, and sexual deprivation. While many of these factors contribute to the "stress components"of suffering it should be recognized that physiological measurements of welfare are limited by lack of knowledge of what to measure and difficulty in relating what is measured to whether the animal is suffering.[19]

IV. HUMAN/ANIMAL PERCEPTION

It is suspected that man and animals share many of the same reactions to noxious stimuli. A general consensus of the literature is that if humans interpret a stimulus as painful, then it should be assumed that the same stimulus would elicit pain in animals.[4,12,13,16,22,24] In other words, if it hurts humans, it probably will hurt animals.

It must be remembered also, that while man and animals may share the same sensations in response to various stimuli, there is a distinct difference of pain perception in quadrupedals and man. This perception is seen in postsurgical observations. Abdominal surgery, for example, is far more tolerable in quadrupedals than in man. This is most likely due to the fact that minimal use of abdominal muscles is needed for ambulation in animals, compared with the upright posture of humans. Another comparison of similar surgical procedures but dissimilar response is seen in the lateral approach to the chest cavity. This is possibly due to the diaphragmatic ventilation of animals, whereby chest expansion is reduced. Trauma to the rib cage of man is generally quite painful,[21] while in animals it is apparently less so.

While animals may be unable to communicate their sensitivity in actual words, many of their reactions can be compared to those of humans. Both species cry out, squeal, withdraw from the source of the noxious stimuli when possible, wince, and often become aggressive. For these reasons, pain in animals should be treated as subjectively as that in humans. The consideration of one's own experiences and reactions to pain is necessary in order to understand an animals reaction to pain.

V. ASSESSMENT OF PAIN IN THE LABORATORY SETTING

A fundamental approach to assessment of pain in laboratory animals does not begin with chemical or biological evaluations. The key to adequate assessment lies in the hands of the animal care personnel: technicians, laboratory specialists, and researchers. It is here that clinical observations and abnormal behavior should be recognized as possible identifying factors of pain in laboratory animals. It is therefore essential that all personnel involved in the care of animals be well versed in normal animal behavior patterns and that they recognize any deviation from the normal or usual pattern. The conscientious laboratory animal personnel performing daily routine functions should identify changes in personality, eating habits, physiological functions, etc. Such observations should be reported quickly to the clinical veterinarian or appropriate animal health care official. Good communication among all animal health care personnel is essential. Early recognition of abnormal signs, or any deviation from usual daily animal performance can mean the difference between mild, moderate, or severe pain.

Anticipating when signs of pain may occur is an important part of minimizing and preventing unintended suffering in animals.[9] This can be accomplished by a thorough knowledge of expected results of all experiments which are known or are likely to produce pain and suffering. In recent years, the field of laboratory animal medicine has established a protocol system that describes each animal experiment prior to its initiation. Clinical veterinarians should review each protocol for assessment of research which may cause pain, stress, distress, discomfort, or suffering to animals. This review also will reveal proposed drug usage which could interfere with or react with postprocedural pain medication. Review of protocols prior to performance and review of drug literature and analgesics known not to interfere with the experimental design or protocol can enhance treatment of postprocedural pain. Knowledge of the general responses of animals to a given procedure is important in the assessment and management of pain.[23]

Knowledge of an animal's disposition and normal physiological functions prior to execution of experimental protocols is extremely helpful in determining whether an animal is in pain. Aggressiveness, attempting to bite, hissing, and/or withdrawal can be interpreted as signs of pain. However, if such behavior was present prior to manipulation and is characteristic of the animal in question, then these indices are not necessarily indicative of pain or suffering. It cannot, however, be assumed that the animal is not in pain, and a thorough assessment for postprocedural pain should be performed. Comparison of pre- and postprocedural behavior may indicate that the animal is still growling, hissing. or attempting to bite, but movements or attempts to escape may be minimum to none. The importance of being aware of preprocedural traits cannot be overemphasized.

VI. FACTORS CONTRIBUTING TO PAIN RESPONSE

There are a variety of factors that contribute to an individual's response to pain.[8] Age response can be seen when comparing conditioned reactions. The younger animal may have a lower tolerance to acute pain, but because of lack of learned or conditioned control, may have more tolerance for stress resulting from emotionally charged or anxious situations. An acute or chronically ill animal has a low pain tolerance threshold. Because a debilitated or very ill animal shows less response to painful stimuli, care must be given to not underestimate the degree of pain which this animal is experiencing. The pain tolerance threshold is high in animals with stoic or apathetic temperament, while considerably lower in the high-strung, nervous type.

VII. CONCLUSIONS

Since the passage of the Cruelty to Animal Act of 1876 in England, and the Animal Welfare Act of 1966 in the U.S., there have been many definitions and attempts at assessment of pain, distress, stress, discomfort, and suffering. A universal definition for each has yet to be found. Advancement in knowledge of the peripheral and neurophysiological mechanisms has helped to establish a better understanding of these various responses in many animal species. Until recently, much of this information was not utilized by the people involved with monitoring and assessing pain in animals. As personnel in the field of laboratory animal medicine take a closer look at alleviation of pain, the material compiled for assessment of pain, discomfort, distress, stress, and suffering can be seen in situations where previously it was not seen or, perhaps, was ignored because "it did not have a human face."[13]

In order to assess specific species, the guidelines provided in this paper should be expanded as they are used. It has been shown by Morton and Griffith and in other publications that it may be possible to use a system of measurement to assess intensity or severity of pain.[6,9] Once definite parameters are established for each species with which we work, such a system may be a great help. Until such parameters are developed, however, a wide variety of factors, such as the ones presented in this chapter, must be utilized.

REFERENCES

1. **Seamer, J.,** Historical basis of British Veterinary Association policy on animal experiments, *ILAR News,* XXVII (1), 11, 1983.
2. **Sanford, J., Ewbank, R., Molony, V., Tavernor, W. D., and Uvarov, O.,** Guidelines for the recognition and assessment of pain in animals, *Vet. Rec.,* 118, 334, 1986.
3. **Wright, E. M., Jr., Marcella, K. L., and Woodson, J. F.,** Animal pain: evaluation and control, *Lab. Anim.,* May/June, 20, 1985.
4. **Kitchell, R. L. and Johnson, R. D.,** Assessment of pain in animals, in *Animal Stress,* American Physiologic Association, Bethesda, MD, 1985, 113.
5. **Lineberry, C. G.,** Laboratory animals in pain research, in *Methods in Animal Experimentation,* Vol. 6, Gay, W., Ed., Academic Press, New York, 1981, 237.
6. **Duncan, I. J. H. and Molony, V.,** Eds., Assessing pain in farm animals, Commission of the European Communities, 1986.
7. **Merskey, H., Albe-Fessard, D. G., Bonica, J. J., et al.,** Pain terms: a list with definitions and notes on usage, *Pain,* 6, 249, 1979.
8. **Zimmerman, M.,** Behavioural investigations of pain in animals, in *Agriculture: Assessing Pain in Farm Animals,* Duncan, I. J. H. and Molony, V., Eds., 1986, 16.
9. **Morton, D. B. and Griffiths, P. H. M.,** Guidelines on the recognition of pain, distress, and discomfort in experimental animals and an hypothesis for assessment, *Vet. Rec.,* 20, 431, 1985.
10. *The American Heritage Dictionary,* 2nd college ed., Berube, M. S., Ed., Houghton Mifflin, Boston, 1982.
11. *The Reader's Digest Dictionary,* Londau, S. I., Ed., Reader's Digest Association, 1966.
12. **Spinelli, J. S.,** Clinical recognition and anticipation of situations likely to produce suffering in animals, presented at AVMA Colloq. on Recognition and Alleviation of Animal Pain and Distress, Chicago, May, 1987.
13. **Dawkins, M. S.,** *Animal Suffering: The Science of Animal Welfare,* Chapman and Hall, London, 1980.
14. **Coffey, D. J.,** The concept of stress, *Br. Vet. J.,* 130, 91, 1971.

15. **Wood-Gush, D. G. M., Duncan, I. H. H., and Fraser, D.,** Social stress and welfare problems in agriculture animals, in *The Behaviour of Domestic Animals,* Hafez, E. S. E., Ed., Bailliere Tindall, London, 1975, 182.
16. **Moberg, G. P.,** Biological responses to stress: key to assessment of animal well-being, in *Animal Stress,* American Physiologic Association, Bethesda, MD, 1985, 28.
17. **Vierck, C. J.,** Extrapolations from the pain research literature to problems of adequate veterinary care, *JAVMA,* 168(6), 510, 1976.
18. **Spinelli, J. S.,** Pain in laboratory animals, presented at PRIM & R Conf. on the Standards for Research With Animals: Current Issues and Proposed Legislation, Boston Park Plaza Hotel, October 1983.
19. **Spinelli, J. S.,** Reducing pain in laboratory animals, *Lab. Anim. Sci.,* January (Spec. Issue), 65, 1987.
20. *Dorland's Illustrated Medical Dictionary,* 26th ed., Friel, J. P., Ed., W. B. Saunders, Philadelphia, 1985.
21. **Soma, L. R.,** Assessment of animal pain in experimental animals, *Lab. Anim. Sci.,* January (Spec. Issue), 71, 1987.
22. **Kitchell, R. L. and Erickson, H. H.,** Pain perception and alleviation in animals, *Fed. Proc. Fed. Am. Soc. Exp. Biol.,* 43(5), 1307, 1984.
23. **Carstens, E.,** Descending control of spinal nociceptive transmission, in *Animal Pain, Perception, and Alleviation,* Kitchell, R., Erickson, H. H., Carstens, E., and Davis, L. E., Eds., American Physiological Association, Bethesda, MD, 1983, 83.
24. **Soma, L R.,** Behavioral changes and the assessment of pain in animals, in *Proc. Second Int. Congr. Veterinary Anesthesia,* Veterinary Practice Publishing, Santa Barbara, 1985, 38.

Chapter 14

AN ANALYSIS OF SUFFERING

A. F. Fraser

EDITOR'S PROEM

We have suggested that the notion of animal pain and its control received little attention from the research community until recently, when it was thrust into prominence by escalating social-moral concern about the treatment of animals. This is *a fortiori* the case for the concept of suffering. It is patent to common sense that animals can suffer in a variety of ways, physically and psychologically, but no precise, rigorous, universally accepted account of suffering is as yet commonplace in the scientific community. Effecting such an account, given traditional neglect of the concept, will require a combination of conceptual analysis and empirical synthesis. In his paper, Professor Fraser, credentialled in both veterinary medicine and ethology, presents an account of suffering which not only combines the empirical and conceptual, but also realizes that animal research is not value free and does not avoid the ethical implications of his analysis.

TABLE OF CONTENTS

I. INTRODUCTION

The existence of suffering in animals is assumed by deduction based on private human events, viewed comparatively. It has been said that the strongest and most respected of the anthropomorphic tenets is that animals can and do suffer. Animal suffering involves the manifestly perceptible experience of a disturbing condition of the body or the senses. Suffering is a negative emotive state when invoked as the affective component of a bodily dysfunction. Clinical conditions create manifest suffering. Disturbed or affected equanimity is a definitive characteristic of suffering which is discernible in behavior. Suffering is becoming behaviorally identifiable in animals through progress in applied ethology and animal welfare science.[1] The condition, as an entity, has been traditionally ignored in textbooks which are otherwise relevant to laboratory animal medicine, but special attention has recently been focused on animal pain and distress.[2] It is common in ethological study for activities to be taken out of their context, out of the behavioral continuum, and then considered as isolated, independent entities. This is done to improve understanding not only of the part, but also of the whole. Unfortunately, suffering is holistic in nature. In the analytical approach taken here, parts are studied to arrive at global understanding. Discrete data accumulated in research do not necessarily improve understanding, for factual knowledge and understanding can be separate matters. Perhaps understanding comes more readily from long-range hindsight than from frontline study. The global condition of suffering is a pointed example; its appraisal requires reflective reviews of many cases and the summation of current opinion and knowledge.

Suffering is seen as a component in clinical conditions characterized by pain, agitation, or depression. Its detection is therefore dependent on a process of diagnosis. It may be self-evident in circumstances of continuing acute pain and protracted agitation, but the diagnostic discipline is required for its confirmation. Until now, the condition, however, has seldom been sought out as an isolated feature requiring separate appraisal, except in conditions in which welfare is in question or when euthanasia is an evident course of action. Humane death is often warranted in veterinary case management when animal suffering is confirmed and its prompt or adequate control is not possible or feasible.

II. TERMINOLOGY

The word "suffer" comes from sub — under, and ferre — to bear. It therefore indicates the carriage of a burden. To undergo an imposition or to endure a condition which is painful, distressing, or injurious is the fuller denotative meaning of the word. Endurance implies an element of continuity. To sustain a pathological condition is the medical connotation. A degree of continuing mental distress is assumed as an added burden to the pathological one. It is this comprehensive meaning which is taken as the central understanding in the context of this discourse. Connotation gives breadth and diversity to the meanings of many common terms and the word "suffer" is one which has acquired extensive connotation.

In the preceding section of this chapter, the meaning of the term, as it can be properly put to use in relation to animals, has been implicitly outlined. This represents the veterinary application of the term in its medical role. It serves as an umbrella term, covering all the likely affective features which may coexist with any clinical condition. In this there is a given understanding that suffering constantly coexists with clinical conditions which are substantive. The notion of animal suffering is therefore broad in the clinical realm, but is ubiquitously acknowledged. It is thus taken for granted, something as a matter of fact, but having high priority in the background of clinical considerations.

III. NEURAL BACKGROUND

It is a tenable belief that suffering has a neural basis. This can be better appreciated with progress in brain science. Neuroethology is now an established discipline.[3,4] Other relevant literature is provided by Satinoff and Teitelbaum[5] and by Kandel and Schwartz.[6] The precise area to which current neuroethology is devoted is that which is concerned with the physiological processes involved in the release and control of behavior. Neuroethology deals with neuron function, stimulation, neural network chemistry, neural circuits, sensory discrimination, motivation, learning, the inheritance of neural traits, release, and control of action patterns.

Sufficient progress has been made in some of these areas to allow neurophysiological knowledge to mesh with behavioral knowledge. For example, studies on the neural determinants of sleep and trophism rationalize the bulk of behavioral observations on these features. The nature of motivated actions such as ingestion, aggression, and courtship also rationalize the behavioral studies which have gone into these areas. Neurophysiological studies reveal a complex interplay between sensory information and the action of localized groups of cells and their broadcast fibers in the central nervous system. In the course of explaining the basis of behavior which is normal, neuroethology begins to reveal the probable basis of behavior which is aberrant and relates to suffering.

The limbic system is the paleomammalian brain and represents a device for providing the animal with better means of coping with the environment.[7,8] Parts of the limbic system are concerned with primal activities related to feeding and breeding sex; others are related to emotions and feelings; and still others combine messages from the external world. One way to conceptualize the limbic system may be to see it as a regulator of mood and action in survival and maintenance. Each limbic structure may be sensitive to its own effect on the whole manifestation of the system. Each of the limbic system structures is highly specialized and is tuned to specific changes of the internal or external environment. However specialized the structures of the limbic system may be, their end product is the regulation of basic, even primitive, feelings and activities. The hypothesis is offered that suffering may be generated here.

The frontal cortical areas may play a role in coordinating the signals from the limbic system that are to be integrated with the activities of the "cognitive brain", that is, the majority of the neocortical surface.[9] Portions of the neocortex may speed up the processing of information, at least some types of information. Experimental stimulation of the temporal neocortex in animals can alter the activity from disinterested to attentive. By implication, this temporal neocortex can be thought of as regulating the excitability and the processing time of the primary sensory systems. In such arrangements the neural background to states of suffering can be imagined.

IV. PAIN

Since pain is the essence of severe suffering, any conception of suffering must include an appreciation of the variety of features which relate to the experience of pain. Rollin[10] has addressed the dilemma concerning the recognition of pain in laboratory animals by the scientific community. He describes an ambivalence displayed regarding the reality and "knowability" of animal pain. He points out that scientists are loath to speak of animals experiencing pain or pleasure, or any other mental state, but must nevertheless presuppose pain and various mental states in the pursuit of their work on animals. He further points to the use of various tests of pain (such as withdrawal tests, pressure tests, and electrical stimulation tests) to screen substances for analgesic potential in humans. These methods are well recognized,[11] although it is also recognized that different species of animals may express pain in differing ways.[12]

The problem of recognizing the characteristics of pain among various animals has been acknowledged by a working party of the Association of Veterinary Teachers and Research

Workers in Britain. This group set out some guidelines for the recognition and assessment of pain in animals.[13] While these guidelines are a somewhat perfunctory document, it is clear that the recognition and assessment of pain in experimental animals can be achieved. However, a major admission of difficulty is made as follows: "The detailed information readily available on species specific signs is limited and it has been particularly difficult to obtain such signs for the smaller mammals and for the non-mammalian species." The document concludes with the observation that to describe manifestations of pain specific terms are needed: "to describe mental state, vocalization, posture and gait, and to rank these, where appropriate, in order of associated severity."

Pain is a highly variable and subjective experience, but its existence in animals is abundantly evident in typical reactions.[14] Flecknell[15] discusses the issue of pain perception in animals and describes behavioral features which are responses to pain, such as inactivity, tucked-up posture, restlessness, flailing or rigid limbs, writhing, self-inflicted bites, reduced ingestion, and abnormal vocalization. He adds that all of these criteria must be considered in conjunction with the nature of any surgical procedure undertaken or of any disease process present.

According to Melzack,[16] qualities of pain experience include sensory, affective, and evaluation properties. Melzack defines pain broadly as experiences having somatosensory and negative-affective components which elicit behavior aimed at stopping the conditions that produce them. It is stressed by Melzack that pain is characterized by noxious input (into the subject) which evokes negative-affect and aversive motivation. Pain certainly commands the attention of the subject. It is the high priority signal which normally arrests all concurrent behaviors, with the exclusion of responses to the painful stimulus. The behavior resulting from pain in animals with its characteristic vocalizations, struggles, and agitations can be easily observed, recognized, and interpreted by anyone with appropriate experience and knowledge, such as a veterinary clinician familiar with the given species in normal states.

V. HOMEOSTASIS

Much positive function and endeavour go into homeostasis, or physiological stability, and the loss of the latter is associated with negative feedback. Any force of destabilization may disturb the function of the animal and act as an initiator of suffering. Suffering is therefore a potential component of any dysfunction affecting the body, the behavior of the animal, or the group behavior in a collection of animals. In this alliance with dysfunction it exists at a sensory level. Its association with nociception is therefore implicit, although it has its own status as an affective condition. In addition to painful association, suffering is a component of more amorphous conditions such as stress or distress by pertaining to the mental state of the subject in such conditions which appear to create anguish. The nature of anguish in animals is intangible. It still has to have its case proven to the scientific jury, although substantial anecdotal evidence attests to its existence. Animals can be made to suffer in very real ways with which we can identify. If this is anthropomorphism, perhaps this is a justification for this remarkably common form of empathy.

In addition to intra- and extracellular phenomena, physiological homeostasis also takes place as a balancing arrangement at the levels of organs, or whole systems, of multisystems, and behavior. Behavioral homeostasis typically invokes much self-determination and maintenance[17] (Tables 1 and 2). Arrest of behavioral homeostasis is assumed to be linked with suffering, in the opinion of veterinary ethology. While defensive reflexes may be complex in their organization but limited in function, other self-protecting activities are of a highly complex nature and involve conation, e.g., seeking associates, selecting food, choosing sites for rest and excretion, and developing social organizations. It is precisely such behavioral work which may be diminished or eliminated in manifestations of suffering.

TABLE 1
Self-Determination in Relation to Behavioral Homeostasis in Laboratory Animals

Homeostatic factor	Tactics	Strategy	Functional outcomes	
			Short-range	Long-range
Feed	Search, selection and ingestion of food, consumption of water; self-sufficient maintenance	Self-maintenance in respect of visceral needs	Ingestion to repletion	Building up body reserves and storage
Fight	Aggression, agonistic acts mock contests	Self-assertion	Suppression of competitor	Control over resources of varied kinds
Flight	Submissive withdrawal; discriminate social movement and contact	Avoidance of threatful situations	Self-protection	Security of individual and stabilization of groups

TABLE 2
Features of Behavioral Homeostasis Relating to Maintenance

Behavioral category	Primary features	Secondary features	Role in maintenance
Body care	Grooming; comfort-seeking	Evacuation; modes of excretory behavior	Three homeostatic functions of surface hygiene, excretion, and comfort-state are basic to health
Motion	Changes of position and posture; exercise movements	Playing and stretching	Certain processes of movement require regular expression to maintain fitness in the animal
Exploration	Investigative efforts and attentiveness	Empirical activities in general	Stimulation, with quantity and variability needed for sensory satisfaction and learning
Territorialism	Using individual space for basic functions	Proprietary behavior in core of home and feeding area	Spatial needs are both quantitative and qualitative for variety of self-maintaining requirements, e.g., nutrition, shelter, defense, reproduction
Rest	Drowsing, resting, and sleeping in recumbency in diurnal phases	Idling in various postures	Physical self-conservation; physiological restoration
Association	Bonds; positive social acts, e.g., affiliation	Socialization; group affinity	Social group membership and stability of

Although the body and its nervous system have many parts, they function in unity, and behavior is the manifestation of this composite function. Seemingly the great, broad ethological purpose is the integration of the animal with the environment through homeostasis. In addition to the basic behavioral processes in each animal utilized in the laboratory is the incarcerated condition as the common problem for them all. This problem is acute for the animals so used. Their chief behavioral characteristics, either singly or as a group, appear largely homeostatic in purpose, and the variability in these, from animal to animal and from time to time, bears some relationship to "feelings" or mental states relating to the purpose.

VI. SENTIENCE

With suffering taken as an affective (mental) aspect of a clinical state, it follows that it will express itself in the illnesses and abnormalities of the sentient animals with which veterinary medicine is concerned. Rollin[18] draws attention to the essential role of the veterinarian in regard to the health and welfare of domestic and experimental animals. It could be said that veterinary medicine exists to address suffering in these sentient animals. This calls for acknowledgement of sentience as a defined property, for clearly the animals with the capacity to experience suffering possess this character.

The notion of sentience has been vague, although sentience is commonly referred to, usually as a factor considered to motivate behavior. It is widely held also to be a property of many animals, in addition to man, and is defined simply as the power of sense perception. Few would now dispute that animals can feel pain and pleasure, but some would concede only these and no other feelings to animals. Since pain and pleasure are not omnipresent in human experience, this can impart a false impression of sense perception as an occasional property. However, sentience is not a capacity to experience merely pleasure or pain; neither is it simply an ability to respond to stimuli. More than simple responsiveness to stimuli, sentience includes sensory experiences which have various values. It involves "open awareness" and is not restricted in time to only those moments when painful or pleasurable situations impose immediately on the subject's consciousness.

Sentience in its comprehensive connotative meaning is holistic reactivity which is more than reflex responses to stimuli. It is responsiveness which is more psychological than physical. Ethologically, it is coordinated, sense-motivated behavior that is in appropriate response to complex circumstantial stimulation. Sentience is intelligence in sensory terms. More subtle episodes of experience than elemental ambient factors motivate sentient animals. Behavioral heterogeneity in general is a reflection of the open awareness which results from the property of sentience. Featured in sentient animals is a capacity to possess this awareness as a sensory faculty and reflect it in behavior. The accumulation of experience in sentient animals permits them to acquire cognitive sensory ability which controls and supports increasingly variable and complex behavior. Cognitive sensory ability, whether innate or acquired, is revealed in behavior which is subjectively appraised as being environmentally interpretive.

A catalogue of sentient behaviors would feature maternal phenomena such as neonate recognition, nursing, suckling, and care-giving so prominently that sentience would appear to be only a mammalian property; but epimiletic (care-giving) behavior in non-mammalian subjects prevents such a limited allocation of the term. The concession must therefore be made that there evidently exists an open-ended gradient of sentience covering bridges of mutation between classes of animals in the upper phyla, from warm-blooded mammalia to birds, from cold-blood reptilia to fish. All such animals certainly possess limbic systems in their brains. With this property of sentience, in each presumed degree, goes the equivalent capacity to suffer — in the full connotative meaning of the latter term.

VII. DEPRESSION

Among the identifying signs of many clinical conditions in laboratory species of animals, mention is made of depression in standard veterinary literature. For example, depression is described as a major diagnostic clinical finding in such conditions as panleucopenia in cats, myxomatosis in rabbits, amyloidosis in hamsters, and Newcastle disease in poultry. By the clinical use of the term is meant a marked reduction in general activity, diminished responsiveness to exteroceptive stimuli, and an appearance of reduced awareness in a generalized behavioral atony. It is taken as a major display of suffering.

The typical behavioral picture of the depressed animal is a passive one. The bulk of the depressed animal's activity is likely to be passively derived from aversive stimulation rather than through spontaneous relationship with the environment. In depression, the suffering animal shows a depletion of the behavioral repertoire characteristic of the normal animal. The principal features of maintenance behavior such as trophic activities and restorative functions, together with collateral social behavior, become diminished, adding to the picture of suffering. Loss of maintenance priorities appear to be the essential criteria of that general aspect of animal illness referred to clinically as depression.

The established concept of adjunctive depression in animal illness recognizes the behavior of the animal as globally changed rather than regionally modified. In this, the main significant measure is behavioral frequency. It shows most often in the reduced frequency of maintenance activities involved in behavioral homeostasis. Frequency can also measure high incidences of irregular acts such as agitations or stereotypies, and apparently these can also indicate a form of suffering. Agitated states of depressed behavior in animals can thus be recognized as forms of suffering.

Depression can occur by itself in animals as a behavioral response to certain psychologically disturbing experiences. Notable among these is separation of the animal from its home base or from its immediate kin. McKinney and Bunney[19] report that a variety of separation experiments in animals have relevance to depression. They give an account of animal depression in a variety of species; cases are recorded among non-human primates, in particular, and also in dogs and in some birds. An extensive collection of case histories is given in evidence by these authors. In many of these, the condition endured for periods ranging from several days to months. In one typical case, the subject, a chimpanzee, was reported to "sit unresponsive to the environment for long periods of time and never participate in grooming or play." Other cases showed agitation, huddling, screaming, unresponsiveness, anorexia, self-mutilation, social withdrawal, or arrest of body care. Reduction in appetite, in motor activity, in exploratory behavior, and the adoption of huddled posture are presented as the most constant signs of the condition. In the majority of these cases, the "separation", as the presumed cause of the depression, related to a major bond disruption. In all cases cited depressant suffering was behaviorally self-evident.

VIII. CLINICAL SUFFERING

Physicopathological conditions commonly provide *prima facie* evidence of animal suffering on veterinary examination. Numerous clinical conditions in animals, including disease processes and traumatic incidents, first become apparent to observers through a set of behavioral indicators of suffering. These include unusual forms of conduct, unsoundness of bodily movements, reduced activity, loss of appetite, and sundry abnormal behaviors. Such signs are used for diagnostic purposes when cases of illness and injury are being appraised. They help to determine the nature and the extent of the dysfunction of the animal and the appropriate treatment. Evidence of clinical suffering is so variable and vast, however, that its detection and

TABLE 3
Common Sources and Forms of Unnecessary Suffering

	Common sources	
Forms	**Husbandry**	**Experimentation**
Direct	Careless handling	Unauthorized or unwarranted experiments
Indirect	Neglect of provisions; poor husbandry environment	Continued close confinement during inactive projects
Acute	Inflicted trauma, hunger, thirst, severe housing conditions	Gross invasiveness in procedures (e.g., Category E in Table 6)
Subacute	Mismanaged malnutrition, infestation, fatigue	Failure to provide tranquilization during stress
Chronic	Unattended persistent clinical conditions	Protracted restraint beyond initial experimental procedure

evaluation require veterinary expertise. Among features of clinical suffering are abnormal vital signs, physical and behavioral changes, pathological lesions, and altered mood.

Since laboratory animals are not subjects normally considered for return to a natural or normal life, treatments to carry them through episodes of clinical suffering for a future existence, free of noxious experience, are not always as justified as with domesticated animals. On the other hand, the experiment imposed on the animal may be, by itself, the reason for the suffering, and this might mitigate against any decision to eliminate the suffering in the laboratory setting. Acceptable suffering therefore becomes a challenge for modern bioethics to define. Veterinarians, however, are constantly bound to a code which obliges them to provide relief from suffering in animals in their charge. In the laboratory situations, this decree relates to unnecessary suffering (Table 3).

Adjunctive suffering is usually addressed indirectly when the clinical dysfunction itself is dealt with by case management. In a number of instances, however, such as in trauma, the relief of actual and acute suffering may be an immediate objective before the physical or bodily dysfunction receives specific attention. Certain behavioral manifestations are unequivocal evidence of suffering through being outward expressions of mental states in clinical conditions accompanied by pain, distress, or fear. Intense vocalizations, struggling, trembling, passively depressed behavior, and agitated behavior certainly reflect states of suffering. If physicopathological correlates are absent, evidence of suffering may relate to a subclinical or functional condition.

IX. FUNCTIONAL SUFFERING

Many syndromes of anomalous behavior are not associated with recognizable organic dysfunctions. The major etho-anomalies are abnormal behaviors in syndromes characterized by frequent recurrence and consistency of characteristics with some degree of harmful consequence. Although noxious, they essentially represent a functional type of suffering, since the affected animal may function basically normally in all other physical respects.[20] Sambraus[21] has referred to this as "immaterial" suffering. Examination of the nature of these syndromes shows that they fall into five broad categories (Table 4). Some relate to abnormal reactivity. Others are anomalous forms of reproductive behavior. In many cases, the behavior typically involves somatic movement, often of a stereotyped nature. A large class of behavioral disorders is mouth-based, taking the form of energetic mouthing activities. The term "orosthenia" has been applied to these. The remaining large category contains those in which the behavioral abnormality concerns ingestive behavior (paraphagia), with the abnormality relating mainly to manner of ingestion and the nature of the material ingested.

TABLE 4
Major Classes of Etho-Anomalies Involved with Functional Suffering in Laboratory Animals

General classification	Alternative terminology	Identified component syndromes	Related ethological systems
Reactive	Reactive anomalies; problems of temperament	Exaggerated aggression, or alarm (threatening, biting); hypotonia	Reactivity
Reproductive	Sexual/maternal/ neonatal anomalous behavior	Neonatal rejection; refusal to nurse/suck; maternal cannibalism	Coition and nursing
Somatic	Body-based anomalies; stereotypies	Rapid circling; stereotyped pacing	Kinesis
Oral	Orosthenia; oral stereotypies; mouth-based anomalies	Self-mutilation by chewing; over-grooming; vacuous chewing, barbering	Prehension
Ingestive	Paraphagia; pica; eating anomalies	Coprophagia; trichophagia; hyperphagia; polydipsia; pica; litter-eating	Prehension-ingestion

The diagnosis of "suffering" in such behaviorally affected animals entails clinical and ethological judgement. There is little doubt that much suffering is evident in many abnormal behaviors, in which the subjects are evidently free of a detectable physical disease. In applied ethology, functional suffering is seen as a condition which exists without concurrent organic pathology and which, in varying degrees, is associated with disturbed homeostasis.

Not all etho-anomalies are indicative of functional suffering. Major and minor forms must be recognized. Some minor forms are innocuous, while the major ones are noxious. Many of the latter entail a state of depression. Those syndromes which do possess implicit suffering, and are harmful or unwholesome, usually have their etiology in the circumstances of stressful husbandry. This can present any of five pathogenic features, viz., confinement, crowding, isolation, frustration, and restriction. As indicators of suffering from aversive, restrictive husbandry, noxious etho-anomalies are at once obvious and reliable. Husbandry which produces them at a biologically significant level in a common population is clearly behaviorally pathogenic through its capacity to induce states of suffering.

In a general way, suffering can also emanate from the deprivation of significant, continuing, basic needs in the animal. Today we can recognize the significant behavioral needs;[22] these relate to the behaviors of maintenance (Table 5). The behavior of association, for example, shows the major role of social activities in the total system of group behavior with such other categories as feeding, resting, exploring, moving, grooming, communicating, aggression, territorialism, dominance, and symbiotic social relationships. Failure to comprehend the rich fabric of such behavior in animals may be fostered by a Cartesian attitude, which sees the laboratory animal as a biological machine, research tool, or reagent. Even in the confined conditions of the laboratory, these animals show analogues of the classes of maintenance behavior exhibited by free-living animals.

X. EXPERIMENTAL SUFFERING

Closely related to the assumption of suffering, as a condition associated with diseases and pathological states, is the concept of its similar relationship with any experimental stressor. Even the impounding of animals formerly free (or comparatively so) can induce depression and suffering. Just as increasing levels of suffering are assumed with greater degrees of clinical

TABLE 5
Behavioral Needs with their Homeostatic Functions and Forms

Categories of behavioral needs	Principal sources of stimulation	Types of homeostatic function	Homeostatic roles
Reaction	Reception of releasers; environmental stimuli through social/physical contacts	Attentive	Generation of orienting responses to novel stimuli and their coordination with states of readiness and response
Ingestion; rest	Visceral signals; blood chemistry	Vegetative	Appetitive and restorative processes; elaboration of trophotropic responses
Motion; association; body care	External, nonvisceral, sensory stimuli, e.g., sight, sound, touch	Expressive	Expressions of neurophysiological build-up into receptive settings
Exploration; territorialism	Telereception of releasers and stimuli from environment	Effective	Effectors of complementary/ excitatory/receptive states according to sensory evaluation of environment

illness, so there is held to be more suffering with greater experimental insult. Collectively these insults, though very widely varied in their forms, are described as invasions. It is the laboratory animal's homeostatic barrier to suffering which is conceptually invaded, as much as its corporal form. Degrees of invasiveness are now assigned to experimental procedures according to the perceived levels of distress, or suffering, which they are considered to generate. There is an anthropomorphic thread in the fabric of this method, but it is one which is apparently strong enough to base a respected system upon.

Degrees of invasiveness in the use of laboratory animals have been rationalized in several documented systems. Notable among these is the plan which currently prevails in North America and which has been developed and promoted by the Scientist's Centre for Animal Welfare (SCAW). A modified version of this is given in Table 6, as it is in use in this author's unit. It will be apparent that the potential suffering in Category E is bioethically unacceptable. Ethically acceptable suffering is assumed in Categories C and D, provided that the appropriate body of approval has sat in judgement of the proposal. It is likely that limiting conditions would be included in the approval of a project falling within Category D in any institution practicing bioethics.

XI. BIOETHICS

Wherever laboratory animals are in use in order to address experimental and iatrogenic suffering, it is believed that increased pressure of opinion will oblige institutions to adopt a charter of bioethics. This is the study of the ethical implications in biomedical research. In this there will be a greater responsibility for veterinarians to represent the animal with a viewpoint relevant to the concerns of welfare philosophy and society.[23]

Increasingly it becomes evident that ethically acceptable, or tolerable, laboratory animal suffering must be recognized. It lies more in the domain of biomedicine than in pure psychology. In this regard it is notable how the higher categories of invasiveness mainly contain procedures

TABLE 6
Categories of Invasiveness: Based on a Code of the Scientist's Centre for Animal Welfare

Category	Examples and comments
Category A Experiments involving either no living materials or use of plants, bacteria, protozoa, or invertebrate animal species; NO SUFFERING	Microbiological, or invertebrate animal studies; invertebrate animals have nervous systems and respond to noxious stimuli
Category B Experiments on vertebrate animal species that are expected to produce little or no discomfort; NO SUFFERING	Holding of animals captive for experimental purposes; simple procedures such as injections of nontoxic substances and blood sampling; observing natural behavior; behavioral testing without significant restraint or noxious stimuli; experiments on completely anesthetized animals which do not regain consciousness; standard methods of euthanasia that induce rapid unconsciousness, such as anesthetic overdose
Category C Experiments that involve some discomfort (short-lasting pain) to vertebrate animal species; NO LIKELY SUFFERING	Exposure of blood vessels and implantation of chronic catheter performed under local anesthesia; gonadectomy by standard methods; behavioral experiences on awake animals that involve restraint, with or without food/water deprivation for short periods (a few hours); noxious stimuli from which escape is possible; social isolation or crowding; maternal deprivation; restricting normal instinctive behavior or functioning; relatively minor surgical procedures under anesthesia with subsequent recovery of consciousness and some minimal postoperative discomfort; Category C procedures incur additional concern in proportion to the degree and duration, of unavoidable suffering
Category D Experiments that involve significant but unavoidable distress or discomfort to vertebrate animal species; SUFFERING WHICH MAY BE UNNECESSARY	Major surgical procedures conducted under anesthesia and with postsurgical recovery of consciousness and discomfort/pain or functional deficit; application of noxious stimuli from which escape is impossible; prolonged periods (several hours or more) of physical restraint; induction of behavioral stress; maternal deprivation with substitution of punitive surrogates; induction of aggressive behavior leading to self-mutilation or intraspecies aggression; procedures that alter the perceptual or motor functioning of an animal to test consequent behaviors; Category D experiments present an explicit responsibility on the investigator to explore alternative designs to ensure that animal suffering is minimized
Category E Experiments that involve inflicting severe pain on unanesthetized, conscious vertebrate animals; UNETHICAL SUFFERING	Procedures involving severe pain or deprivation; burn or severe trauma infliction on *unanesthetized* animals; burn or severe trauma infliction on *anesthetized* animals allowed to recover consciousness; attempts to induce psychotic-like behavior, such as approach-avoidance conflict, inescapable stress, or terminal stress; use of painful shock to modify or stimulate behavior; use of muscle relaxants or paralytic drugs for surgical restraint without use of anesthesia in sufficient dosage to produce loss of consciousness; Category E experiments are considered to invoke suffering and to be unethical or unacceptable

employed in psychological research and in surgery. Future justification for all such work will surely be subject to bioethical judgement and evaluation of the applied content of the proposed laboratory animal usage. Indeed, full respect for suffering among these animals will force their approved use to be chiefly in the nature of applied research in which a clear biomedical problem is being directly addressed. This should lead to a substantial restriction on experiments relating to such psychological studies as prey killing and learning studies which use any toxic substance to a lethal level. The creation of states of severe mental disorder, such as learned helplessness, would not meet the requirements of bioethically acceptable suffering in a more ethical era than has previously persisted.

The level of implicit suffering in experiments of the above types may not be viewed as acceptable if the applied content in the proposal is inadequate to elevate its status. Bioethically tolerable suffering, in all instances, warrants an umbrella of intensive animal care in which the individual provision of analgesia, tranquilization, or euthanasia would be a prompt option. Individuality must be taken into account so as to cover the behavioral heterogeneity which characterizes variable disposition in animals. The latter was not a feature in earlier behaviorist research, but has become a common finding in the contemporary work of applied animal ethology. Heterogeneity modifies manifest suffering among animals in the presence of a common pathogenic (or invasive) circumstance. Veterinary clinicians are all aware of this fact, and this alone warrants veterinary involvement in the provision of intensive animal care in order to control unnecessary or ethically unacceptable suffering in laboratory animals.

Suffering deemed acceptable does not extend to unnecessary illness, and animal care should be organized as the practice of preventive medicine applied to laboratory animals. Full arrangement for this should be in place as a foundation for a bioethical charter in any institution using laboratory animals, particularly if experimentation beyond Category B is to be performed. Programs of preventive medicine represent another feature of intensive animal care. Laboratory animals should be kept entirely free of their natural diseases. These are extensive in range and carry the inevitable burden of clinical suffering to the subject. Bioethics dictate that animals should be secure against such intercurrent suffering in the laboratory environment which usually provides a constant background of some confinement stress in addition to the scheduled invasiveness. Such incremental stressors are considered capable of contributing to experimental suffering. Control of the latter can be seen as the purpose in a bioethical system.

It is widely understood that cruelty, with reference to laboratory animals, is infliction of unnecessary pain or suffering. Unnatural suffering might also be included, to contrast with much natural suffering in animals which lies outside man's province. Animals are seen to merit freedom from suffering, particularly under certain conditions of intensive husbandry, including the laboratory. To a large extent, this can be addressed by recognizing basic freedoms, in addition to behavioral needs. Webster[24] has listed several basic freedoms that should be accorded the intensively maintained animal, as follows: freedom from hunger, freedom from thermal or physical discomfort, freedom from pain, injury or disease, freedom from fear and distress, freedom to indulge in most normal socially acceptable patterns of behavior.

XII. PUBLIC CONCERN

When criminal cruelty occurs, it implies imposed suffering and stems, by definition, from the actions of mankind. Society requires this to be policed and suppressed. Contemporary society has a relatively low tolerance of manifest cruelty, in any form, perhaps for its own protection. Laboratory animals are characteristically perceived to be particularly defenseless against imposed suffering. Actions against animal abuse have usually been pressed according to the administrative resourcefulness of public societies for the prevention of cruelty to animals. Today there is an international legal architecture to supply legal statutes and standards of propriety in laboratory animal care.

Increasing forces for animal welfare are reacting to such utilization of animals as they perceive as unjust exploitation. Progressive attitudes among users of laboratory animals must develop in accordance with public concern. This progress can lead to a great improvement in appreciation of the importance of dynamic, behavioral mediation of the animal with the environment. In the matter of laboratory animal well-being, the prevention of suffering is paramount. On such prevention may rest, in the future, the public approval of laboratory animal use.

REFERENCES

1. **Fraser, A. F.,** The behaviour of suffering in animals, *Appl. Anim. Behav. Sci.,* 13, 1, 1984.
2. AVMA Panel, Report on the Colloquium on Recognition and Alleviation of Animal Pain and Distress, *JAVMA,* 191(10), 1186, 1987.
3. **Ewart, J. P.,** *Neuroethology: An Introduction to the Neurophysiological Fundamentals of Behaviour,* Springer-Verlag, Berlin, 1980.
4. **Guthrie, D. M.,** *Neuroethology: An Introduction,* John Wiley & Sons, New York, 1980.
5. **Satinoff, E. and Teitelbaum, P.,** *Handbook of Behavioral Neurobiology,* Vol. 6, Plenum Press, New York, 1983.
6. **Kandel, E. R. and Schwartz, J. H.,** *Principles of Neural Science,* Edward Arnold, London, 1981.
7. **Heimer, L.,** The olfactory cortex and the ventral striatum, in *Limbic Mechanisms: The Continuing Evolution of the Limbic System Concept,* Livingston, K. E. and Hornykiewicz, O., Eds., Plenum Press, New York, 1978, 95.
8. **Isaacson, R. L.,** *The Limbic System,* 2nd ed., Plenum Press, New York, 1982.
9. **Macphail, E. M.,** *Brain and Intelligence in Vertebrates,* Clarendon Press, Oxford, 1982.
10. **Rollin, B. E.,** Animal Pain, in *Advances in Animal Welfare Science, 1985,* Fox, M. W. and Mickley, L. D., Eds., Martinus Nijhoff, Boston, 1986, 91.
11. **Lineberry, C. W.,** Laboratory animals in pain research, in *Methods of Animal Experimentation,* Gay, W., Ed., Vol. 6, Academic Press, New York, 1981.
12. **Davis, L.,** Species differences in drug disposition as factors in alleviation of pain, in *Animal Pain: Perception and Alleviation,* Mitchell, R. and Erickson, H., Eds., American Physiological Society, Bethesda, MD, 1983.
13. **Sanford, J., Ewbank, R., Molony, V., and Tavernor, W. D.,** Guidelines for the recognition and assessment of pain in animals, *Vet. Rec.,* 118, 334, 1986.
14. **Iggo, A.,** Somesthetic sensory mechanisms, in *Duke's Physiology of Domestic Animals,* Swenson, M. J., Ed., Comstock, Ithaca, NY, 1984, 640.
15. **Flecknell, P. A.,** Recognition and alleviation of pain in animals, in *Advances in Animal Welfare Science, 1985,* Fox, M. W. and Mickley, L. D., Eds., Martinus Nijhoff, Boston, 1986, 61.
16. **Melzack, P.,** *The Puzzle of Pain,* Basic Books, New York, 1973, 41.
17. **Fraser, A. F.,** Processes of ethological homeostasis, *Appl. Anim. Ethol.,* 11, 101, 1983.
18. **Rollin, B. E.,** *Animal Rights and Human Morality,* Prometheus Books, Buffalo, 1981.
19. **McKinney, W. T. and Bunney, W. E.,** Animal model of depression, in *Origins of Madness: Psychopathology of Animal Life,* Keehn, J. D., Ed., Pergamon Press, Oxford, 1979, 79.
20. **Fraser, A. F.,** Background to anomalous behaviour, *Appl. Anim. Behav. Sci.,* 13, 199, 1985.
21. **Sambraus, H. H.,** Abnormal behavior as an indication of immaterial suffering, *Int. J. Study Anim. Prob.,* 245, 1981.
22. **Fraser, A. F.,** Behavioural needs in relation to livestock maintenance, in Proceedings of a Workshop on Behavioural Needs, *Appl. Anim. Behav. Sci.,* 19, 339, 1988.
23. **Soulsby, E. J. L.,** Animals in Society: A Veterinary Viewpoint, Hume Memorial Lecture, published by Universities Federation for Animal Welfare, London, 1985.
24. **Webster, A. J. F.,** *Calf Husbandry, Health and Welfare,* Collins, London, 1984, 15.

Chapter 15

PREVENTING SUFFERING IN LABORATORY ANIMALS

Joseph S. Spinelli

EDITOR'S PROEM

The moral obligation to recognize and control suffering detailed in Dr. Fraser's paper is echoed in this contribution by Dr. Spinelli, a pioneer in developing both moral concern for pain and suffering and mechanisms for minimizing them in research contexts. In his paper, Spinelli outlines criteria for identifying physical and emotional pain, suffering, and distress. In addition, he explores a variety of methods for diminishing and controlling these noxious modalities of experience. Especially interesting to researchers is Spinelli's emphasis on nonpharmacological ways of preventing suffering, particularly his discussion of environmental enrichment and acculturation of animals to humans. Like Dr. Fraser, Dr. Spinelli urges a new, morally based emphasis on the control of suffering as a keystone for scientific work with laboratory animals.

TABLE OF CONTENTS

I. INTRODUCTION

This chapter is based on the following premises:

- There is no contradiction between providing high-quality animal care and conducting good science. If there were, most of us would probably leave this profession, as we are typically committed to both.
- If "laboratory animal science" is a science, then we have a moral and scientific responsibility to investigate and then implement better ways of caring for and using laboratory animals.
- To admit that we in the field need to improve does not imply that what we have been doing is evil; it merely reaffirms the obvious point that humans can usually improve upon whatever they do.

Among the many ethical issues current in both human and veterinary medicine is a debate as to whether it is proper to use animals in biomedical research. Surveys in the U.S. show there is wide public support for using animals in research. However, that support could deteriorate if people became convinced that laboratory animals suffer. There is no constitutional right to use animals. Local jurisdictions, state legislative bodies, and/or the U.S. Congress have the power to pass laws which restrict or even eliminate the use of animals in research.

If the antivivisection movement has a rallying cry, it probably is "laboratory animals never have a nice day." Both from a moral perspective and in order to continue to have public support, we should see that laboratory animals have many nice days. With current technology it is probably impossible to guarantee that laboratory animals will not endure some level of pain or discomfort — conditions also experienced by both domestic and wild animals. However, we should assure that we minimize pain and discomfort and that laboratory animals never have to suffer. If we believe that laboratory animals shouldn't suffer, then we have to be very clear about the definition of that term.

A. SUFFERING

Dorland's Medical Dictionary[1] neither defines "suffer" nor "suffering". Another dictionary[2] defines "suffering" as "To feel or endure pain or distress." The problem with equating all levels of pain with suffering is that it leaves us without a word to differentiate between pain that one can — and even would be willing to — tolerate, and pain that one cannot tolerate.

In order to differentiate between pain, discomfort, and suffering, we define suffering as "a severe emotional state, extremely unpleasant, which results from one or more of the following: physical pain, mental pain, and/or discomfort at a level not tolerated by the individual, and which results in some measurable level of physiological distress."

If you accept the above definition of suffering, then any circumstance that causes intolerable pain or discomfort in animals can cause an animal to suffer. The list of such items could go on forever. It would include a failure to properly use anesthesia during surgery, not using postsurgical analgesics when indicated, not getting proper medical attention for a severely ill animal, or proceeding with toxicity testing to a point that would cause an animal to experience intolerable sickness and discomfort.

In order to effectively prevent suffering in laboratory animals, we must first believe that they can suffer. Once we make that commitment, the general ways that we can prevent suffering in animals are by reducing the stimuli that may result in pain or discomfort and encouraging those conditions which allow the animals to better tolerate the pain or discomfort that they must endure.

For the remainder of this chapter we will define and characterize those conditions which are likely to result in suffering, discuss means of diagnosing the presence of suffering, and describe some ways of reducing or eliminating the causes of suffering.

II. CHARACTERISTICS OF PHYSICAL PAIN

Physical pain has been defined as a perception invoked by stimuli that injure or threaten to injure tissue (noxious stimuli), exciting specialized nerves, and which each person introspectively designates as that which hurts.[3]

Because pain is a perception, we can never know how pain is perceived by other individuals, much less by other species. Throughout the history of thought, some Western philosophers and psychologists have believed that animals were incapable of perceiving pain. There is no evidence to support this and, indeed, there is significant evidence to the contrary. Vertebrate animals have central nervous systems, peripheral nerves, and deep and superficial nociceptors similar to humans. Also, the reaction of animals to noxious stimuli is typically similar to that of people.

A. TOLERANCE OF PAIN

To prevent suffering, the pain that we should avoid causing in laboratory animals is that above the pain tolerance threshold. The pain tolerance threshold is defined, from an experimental point of view, as the highest intensity of a noxious stimulus that a subject will permit an experimenter to deliver. This threshold — perhaps better called the pain intolerance threshold — is an expression of unwillingness to receive more intense stimulation. The pain tolerance threshold is extremely variable between individuals and even within an individual at different times. Factors such as anxiety tend to reduce the pain tolerance threshold. By use of drugs or by use of training and rewards, we can increase the pain tolerance threshold.[4]

Clinically it is far more difficult to determine when the pain tolerance threshold has been reached than it is when one is performing pain research and has specific parameters enabling one to objectively make that determination. However, if a procedure has been performed that is likely to cause pain and if the animal shows signs of pain then, in our view, one should assume the pain tolerance threshold has been reached and steps should be taken to alleviate the pain.

B. DIAGNOSIS OF PHYSICAL PAIN

The first thing one should do in evaluating an animal for pain is to know its history. One should consider whether the procedure being performed on the animal has a high probability of producing intolerable levels of pain. Then one should determine whether the animal shows signs of pain.

1. Examples of Painful Procedures

Soma[5] has listed surgical procedures in dogs and cats that are most likely and least likely to be painful. Surgery of the ears, eyes and orbit as well as orthopedic procedures of the femur or humerus or invasion of large muscle masses may be quite painful.

While all thoracotomies are likely to be painful, the intercostal approach in dogs and cats is not as painful as the sternal approach. While this may not be the case in humans, dogs and cats will resume normal activity more quickly if the intercostal approach is used. If possible, this is the approach that should be used.

Abdominal procedures do not appear to be as painful in the immediate postsurgical period in dogs and cats as in people. However, if the procedure is extremely traumatic or if peritonitis develops, the animal will be reluctant to move about and when forced to do so will tuck in its abdomen and walk with an arched back. Thus, the animal showing severe abdominal pain following a laparotomy needs careful evaluation.

2. Signs of Pain

One must rely on signs rather than symptoms in performing diagnostic work on animals. The behavior of one individual may be quite different from the normal behavior of another individual. If an animal is acting in a way that is considered to be normal for the species, but that behavior is a radical change from its own normal behavior, then pain may be the cause. Therefore, it is advantageous to know as much as possible about the behavior of an individual animal prior to the time that it is exposed to a potentially noxious stimulus. If there is a radical change in animal behavior following a procedure that is likely to cause pain, then it is likely that pain is the cause. So one of the most important aspects of evaluating an animal for pain is to observe the animal before and after it is exposed to a noxious stimulus and determine if there is a difference in the animal. Any difference consistent with pain may lead to a presumptive diagnosis of pain.

First one should observe the animal's behavior and ask the following questions: Is is alert? Does it move about normally? For example, if a dog that normally comes out of its cage and jumps up on a person and barks, instead slowly leaves its cage and walks over to an individual but does not jump up, it might be in pain. Is it eating and drinking? Animals with pain may lose their appetite. Next, one should perform a physical examination. As an attendant gently restrains the animal, one should palpate those areas that are likely to be painful, and note responses indicating sensitivity or unusual response.

Soma[6] has listed a variety of responses to pain in animals. One is guarding, or an attempt to protect, move away, or bite at a painful area, particularly when one is palpating the area. If an animal limps when it moves, one can assume that the limping is due to pain if one has done a procedure that is likely to cause pain in one or more of the legs. Crying out can be a sign of pain, especially when a painful area is palpated or when an animal is forced to move. Animals frequently will self-mutilate painful areas. This may include licking the area, biting, or scratching. A normally calm animal may become restless, as evidenced by an abnormal amount of pacing, then laying down, getting up, and laying down again in an apparent attempt to become comfortable. Other animals may become recumbent for an unusually long length of time. The animal may also assume unusual body postures such as tilting its head, headshaking, or pawing at the ears or eyes. An animal with neck pain may show reluctance to move and may hold its head lower than normal. If the pain is severe, animals may show typical signs of shock, including an increased pulse rate and pale mucous membranes.

A presumptive diagnosis of pain can especially be made when the above signs are alleviated by the use of an appropriate analgesic drug or by the use of a local anesthetic.

C. KEEPING PAIN BELOW THE PAIN TOLERANCE THRESHOLD

In order to keep pain at a level that the animal can endure, we need to either reduce the level of the noxious stimulus or increase the animal's ability to tolerate pain.

1. Pain Control without Drugs

By familiarizing animals with the project environment prior to initiating work with the animal, one can reduce anxiety in the animal. This will increase the degree of pain the animal will tolerate. One should also train animals using positive, rather than negative, reinforcers.

By having highly competent surgical teams, one can minimize pain associated with surgical procedures. Those who administer the anesthetic, do the surgery, and provide the postsurgical care must all be highly competent.

An animal is likely to experience less pain following surgery if the person who performs the surgery is highly skilled. A skilled surgeon will induce less trauma and, by using proper aseptic technique, will reduce the chance for infection. Each institution should develop strict standards regulating who is allowed to perform survival animal surgeries.

The level of pain following surgery is also likely to be reduced if the animal is provided with a high level of postsurgical care. As with the human patient, it is important to provide supportive therapy to counteract any pathophysiology brought on by the surgery, prevent hypothermia, and design a regimen for the treatment of pain.

2. Analgesics

If a procedure has a high probability of resulting in pain, postsurgical analgesic drugs should be administered to the animal. Occasionally even veterinarians are reluctant to administer such drugs unless it can be proven that the animal is in fact experiencing pain. While this might be wise relative to the use of drugs for other conditions, for pain it is probably better to treat an animal that doesn't need treatment than to withhold treatment from one that does. Dr. Lloyd E. Davis, a veterinary pharmacologist, has for a long time warned veterinary practitioners against the overuse of drugs in managing diseases in animals. However, regarding the clinical management of pain, he has stated the following: "One of the psychological curiosities of therapeutic decision-making is the withholding of analgesics drugs because the clinician is not absolutely certain that the animal is experiencing pain. Yet the same individual would administer antibiotics without documenting the presence of a bacterial infection. Pain and suffering constitute the only situation in which we believe that, if in doubt, one should go ahead and treat."[7]

The appropriate analgesic drugs will depend upon the species, the procedure that caused the pain, and the degree of pain. Some experiments will also preclude the use of certain analgesics. The design of the analgesic regimen is a therapeutic decision that has to be made on the basis of many circumstances. It is probably best developed by joint decision of the investigator and the clinical veterinarian. A treatise on specific analgesic regimens is beyond the scope of this chapter (see Chapter 17).

There are several things to be kept in mind by those designing anesthetic and analgesic regimens for laboratory animals. As in humans, the average dose and reaction to a drug(s) may vary in a given individual from what is normally expected. Differences in dose, frequency of administration, and even the physiological effects frequently vary greatly between species. If one knows the dose and frequency of administration in one species, one should not assume that that will be appropriate for another species. Also, the experimental procedure one performs on a given animal may have major effects on the animals' reaction to an anesthetic or analgesic drug. This is especially true if anything is done to the liver, kidneys, brain, or to nerves that supply the respiratory or cardiovascular system.

Under the provisions of the Animal Welfare Act, each institution is now required to provide information regarding appropriate use of analgesic drugs in laboratory animals.

Some veterinarians use tranquilizers, particularly phenothiazine derivatives alone or with other drugs to relieve pain. While the promazine tranquilizers probably do not relieve pain as such, they appear to increase an animal's ability to tolerate pain. Thus, in addition to an appropriate use as a preanesthetic drug, such tranquilizers may also have some benefits as an animal is recovering from traumatic procedures as well.

The use of analgesics, tranquilizers, and anesthetics must be carefully designed and monitored as these drugs can cause some life-threatening physiological reactions. Cardiac and respiratory functions should be closely evaluated.

3. Euthanasia

An opportunity is available in the practice of veterinary medicine that is not available in the practice of human medicine. When pain is excessive and cannot be controlled by routine measures, euthanasia is an appropriate way to put an end to the animal's pain. It is sometimes the only practical measure, particularly in animals experiencing intolerable chronic pain.

III. EMOTIONAL PAIN

If some would argue that animals do not experience physical pain, then they would also probably state we are on very shaky ground to discuss emotional pain in animals. While there is general agreement that animals exposed to certain unpleasant environments may develop abnormal behavior, some would state that it is anthropomorphic to attribute mental states to animals. However, clinically one sees behavior in animals that is suggestive of behavior in people that are experiencing anxiety, depression, or frustration.

Our definition of emotional pain in animals is an unpleasant emotional reaction to external or internal stimuli which results in a state like anxiety or frustration. Emotional pain can be caused by various environmental conditions to which the animal is exposed.

A. CAUSES OF EMOTIONAL PAIN

We believe that some of the ways we cage laboratory animals may cause them distress. If that is true, then it is our responsibility to devise new methods of housing laboratory animals so that it will not distress them.

There are many differences between living in cages or pens and living in nature. Some of these differences may benefit the animal and some may be harmful. Among the benefits are

- The absence of predators (except humans).
- The animals' health needs are attended to.
- Animals receive a quality diet.
- Animals do not experience extremes of temperature.

Some aspects that may be detrimental for the animals are

- Noise levels.[8]
- Light levels, especially for albino rats and mice.[9]
- Consistent temperatures.[10]
- Lack of variety of diet.
- Caging, its attendant lack of activity, and its social isolation, which can lead to pathological behavior and physiology. For example, nearly 30 years ago Cross and Harlow[11] reported on some of the prolonged and progressive effects of partial isolation on the behavior of macaque monkeys.

B. SIGNS OF EMOTIONAL PAIN

Erwin and Deni summarized abnormal behaviors commonly found in non-human primates kept in caged environments.[12] They distinguished between two types of abnormal behaviors, qualitative abnormalities and quantitative abnormalities. Qualitative abnormalities are those that occur in captivity but not in natural settings, while quantitative abnormalities are those that occur more often or less often in captivity than they do in natural settings. Among the qualitatively abnormal behaviors, Erwin and Deni listed bizarre postures (e.g., floating limbs, self-biting, self-clasping, and self-grasping), saluting (eye poking), and stereotyped motor acts (e.g., pacing, head tossing or weaving, bouncing in place, somersaulting, and rocking). Examples of some quantitatively abnormal behavior patterns are apathy and depression as well as appetite disorders such as hyperphagia (uncontrollable eating), hypophagia (insufficient eating), and polydipsia (frequent drinking).

C. PREVENTION AND TREATMENT OF EMOTIONAL PAIN

How can these problems be addressed? Research programs should be funded in order to

- Determine which noise levels harm animals and then develop strategies for reducing noise levels. (For example, one could use materials in laboratory animal facilities that would allow for good sanitation, but which would also reduce noise levels as animals are serviced.)
- Determine whether it is physiologically and/or psychologically sound to house various species on grid-type floors.
- Determine the ideal light levels for laboratory animals.
- Determine the need for exercise for animals' health.
- Determine how providing the animal with opportunities for voluntary activity affects the animals' physical and psychological health.

IV. DISCOMFORT

We define discomfort as a state of being in which one feels poorly, even in the absence of physical or emotional pain. Perhaps the best example in humans is a hangover. Another example in both humans and animals is motion sickness. Induction of many disease states in animals as well as some toxicity testing may not result in physical nor emotional pain. However, some procedures may cause discomfort.

A. SIGNS OF DISCOMFORT

Commonly, animals that experience severe discomfort will become depressed, eat less, and groom less than normal. Again, knowing how an animal behaved prior to the induction of a diseased state that causes discomfort will give one good clues as to whether the individual animal is acting normally or abnormally. In addition, other signs typically associated with illness may be present. However, because normal ranges are different in animals than in people, one needs familiarity with normal physiological parameters in the species with which one is dealing. Animals may develop fever or hypothermia, develop abnormal blood profiles of the circulating white and red cells, have abnormal serum chemistries, become dehydrated, lose weight, vomit, or develop diarrhea or constipation.

B. CONTROL OF DISCOMFORT

Generally, to control discomfort, one should provide treatment that will alleviate the signs or cure the illness. Such a regimen should be designed in concert with the facility veterinarian. However, if the project requires that the disease progress, then one should perform euthanasia before the animal suffers. Admittedly, it is difficult to know when this occurs, but many would draw the line at the point where the animal becomes uninterested in its surroundings, stops eating, stops drinking, and appears either unwilling to move for prolonged periods or becomes extremely restless. One should design experiments where such a state, rather than death, becomes the end point.

V. ENVIRONMENTAL CONSIDERATION

While the proper use of anesthetics, analgesics, and veterinary care will do much to prevent suffering in laboratory animals, in the past there has been little emphasis on manipulating the environment to prevent suffering. We believe this to be one of the great challenges in laboratory animal science. We believe that manipulation of the environment, if done properly, can reduce the probability that laboratory animals will suffer. However, to accomplish this we will need to change much about the way we have traditionally organized and operated our centralized animal resources.

A. STAFFING

If we are to successfully improve the environment of laboratory animals, a multidisciplinary team consisting of veterinarians, animal behaviorists, engineers, anthropologists, physiologists, and others will we needed. We believe that in order to address the problem of reducing environmentally imposed stress in our own laboratory animal facilities, such individuals should be on the staff of, or consultants to, centralized animal resources. Since this will be expensive, one needs to ask how much of the animal behaviorist's time is going to be spent diagnosing, treating, and preventing problems, and how much of their time is going to be spent on research and development. The effectiveness of many of the "treatments" has not been proven. So if we ask the behavior specialist to spend most of the time on treatments there may be little or no benefit for the animals. On the other hand, if we ask the animal behaviorists to spend most of the time on developing and verifying effective treatment, the long-term benefits for the animal may be substantial. However, the short-term benefits for the animals in the institution may be small. The job of justifying the expense will be enormous, especially if that expense is being charged to animal users' research grants. We believe that determining the type of staffing needed to work out such modification of laboratory animal environment and how this work will be funded is one of the major issues facing our field today.

B. ENVIRONMENTAL ENRICHMENT

We hypothesize that by improving one aspect of the environment, one may be able to compromise on other aspects of the environment without harming the animal. For example, if the animal is provided with activities, it may be able to tolerate less cage space, less exercise, and less intraspecies interaction than if it did not have an opportunity for activities. If this is true, we could keep animals reasonably content in an enriched urban environment.

What do we mean by environmental enrichment? There is unfortunately no uniform definition of this popular term. Here are some examples of definitions.

One definition could be, "changing the environment of captive animals so that the frequency and duration of abnormal behavior is reduced or eliminated." This definition focuses on a change in the abnormal behavior of the animals. Did they have abnormal behavior before one enriched the environment? Was this measured? Was it reduced by the enrichment? Another definition could be, "providing captive animals with some kind of device that they are willing to manipulate during periods of time when they would have just been sitting and doing nothing." Here one is focusing on the activity of the animal. Does the animal that previously sat in its cage now have some activities that it participates in?

A third definition could be, "placing something in the cage for the animal to sit on or changing the animal's environment in any way." Does buying the animal a new cage constitute environmental enrichment? Does painting the walls of the animal rooms constitute environmental enrichment? The focus here is on a change. The hypothesis is, "I have changed the environment; therefore it is enriched."

We believe that environmental enrichment consists of all of the above factors. To qualify as environmental enrichment

1. There must a change in the animal's environment — something has been done to the environment to make it different than the way it was before it was enriched.
2. The animal should respond to that change in a positive voluntary way. It should become more active than it was prior to the enrichment. So from our perspective, there is some activity on the part of the animal associated with an enriched environment.
3. Some measure should be made of the animal's behavior to validate that there is indeed a positive change in the animals behavior.

So the definition that we would propose is that environmental enrichment is "changing the animal's environment in such a way that the animal voluntarily becomes more active than it was, and there is a measure of a reduction or elimination of abnormal behavior."

Among the things done to enrich the environment of non-human primates are the following: providing animals feed like raisins, nuts, and seeds in deep bedding; providing animals with puzzle feeders; and providing the animals with electronic devices that they can manipulate and respond to.

There is in fact evidence that animals will participate in such voluntary activity:

- Animals in zoos will participate in such activities when rewarded with food, even when the same food is supplied by free choice.[13]
- It has been reported that when a limited number of pigs housed in bare pens were provided with a trough with dirt in it those pigs were more active than those with a bare environment.[14]
- A group reported that when Asian small-clawed river otters were given the opportunity to hunt and capture live crickets, even though the crickets were an insignificant part of their diet, the otters were highly motivated to hunt them. The otters would even work to "capture" gelatin capsules. The authors conclude that "any active opportunity to produce a change in their environment was rewarding."[15]

There are also results which show that enriched environments benefit animals:

- It has been reported that young rats raised in an enriched environment with toys have larger forebrains than rats raised in isolation.[16] Also, rats raised in isolation who are later maze-trained develop larger brains than animals raised and kept in isolation.[17]
- Investigators showed changes in the brain RNA/DNA ratio in rats raised in enriched vs. impoverished environments.[18]
- It has been shown that learning in mice occurs more quickly in enriched groups than in impoverished groups.[19]
- Other studies indicate that enriched groups of rodents have more developed brain structure than impoverished groups.[20]
- Hyperaggressiveness was reduced in monkeys in a zoo environment when they were provided with activities.[21]

C. SOCIALIZATION TO HUMANS

Early socialization of animals with humans can reduce fear in animals. By reducing fear in animals we reduce the chance that they will suffer.

What kind of human contact do many laboratory animals have? Our animal technicians come into the room and in the process of caring for the animals make a lot of noise as they manipulate stainless steel cages. In addition, they may hose down the animal room. Even though they get fed at the same time, animals may associate the presence of animal technicians with unpleasant experiences. Also, investigators may come in and take the animals away and do surgery on them. Then the animal wakes up and it is in pain.

To make animals comfortable around humans, we must provide them with human contact that is nonthreatening. Examples of nonthreatening human contact are humans gently handling rodents before the animals are used in research protocols, sitting quietly in rooms in which non-human primates are housed, or quietly presenting some especially palatable feed to animals.

One disadvantage of providing nonthreatening human contact is that it is expensive. Humans want to be paid. However, we still believe that it is an important source of environmental enrichment. Perhaps we need to consider using volunteers for this activity. Some animal facilities are currently utilizing such volunteers to exercise and socialize dogs.

D. SOCIAL ORDERS

We believe we need to explore giving the animals an opportunity to develop some social orders. Perhaps this is difficult in a medical school animal facility. However, preliminary information from Victor Reinhart at the Wisconsin Primate Center indicates this is possible with rhesus monkeys.[22]

While developing social orders for animals may reduce frustration in the long term, one must be extremely careful how this is done. Putting monkeys together that have not had a chance to develop a normal social order may result in fights, severe physical injury, and even death for some animals. However, in other species such as cats, housing animals in a large room and giving them a chance to interact socially may be relatively easy.

E. ASSESSMENT OF HEALTH AND WELL-BEING

There are various ways of looking at the roles of the health science professionals. One is to consider that our primary role is to minister to the needs of the ill. In fact, we are often referred to as healers. Of course, another view is that we should focus on the prevention of disease — the public health perspective. Perhaps a step beyond the treatment and prevention of disease is the promotion of health and well being.

According to Fox,[23] the World Health Organization defines health as follows: "Health is a state of complete physical, mental and social well-being and not merely the absence of disease or infirmity." To paraphrase this, we would define well-being as "a state of complete physical, mental and social health and not merely the absence of suffering." What we are suggesting is that while well-being involves more than a lack of suffering, certainly for a given animal there is no state of well-being in the presence of suffering.

So far we have discussed animal behavior as a reflection of the absence of well-being due to suffering. Now we would like to attempt to discuss the use of environmental enrichment devices as a means of evaluating the "wellness" of an animal. Therefore, on the one hand, the behavior of an animal can be used to evaluate an individual for the lack of well-being, and on the other hand, for the presence of well-being. An animal's behavior is not proof that it is well or not well; but behavior is an indication of "wellness" and, combined with other observations, may assist in an assessment of the animal's status.

Environmental enrichment devices provide us with one opportunity to give animals in captive environments, such as zoos and lab animal facilities, a chance to behaviorally express wellness. In fact, such devices can also give us an opportunity to determine illness that otherwise might have gone unnoticed. This hypothesis was discussed in a paper[24] reporting behavioral enrichment experiments conducted in a zoo, but the examples therein may also act to stimulate us to develop enrichment devices that could help to establish wellness for laboratory animal species as well.

Animals given voluntary opportunities to perform species-specific behavior may demonstrate wellness by voluntarily participating in such activities. Also, animals that had previously voluntarily participated in such activity and who abruptly stop that activity may be giving the clinician a clue to a pathological condition and they may be demonstrating it much earlier than they would have if such voluntary activities had not been available. Two examples follow.

Servals in nature can catch birds on the fly. The beauty of this act is usually not present in zoo displays because such activity would not be accepted by zoo visitors. As a result, servals in zoo situations may demonstrate stereotyped pacing and inactivity. Some servals in a zoo were provided with flying meatballs, which they learned to catch in midair in a manner similar to wild servals catching birds in flight. This voluntary species-specific behavior was perhaps an indication of wellness. However, one male serval abruptly resigned from the task. This prompted an examination by the zoo veterinarian and the detection of a hernia — a diagnosis that the veterinarian believed would have almost certainly gone unnoticed and untreated for a much longer period under normal cage conditions. The authors also reported that they had seen significant changes in behavior relative to females catching meatballs prior to a diagnosis of pregnancy.[24]

When testing the memory of three elephants regarding some discriminatory tasks the elephants had learned years earlier, the research team noted that the first elephant tested remembered the tasks very well but the other two had problems even though they appeared to be very cooperative. This led the veterinarian to examine the elephants' eyes, which revealed a vascular deficiency in the areas of the optic disc in the two elephants who had struggled with the tasks and a richer vascularization in the first animal.[24]

The points to be made are that (1) a trained observer may use behavior as an indication of wellness or illness or even suffering; (2) environmental enrichment may result in an additional milieu of behaviors that assist the clinician in evaluating the condition of an animal.

IV. CONCLUSION

Traditionally, veterinarians concerned with laboratory animal medicine have focused their attention on the prevention and treatment of disease and, as a component of that, on good husbandry practices. While both laboratory animal medicine veterinarians and scientists need to continue to address the wide variety of issues that constitute laboratory animal science, in our view, we also need to place a new emphasis on the prevention of suffering in laboratory animals.

REFERENCES

1. **Friel, J. P.,** Ed., *Dorland's Illustrated Medical Dictionary,* 26th ed., W. B. Saunders, Philadelphia, 1985.
2. **Berube, M. S.,** *Second College Edition of the American Heritage Dictionary,* Houghton Mifflin, Boston, 1982.
3. **Kitchell, R. L. and Johnson, R.D.,** Assessment of pain in animals, in *Animal Stress,* Moberg, G. P., Ed., American Physiological Society, Bethesda, MD, 1985, 113.
4. **Moberg, G. P.,** Ed., *Animal Stress,* American Physiological Society, Bethesda, MD, 1985.
5. **Soma, L.R.,** Behavioral changes and the assessment of pain in animals, in *Proc. Second Int. Congr. of Veterinary Anesthesia,* Grandy, J. et al., Eds., Veterinary Practice Publishing, Santa Barbara, CA, 1985, 38.
6. **Grandy, J. et al.,** Eds., *Proc. Second Int. Congr. of Veterinary Anesthesia,* Veterinary Practice Publishing, Santa Barbara, CA, 1985.
7. **Davis, L. E.,** Species difference in drug dispositon as factors in alleviation of pain, in *Animal Pain, Perception and Alleviation,* Kitchell R. L. and Erickson H. H., Eds., American Physiological Society, Bethesda, MD, 1983, 175.
8. **Patterson, E. A.,** Noise and laboratory animals. II. *Lab. Anim. Sci.,* 30(2), 422, 1980.
9. **Bellhorn, R. W.,** Lighting in the animal environment, laboratory animal environment. II. *Lab. Anim. Sci.,* 30(2), 440, 1980.
10. **Weihe, W. H.,** The effects on animals of changes in ambient temperature and humidity, in *Control of the Animal House Environment* (Laboratory Annu. Handbk. No. 7), McSheehy, T., Ed., Laboratory Animals Ltd., London, 1976.
11. **Cross, H. A. and Harlow, H. F.,** Prolonged and progressive effects on partial isolation on the behavior of macaque monkeys, *J. Exp. Res. Pers.,* 1, 39, 1965.
12. **Erwin, J. and Deni, R.,** Strangers in a strange land: abnormal behaviors or abnormal environments?, in *Captivity and Behavior,* Erwin, J., Maple, T. L., and Mitchell, G., Eds., Van Nostrand Reinhold, New York, 1979, 1.
13. **Markowitz, H.,** *Behavioral Enrichment in the Zoo,* Van Nostrand Reinhold, New York, 1982.
14. **Wood-Gush, D. G. and Beilharz, R. E.,** The enrichment of a bare environment for animals in confined conditions, *Appl. Anim. Ethol.,* 10(3), 209, 1983.
15. **Foster-Turley, P. and Markowitz, H.,** A captive behavioral enrichment study with Asian small-clawed river otters, *Zoo Biol.,* 1(1), 29, 1982.
16. **Bennett, E. L., Rosenzweig, M. R., and Diamond, M. C.,** Rat brain: effects of environmental enrichment on wet and dry weights, *Science,* 163 (3869), 825, 1969.

17. **Cummins, R. A., Walsh, R. N., Budtz-Olsen, O. E., Konstantinos, T., and Horsfall, C. R.,** Environmentally-induced changes in the brains of elderly rats, *Nature (London),* 243 (5409) 516, 1973.
18. **Vitvitskaya, L. V., Bikbulatove, I. S., and Vivitsky, V.,** Changes in the content of nucleic acids and proteins in different brain parts of rats raised in enriched and impoverished media, *Zh. Vyssh. Deyat. Nosti*, 32(3) 455, 1982.
19. **Bouchon, R. and Will, B,** Effects of post weaning rearing conditions on learning performance in 'dwarf' mice, *Physiol. Behav.,* 28(6) 971, 1982.
20. **Wahlsten, D.,** Deficiency of corpus callosum varies with strain and supplier of the mice, *Brain Res.,* 239, 329, 1982.
21. **Yanofsky, R. and Markowitz, H.,** Changes in general behavior of two mandrills (*Papio sphinx*) concomitant with behavioral testing in the zoo, *Physiol. Rec.,* 28, 369, 1978.
22. **Reinhardt, V., Eisele, S., and Houser, D.,** Environmental enrichment program for caged macaques at the Wisconsin Primate Research Center: a review, *Lab. Primate News,* 27(2), 5, 1988.
23. **Fox, M. W.,** *Laboratory Animal Husbandry,* State University of New York Press, New York, 1986, 147.
24. **Markowitz, H, Schmidt, M.J., and Moody,** Behavioral engineering and animal health in the zoo, *Int. Zoo* (New York), 18, 190, 1978.

Chapter 16

BOREDOM AND LABORATORY ANIMAL WELFARE

Françoise Wemelsfelder

EDITOR'S PROEM

As we have indicated, the scientific acceptability and talk of pain and suffering in animals is a very recent phenomenon, and much work, both conceptual and empirical, needs to be done to clarify the many forms of noxious experience which animals can undergo. At the same time, such clarification is absolutely essential as a basis for establishing human obligations to the animals. In this essay, Dr. Wemelsfelder grapples with providing some precision to the intuitively plain insight that animals can suffer as a result of boredom and frustration growing out of being placed in sterile environments where they cannot perform as their *telos* dictates. Drawing upon a literature primarily directed at zoo animals and farm animals reared in confinement, Wemelsfelder attempts to apply these notions to laboratory animals, whose environments are typically extremely impoverished in terms of opportunities for expressing natural behaviors.

TABLE OF CONTENTS

I. INTRODUCTION

Over the past 10 or 15 years, concern for the welfare of laboratory animals has been growing. Anyone who lives in close relationship with animals or has had the privilege of watching animals in their wild surroundings knows how active, playful, and inquisitive most of these animals naturally are. Yet in our scientific laboratories, animals are kept mostly in small, barren cages, without companions whom they might touch or play with, without any means to perform natural, spontaneous behavior and, even without the most basic provision of all, space to move around and exercise their bodies.

The question is, does this matter to the animal? Does an animal experience something like "unhappiness" in its restricted and barren environment if it has never experienced anything else? Or, in other words, can an animal miss environmental stimulation which it has never known, and does it consequently suffer from this experience?

Formulated this way, the problem concerning the suffering of animals in laboratory environments can be regarded as equivalent to the problem of boredom. Boredom, as it is considered by psychologists, has been described as "the unpleasantness of monotony",[1] or as an adverse state induced by sensory deprivation on a short-term scale as well as a long-term scale.[2] So the question becomes, then, "can laboratory animals be bored?"

The term boredom derives its meaning from human experience, as do all concepts concerned with psychological states. In daily life, we use these terms on the basis of a general intuitive consensus, presuming that we as human beings share similar inner experiences. This intuitive consensus arises from the observation of behavior, combined with verbal communication. In the case of our relationship with animals, we lose the possibility of verifying the intuitive interpretation of their behavior by verbal confirmation. Despite this difficulty, however, we generally extend the application of psychological states to animals (at least to the higher vertebrates) as well, thereby assuming the inseparability of behavior and psychological experience. This assumption is enhanced by the idea that, according to evolutionary theory, the neural structures underlying behavior in higher vertebrates are similar to ours to a large degree.[3]

Based on such an interpretation of animal behavior, satisfactory "working-relationships" with animals have been developed in daily life, in training situations,[4] in zoos,[5] and in science. After some 50 years of skepticism, some experimental psychologists and ethologists are beginning to attribute to animals such states as fear, aggression, frustration, satisfaction, curiosity, etc. As a result of the extension of psychological concepts to animals in a scientific context, these concepts have undergone considerable refinement. In working with animal subjects it is possible to create experimentally controlled circumstances in which reproducible, detailed results can be obtained. It would be very difficult, if not impossible, to obtain such results with humans subjects. Theories arising from the study of animals can then be reapplied to the understanding of human behavior. The successful application of this procedure confirms the validity of behavioral analogy between humans and animals, and has improved the scientific status of psychological concepts. The fruitful interaction between scientific research programs into human and animal behavior thus provides evidence for the existence of real psychological states in animals.

Within this context, it appears to be justified to investigate boredom in animals which is supposed to occur as the result of environmental deprivation. Two things would have to be done. First, the behavior of confined animals would have to be compared with the behavior of their conspecifics in more enriched environments. Second, an ethological characterization of boredom, derived from human psychology, would have to be formulated. By evaluating these two aspects, the extent to which laboratory animals experience a state of boredom could be elucidated.

This approach is followed in the first two sections of this chapter. In the first section, a rough outline is given of the behavioral response of animals to long-term confinement and sensory deprivation. In the second section, the meaning of the boredom concept is considered as it is applied in experimental psychology and ethology. No consensus as yet exists among ethologists on the question of whether boredom is the most appropriate concept to describe the effects of environmental deprivation on the animal. Other concepts, such as frustration or helplessness, are used in a similar context and seem to eliminate the need for a boredom concept. Suggestions for an unambiguous redefinition of these concepts are given.

Although the boredom concept is used in experimental context by several authors, resistance towards its application to animals is expressed by other scientists. Many authors explicitly state that boredom is a "human" term or put the word boredom between quotation marks. By this, they mean to indicate that the analogy between human and animal behavior is questionable in the case of boredom. It is striking that these authors seldom make such remarks in regard to concepts like fear, frustration, or helplessness, although these concepts play a prominent role in the description of the behavioral problems animals face in confined environments. Why then does the boredom concept give rise to more doubt concerning the application of behavioral analogy than other concepts do? This question is dealt with in the third section.

As widespread as the scientific use of psychological concepts may be, a large part of the ethological community still insists on the metaphorical nature of these concepts, when used in relation to animals. It is said that seriously attributing to animals the actual emotional experience that these concepts imply would be making the mistake of so-called "anthropomorphism".[6] Since human language provides the framework for the description of animal behavior, however, ethologists have grown accustomed to applying psychological concepts to animals in practice. They assume at the same time, though, that the emotional experience might be absent, or at least different from the human experience denoted by these concepts.

A crucial differentiating factor which apparently causes ethologists either to accept or to reject a concept is the degree to which a concept suggests that the animal might be *consciously aware* of its own behavior.[6] For example, hardly any scientist will have trouble with the attribution of pain to laboratory animals. In fact, many experiments are done on animals to understand the functional organization of pain. In discussions on the well-being of these animals, however, much resistance can be seen towards the idea that animals might *suffer* from their pain or, in other words, towards the idea that animals might to a certain extent be aware of their pain. It could be possible, therefore, that the boredom concept implicitly refers to a higher degree of awareness in animals than ethologists are willing to accept.

To presume that animals actually experience boredom implies that the analogy between man and animal is drawn not only in relation to active behavior, but also in relation to passive behavior. As will be shown in the third section, the attribution of a psychological state to an animal wherein this state is not directly linked to an active behavioral sequence poses several problems to cybernetic models which are being developed in the field of animal welfare. Such an attribution does suggest that an animal has a certain awareness of itself, even when there is no sign of outwardly visible behavior. This suggestion causes several authors to reject the boredom concept as being "mentalistic" and therefore unscientific.

Suggestions for relevant approaches to these problems will be given.

As long as the underlying theoretical problems are not dealt with explicitly, the use of the term boredom can only indicate what the situation might be like for animals in restricted housing environments. It will certainly not lead to a clear diagnosis, nor to adequate remedies which might diminish this form of suffering in laboratory animals. It is argued that in spite of theoretical unclarity, a sufficient basis is available to alleviate the worst signs of apparent boredom.

Articles on boredom which are mentioned in this chapter are drawn to a large extent from fields other than the field of laboratory animal science. The reason for this lies in the fact that experimental work on animal welfare and environmental deprivation has mainly been done in relation to husbandry systems for farm animals. In zoos, on the other hand, more attention has been paid to possible ways of enriching the animal's environment and the effect this has on its behavior. Now that the interest in laboratory animal welfare is rapidly growing, it seems only reasonable to draw from the experience gained in other fields.

For in all cases where animals are produced and confined for the sake of human advantage, the same central question presented in the beginning can be posed: does the hindrance of the species-specific behavior of an animal affect its welfare adversely?

This chapter is meant as a contribution to the general understanding of this problem.

II. BEHAVIORAL RESPONSES TO ENVIRONMENTAL DEPRIVATION

A. THE THWARTING OF EXPLORATORY BEHAVIOR

It is important to know whether domesticated or laboratory bred animals are still capable of as full a behavioral range as their wild conspecifics before the impact of the thwarting of exploratory behavior on their welfare can be evaluated. This has been investigated for several species. For example, albino Norway rats,[7] rabbits,[8] and pigs[9] can display behavior patterns which are as complex as those of their wild relatives. Price,[10] in a review article on domestication, concludes that "it is apparent that, with respect to animal behavior, domestication has influenced the quantitative rather than the qualitative nature of the response." It can be assumed, therefore, that adaptation to the captive environment did not fundamentally alter the behavioral capacity of domesticated animals.

In their natural, wild environment, the relatives of our domesticated animals have a large array of behavioral strategies which eventually contribute to the fulfillment of their basic needs. Locomotion, social interaction, and interaction with the physical environment constitute the major parts of these strategies. Wild pigs, for example, spend 6 to 7 h/d seaching for food;[11] wolves hunt for prey in groups, using cooperative strategies.[12]

To a large extent, such behaviors can be qualified as "exploratory behavior". Exploration, in a broad sense, refers to all activities concerned with gathering information about the environment.[13] Exploratory activities, although not specifically aimed at the fulfillment of basic needs continuously, nevertheless form a vital part of the survival strategy of the animal.[14]

In the laboratory environment, the animal is provided with its basic physical needs by its caretakers. Thus, its physical survival is ensured. Laboratory housing conditions, however, deprive the animal of the possibility of performing most of the above-mentioned exploratory behavior patterns. Animals which maintain large home ranges in the wild, like monkeys, and who displace themselves over large distances each day, are kept in cages which are too small even to stretch themselves to their full body length. Social animals whose natural behavior is adjusted to interdependence between conspecifics, like dogs, cats, monkeys, rabbits, rats, etc. are often kept alone. None of these species are able to spend any time searching for food; it is delivered to them at fixed times.

In general, laboratory housing conditions lack the presence of stimuli which might elicit exploratory behavioral sequences. Animals need specific stimulation to fulfill their behavioral needs; when this is denied to them, they are subject to so-called sensory deprivation.

Under the influence of the growing concern for animal welfare, the question of whether the thwarting of exploratory behavior affects the health and well-being of animals has received increasing interest. A series of studies has been generated in different fields which consider the effect of sensory deprivation on the behavioral repertoire of the animal. Generally speaking, animals show the following types of response to the deprivation of the possibility of performing exploratory behavior:

1. Active:
   a. Redirected behavior
   b. Stereotyped behavior
      i. Reversible
      ii. Irreversible
2. Passive:
   a. Lying, nonbehaving
   b. Drowsy posture

Several examples of studies which have described these different types of response will be given below.

B. SOME EXAMPLES OF RESPONSES TO SENSORY DEPRIVATION

1. Active and Passive Responses

First, two examples will be given of studies which report both active and passive responses to sensory deprivation.

Wood-Gush and Beilharz[15] start their article, "The enrichment of a bare environment for animals in confined conditions," by suggesting that "the lack of stimulation in bare environments may lead to ... the development of stereotypies and abnormal behavior." They tested this assumption by comparing the behavior of control piglets in a flat-deck cage (normal rearing environment) to the behavior of piglets in a flat-deck cage enriched with a trough containing sterilized earth.

Their main results were that (1) the presence of the trough greatly reduced the amount of lying and reduced the amount of biting and sucking the cage and food-hopper; (2) the amount of rooting (using the snout on the cage-floor) was greatly increased, as well as the amount of feeding (and, to a lesser extent, drinking).

Considering that the earth trough was always attended to before feeding, and that the amount of rooting was increased, Wood-Gush and Beilharz suggest that the presence of the earth trough mainly results in the elaboration of appetitive behavior. Regarding the fact that the amount of food intake itself was not increased, they postulate that the motivation behind this elaboration in rooting was not hunger, but exploration.

In this study, the thwarting of exploratory behavior in the bare control environment led to an increased amount of lying down and to seemingly harmless redirected behavior like cage-biting, etc. The authors warn that the results of deprivation of exploratory behavior have been reported to be much more severe: lack of exploratory possibilities can lead to tail biting in piglets and fattening pigs, resulting in severe production loss.[16,17]

Stolba et al.[18] studied the development of behavior patterns of pregnant sows over 15 weeks from the moment the sows were brought into severe confinement; they were put in small stalls and tethered around the neck.

After a few weeks, "an increasing number of patterns of investigative and feeding quality become fixated in form and orientation and are performed in more and more frequently repeated stereotyped sequences." It is suggested that these stereotypic patterns (vacuum chewing, bar biting, head weaving, etc.) originate from thwarted exploratory behavior and serve to increase sensory input. Stereotyped behavior patterns would in this way compensate for the lack of sensory input resulting from the confinement of the sows. The form of these patterns indicates that it is mainly food-related exploratory behavior which is thwarted. This is in accord with the results obtained by Wood-Gush and Beilharz[15] which show that environmental enrichment leads to the elaboration of food-related exploratory behavior.

Stolba et al. also report "prolonged drowsy stances" and periods of "alert inactivity" preceding the gradual development of stereotyped behavior patterns.

These two studies show that pigs, when placed in an understimulating environment, develop active as well as passive responses as a reaction to sensory deprivation. Piglets perform redirected activities and lie down more frequently; pregnant sows develop stereoptyped behavior patterns, preceded and accompanied by drowsy, inactive periods.

Redirected behavior can be defined as behavior which occurs when, because of the absence of adequate stimuli, desired behavior is performed on substitute stimuli.[19]

Stereotyped behavior can be defined as "aberrant behavior, repeated with monotonous regularity and fixed in all details."[20]

More evidence for the occurrence of these responses in deprived environments will now be given.

2. Active Responses

a. Redirected Behavior

Tethered sows do not only perform stereotyped behavior to satisfy their natural tendency to be active and inquisitive, but also redirected behavior patterns.[21] Up to 17.5% of their behavioral range consists of behavior such as licking bars, trough, floor, and chain, play drinking, and vacuum behaviors such as teeth-grinding, air-chewing, tongue-rolling, and rooting.[21-23] The provision of straw prevents such behavior to a large extent.

For laying hens in battery cages as well, the absence of litter is the most important factor contributing to feather pecking and vacuum dust bathing.[24,25] Piglets, veal calves, and dairy cows suck different parts of their peer's bodies or drink their urine.[26] These behaviors can develop into cannibalism, where the animal actually starts eating tissue of other animals.[26] In this way, animals can cause serious bodily damage to each other. When an animal is housed alone in a cage, however, it will compensate for the lack of stimuli by directing its behavior at its own body, or its own excrements. In laboratory and zoo animals, many cases are known of self-mutilation, the eating of feces (coprophagia) and the eating of vomit.[27,28] Self-mutilation involves biting of limbs, tail, or parts of the body, or plucking out hair or feathers.[28] Isolated parrots are reported to pull out their own feathers,[28,30] but such activities are especially known to occur in monkeys who are housed alone in small, bare cages. They might pull out their hair in compulsive bouts,[29] or gnaw at their tail, penis, scrotum, or testes.[27] The eating of feces is also frequently observed in captive primates, particularly the great apes.[28,31] Chimpanzees are not only known to eat their feces, but also to mix these with saliva and smear it over the glass cage front with lips and tongue.[32] These behaviors are not observed in the wild.[27]

Such self-directed behaviors are presumably related to foraging behavior. In addition to these, self-directed social behaviors have been reported. Goosen[33] found that several species of individually housed macaque monkeys show self-aggression, self-sexual behavior (masturbation), self-holding, thumb-sucking, etc. These behaviors were not observed when the monkeys were housed in groups.

Besides performing abnormal behavior patterns to compensate for a lack of adequate stimuli, animals might prolong and exaggerate the performance of normal behavior patterns. Excessive feeding (hyperphagia), hyperaggression, and hypersexuality may develop.[29] Morris[32] mentions "super-normal responses to normal stimuli, as a compensation device for the enforced absence of responses to missing stimuli." Animals can, for example, over-groom or over-clean themselves. Minor injuries might enlarge and become serious by an excessive amount of rubbing and licking. Rodents are known to damage their young by over-cleaning them, which results in the eating of the ears and tails of their offspring.

b. Stereotyped Behavior

In a similar context as Stolba et al.,[18] Rushen[34] observed the behavior of tethered sows 1 h before and 1 h after the delivery of food. He found that stereotypies like head waving, bar biting, and rubbing occurred mostly before feeding, and excessive stereotyped drinking and rubbing after feeding. Vacuum chewing and chain playing occurred evenly before and after feeding. In a follow-up experiment,[35] where he observed tethered sows for 9 h during the day, he found that these stereotypies did occur almost exclusively before and after feeding, except vacuum chewing, which was distributed more evenly over the day.

Keiper[36,37] studied the development of stereotypies in caged canaries. He suggests that stereotypic route tracing is "a cage-related stereotypy," caused by the restriction of locomotory activity, whereas stereotyped spot-pecking "seems to develop as a result of deficiencies associated with laboratory feeding conditions." Birds, provided with the opportunity to "work for their food," did not develop this stereotypy. He considers stereotypies to "occur primarily to provide a source of stimulation to birds confined to the monotonous, stimulus deficient environment of a cage."

Bareham[38] described how laying hens in battery cages perform stereotypic head flicks. Like Keiper, he suggests that these activities "function to increase sensory input to a bird living in a monotonous environment with restriction of external stimuli." Berkson et al.[31] found that chimpanzees confined in bare cages performed almost six times as many repetitive stereotyped movements (rocking, swaying, head shaking) as chimpanzees in an outdoor enclosure and provided with a playground. Cage size did not seem to influence the performance of stereotypies in this experiment; the effect was mainly produced by the lack of exploratory opportunities.

Kiley-Worthington,[20] however, reports several cases in which movement restriction by small cage size appeared to be the major cause of stereotypic activities. Horses housed in stalls perform activities like weaving, crib biting, and wind sucking. The pacing up and down of many caged felids, canids, and ruminants also appears to be related to restricted size of cage. Morris[32] considers pacing and weaving in ungulates and carnivores, as well as vertical circling in chipmunks and squirrels, as indicative of the need for patrolling in a great territorial space.

Several studies have shown that social isolation can also result in "severe autistic-like behavior."[29] Stereotyped behavior, such as rocking, has been reported for chimpanzees who were denied contact with peers.[31,39] Dogs who are raised in isolation up to 16 to 32 weeks, show whirling, pacing, and bizarre postures.[40] Isolated veal calves perform stereotyped tongue playing, which does not occur when the calf stays with its mother. Dairy cows and fattening bulls perform this stereotypy in confinement as well.[26]

Although quite a few authors have attempted to differentiate between several causal factors regarding stereotypies, it is more likely that these factors are intertwined.[20,28,29] Considering stereotyped head flicks in laying hens, Bareham[38] concludes that "social isolation, movement restraint and monotony, individually or together, lead to reduced sensory input, which is the suggested main causal factor in this study."

These are a few examples of the relation between environmental monotony and the performance of redirected behavior as well as stereotypic behavior.

Stereotyped behavior (and some self-directed behavior) seems to have lost much of the flexibility which most redirected activity appears to possess. When animals who perform redirected behavior are provided with adequate stimuli through environmental enrichment, they will often shift their behavioral attention to these new stimuli and stop performing redirected activities.[28,32] However, with stereotyped behavior, this appears not always to be the case. In some cases, stereotyped behavior can be "cured", and disappears or decreases when adequate stimuli are provided; in other cases the animals do not react to new stimulation anymore, but go on performing their stereotyped movements.

This difference in "reversibility" might be related to the length of time the animal has been housed in the deprived situation.[18,20,32] When stereotyped movements start to develop, they do not become fixed instantaneously, but remain flexible for a while. After a prolonged period of sensory deprivation, however, the animals are not likely to react to new stimuli anymore. This was shown by Wood-Gush et al.,[9] who introduced a dangling sack into the stalls of sows, with varying degrees of developing stereotypies. The number of reactions to the sack was inversely related to the degree of stereotypy of the sows. Sows with fixed stereotypies interacted very little with the sack compared to sows with less fixed stereotypies; also their response to the sack was always of low intensity.

Kiley-Worthington[20] reports that similar results were obtained with autistic children; a rise in environmental complexity *in*creased the performance of established stereotypies, whereas this caused a *de*crease in the performance of children whose stereotypies were not so well established.

Whether any form of environmental enrichment will be successful in the reduction of stereotypies depends, therefore, on the stage of development of the stereotypies at the time the enriching factors are introduced.

Although hardly anything is known at the moment about the mechanisms which influence these processes, several examples can be given of cases in which the reduction of stereotypies by environmental enrichment was succesful. Berkson et al.[31] found that stereotyped activities performed by chimpanzees in small cages could be reduced by the provision of objects which could be manipulated. Fox[40] showed that frequent change of the cage of his dogs reduced the performance of stereotypies. Stalled horses who perform stereotypies do not do so anymore when they are moved outdoors on grass.[20] Stevenson[28] reports several enrichment projects carried out in zoos. By providing polar bears, servals, and orangutans with devices which produce food at unexpected times in such a way that the animals had to catch the food actively, stereotypic pacing was significantly reduced. The presence of straw greatly reduces the performance of stereotypic movements by tethered sows. Vestergaard[41] found that these movements increased a few days after the removal of straw, and decreased as soon as straw was present again.

Stolba et al.,[18] however, found that the provision of straw to tethered sows did not prevent the development of fixed stereotypies over a time period of 15 weeks. They suggest that after a few weeks, the novelty of the stimulation provided by the straw might wear off. Along the same lines, Keiper[36] found that the view of a neighbor for a short period of time did decrease stereotyped spot-pecking in canaries, while housing canaries permanently together did not. Like Stolba et al., Keiper suggests that habituation to the presence of other canaries might reduce their stimulation value and, hence, causes the stereotypies to remain.

These studies show that the enrichment of a partial aspect of the housing environment might not be sufficient to cancel the overall effect of sensory deprivation on the animal by the environment. Observations over a longer period of time are necessary to observe the impact of an environment on the animal.

In accordance with this conclusion, Hutchins et al.[42] comment on the enrichment of the zoo environment by artificial devices. They concede that such devices might initially reduce stereotypies like pacing by providing the animals with a chance to exert more foraging behavior. However, they warn that in the long run, the animals might habituate to these devices and might not be sufficiently stimulated by them anymore. The apparatus could then become an outlet for abnormal behavior "in much the same way as an animal might use its own body in the case of pacing or masturbation." Signs of this process have been observed in the Portland Zoo, where "gibbons would pull, shake, and even attempt to copulate with portions of the apparatus when it is not functioning (Markowitz, 1975b)." Hutchins et al. argue that only the provision of stimulus *variation* through the creation of a situation which resembles the natural habitat of a species would solve severe behavioral problems permanently.

Despite these examples of succesful reduction of stereotypic behavioral patterns, there are also cases in which the introduction of new, enriching stimuli does not lead to a decrease in stereotypic performance.

Sometimes an animal continues its repetitive movements without paying attention to the new stimulus, as Wood-Gush et al.[9] have described for tethered sows when a dangling sack was introduced into their cage. At other times, the performance of stereotypies actually increases, as Hutt and Hutt[43] found for autistic children when they were brought into an enriched environment.

Berkson et al.[31] reported that novelty increased stereotypic behavior in chimpanzees, as did Mason and Green[44] for Rhesus monkeys, which were separated from their mother at a very early age and raised individually.

Meyer-Holzapfel,[27] working with bears, found that after acquisition of a stereotype, a change in the environment, such as the entrance of a caretaker into the cage, resulted in an increase of stereotypic performance. She also reports that the housing of former circus polar bears in large enclosures did not eliminate stereotypic pacing and weaving movements. One polar bear would create an oval circuit the size of its old circus wagon, and pace along this circuit continuously. Another bear would stand in a fixed place at the border of the enclosure and weave for long periods of time. Monkeys are also known to continue their stereotyped movements in large enclosures.[32] Once self-directed abnormal behavior patterns like self-mutilation, coprophagia, and the eating of vomit are acquired, they appear to be very difficult to eliminate.[28,32]

These examples show that once animals have been deprived long enough of sensory stimulation to establish stereotyped behavior patterns within their behavioral repertoire, this process can hardly be reversed; on the contrary, new stimulation is not assimilated normally, but is often counteracted by increased stereotypical performance.

Animals do not respond to understimulating environments only by performing abnormal compensatory behavior, nor do all species develop forms of stereotyped behavior. Another major way of responding to the lack of key stimuli is to become passive. Several examples are given below.

3. Passive Responses

Inactivity is often observed in an indirect way by the observation that the overall activity of an animal increases considerably when enriching stimuli are introduced, or that the overall activity in an environment is low compared to activity in a more enriched environment.

Stevenson[28] reports concerning enrichment projects in zoos with chimpanzees, orangutans, and servals who showed high levels of inactivity (sleeping and lying). The provision of exploratory devices increased the overall rate of activity of these animals, not only of exploratory behavior, but also of social interaction. Wood-Gush and Beilharz[15] (see Section II.B.1) showed that the introduction of an earth trough in a flat-deck cage greatly reduced the amount of lying in piglets, while exploratory behavior patterns were elaborated. Stolba et al.[18] found that deprived piglets showed a greater response to a new stimulus than nondeprived piglets. The deprived animals performed a whole range of exploratory behavior patterns in reaction to a round tire, while nondeprived animals were hardly interested.

It is known that animals often prefer to "work" for their food, even if they can get it for free. Duncan and Hughes[45] showed that chickens in a Skinner box prefer to perform an operant response for their food, even when identical food is available. Markowitz[46] found the same result for ostriches. Fox[29] reports of research which "demonstrated that animals raised in an environment in which they learned to exercise control over feeding, access to water, and amount of light by pressing appropriate levers grew up to be more exploratory, self-confident, and less anxious than animals that received the same food, water, lighting changes, but had no control over their environment."

Besides these indirect ways in which the lack of activity in bare environments can be inferred, overt signs of passivity are also reported. In laboratory housing conditions, monkeys, dogs, and cats develop signs of passivity and lack of interest in the external environment. These animals have difficulty learning that responses can produce relief and show lack of aggression and loss of appetite.[29] Laboratory-reared Rhesus monkeys quite commonly stay passive to their babies,[33] so that these have to be removed and nursery raised.

Tethered sows and fattening pigs spend long periods, up to 6 h, of motionless "sitting",[22] often with their heads hanging down or pressed against the stall divisions.[30,41] Standing, which occurs for long periods, may be regarded as a conflict between the desire for activity and the impossibility to achieve it.[22] The motionless postures the animals assume in these cases appear to be fixed in a similar way as stereotypies. The animal's posture is often described as "drowsy or apathetic", its head droops down, the eyes are not alert or are half closed, limbs are bent in abnormal ways. These positions are also often taken while the animal is performing stereotypic behavior.

The occurrence of drowsy and apathetic postures has mostly been described for tethered sows and fattening pigs,[26] but it is likely to be equivalent to, for example, the prolonged self-clutching and rocking of isolated monkeys.

These examples show that in understimulating environments, animals fall back into a general passivity, many different behavioral capacities being inhibited. This inhibition is of general rather than of specific nature, since a few specific stimuli can induce a diversity of behavior patterns. If the deprivation continues over a long time, however, animals might lose their ability to give up their passivity in response to new stimuli,[29] as is the case with stereotyped behavior patterns as well (see Section II.B.2.b).

In the first section, the response of animals growing up and/or living in an understimulating environment has been described. Active as well as passive reactions occur, and, under the pressure of prolonged deprivation, acquired behaviors from both categories appear to be non reversible by enrichment.

The question now arises as to how these responses relate to the well-being of the animals. Does the behavior animals perform in deprived environments constitute evidence for a state of boredom, subjectively experienced by the animal? In the next section, the concept of boredom, as it is used by several experimental psychologists and ethologists, will be evaluated.

III. PSYCHOLOGICAL CONCEPTS RELATED TO ENVIRONMENTAL DEPRIVATION

In this section, relevant ethological and psychological literature will be examined in search of an operational definition of boredom. Reasonably satisfying operational definitions of other psychological concepts, such as fear and aggression, have been provided in the past. These states are accompanied by rather specific behavioral patterns, while boredom seems to be expressed by a diversity of behavioral patterns. This makes it quite difficult to give an operational definition of boredom which does not overlap with definitions of other psychological concepts, such as frustration and helplessness, which are also applied to the performance of stereotypies. A close look will therefore be taken at the context in which the concepts of boredom, frustration, and helplessness are used. Suggestions concerning operationally meaningful differences between these concepts will be given. It will be proposed that the kind of behavioral response of an animal to an introduction of novel stimuli in the environment might serve to differentiate between the three concepts.

A. BOREDOM

The term boredom appears in an experimental context in relation to the concept of "exploratory drive". Exploratory behavior has long presented difficulties for theories of motivation. For a time, exploration was considered as a form of general activity, a view which changed during the 1950s. Once it was realized that the opportunity to explore stimuli is itself reinforcing, exploration was considered by many to be an autonomous drive.[13] The reinforcing value of exploratory behavior can be explained by assuming that inquisitive exploration and novelty seeking are attempts to reduce boredom by self-produced exposure to change. Different formulations of such boredom-drive theories were developed by authors like Myers and Miller, Berlyne and Fowler.[47]

When the animal is free to move around in an environment which is sufficiently stimulating, the actual psychological experience of boredom would most likely not occur. As soon as a certain amount of stimulus satiation would have been reached, the animal would move on to a new situation in the environment, which would bring forth new stimulation. In such a context, boredom serves mostly as a hypothetical construct to account for a highly variable and complex set of behavioral patterns. Furthermore, to assume that an animal will experience boredom when it does *not* have the opportunity to explore, implies falling into the trap of circularity.[13,47] Boredom and exploration are defined in terms of each other without independent specification of either of these concepts.

To avoid this fallacy, it is necessary to consider passive behavior as not merely a complementary to active behavior. The question what animals experience in confined, unchanging environments must be considered seperately.

An early experiment on boredom in human beings was done in 1938 by Barmack.[48] He tested the response of people to the forced prolongation of a relatively simple task and asked them to report on their feelings during this task. The subjects reported conditions of strain, restlessness, irritation, fatigue, sleepiness, day-dreaming, and inattentiveness. Barmack[49] concluded that "according to our hypothesis, then, after the exploratory drive is gratified (of which the relative novelty of the task may be the incentive), ... a condition develops which is unfavorable to the maintainance of the alert state. Out of this tendency stem two main effects, a desire to get away from the task or other environmental condition, which, in a sense, had produced it, or a desire (usually unconscious) to correct the unpleasant state by introducing new methods of work, thinking of more strongly affective situations, which are, in a way, means of staying alert."

Thus, according to Barmack, the experience of sleepiness, inattentiveness, fatigue, etc. or, in short, a condition "which is unfavorable to the maintainance of an alert state," is followed by sensations of irritability, restlessness, and strain, resulting from the desire to get away from or change the situation. Together these symptoms make up the experience of boredom.

In accordance with these results, Berlyne[2] reports of studies done by Karsten and by Bexton et al. Karsten had her subjects perform monotonous, repetitive, uninteresting tasks. As the experiment proceeded, the quality of the performances deteriorated; subjects reported distaste toward the task, aggression towards the task, the experimenter, and themselves, and they would invent all kinds of ways of bringing variation into the task or fall back into day-dreaming. Bexton et al. housed their subjects in an isolated cubicle with as little sensory stimulation as possible. These people would fall asleep after being placed in the cubicle. However, after their need for sleep was satiated, they would become increasingly irritated and stressed and would try to provide themselves with some stimulation in whatever way they could. Hallucinations would also occur.

According to these studies, the boredom concept refers to a state induced by the lack of sensory stimulation in monotonous, restricted environments. It is characterized not by a single unvarying component, but by several factors which interact and cause a certain progression in emotional experience over time. After exposure to understimulation for a while, decreasing alertness will manifest itself by a decrease in concentration, intellectual capacity, and general performance. Sleepiness and fatigue result. However, if not in need of sleep, the person will then experience an increase in restlessness, and a dissatisfaction with the situation. If escape appears to be impossible, means to change the aversive situation will be invented. The subject will try to bring variation into its nonstimulating tasks in as many different ways as possible. In addition, new stimulation may be created by day-dreaming or hallucinating.

This rather complex conceptualization of boredom was made possible by detailed information provided by the verbal reports of the human subjects on their emotional experience. How have animal ethologists used or defined the boredom concept? Several examples of the context in which ethologists use the term boredom will be given below.

In an early article on animal welfare, Wood-Gush[50] refers to research which has shown that animals prefer to work for their food and that they develop stereotypies in bare enviromments. He states that "thus it seems likely that in a barren but light environment the fowl, rather than staying inactive, will compensate for lack of external stimulation by performing some activity... It is often suggested that some of the trouble found in poultry industry, such as feather picking and cannibalism, is due to 'boredom'." Duncan[51] similarly notes that "in a confined and barren environment the animal is restricted in the amount of exploration it can do and this may lead to what, in human terms, would be called 'boredom' "; the animals "may try to increase their general stimulation" by performing stereotypies or by working for their food. Kiley-Worthington[20] also comments that "insufficient environmental stimulation (boredom) may result in the performance of stereotypies, which may lead to self-generated increased sensory input." Murphy[52] says that "boredom resulting from an unchanging environment has been proposed as an predisposing factor to vice...or stereotyped movements."

Considering the results of an experiment on stereotyped behavior in tethered sows, Stolba et al.[18] comment that "the longer the sows stay, the more boring the ever unchanging environment will become. So...the sows would try to increase sensory input by investigative sniffing and licking, which in the long run however is ineffective in the unvarying stall." Fixating movement sequences into stereotyped patterns would be a means to occupy the perceptive channels more permanently. "The lack of stimulation of animals in bare environments may lead to boredom, which in turn may lead to the development of stereotypies and abnormal behavior..." Thus, do Wood-Gush and Beilharz[15] introduce their report of experiments concerning the welfare of piglets in confined conditions (see Section II.B.1). The piglets are also reported to show higher frequencies of laying inactive as a result of boredom; "as if they have learnt that there is nothing else to do."

Stevenson[28] as well relates boredom not only to compensatory behavior, but also to passivity. "Captivity may therefore provide too little novel stimulation, so that the animal is underaroused, inactive and 'bored'." As a result of an abnormally low sensory input, "the animal may develop various types of abnormal behavior." Stevenson considers self-mutilation, coprophagia, and the exaggeration of foraging behavior a direct reponse to boredom.

In summary, ethologists who use the term boredom agree that boredom is induced by a *general lack of environmental stimulation*. Some authors explicitly suggest that lack of sensory input initially causes a state of inactivity or lethargy, which in turn causes the performance of various abnormal activities. Most authors, however, do not provide a further qualification of this process, but just state that boredom results from sensory deprivation, inducing the performance of abnormal behavior.

The limited input of environmental stimulation which is forced upon animals in laboratories, zoos, or industrialized agriculture is comparable to the highly monotonous, repetitive tasks which human subjects were asked to perform. Thus, the boredom concept for animals might be similar to the boredom concept as developed for human beings by psychologists.

In the first section it was shown that animals, like humans, show signs of decreasing alertness and activity; they also perform a large variety of abnormal behavior, which is interpreted as serving a self-generated increase in sensory stimulation, compensating for environmental monotony. Normal behavior patterns are exaggerated or redirected towards inadequate stimuli; new behavior patterns such as stereotyped movements are created.

These responses can all be considered as symptoms of boredom within the broad conceptualization of ethologists and psychologists. Not all ethologists concerned with animal welfare would agree with this, however, since other emotional states have been connected with similar types of response, especially the performance of stereotyped movements. Besides boredom, central concepts in relation to long-term sensory deprivation are frustration and helplessness. The meaning attributed to these concepts in relation to animal welfare will be discussed briefly.

B. FRUSTRATION

In animal welfare research, the concept of frustration has first been used for domestic fowl. The Brambell report,[53] written by an advisory committee on animal welfare, stated that "Much of the ingrained behavior pattern is frustrated by caging. Normal reproductive patterns of mating, hatching and rearing young are prevented and the only reproductive urge permitted is laying. They cannot fly, scratch, perch or walk freely. Preening is difficult and dust-bathing is impossible."

Duncan[54] and Duncan and Wood-Gush[55] attempted to operationalize the concept of frustration by experimentally frustrating chickens and recording their reactions. They defined frustration as "the state of an organism placed in an objectively defined frustrating situation..., (A) frustrating situation will be restricted to those situations in which there is interference with a behavior sequence normally leading to a goal response." This definition is in agreement with Hinde,[56] who comments that frustration occurs when an animal engaged in a sequence of behaviors is unable to complete it because of a physical barrier or the absence of an appropriate stimulus link. Duncan and Wood-Gush[55] deprived chickens of food for 24 h and then put them in an experimental cage with food present under a Perspex® cover and observed their behavior. They found that the birds showed escape behavior which developed into stereotypic back-and-forward pacing over time, when escape appeared to be impossible; they also showed displacement preening and increased aggression if other birds were present. These responses are related to the thwarting of the hunger drive, but they have also been shown to occur when nesting, incubating, and brooding behavior are frustrated. Stereotyped pacing would only develop, however, if the bird was highly motivated for a certain behavior pattern. If the motivation was weaker, for instance, as a result of a shorter deprivation time, the bird would show displacement preening, but no stereotyped movements. On the other hand, it was found that once stereotyped pacing was firmly established, it would not disappear when the source of frustration was removed.

Along similar lines, Dantzer et al.[57] investigated frustration in pigs. They defined frustration as "a hypothetical state elicited by the omission of an expected reward, e.g., withholding positive reinforcement after a history of reinforcement contingent on emission of an operant." After exposing pigs to such conditions, the animals, if housed alone, would become very restless, attempt to escape, and rub and scratch the floor with their feet. If housed in pairs, aggressive biting, pushing, and fighting would ocurr. Aggressive encounters would not take place, however, if the two pigs had established a social bond before the experiment.[58]

These experiments indicate that a state of frustration apparently leads to displacement activities, escape behavior, and aggressive behavior, and to stereotyped pacing when prolonged frustration occurs.

Defining frustration as a state caused by the blocking of, or interference with, behavioral sequences which normally lead to expected goal responses, several authors suggest that the restrictive housing environments of intensive husbandry are likely to give rise to frustration in animals. "This psychological state is therefore believed to play a key role in the development of abnormal physiological and behavioral reactions observed in farm animals."[58] Duncan and Wood-Gush[55] argue that stereotyped pacing in hens resembles the repetitive stereotypies commonly observed in zoos and pet shops, thereby suggesting that general welfare problems might be related to frustration.

Duncan,[51] however, remarks that the specific behavioral expressions of frustration described above, *do not occur* in battery caged hens, with one exception: during the pre-laying phase, hens cannot find a suitable nesting site and perform stereotyped pacing. "It can be concluded," so he states, "that generally speaking, caging per se does not lead to frustration." It is clear that Duncan does not define frustration in relation to environmental circumstances, as Arnone and Dantzer do,[58] but in relation to the performance of specific behavior patterns. Such a difference in opinion concerning the definition of frustration in turn leads to differences in the interpretation of general stereotypic performance by understimulated animals.

Laying hens in battery cages show several forms of redirected and stereotyped movements, such as ground picking and head flicks.[38] Bareham remarks that "it is unclear whether headflicks and groundpicking are a result of frustration, or due to a lack of environmental stimulation necessary to elicit 'normal' behavior." Because the hens do not perform displacement preening, however, he chooses to interpret the performance of stereotypic activities as a means of increasing sensory input in a monotonous environment, and not as a result of the frustration of behavioral drives by caging. For this reason, the results of Bareham are referred to as evidence of boredom by several authors.[51,52]

Rushen,[34,35] on the other hand, does interpret the performance of stereotypies like bar biting, head rubbing, and head weaving in tethered sows as a result of frustration, referring to the work of Duncan and Wood-Gush to justify this interpretation. These stereotypies occur mainly before feeding, and Rushen proposes they are therefore caused by frustration of the hunger drive, and not by a lack of adequate sensory stimulation or boredom.

The behavior shown by the sows is not as specific as the response of the chickens to a frustrating situation, the latter showing escape behavior and stereotyped pacing. Head weaving is interpreted as a stereotyped form of escape behavior in bears and horses[27] and might therefore be equivalent to the stereotypic pacing of frustrated chickens; but bar biting and head rubbing do not appear to be related to escape behavior. Rushen apparently justifies his use of the term frustration on the basis of the similarity between the restrictive situation of the sows and the chickens, and not on the similarity of their behavioral responses to the frustrating situation. Furthermore, Rushen reports of stereotypies such as vacuum chewing, drinking, and mouth running, which occur after feeding or are more evenly distributed over the day. It would be difficult to interpret these stereotypies as all resulting from the frustration of the hunger drive. As Stolba et al.[18] observe in regard to their own similar reports on stereotypies in tethered sows, "most of the observed stereotypies...seem not to be direct derivatives from the early flight attempts, but are more likely to originate in the subsequently thwarted exploratory behavior." They go on to say, however, that the stereotypies "thus appear to be products of boredom or frustration (cf. Duncan and Wood-Gush, 1972), as the sows cannot actively alter the features of their environment and cannot influence the kind and occurrence of the situations of daily activities, as they would under unrestricted conditions."

Stolba et al., like Rushen, define frustration in relation to the restrictive effect of the environment and not in relation to the performance of specific behavior. The concept of frustration thereby loses its specific experimentally defined meaning and the distinction between boredom and frustration diminishes, both concepts indicating the general effect of the environment on the welfare of the animal. As Murphy[52] argues, "Animals, or people, which receive little sensory input, and have little opportunity for behavioral output may be said to be experiencing frustration of the general activity drive. Or, in every day language, they are bored."

Such a formulation, however, denies that an animal might experience a differentiation in its psychological state in response to different aspects of the environment. Would an animal be able to experience the difference between the blocking of its behavior by the environment, and the lack of general stimulation from the environment? In Section III.D, a redefinition of the concepts of boredom and frustration is proposed which does justice to the differentiation between these concepts while also indicating the way in which they are related.

C. HELPLESSNESS

The concept of helplessness is derived from the concept of "learned helplessness". The term "learned helplessness" has been introduced in the context of psychological learning theories. It appeared that dogs, who had been given unavoidable electric shocks, showed poor motivation and a decrease in the ability to learn a simple avoidance task, such as jumping over a barrier to avoid shock after a warning signal is given, compared with dogs who received escapable shock or no shock.[59] These dogs sit passively and take the shock, having learned apparently, that irrespective of any response they might give, the unpleasant event will follow. This lack of a relationship between response and outcome causes the animal to experience a complete lack of control over its situation, without any predictability of future events. Such a state of experimentally induced learned helplessness has been observed in humans, monkeys, cats, dogs, rats, gerbils, mice, and goldfish.[29,60]

The concept of helplessness has mostly been investigated in research aimed at developing an accurate model for clinical depression in humans. Notwithstanding the fact that in this experimental context, the meaning and value of the concept is still in discussion, the terms helplessness and depression have been taken by some ethologists in the field of animal welfare to describe the psychological state animals might come to experience in long-term confinement.

In an evaluation of intensive housing conditions, McBride and Craig[61] write that "absence of opportunity for the animal to control recurring aversive events may lead to the condition of 'learned helplessness' (see review by Seligman, 1975), which may be an integral part of the failure to adapt....The construct of learned helplessness is well established among psychologists. The parallels with conditions in animal husbandry are too close for applied ethologists to ignore them further."

Considering the suffering of confined animals, Fraser[62] writes that "the typical behavioral picture of the depressed animal is a passive one. Positive, reinforcing stimuli lack influence. The bulk of the depressed animal's activity is likely to be passively derived from prompts, commands, and aversive stimuli rather than through spontaneous relationship with the environment." Besides symptoms of passivity, the animal may also express its helplessness by some form of activity: "The stereotyped continuation of an activity disorder could be likened to an agitated state of depression..." Rushen[35] as well suggests that "some element of 'hopelessness' from continued frustration may be involved in the formation of stereotypies." He considers this to be equivalent to the state of learned helplessness.

Fox[29] considers that learned helplessness in laboratory animals, resulting from lack of control and predictability, may lead to movement stereotypies, self-mutilation, sexual and social deficiencies, as well as to a lack of aggression, appetite loss, lassitude and weakness, lack of interest in the environment, and a general passivity. The experience of lack of control might be caused by the the inability of the animal to change its adverse environment, as well as by the unpredictable imposition of test procedures upon the animal.

In a context apparently similar to learned helplessness, the term apathy is also used. Ödberg[63] characterizes apathy as lasting immobility as an abnormal response to long-term restriction. He adds that "a difficulty is the exact description of apathy. An immobile animal can be sleeping, resting, freezing, simply watching, either relaxed or intensively seeking stimuli. All this is different from apathy." van Putten[21] gives an interpretation of apathy which avoids this problem: "There are housing conditions, which restrict the animals in their behavioral expressions to such an extent, that they become apathetic. Typical for this state is that these animals do not react to stimuli to which conspecifics in less extreme conditions do react very lively." A state of apathy might be accompanied by sitting with a drooping head and with the eyes half closed.

Helplessness, depression, and apathy appear to be fairly equivalent concepts, indicating a state which results from the long-term experience of lack of control over the environment and of the inability to change this aversive situation. The helpless animal can react by performing abnormal activities like stereotypic behavior or self-mutilation, or it can become passive, sometimes maintaining motionless, so-called "drowsy" postures for a long time.

Characteristic of a state of helplessness, whatever response an animal shows, is that the animal apparently has cut itself off from its adverse environment, because it does not react normally anymore to new stimuli or to changes in the environment. Although the behavioral expressions of helplessness may be very similar to those of boredom, this absence of the motivation to react to environmental change may provide a useful differentiation between the two states.

D. SUGGESTIONS FOR INTEGRATION AND OPERATIONALIZATION

In the previous discussion, it has become apparent that attempts to describe the states of boredom, frustration, and helplessness take into consideration (1) the environmental conditions which might give rise to either of these states, and (2) the behavioral symptoms a state produces.

Comparing the results from experimental work with the intuitive interpretation of ethologists, the following general consensus on the environmental circumstances which induce a certain state can be given:

1. **Boredom** — A state which arises because of a general lack of sensory stimulation in the environment.
2. **Frustration** — A state which arises because of the blocking of certain behavior by the environment.
3. **Helplessness** — A state which arises because the environment cannot be controlled in its adverse aspects by the animal.

Within the context of these environmental conditions, ethologists consider these states to be accompanied by the following behavioral categories:

1. **Boredom** — Passivity, redirected behavior, stereotyped behavior.
2. **Frustration** — Escape behavior, stereotyped behavior (notably stereotyped escape behavior).
3. **Helplessness** — Passivity, self-directed, or stereotyped behavior.

It is clear from these results that a differentiation between the three states, based on behavioral performance alone, cannot easily be made. A considerable overlap exists between the behavioral symptoms the three states are believed to produce. It will therefore be necessary to take the environmental conditions into account. Presently, no clear analysis concerning the question whether different aspects of the environment might have a different effect on animal welfare is available.

The question therefore arises how the effect of the environmental conditions on the animal could be clarified experimentally. How can we determine whether an animal performs a stereotypy out of frustration because the environment blocks a certain behavior pattern or out of boredom, in order to compensate for a general lack of stimulation; or, in the case of helplessness, because it has given up any sense of control over the environment?

In regard to helplessness, experimental data exist which clearly differentiate this state from other states. As has been pointed out by several authors, helplessness and apathy can be characterized by the fact that the animal does not respond normally to new test stimuli, but with a much lower intensity than nonhelpless animals, or not at all. This occurs because the animal has learned that no response to the environment will have the desired effect. It can be said, therefore, that the psychological state the animal experiences cannot be observed directly, but is expressed when changes in the environment are introduced by the experimentor. Could such a criterion be used also to distinguish frustration from boredom?

In the case of frustration, when certain behavior patterns are blocked, it would be expected that when this blockage is removed by a specific environmental change, the animal would respond actively and its frustration-induced behavior would disappear. In case of boredom, however, it would be expected that a specific change in the environment would not be able to eliminate abnormal behavior in the long run, since the general lack of sensory stimulation would not be cancelled by such specific changes. The animal would initially be expected to respond with much interest to the new stimulus, and its abnormal behavior might disappear. Over time, however, habituation to the stimulus would occur, and the animal would fall back into its behavioral expressions of boredom.

Thus, the introduction of a specific environmental stimulus would be expected to have the following effect on abnormal behavior: nonreversible in the case of helplessness, temporarily reversible in the case of boredom, and reversible in the case of frustration.

Evidence for such a differentiation in the response animals give to the introduction of a new stimulus in an understimulating environment is given in Section II.B.2.b. Several examples are given in which stereotyped behavior could not be eliminated by new stimuli, not even by bringing the animal into a completely different, enriched environment, as in the case of zoo polar bears or laboratory monkeys. Other examples are presented in which stereotypies could be reduced by the introduction of devices or manipulative objects to extend foraging behavior. Few authors, however, have observed their animals long enough to be able to report the effects of habituation to the novel stimuli and the reappearance of abnormal behavior. Those who have warn that initially successful enrichment might lose its effect in the long run, because abnormal behavior, passive as well as active, is seen to return.

On the basis of this evidence, I suggest that the concepts of boredom, frustration, and helplessness might be operationally distinguished by observing the behavioral response of the animal to the introduction of novel stimuli over a sufficiently long period of time. Within this context, an animal can be said to be bored when abnormal behavioral reponses to a monotonous environment can be temporarily reversed by the introduction of novel stimuli. In the case of permanent reversal or absence of reversal, the animal would be frustrated or helpless, respectively.

Only limited research has been done in this direction and has not been sufficiently specific to provide information of any detail. It would be important to test the correlation between different types of environmental stimulation and different types of stereotyped and redirected behavioral patterns. Some abnormal behaviors might be indicative of the frustration of specific behavioral patterns, others might be expressions of general boredom. Route tracing might be related to the frustration of locomotory behavior, since enlargement of the cage can eliminate this stereotypy in canaries.[36] On the other hand, stereotypies which are apparently related to exploratory foraging behavior like bar biting or vacuum chewing do only temporarily disappear with the provision of extra foraging devices and might be signs of boredom.

Not only is the stimulus substrate in regard to the reversal of abnormal behavior important, but so is the timing of the stimulus introduction; duration, variation, and predictability are factors which might induce change in different abnormal behaviors in different ways.

By setting up such experiments, a fairly detailed impression might be obtained of the differentiation between psychological states experienced by animals which are housed in restrictive, monotonous environments.

Now that it might be possible to apply the concepts of boredom, frustration and helplessness to animals in confinement, since their behavioral reaction to controlled enrichment is known, something more needs to be said about the interrelationships of these concepts.

Duncan and Wood-Gush[55] found that stereotyped pacing resulting from the frustration of feeding behavior would become fixed and irreversible after prolonged frustration. Wood-Gush et al.[9] found that different stereotypies in tethered sows became more and more fixed over time; where the sows would respond to new stimuli in early stages of the development of stereotypies, they would not do so anymore in later stages. These results show that both frustration and boredom can develop into helplessness if the environment continues to be adverse over a longer period of time. How long this development would take might depend on the characteristics of the animal species and the individual animal, and on the variety of sensory stimulation present in the environment. Very little is known at present about these processes. Insight in this matter would be crucial, however, when our aim is the prevention of the development of helplessness in confined animals.

Helplessness can be considered as the final stage of the "struggle for life" between the animal and its environment. The animal has withdrawn its attention from the environment, not responding anymore to any impulse coming from it. The terms helplessness, hopelessness, and apathy indicate that the animal no longer exerts any psychological interaction with the environment anymore; it has cut itself off. Both boredom and frustration are states preceding this final cut-off. How do they relate to each other?

As discussed in Section III.B, frustration is defined as a state caused by the blockage of behavior patterns. The animal is frustrated because it is motivated to perform a certain behavior and is blocked in doing so. In other words, the animal has a specific goal and is oriented towards the environment to reach this goal. In the case of boredom, however, the animal has no specific goal, but rather experiences the lack of a goal, because of too little environmental stimulation. It cannot satisfy its general need to behave and explore because the environment provides no substrate. The concept of boredom appears to indicate a state in which the animal suffers from too little opportunity to focus its attention towards the environment, rather than from the blocking of attention already focused towards the environment. Saying that an animal is bored, therefore, implies that the psychological interaction between the animal and the environment is low.

Thus, it appears that the differentiation between boredom, frustration, and helplessness might imply a differentiation in the degree of interaction with the environment. In the case of frustration, the interaction is intense, in the case of boredom it is low, and in the case of helplessness, the interaction is cut off.

By introducing a new stimulus into the environment, this differentiation is brought to the surface. When the stimulus is adequate, a frustrated animal will actively respond and resume normal interaction with the environment; a bored animal will temporarily reverse its low interaction, but will eventually fall back into it; and a helpless animal will not respond. The extent to which an animal interacts with its bare, restrictive environment can thus be tested by observing the extent of its interaction with a new stimulus.

Within this context, it is possible to consider frustration, boredom, and helplessness as stages of withdrawal from the environment. When the animal is introduced into monotonous surroundings, it will attempt to perform its natural range of behavior. When it does not succeed, it will show escape behavior, and perhaps signs of fear and aggression. After a while, the animal will quiet down and become more passive; it will also, however, go on attempting to exert its normal behavior by performing redirected behavior. Gradually, the animal will develop stereotyped behavior patterns, which eventually become fixated, and the animal will give up its normal behavior patterns almost entirely. Stolba et al.[18] describe such a process for sows which are brought into close confinement. They comment that the emerging stereotypies do not seem to be direct, fixed derivatives from early flight attempts, but more likely originate from the lack of sensory stimulation caused by the thwarting of exploratory behavior over a longer period of time. This formulation indicates a succession of different states from frustration to boredom and from boredom to helplessness.

The recognition of such a process provides an answer to the problem indicated at the end of Section III.B. The problem was that if frustration is broadly defined as resulting from the blocking of general behavior by severe restriction, it becomes quite difficult to distinguish frustration from boredom. The difference between the blocking of behavior by restriction and the lack of sensory input because of restriction might seem rather arbitrary. However, the line of thought discussed above has indicated that while the environment itself does not change, the impact of the environment on the animal does change over time. Thus, the animal does not experience either frustration or boredom, but both in succession.

Although it is outside the scope of this chapter, it can be noted that in case the level of sensory stimulation of the environment is relatively normal, but only certain key stimuli are lacking, frustration in the animal may develop into helplessness without the intermittent experience of boredom. Strongly motivated behaviors, like long-distance patrolling, mating, or nest building, may lead to the development of stereotypies when adequate substrate is lacking; the presence of these stereotypies within an otherwise normal behavioral range indicates that the prolonged deprivation of certain needs has influenced the well-being of the animal irreversibly.

In this section, the discussion of the concepts of boredom, frustration, and helplessness, based on experimental work in psychology as well as ethology, has led to the suggestion that these concepts can be given operational meaning by observing the effect of the introduction of new stimuli on the performance of abnormal behavioral patterns. The degree of interaction between the animal and the understimulating environment which these concepts represent thus comes to expression.

IV. BOREDOM: DO ANIMALS SUFFER FROM LONG-TERM INACTIVITY?

The purpose of the first two sections has been to integrate separate pieces of research on the concepts of frustration, boredom, and helplessness in order to eliminate some of the confusion concerning the experience of animals in understimulating environments. It would be hoped that suggestions for a clear, operationally valid demarcation of the concepts of boredom, frustration, and helplessness would benefit the improvement of the well-being of animals in confinement.

However, as was mentioned in the introduction, the application of the term "boredom" to understimulated animals meets with considerable resistance among ethologists. The fact that the term boredom in particular is subject to resistance, while concepts like frustration, fear, and aggression are not, suggests that the boredom concept might ascribe characteristics to animals which the other terms do not and which are considered uniquely human. This section will therefore investigate some theoretical implications of the ascription of a state of boredom to animals.

Are animals "really" capable of experiencing an emotion which we, human beings, call "boredom"? Duncan remarks that "In a confined and barren environment the animal is restricted in the amount of exploration it can do, and this may lead to what, *in human terms*, may be called 'boredom'"[51] (emphasis added). He also says, however, that "severe confinement may lead to frustration," without any additional remarks. The doubts expressed towards boredom apparently do not apply to other concepts which are used to discuss the welfare of animals. Wood-Gush[50] asks in relation to boredom, "can one apply such a concept to an animal as primitive as a fowl, particularly if its behavior is largely governed by releasers?" Birke and Archer[13] comment that "in every day language, we talk of animals being bored, or conversely, being curious, implying that they need or seek a certain level of stimulation. Although scientifically, we may prefer to avoid such mentalistic terms, we can study the behavioral outcome of 'curiosity' or 'boredom'...." Toates[64] concludes that "boredom is an unhelpful mentalistic construct."

Apparently there are reasons to think that the boredom concept has mentalistic connotations which other concepts do not have. When an animal is lying down for long periods, but does react normally to new stimuli, does it miss some form of stimulation? Is it bored, or just resting?[63] And when an animal redirects its behavior to inadequate stimuli by, for instance, licking and sniffing the cage bars, the floor, or the body of its mates, does it suffer from the lack of adequate stimuli? Or, as Murphy[52] put it, "as long as they are doing something, does it matter to them what it is?...Do animals have a sense of 'the quality of life'?" Thus, it seems that boredom refers to some form of awareness in the animal of the inadequacy of its own situation.

On what grounds, however, could it be decided whether or not animals might posses such capacities? The preceding analysis of welfare-related concepts has been justified, as was explained in the introduction, by the behavioral analogy between humans and animals. If animals show behavior which in humans is known to imply a certain emotion or awareness, it is therefore assumed that animals will have this subjective experience also. There seems to be no reason to exclude the concept of boredom from this principle. If the neurophysiological organization of an animal is complex enough to induce behavior indicative of boredom, there is no reason to assume that the actual experience is lacking, despite the fact that a boredom experience might not be accompanied by very conspicuous symptoms.

However, the correlation between animal behavior and conscious experience, expressed in concepts such as awareness, self-awareness, or intentionality, has been subject to fundamental criticism for a long time.[3] Only recently have cognitive scientists started to discuss the possibility of intentionality in animal behavior (see Reference 65). In the field of animal welfare, however, the use of intentional terminology for the description of animal behavior has not yet been accepted. Concepts which imply elements of awareness will therefore meet with considerable resistance. Evidence would have to be provided that explanatory models for animal welfare gain explanatory power by the use of such concepts.

In this section, it will be discussed whether the use of the boredom concept can be justified within the framework of cybernetic models. The application of cybernetic models in animal welfare research has recently been introduced.[66] The science of cybernetics had originally been developed in the field of technical engineering. The insight has been growing, however, that cybernetic models are applicable to a wide range of self-regulatory (i.e., living) systems. They attempt to integrate internal and external influences on the behavior of an animal, regarding the animal as an information-processing system. Within these models, emotions can be assigned an evaluating function regarding the adequacy of behavioral programs. For this reason, cybernetic models are considered to be particularly useful in animal welfare research.

In human psychology and psychiatry, cybernetic models are applied to understand and alleviate suffering which arises from inadequate behavior. In these fields, however, a cognitive framework is used to formulate discrepancies between, for instance, the internal expectancies of a patient and the actual environmental circumstances. This implies that the expectations of a patient are described in intentional terms, expressing what the patient wants, desires, or plans to happen. Such intentions are considered to be mental events, complemented with outwardly visible behavior (see Reference 67).

Since an intentional framework is not accepted in the field of animal welfare, ethologists who are developing cybernetic models for the explanation of animal suffering attempt to do so within a physiological framework; they regard behavior as a function of the physiological system, which includes the brain. It is questionable, however, whether such a physiological framework would have explanatory power for all levels of suffering an animal might be capable of. Human beings certainly suffer when their physiological system is threatened with disintegration, but discrepancies on other levels of organization, be it psychological, social, or cultural, can also lead to suffering when the physiological system is still intact.

It will be discussed below whether cybernetic models on a physiological level can account for suffering in animals in the case of boredom.

A. BOREDOM IN RELATION TO CYBERNETIC MODELS FOR WELL-BEING

In cybernetic models for well-being in animals, the animal brain is supposed to possess certain set-points or expectancies,[68] or so-called "Soll-werte",[69-71] regarding its environment. "These Soll-werte are neural configurations or representations of various 'Umwelt'-aspects and determined by phylogenetic, ontogenetic and experiential processes."[69] In order to compare incoming sensory input ("Ist-werte") with these Soll-werte, the animal possesses a "comparator".[68] If there is a mismatch between Soll-werte and Ist-werte, the comparator initiates a behavioral program which is aimed at the removal of the observed mismatch. "In this type of model, the motivation of a given behavior program is defined as the difference between Soll-werte and Ist-werte on which the program is directed."[69]

The attainment of Soll-werte can involve the performance of different flexible behavior programs. An animal must be able to evaluate the contribution of a certain behavior pattern to the realization of a given Soll-wert. This evaluation results in the experience of positive or negative feelings. If a behavior pattern results in the desired environmental change, positive feeling will result, which strengthens the value of that particular behavior pattern. If an observed mismatch is not reduced, negative feelings will cause the current behavior to stop or will instigate alternative programs.[68,69] These positive and negative emotions can be compared with the maintenance or loss of control by the animal over its environment.[66,69,71] The experience of control over the environment is considered to be the essential motivating force of behavior in cybernetic models.[69] If the animal would lose its sense of control, the behavioral system would be out of balance.

Authors who have taken the effort of elaborating on these cybernetic models, like Wiepkema and Inglis, make a point of qualifying their models as "cognitive". This implies that the goal (or Soll-werte, expectancy) the animal is trying to attain is not fixed, but is in integral part of an information-processing device. The presupposition of the presence of such a device means that internal set-points can themselves be subject to the influence of the external environment by negative feedback loops.[66]

For instance, if an environment is bare and does not provide enough stimulation to satisfy the exploratory needs of an animal, eventually the expectation of the animal will adapt to a new, much lower, value. The expectancy structure will slowly degenerate, and the so-called "assimilation efficiency" will decrease greatly. A new balance point will be set, which is in "harmony" with the low stimulation value of the environment. As a result, the animal's motivation to explore will decrease substantially. "If the animal is given the opportunity to explore, then such behavior will be at a much lower intensity than that shown to identical test stimuli before the environmental change."[68]

By adapting its expectation to a lower value, the animal is able to restore its sense of control over the environment. In case the external environment cannot be controlled or actively influenced, the animal will strive instead for a maximum of predictability in order to simulate a sense of control. This is accomplished by changing flexible programs into routines or habits.[71]

Stereotyped behavior patterns as shown by confined animals are considered to be such a routine. "The first time farm animals find themselves in such an abnormal housing situation, they test different behaviors to rid themselves of that situation. They try to escape, to break out, they redirect their behavior and may even start to perform sham behaviors (dust-bathing without dust, chewing with an empty mouth...and so on). In terms of the behavior model given above, farm animals will endure a disturbed welfare and experience negative emotions (suffer) in this phase. When no alternative brings about the wished solution, parts of the disturbed behavior become ritualized."[69]

Although stereotyped behavior can therefore be regarded as an indicator of a stressful environment which caused suffering in the past, it can no longer be regarded as a sign of acute suffering. The predictable character of the feedback produced by the ritualized movements reduces the probability of experiencing negative emotions. "The conclusion might be then that routines and habits, because of their highly predictable/controllable outcomes, are no longer associated with emotions."[71]

Research has shown that stereotyped behavior reduces corticosteroid levels in chickens[54] and pigs;[72] a negative correlation between stereotyped behavior and abomasal ulcers has been found in veal calves.[71] These results provide additional evidence that the performance of stereotyped behavior serves the maintenance of physiological equilibrium and, therefore, of control. The appropriateness of regarding stereotyped behavior as a parameter for some form of suffering is therefore doubted by many authors; rather, this behavior could be regarded as a succesful *adaptation* to an adverse environment.[51,52,69,70,72,73]

By defining emotion in terms of its evaluating function regarding physiological equilibrium, it follows that the experience of negative emotion disappears when such equilibrium has been restored at a lower level of behavioral flexibility. Physiological homeostasis must necessarily be attained by the animal, since it would otherwise disintegrate and die. Thus, prolonged negative emotions (suffering) cannot occur; the homeostatic state will be restored by the system as soon as possible.

Boredom and helplessness are both concepts which indicate states of suffering which result from long-term understimulation. Helplessness is specifically defined as the permanent loss of any sense of control over the environment. Within physiological cybernetic models, however, a decrease in behavioral flexibility would mean that the Soll-wert of the animal was in the process of adapting itself to a low level of stimulation and that the experience of control was being restored. The experiences of boredom and helplessness therefore have no place in such models.

Cybernetic models such as have been developed by the authors discussed above can be criticized, however, in regard to the relationship they presuppose between emotional experience and physiological experience.

The fact that emotional disturbance occurs when there is a disturbance in physiological equilibrium does not necessarily mean that emotional disturbance occurs then, and *only* then when the physiological system is out of balance. Assuming this would imply being guilty of the so-called "reductive fallacy".[74,75] This fallacy implies the unjustified reduction of one organizational level to a lower organizational level by claiming that the explanation of certain phenomena in terms of lower-level phenomena fully accounts for the higher-level phenomena. In other words, it would be wrong to claim that if negative emotion occurs in the case of physiological discrepancy, it would never occur apart from such discrepancy.

So it might be true that some emotions only occur in case of physiological discrepancy, but for other emotions this might not be true. In human beings, it is accepted that long-term confinement leads to suffering, despite the fact that the physiological system might be functioning well on a low set-point. Long-term prisoners may adjust their information-processing system to a very low level, but no one would claim that these people do not suffer under their situation. In fact, they are meant to suffer.

B. QUALIFICATIONS FOR WELFARE MODELS WHICH INCLUDE LONG-TERM SUFFERING

The question thus arises whether animals can have intentions or goals parallel to the maintenance of physiological homeostasis, and whether they would suffer if such goals would not be satisfied in the environment. Would it be possible for animals to attribute "meaning" to incoming information, not in terms of their physical survival as a physiological system (control), but in terms of a subjective evaluation of the relationship between themselves and the environment? This would imply that animals would be capable of having desires other than mere physical survival or reproduction.

Such desires would be related to the exertion of behavior *in its own right,* and not to the functional role of behavior regarding physiological homeostasis. Animals might possess the desire, and the awareness of this desire, to perform the full range of their behavioral capacities. The possibility to relate to conspecifics, to rear their young, or to patrol their territory might fundamentally matter to them. And if such behaviors are made impossible by the environment, the memory of pleasant experiences in the past related to these behaviors might cause them to suffer, even if no present stimulation may remind them of the possibility of a more active life.

Very little is known of the extent to which memory affects the well-being of animals. Anecdotal evidence exists concerning the capacity of elephants, dogs, and cats to remember previous environments for years or for their entire lives. The memory of their wild home environment might cause unrelievable suffering in wild-caught laboratory primates. For laboratory-bred animals, the memory of close intimacy with their mother and their brothers and sisters after they were born might induce suffering when they are isolated. Such memories might serve as cognitive set-points in the awareness of animals and might cause suffering for long periods of time. It would therefore be very important for animal welfare research to investigate the degree to which (long-term) memories influence the animal's awareness of its own situation.

Any subjective goal animals might have in regard to their environment could theoretically become apparent in their behavioral interaction with the environment. It might be possible to clarify the nature of the subjective relationship between animal and environment by introducing frequent environmental change and observing the diversity and the flexibility of the animal's response. If it is kept in a bare environment, which provides only for its most basic behavioral needs, the animal will not be able to express any goal or desire other than staying physically alive. The provision of a more complex and demanding environment may evoke the expression of subjective goals and desires which might be essential for its emotional experience of the environment.

The best conditions to study the behavioral goals of animals would, logically, be found in their wild, natural surroundings. There, the animal could spontaneously express its subjective capacities in ways which would never be possible in any captive environment. It is therefore appropriate that the discussion on the degree of awareness, self-awareness, or consciousness in animals is led by cognitive ethologists. Ethologists, more so than cognitive psychologists, are aiming at the provision of an appropriate conceptual framework for the characterization of the natural behavioral capacities of animals. Griffin[3] proposes that it might be possible to gain insight into the mental capacities of animals by intervening with their communicatory systems and observing the flexibility of their response. This could be done in the wild, but also in seminatural conditions. This approach is being taken with several species, such as bees,[76] parrots,[77] dolphins,[78] sealions,[79] vervet monkeys,[80] and chimpanzees,[81] and in the sign-language projects with chimpanzees and gorillas (for a review, see Reference 82). As a result of such research projects, a lively debate has been generated, discussing the degree of complexity the subjective goals of an animal might attain and the degree to which an animal is aware of its own goals and actions.

This debate is of central importance to cognitive psychology and animal welfare research, since it may provide insight into the nature of the *spontaneous* goals and desires animals might have with regard to the exertion of their behavior. This knowledge will be needed when the response of animals to *experimentally imposed* goals or purposes, as is the case in cognitive psychology and animal welfare research, is to be interpreted correctly. The systematic imposition of environmental conditions on the animal might bring physiological or behavioral mechanisms to the surface. The subjective evaluation of disturbances in these mechanisms, however, depends upon the intentions the animal might have. Such intentions are not apparent within the experimentally controlled conditions; they must be evoked by circumstances which provide the animal with the opportunity for spontaneous behavioral expression.

This principle was applied in Section III.D, where it was proposed that a differentiation between the states of boredom, frustration, and helplessness could be made by observing the animal's response over time to a new stimulus. The existence of states of long-term suffering such as boredom or helplessness can only be accepted if the animal is credited with long-term subjective goals; it can express that it has these, and that the distortion of these goals matter to it, only when provided with new stimulation. An animal can show that it experiences its environment as boring by vigorously responding to new stimuli initially, but falling back into passivity or abnormal behavior after the novelty of the stimulation has worn off. In the case of helplessness, the animal remains abnormally passive to new stimuli. This does not necessarily imply that the animal does not have any goals anymore; it only shows that the animal does not expect that its goals will be gratified by the environment (see Section III.C).

The understanding of such processes might be deepened by fundamental research which studies the effect of different levels of environmental complexity on the capacity of animals to maintain and exert subjective goals.

Whether such capacities could be attributed to physiological systems, or whether the system then ceases to be physiological and must be qualified differently, is a fundamental question which needs consideration on different levels.

This question is in fact a formulation of the so-called "mind-body problem". This is a philosophical problem which has been debated vigorously for centuries which concerns the nature of the relationship between mind and body, or psyche and physiology. A broad range of positions on this problem can be taken, and so far, no conclusive evidence on either of these positions has been provided.

Such unclarity has led to theoretical confusion not only in animal welfare research, but also in cognitive biology and cognitive psychology. In comparative psychology, the study of the "mind" in animals has become equivalent to regarding animals as information-processing systems (see References 83 and 84). This point of view denies the existence of a separate psychological entity which might "use" the information processing system, since such a system "is" the mind.[85-87]

An ethologist like Griffin,[88] on the other hand, argues that "information processing is doubtless a necessary condition for mental experience, but is it sufficient? Human minds do more than process information, they think and feel....Contemporary behaviorists...almost never mention the possibility that the animals might have feelings, memories, intentions, desires, beliefs, or other mental experiences." Griffin qualifies "information processing" as "non-conscious", and "having feelings, memories, intentions, etc." as "conscious". He does not deny, however, that such conscious capacities could be generated by a neurophysiological substrate.

The conflict between such points of view, which represent different stances in the mind-body problem, can be approached from two perspectives. First, from an epistemological perspective, which implies that the mind-body differentiation would represent the differentiation between two modes of observing the world; a subjective and an objective framework, respectively. Mind and body would then be different aspects of the same reality, being differentiated only by the perspective of the observer. Second, the mind-body problem can be approached from an ontological perspective, which implies that the mind-body distinction actually indicates the existence of two separate entities. In the first case, the "problem" of the mind-body relationship concerns the relevance of different observational perspectives for human existence, and in the second case, it concerns the nature of the causal relationship between mind and body.

For both perspectives on the mind-body problem, however, it would be useful and important to develop a conceptual framework for animal behavior on an intentional or psychological level of description. The scientific status of such a cognitive framework might be unclear, but it must be realized that such unclarity equally applies to a physiological framework.

Explanatory models such as the cybernetic models discussed in this section, which presuppose a functional relationship between emotional experience and physiological homeostasis, implicitly suggest that the causal relationship between emotional and physical events is unproblematic. This, as has been indicated, is not the case. For instance, it could be asked why, in a physiological cybernetic system, it would necessarily have to be emotional experience which fulfilled the function of the evaluating mechanism. Some form of neurochemical mechanism emitting chemical or electrical control signals might do the job as well. The animal could then be described as a causally closed physiological system, without the need for presupposing the existence of any form of subjectivity in the animal. However, if it would be considered desirable to incorporate some form of subjective awareness into explanatory models for animal welfare, it would not be justified to attribute to such awareness merely the role of an evaluating "Deus ex Machina", for the sake of physiological homeostasis. It must then be asked what the contents and the depths of subjective awareness in animals are and how such awareness generally relates to the physical system of the animal.

V. SUMMARY AND CONCLUSIONS

Can animals be bored? Most laboratory animals, as well as many farm animals and zoo animals, are housed in environments which would be considered as unbearably monotonous by human beings. Do animals suffer from their confinement in these bare, depriving surroundings, like humans would? This chapter has attempted to provide some insight into this question.

In the first section, it was shown that the behavioral response of animals to understimulating environments differs considerably from the performance of animals in more complex, demanding environments. First, understimulated animals show abnormal behavior patterns (stereotypies and redirected behavior) which are largely absent in enriched environments; second, they show a higher level of passivity, and the diversity of their behavioral range decreases considerably. Can these responses be regarded as signs of boredom? Animal welfare scientists have not been able to agree on this matter for two reasons, which were discussed in the second and third section.

In the first place, little conceptual clarity exists concerning an operational definition of boredom. The concepts of boredom, frustration, and helplessness are alternately applied to the behavioral response of confined animals. In the second section, therefore, a detailed analysis is made of the context in which these concepts are used. It is proposed that the range "frustration-boredom-helplessness" represents a decrease in the intensity of interaction of the animal with its adverse environment. The different states might be regarded as different stages in the process of withdrawal from the environment, while the animal gradually loses its sense of control over the environment. The stage in which the understimulated animal might be could be tested by providing it with a variety of new stimulation. The intensity of its response and the degree to which its abnormal behavior appears to be reversible would indicate the present emotional state of the animal. Thus, the otherwise invisible differentiation between different emotional states would be brought to expression.

In the third section, the concept of boredom is discussed in relation to cybernetic models for animal welfare, since such models are considered to provide an adequate explanatory framework for the functional relationship between emotional experience and behavior. The cybernetic models which are currently being developed, however, aim to explain behavior on a physiological level of organization. It was shown that as a consequence, a gradual decline in behavioral flexibility is not interpreted as an indication of boredom, but as an adaptive process to understimulation, restoring physiological balance (and, at the same time, the animal's sense of control) at a lower level of stimulation. Such a physiological interpretation of cybernetic models, therefore, cannot provide us with insight concerning states of long-term suffering such as boredom and helplessness. The suggestion that animals might suffer despite renewed physiological balance is even discarded as mentalistic and unscientific by several authors.

Contrary to such opinion, it is argued that although the existence in animals of psychological capacities such as boredom cannot be derived from theories formulated at a physiological level, such capacities might exist on a higher level of organization. Particularly ethologists might contribute to the development of a conceptual framework which does justice to the behavioral flexibility and complexity of animals in a natural environment, where they show spontaneous, unhampered behavioral expression. Such a framework might provide us with insight into the subjective awareness animals might have regarding the expression of their own behavior. Furthermore, research would be needed which studies the effect of the level of environmental complexity on their behavior in order to gain understanding of the psychological processes taking place in understimulated animals.

How the physiological and the psychological level of organization are causally related to one another and whether a distinction between the two levels is scientifically meaningful is a question which directly refers to a fundamental philosophical problem: the so-called mind-body problem. This problem has been debated for centuries and implicitly influences the explanation of behavior in the cognitive sciences. Ethologists, concerned with the development of explanatory models for animal welfare, might therefore deepen their understanding of current theoretical problems by the study of literature dealing with the the mind-body problem.

Does the recognition that laboratory animals can suffer because of boredom have to wait until clarity on such fundamental issues has been attained? The analogy postulate provides us with sufficient evidence (presented in Sections II and III) which indicates the need to start projects which aim at the enrichment of laboratory housing environments. Such projects will not only benefit the well-being of animals, they will at the same time provide ample opportunity for research into questions brought forward in this chapter.

An increasing number of experimentors have started such projects, being aware of the advantages these projects have for the animals as well as for themselves. Bernstein and Gordon,[90] for example, set up a project with monkeys in which improved housing conditions greatly facilitated behavioral research programs. The behavioral research on the social relationships formed by monkeys in these enriched conditions in turn resulted in "specific recommendations for establishing, expanding and culling of nonhuman primate breeding colonies." Other projects report the importance of enriched social and environmental conditions for succesful breeding in, for example, rhesus monkeys,[90] or marmosets and cotton-tops.[91] Behavioral research in this context resulted in improved animal welfare, leading to, for instance, the elimination of aggression-inducing factors[33,92] and also to increasing fundamental insight into social and reproductive behavior, which consequently led to improved breeding success.

The favorable effect of increasing the complexity of the housing environment on overall health and disease susceptability has been reported for cats[29] and for different primate species.[93] For farm animals, Wemelsfelder[94] reports that environmental enrichment has been shown to improve production, reproduction, endurance against disease, and the prevention of physical injury. Concerning rodents, Wallace[95] has designed the so-called Cambridge mouse cage, which is designed to more fully provide for the species behavioral needs; yet the cage design satisfies experimental needs as well. Instead of regarding animal and human needs as necessarily conflicting, Wallace remarks that "I thought of the mouse needs first, and my own concerns afterwards, and was rewarded with a harmonious design, and with bonus features I had not foreseen."

One such bonus point might be that keeping rats and rabbits, for example, in enriched conditions greatly enhances their recovery after experimental operations.[96-98] Environmental enrichment is known to stimulate organization and elaboration of neural structures; areas in the brain which are left intact might therefore have a greater potential to compensate for the inflicted damage.[99] Held et al. suggest that "impoverishment may have effects similar to those of actual brain-damage." If this would be the case, then few experimentors will realize that they are performing their experiments with seriously affected animals which are close to being brain-damaged. Especially where laboratory animals serve as a model for man, the effect of impoverished conditions may seriously confound experimental results.[29]

The suffering which animals have to endure because of a lasting lack of environmental stimulation apparently affects their complete physical health as well. Paying attention to boredom might therefore result in considerable economic advantage.

Besides the importance of care for the physical health of animals, the problem of boredom in animals forces us not to forget their overall integrity. It requires the recognition that animals are sentient beings, possessing psychological capacities which are continuous with those of human beings. Our willingness to accept a fundamental kinship with animals might inspire us to create environments for them which are healthy, both physically and mentally. Changes in this direction will benefit not only the well-being of animals, but the integrity of our own lives as well.

ACKNOWLEDGMENTS

The work for this chapter was supported financially by grant No. 484876 of the Department of WVC of the Dutch government. The author would like to express her appreciation for helpful comments on earlier drafts of this chapter to Prof. Dr. Tj. de Cock Buning, Dr. H. Verhoog, and Prof. Dr. W. J. van der Steen.

REFERENCES

1. **White, R. W.,** Motivation reconsidered: the concept of competence, in *Functions of Varied Experience,* Fiske, D. W. and Maddi, S. R., Eds., Dorsey Press, Homewood, IL, 1961, 278.
2. **Berlyne, D. E.,** *Conflict, Arousal and Curiosity,* McGraw-Hill, New York, 1960.
3. **Griffin, D. R.,** *The Question of Animal Awareness,* 2nd ed., Rockefeller University Press, New York, 1981.
4. **Hearne, V.,** *Adam's Task, Calling Animals by Name,* Heinemann, London, 1986.
5. **Hediger, H.,** *Wild Animals in Captivity: An Outline of the Biology of Zoological Gardens,* Dover, New York, 1950.
6. **Asquith, P. J.,** The inevitability and utility of anthropomorfism in the description of primate behaviour, in *The Meaning of Primate Signals,* Harré, R. and Reynolds, V., Eds., Cambridge University Press, Cambridge, 1984, chap. 8.
7. **Boice, R.,** Captivity and feralisation, *Psychol. Bull.,* 89, 407, 1981.
8. **Vastrade, F. M.,** The social behaviour of free-ranging domestic rabbits (*Oryctolagus cuniculus L.*), *Appl. Anim. Behav. Sci.,* 16, 165, 1986.
9. **Wood-Gush, D. G. M., Stolba, A., and Miller, C.,** Exploration in farm animals and animal husbandry, in *Exploration in Animals and Humans,* Archer, J. and Birke, L. I. A., Eds., Van Nostrand Reinhold, London, 1983, chap. 8.
10. **Price, E. O.,** Behavioural aspects of animal domestication, *Q. Rev. Biol.,* 59, 1, 1984.
11. **Signoret, J. P., Baldwin, B. A., Fraser, D., and Hafez, E. S. E.,** The behaviour of swine, in *The Behaviour of Domestic Animals,* Hafez, E. S. E., Ed., Balliere Tindall, London, 1975, 295.
12. **Fox, M. W.,** Concepts in Ethology, University of Minnesota Press, Minneapolis, 1974.
13. **Birke, L. I. A. and Archer, J.,** Some issues and problems in the study of animal exploration, in *Exploration in Animals and Humans,* Archer, J. and Birke, L. I. A., Eds., Van Nostrand Reinhold, London, 1983, chap. 1.
14. **Wood-Gush, D. G. M. and Stolba, A.,** Behaviour of pigs and the design of a new housing system, *Appl. Anim. Ethol.,* 8, 583, 1982.
15. **Wood-Gush, D. G. M. and Beilharz, R. G.,** The enrichment of a bare environment for animals in confined conditions, *Appl. Anim. Ethol.,* 10, 209, 1983.
16. **van Putten, G. and Dammers, J.,** A comparative study of the well-being of piglets reared conventionally and in cages, *Appl. Anim. Ethol.,* 2, 339, 1976.
17. **Ruiterkamp, W. A.,** Comparative investigations into the well-being of fattening pigs, *Tijdschr. Diergeneeskd.,* 111, 520, 1986.
18. **Stolba, A., Baker, N., and Wood-Gush, D. G. M.,** The characterisation of stereotyped behaviour in stalled sows by informational redundancy, *Behaviour,* 87, 157, 1983.
19. **van Putten, G.,** Quantifying well-being in farm-animals, *Tijdschr. Diergeneeskd.,* 106, 106, 1981.
20. **Kiley-Worthington, M.,** *Behavioural Problems of Farm Animals,* Oriel Press, London, 1977.
21. **van Putten, G.,** Welzijnsaspecten, in *Fokzeugen,* 8th Report of the Study-Committee on Intensive Farming, Ned. Ver. tot Besch. van Dieren, The Hague, 1982, chap. 4.
22. **Buchenauer, D.,** Parameters for assessing welfare, ethological criteria, in *The Welfare of Pigs,* Sybesma, W., Ed., Martinus Nijhoff, The Hague, 1981, 75.
23. **Sambraus, H. H.,** Beurteilung von verhaltens-anomalien aus ethologischer Sicht, in 2nd *GFT-Semin. für angewandte Nutztierethologie,* Bayerische Landesanstalt für Tierzucht, Grub, 1981, 1.
24. **Duncan, I.J . H.,** Overall assessment of poultry welfare, in *Proc. 1st Danish Semin. on Poultry Welfare in Egglaying Cages,* National Committee for Poultry and Eggs, Copenhagen, 1978, 79.
25. **Brantas, G. C.,** The pre-laying behaviour of laying hens in cages with and without laying nests, in *The Laying Hen and Its Environment,* Moss, R., Ed., Martinus Nijhoff, The Hague, 1980, 227.
26. **Wiepkema, P. R., Brown, D. M., Duncan, I. J. H., and van Putten, G.,** Abnormal behaviour in farm animals, in *CEC Report,* Commission of European Communities, Brussel, 1983, 1.

27. **Meyer-Holzapfel, M.,** Abnormal behaviour in zoo animals, in *Abnormal Behaviour in Animals,* Fox, M. W., Ed., W.B. Saunders, Philadelphia, 1968, chap. 25.
28. **Stevenson, M. F.,** The captive environment: its effect on exploratory and related behavioural responses in wild animals, in *Exploration in Animals and Humans,* Archer, J. and Birke, L. I. A., Eds., Van Nostrand Reinhold, London, 1983, chap. 7.
29. **Fox, M. W.,** *Laboratory Animal Husbandry, Ethology, Welfare and Experimental Variables,* State University of New York Press, Albany, 1986.
30. **Sambraus, H. H.,** Ethologische Grundlage einer Tiergerechten Nutztierhaltung, in *Ethologische Aussagen zur artgerechten Nutztierhaltung,* Fölsch, D. W. and Nabholz, A., Eds., Birkhäuser Verlag, Basel, 1982, 22.
31. **Berkson, G., Mason, W. A., and Saxon, S. V.,** Situation and stimulus effects on stereotyped behaviors of chimpanzees, *J. Comp. Physiol. Psychol.,* 56, 786, 1963.
32. **Morris, D.,** The response of animals to a resricted environment, in *Symp. Zool. Soc. London,* 13, 99, 1964.
33. **Goosen, C.,** Recommendations on laboratory housing for Macaques in the Netherlands, in *Standards in Labotatory Animal Management,* Unversities Federation for Animal Welfare, Hertfordshire, U.K., 1984, 245.
34. **Rushen, J.,** Stereotyped behaviour, adjunctive drinking, and the feeding periods of tethered sows, *Anim. Behav.,* 32, 1059, 1984.
35. **Rushen, J.,** Stereotypies, agression,and the feeding schedules of tethered sows, *Appl. Anim. Behav. Sci.,* 14, 137, 1985.
36. **Keiper, R. R.,** Causal factors of stereotypies in caged birds, *Anim. Behav.,* 17, 114, 1969.
37. **Keiper, R. R.,** Studies of stereotypy function in the canary (*Serinus canarius*), *Anim. Behav.,* 18, 353, 1970.
38. **Bareham, J. R.,** Effects of cages and semi-intensive deep litter pens on the behaviour, adrenal response and production in two strains of laying hens, *Br. Vet. J.,* 128, 153, 1972.
39. **Sackett, G. P.,** Abnormal behaviour in laboratory-reared Rhesus monkeys, in *Abnormal Behaviour in Animals,* **Fox, M. W.,** Ed., W.B. Saunders, Philadelphia, 1968, chap. 18.
40. **Fox, M. W.,** Socialisation, environmental factors, and abnormal behavioral development in animals, in *Abnormal Behaviour in Animals,* Fox, M. W., Ed., W.B. Saunders, Philadelphia, 1968, chap. 19.
41. **Vestergaard, K.,** Influence of fixation on the behaviour of sows, in *The Welfare of Pigs,* Sybesma, W., Ed., Martinus Nijhoff, The Hague, 1981, 16.
42. **Hutchins, M., Hancocks, D., and Calip, C.,** Behavioural engineering in the zoo; a critique. II, *Int. Zoo News,* 25, 18, 1978.
43. **Hutt, C. and Hutt, S. J.,** The effects of environmental complexity on steretyped behaviour of children, *Anim. Behav.,* 13, 1, 1965.
44. **Mason, W. A. and Green, P. C.,** The effects of social restriction on the behavior of Rhesus monkeys. IV. Responses to a novel environment and to an alien species, *J. Comp. Physiol. Psychol.,* 55, 363, 1962.
45. **Duncan, I. J. H. and Hughes, B. O.,** Free and operant feeding in domestic fowls, *Anim. Behav.,* 20, 775, 1972.
46. **Markowitz, H.,** *Behavioural Enrichment in the Zoo,* Van Nostrand Reinhold, London, 1982.
47. **Russell, P. A.,** Psychological studies of exploration in animals: a reappreasal, in *Exploration in Animals and Humans,* Archer, J. and Birke, L. I. A., Eds., Van Nostrand Reinhold, London, 1983, chap. 2.
48. **Barmack, J. E.,** The effect of benzedrine sulfate (benzyl methyl carbinamine) upon the report of boredom and other factors, *J. Psychol.,* 5, 125, 1938.
49. **Barmack, J. E.,** A definition of boredom: a reply to Mr. Berman, *Am. J. Psychol.,* 52, 467, 1939.
50. **Wood-Gush, D. G. M.,** Animal welfare in modern agriculture, *Br. Vet. J.,* 129, 167, 1973.
51. **Duncan, I. J. H.,** Animal behaviour and welfare, in *Environmental Aspects of Housing for Animal Production,* Clark, J. A., Ed., Butterworths, London, 1981, chap. 25.
52. **Murphy, L. B.,** A review of animal welfare and intensive animal production, in Report of the Poultry Section, Queensland Department of Primary Industries, Australia, 1978.
53. **Brambell, F. W. R.,** Report of the Technical Committee to Enquire into the Welfare of Animals kept under Intensive Livestock Husbandry Systems, Her Majesty's Stationary Office, London, 1965.
54. **Duncan, I. J. H.,** Frustration in the fowl, in *Aspects of Poultry Behaviour,* Freeman, B. M. and Gordon, R.F., Eds., British Poultry Science, Edinburgh, 1970, 15.
55. **Duncan, I. J. H. and Wood-Gush, D. G. M.,** Thwarting of feeding behaviour in the domestic fowl, *Anim. Behav.,* 20, 1972, 444.
56. **Hinde, R. A.,** *Animal Behaviour,* 2nd ed., McGraw-Hill, Tokyo, 1970.
57. **Dantzer, R., Arnone, M., and Mormede, P.,** Effects of frustration on behaviour and plasma corticosteroid levels in pigs, *Physiol. Behav.,* 24, 1, 1980.
58. **Arnone, M. and Dantzer, R.,** Does frustration induce aggression in pigs?, *Appl. Anim. Ethol.,* 6, 351, 1980.
59. **Seligman, M. E. P.,** *Helplessness: On Depression, Development and Death,* W.H. Freeman, San Francisco, 1975.
60. **Seligman, M. E. P. and Beagley, G.,** Learned helplessness in the rat, *J. Comp. Physiol. Psychol.,* 88, 534, 1975.
61. **McBride, G. and Craig, J. V.,** Environmental design and its evaluation for intensively housed animals, *Appl. Anim. Behav. Sci.,* 14, 211, 1985.

62. **Fraser, A. F.,** The behaviour of suffering in animals, *Appl. Anim. Behav. Sci.,* 13, 1, 1984/85.
63. **Ödberg, F. O.,** Behavioural responses to stress in farm animals, in *Biology of Stress in Farm Animals: An Integrative Approach,* Wiepkema, P. R. and van Adrichem, P. W. M., Eds., Martinus Nijhoff, The Hague, 1987, 135.
64. **Toates, F. M.,** Exploration as a motivational and learning system: a cognitive incentive view, in *Exploration in Animals and Humans,* Archer, J. and Birke, L. I. A., Eds., Van Nostrand Reinhold, London, 1983, chap. 3.
65. **Dennett, D. C.,** Intentional systems in cognitive ethology, the "Panglossian Paradigm" defended, *Behav. Brain Sci.,* 6, 343, 1983.
66. **Toates, F. M.,** The relevance of models of motivation and learning to animal welfare, in *Biology of Stress in Farm Animals: An Integrative Approach,* Wiepkema, P. R. and van Adrichem, P. W. M., Eds., Martinus Nijhoff, The Hague, 1987, 153.
67. **Carson, R. C.,** Selfulfilling prophecy, maladaptive behavior and psychotherapy, in *Handbook of Interpersonal Psychotherapy,* Anchir, J. C. and Kiesler, D. J., Eds., Pergamon Press, New York, 1982, 64.
68. **Inglis, I. R.,** Towards a cognitive theory of exploratory behaviour, in *Exploration in Animals and Humans,* Archer, J. and Birke, L. I. A., Eds., Van Nostrand Reinhold, London, 1983, chap. 4
69. **Wiepkema, P. R.,** On the identity and significance of disturbed behaviour in vertebrates, in *Disturbed Behaviour in Farm Animals,* Bessei, W., Ed., Hohenheimer Arbeiten, Stuttgart, 121, 1982, 7
70. **Wiepkema, P. R.,** Abnormal behaviour in farm animals: ethological implications, *Neth. J. Zool.,* 35, 279, 1985.
71. **Wiepkema, P. R.,** Behavioural aspects of stress, in *Biology of Stress in Farm Animals: An Integrative Approach,* Wiepkema, P. R. and van Adrichem, P. W. M., Eds., Martinus Nijhoff, The Hague, 1987, 113.
72. **Dantzer, R. and Mormede, P.,** Can physiological criteria be used to assess welfare in pigs?, in *The Welfare of Pigs,* Sybesma, W., Ed., Martinus Nijhoff, The Hague, 1981, 53.
73. **Baxter, S. H.,** Ethology in environmental design for animal production, *Appl. Anim. Ethol.,* 9, 207, 1982/83.
74. **Hospers, J.,** *An Introduction to Philosophical Analysis,* 2nd ed., Prentice-Hall, Englewood, NJ, 1967.
75. **Ingle, D. J.,** *Principles of Research in Biology and Medicine, Pitman Medical,* London, no date given.
76. **Gould, J. L.,** The dance-language controversy, *Q. Rev. Biol.,* 51, 211, 1976.
77. **Pepperberg, I. M.,** Cognition in the African grey parrot: preliminary evidence for auditory/vocal comprehension of the class concept, *Anim. Learn. Behav.,* 11, 179, 1983.
78. **Schusterman, R. J., Thomas, J. A., and Wood, F. G.,** Eds., *Dolphin Cognition and Behavior: A Comparative Approach,* Erlbaum, Hillsdale, NJ, 1986.
79. **Schusterman, R. J. and Krieger, K.,** California sea lions are capable of semantic comprehension, *Psychol. Rec.,* 34, 3, 1984.
80. **Cheney, D. L. and Seyfarth, R. M.,** Social and non-social knowledge in vervet monkeys, *Philos. Trans. R. Soc. London Ser. B,* 308, 187, 1985.
81. **Woodruff, G. and Premack, D.,** Intentional communication in the chimpanzee: the development of deception, *Cognition,* 7, 333, 1979.
82. **Ristau, C. A. and Robbins, D.,** Language in the great apes: a critical review, *Adv. Study Behav.,* 12, 141, 1982.
83. **Hulse, S. H., Fowler, H., and Honig, W. K.,** Eds., *Cognitive Processes in Animal Behavior,* Erlbaum, Hillsdale, NJ, 1978.
84. **Roitblat, H. L., Bever, T. G., and Terrace, H. S.,** Eds., *Animal Cognition,* Erlbaum, Hillsdale, NJ, 1984.
85. **Wasserman, E. A.,** Animal intelligence: understanding the minds of animals through their behavioral "ambassadors", in *Animal Cognition,* Roitblat, H. L., Bever, T. G., and Terrace, H. S., Eds., Erlbaum, Hillsdale, NJ, 1984, 45.
86. **Hearst, E.,** Stimulus relationships and feature selection in learning and behaviour, in *Cognitive Processes in Animal Behavior,* Hulse, S. H., Fowler, H., and Honig, W.K., Eds., Erlbaum, Hillsdale, NJ, 1978, 51.
87. **Terrace, H. S.,** Animal cognition, in *Animal Cognition,* Roitblat, H.L., Bever, T.G., and Terrace, H.S., Eds., Erlbaum, Hillsdale, NJ, 1984, 7.
88. **Griffin, D. R.,** *Animal Thinking,* Harvard University Press, London, 1984.
89. **Bernstein, I. S. and Gordon, T. C.,** Behavioral research in breeding colonies of Old World Monkeys, *Lab. Anim. Sci.,* 27, 532, 1977.
90. **Goldfoot, D. A.,** Rearing conditions which support or inhibit later sexual potential of laboratory Rhesus monkeys: hypotheses and diagnostic behavior, *Lab. Anim. Sci.,* 27, 548, 1977.
91. **Evans, S.,** Captive management of marmosets and tamarins, in *Standards in Laboratory Animal Management,* Universities Federation for Animal Welfare, Hertfordshire, U.K., 1984, 250.
92. **Erwin, J.,** Factors influencing aggressive behavior and risk of trauma in Pigtail macaques (*Macaca nemestrina*), *Lab. Anim. Sci.,* 27, 541, 1977.
93. **Chamove, A. S., Anderson, J. R., Morgan-Jones, S. C., and Jones, S. P.,** Deep woodchip litter: hygiene, feeding and behavioural enhancement in eight primate species, *Int. J. Stud. Anim. Prob.,* 3, 308, 1982.
94. **Wemelsfelder, F.,** Boredom in animals, in *Advances in Animal Welfare Science,* Fox, M. W. and Mickley, L., Eds., Martinus Nijhoff, The Hague, 1984, 115.

95. **Wallace, M. E.,** The mouse: in residence and in transit, in *Standards in Laboratory Animal Management,* Universities Federation for Animal Welfare, Hertfordshire, U.K., 1984, 25.
96. **Will, B. E., Rosenzweig, M. R., and Bennett, E. L.,** Effects of differential environments on recovery from neonatal brain lesions, measured by problem-solving scores and brain dimensions, *Physiol. Behav.,* 16, 603, 1976.
97. **Shibagaki, M., Seo, M., Asano, T., and Kiyono, S.,** Environmental enrichment to alleviate maze performance deficits in rats with microcephaly induced by X-irradiation, *Physiol. Behav.,* 27, 797, 1981.
98. **de Vos-Korthals, W. H. and van Hof, M. W.,** Residual visuomotor behaviour after bilateral removal of the occipital lobe in the rabbit, *Behav. Brain Res.,* 15, 205, 1985.
99. **Held, J. M., Gordon, J., and Gentile, A. M.,** Environmental influences on locomotor recovery following cortical lesions in rats, *Behav. Neurosci.,* 99, 678, 1985.0

Part VII
Anesthesia and Analgesia

Chapter 17

ANESTHESIA FOR LABORATORY ANIMALS: PRACTICAL CONSIDERATIONS AND TECHNIQUES

Pamela H. Eisele

EDITOR'S PROEM

The most patent method for control of pain in many procedures involving laboratory animals is anesthesia. Yet, all too often, researchers are not formally trained in anesthesia protocols. As a result, anesthesia which is done in research contexts is often outmoded, the frequent reliance upon ether even today providing an excellent case in point. Yet, in the past few years, great sophistication has developed in the field of laboratory animal anesthesia, as evidenced by a voluminous literature now extant in the field. In this chapter, Dr. Eisele summarizes the state of the art of anesthesia for a variety of laboratory animals, while at the same time stressing significant general principles relevant to a wide variety of situations.

TABLE OF CONTENTS

I. INTRODUCTION

The provision of safe, adequate anesthesia for animals in the laboratory setting presents a challenge to researchers, laboratory personnel, and attending veterinarians. Contributing to this challenge are the wide variety of species used in research, the small size of many of these animals, and special conditions that may be presented by the laboratory environment or dictated by experimental protocol. Nevertheless, a review of the literature provides us with an impressive number of anesthetic protocols that have been developed for use in laboratory animals. Since drugs are rarely, if ever, approved for use in many common laboratory species, current practice is based on general concepts of veterinary anesthesia, current standards of veterinary practice, the experience of those working in the field, and the relevant literature.

Appropriate anesthesia, when required as part of a research protocol, is important for humane and scientific reasons to insure that pain, stress, and distress are minimized in animal subjects. Current animal welfare regulations reaffirm that appropriate anesthesia and analgesia shall be provided for research animals. The attending veterinarian should consult in the planning of procedures that may be expected to cause pain and make recommendations as to how pain and distress can be minimized. Additionally, the Animal Care and Use Committee and the attending veterinarian should provide guidelines for the institutional research community on the proper use of anesthetics, analgesics, tranquilizers, and euthanasia agents.[1,2]

A number of other valuable resources are available to assist investigators in providing appropriate anesthesia for their research animals. Veterinary anesthesia textbooks address principles of anesthesia, species considerations, and equipment and drug use in depth.[3-7] Pharmacology texts discuss drug properties, effects, and interactions.[8,9] References devoted to laboratory animal medicine and surgery describe some of the specific techniques used for anesthesia in species commonly found in the research environment.[10-13] Additionally, specialists in the fields of veterinary or human anesthesiology, pharmacology, and physiology are available at various institutions for consultation in the planning of specific anesthetic regimens.

This chapter is designed as an introduction to some of the practical considerations and techniques for providing anesthesia for the more common species of laboratory mammals. The discussion is focused on the topic of general anesthesia and describes techniques used at our laboratory animal facilities as well as selected techniques from the literature.

II. GENERAL CONSIDERATIONS

A number of factors should be considered in the selection of an appropriate anesthetic regimen for use in laboratory animals. These include considerations related to the animal(s), the procedure, and the anesthetic agent(s) to be used.

Examples of general considerations related to the animal include species, breed, and strain; age; and sex. Overall physical condition, nutritional status, and underlying health problems such as hepatic, renal, cardiovascular, or metabolic diseases should be assessed prior to planning the anesthetic, as these conditions may influence the animal's response to pharmacological agents. A history of previous drug exposures, responses, and sensitivities is also helpful. Standardization of the time of day planned for the procedure may be important in predicting anesthetic dose response and duration of effect and in minimizing experimental variability in rodent and other species with well-defined circadian rhythms.

There may be a marked difference between individual animals in response to anesthesia due to the above described variables. These factors should not only be considered in the planning of an anesthetic but should also be reexamined if anesthetic outcomes are not as anticipated.

A number of general considerations related to the procedure are also important in selection of an appropriate anesthetic regimen. These include the type of procedure to be performed, the projected length of the procedure, and the amount and type of pain or distress that the procedure may cause. While it is difficult to precisely predict the amount of pain that may be associated with a particular procedure, recent guidelines suggest that stimuli that are considered to be painful in humans, threaten to cause or do cause damage to tissues, and evoke aversive behavior in animals are to be considered painful to animals.[14] Steps to minimize pain should be taken accordingly.[15] If general anesthesia is required, the regimen selected should provide loss of sensation with loss of consciousness and adequate muscle relaxation to allow safe and humane completion of the procedure. Rapid recovery at the end of the procedure is generally considered desirable.

Other factors that must be considered include physiological imbalances or dysfunctions that may be caused, either inadvertently or by design, by the procedure such as cardiac arrhythmias, hypovolemia, hypoxia, hypothermia, or compromised organ function. Study goals are of obvious importance, and potential effects of pharmacological agents on study parameters should be anticipated and minimized as much as possible.

Another aspect of the procedure that may be of importance is whether it is to be performed under terminal or survival conditions. Agents associated with prolonged recovery or delayed adverse effects may be satisfactory for studies in which the animal is euthanatized at the end of the procedure or shortly thereafter, while being unsatisfactory for survival studies.

A purely practical aspect of the procedure that may affect the choice of anesthetic has to do with the projected numbers of animals to be anesthetized at one time. This consideration plus the availability of trained personnel and anesthesia equipment may dictate whether inhalational or injectable techniques will be used.

The last group of considerations to be discussed are those related to the anesthetic agents to be used. The anesthetic regimen selected must be appropriate for the animal and for the procedure as discussed above. Agents should be selected based on their desirable properties with an understanding of potential or known side effects. Experience with a particular regimen is helpful as previous problems can be identified and steps taken to avoid them in the future. In many cases, a proven technique familiar to personnel may be the safest approach.

In addition to drug safety, ease of use by laboratory personnel may influence the choice of anesthetic. For example, the use of controlled substances is regulated by the federal government and requires special locked facilities and the maintenance of use logs. Because of these regulations and the abuse potential of these drugs, some investigators may prefer not to use controlled substances for animal anesthesia. Other considerations related to the ease of use include available drug concentrations, volumes of injectable agents required for anesthesia, and preferred routes of administration. For example, some commercially available drugs are too concentrated for safe use in mice. Others may not be concentrated enough for practical administration in large swine. Some agents such as volatile anesthetics require special equipment and additional personnel training for proper use.

Clearly, a thorough understanding of the properties of anesthetic agents and the principles of their use is integral to the success of the anesthetic. General pharmacology and veterinary anesthesia texts should be consulted for relevant details on the various individual agents during the anesthetic planning process.

In summary, an anesthetic regimen should be selected with attention to the above discussed considerations related to the particular animal, procedure, and pharmacological agents. It should provide adequate depth and duration of anesthesia, analgesia, and muscle relaxation with minimal side effects and without untoward effects on important study parameters. It should be appropriate for the animal given its species, medical history, and physical condition. Finally, it should be safe for both the animal subject and for personnel working with the selected agents.

III. PHARMACOLOGICAL AGENTS AND TECHNIQUES COMMONLY USED IN LABORATORY ANIMAL ANESTHESIA

A number of different pharmacological agents may be combined to produce the desired overall effect of general anesthesia. Drug combinations may allow for reduced dosages of individual agents, thus minimizing the undesirable side effects of each. A disadvantage of the use of combination regimens in research is that one often cannot predict the overall effect that drug combinations and interactions will have on study parameters. Therefore, single agent anesthetics may be preferred in some cases where experimental variables must be minimized.

Pharmacological agents can be conveniently described in a number of ways. For example, they may be classified by the way in which they are used. Such categories for agents used in anesthesia might include anesthetic premedications, anesthetic induction agents, agents used for anesthetic maintenance, and adjunctive drugs such as neuromuscular blocking agents. Another way to describe drugs is by their pharmacological action. For example, sedatives produce sedation.

The following section provides a brief introduction to the way in which some of the more commonly used agents are applied in laboratory animal anesthesia. It also briefly addresses techniques that may be used for anesthetic maintenance and provisions that should be made for monitoring and supportive care. Mention of selected attributes and actions of agents used for laboratory animal anesthesia are included for illustrative purposes in this section and in those that follow describing anesthetic techniques for the various species. General veterinary anesthesia and pharmacology texts should be consulted for further information on these drugs.[3-9]

A. ANESTHETIC PREMEDICATIONS

Drugs commonly used as anesthetic premedications in animals include anticholinergics, tranquilizers, sedative analgesics, neuroleptanalgesics, and narcotic analgesics.

1. Anticholinergics

Anticholinergics such as atropine sulfate and glycopyrrolate are used in selective cases to reduce salivary and respiratory secretions when agents such as ketamine, azaperone, and some inhalants are used. An anticholinergic may also be used to prevent bradycardia associated with vagal stimulation and the administration of drugs such as potent narcotics and xylazine. Potential side effects include dilation of the pupil of the eye, reduced tear formation, decreased intestinal motility, and tachycardia. Therefore, use of these agents is contraindicated where preexisting tachycardia is present. Some rabbits and rats are reported to have atropinesterases that allow them to metabolize atropine rapidly.[11] Atropine has only a short duration of action in ruminants as well.

2. Tranquilizers

Tranquilizers are commonly used as preanesthetics in animals to relieve anxiety, reduce stress, and facilitate handling and restraint. While they are not analgesics or anesthetics, the use of tranquilizing premedications reduces the amount of subsequent anesthetic required. Examples of tranquilizers available for use include phenothiazines such as acetylpromazine, chlorpromazine, and promazine; butyrophenones such as droperidol, lenperone, and azaperone; and benzodiazepines such as diazepam and zolazepam. Some, such as acetylpromazine and droperidol, have alpha adrenergic blocking properties which cause vasodilatation. This effect can be used to advantage for blood collection from the ear veins and arteries of rabbits.[16] However, it may also be associated with hypotension and decreased ability to thermoregulate, which can lead to the development of hypothermia.

3. Sedative Analgesics

Xylazine, an agent that has been described as a sedative analgesic with muscle relaxing properties, is commonly used as an anesthetic premedication, in combination anesthetic regimens with agents such as ketamine, for chemical restraint, and for its analgesic properties in a wide variety of laboratory animal species. It is approved for use in dogs, cats, horses, and several species of wild animals. Its use as a premedication allows reduction of subsequent anesthetics administered by as much as 30 to 50%. Thus, careful dosing is required to prevent anesthetic overdosage when combination regimens using xylazine are employed. It is notable that there are wide species variations for safe dosages of xylazine. While an effective dosage for the rabbit may be in the range of 3 to 5 mg/kg, it is much lower (0.05 to 0.2 mg/kg) for sheep and goats. It is also important to know that xylazine comes in two concentrations; an equine preparation containing 100 mg/ml and a small animal preparation containing 20 mg/ml.

The use of xylazine is associated with a number of side effects. Cardiovascular effects include bradycardia, second-degree atrioventricular block, decreased cardiac output, and peripheral vasoconstriction. Other side effects include hyperglycemia and glucosuria, emesis in cats and occasionally in dogs, and depression of thermoregulation. Bradyarrhythmias may be prevented by prior administration of atropine.[17] The alpha$_2$ adrenergic antagonist, yohimbine, is currently being evaluated as a reversal agent for xylazine in a variety of species. Other drugs being investigated for xylazine reversing properties include tolazoline, 4-aminopyridine, and doxapram.[3,18,19]

4. Neuroleptanalgesics

Neuroleptanalgesic agents combine a tranquilizer and an analgesic. A veterinary neuroleptanalgesic commonly used in the U.S. combines the opioid analgesic, fentanyl (0.4 mg/ml), with the butyrophenone tranquilizer, droperidol (20 mg/ml). This combination is used in the laboratory to produce analgesia and sedation in dogs, rabbits, some rodents, swine, and primates. It is not recommended for use in hamsters. While not generally used in cats, it was found to be an effective sedative when used at a dosage of 1 ml/9 kg subcutaneously (SC) in this species.[20] The drug may be used in species such as the rabbit and dog at different dosages to produce effects ranging from light sedation at low doses to heavy sedation and analgesia sufficient for minor surgical procedures at higher doses. Used as a premedication, it allows reduction of subsequent anesthetics by 30 to 50%. Side effects associated with the use of fentanyl-droperidol include bradycardia, which may be prevented by premedication with atropine, hypotension, respiratory depression or panting, salivation, defecation, and hyperresponsiveness to auditory and other external stimuli.[4,21]

5. Narcotic Analgesics

Opioid or narcotic analgesics such as morphine, meperidine, and oxymorphone are sometimes used at low doses as anesthetic premedications in selected species. Narcotic effects vary in different species and with differing drug dosages and potencies. For example, an opioid that produces sedation and analgesia in the dog may produce excitement in cats, horses, ruminants, and pigs. Therefore, these drugs must be selected and dosed carefully with attention to species differences. Potential side effects of these drugs include respiratory depression, histamine release, nausea and vomiting, cough suppression, and decreased gastrointestinal motility. One of the advantages associated with the use of narcotic analgesics is that they may be reversed with specific antagonists such as naloxone.

B. ANESTHETIC INDUCTION AGENTS AND TECHNIQUES

1. Injectable Anesthetic Induction Agents and Techniques

Anesthetic induction in laboratory animals, following premedication when appropriate, is achieved with injectable and/or inhalant anesthetics. Injectable agents may be administered by a variety of routes: intravenously (IV), intramuscularly (IM), intraperitoneally (IP), and subcutaneously (SC). The drug dosage and route of administration selected should be appropriate for the agent, species, and desired effect(s). Drugs such as thiamylal sodium, formulated for IV use, may be damaging to surrounding tissues if inadvertently allowed to leak from a vein or administered by another route. Volumes and concentrations of injectables may need to be adjusted for use in the smaller animals and dilute solutions used for injection into delicate vessels such as porcine and rabbit ear veins.

a. Dissociative Anesthetics

Injectable anesthetic induction agents commonly used in laboratory animal species include the dissociative anesthetic, ketamine hydrochloride, and barbiturates. Ketamine can be administered by a variety of routes for immobilization of animals such as primates and large swine and for induction of anesthesia. The drug has a wide margin of safety and, while it is approved for use in cats and primates, use of the drug has been reported for a wide variety of other species as well.[22] Ketamine produces unconsciousness and somatic analgesia, but animals generally retain high muscle tone and may move spontaneously. Ketamine is not generally considered adequate when used alone for surgical procedures associated with severe pain, especially when of visceral origin. Therefore, it is often used in combination with other agents to achieve better surgical anesthesia, analgesia, and muscle relaxation. Combination with xylazine is often used for this purpose, with duration of effects ranging from 40 to 100 min.[22]

Ketamine anesthesia is characterized by salivation and the persistence of many reflexes including ocular, laryngeal, and swallowing reflexes. Eyes remain open and should be protected from trauma and drying with a bland ophthalmic ointment or artificial tears. Effects on the cardiovascular system are generally supportive or stimulatory with increases in heart rate, cardiac output, and arterial blood pressure observed in normal animals.[23] The drug may additionally cause a shallow, irregular, apneustic pattern of breathing and dose-dependent respiratory depression. The use of ketamine is reported to be associated with seizures in up to 11% of dogs and occasionally in other species.[23] Therefore, it is not recommended for use alone in dogs or for use in animals with a history of epilepsy.

b. Barbiturates

Barbiturates used for induction of anesthesia include ultrashort-acting drugs such as thiamylal, thiopental, and methohexital sodium and the short-acting drug, pentobarbital sodium. These drugs have little if any intrinsic analgesic activity, producing anesthesia and muscle relaxation by depression of the central nervous system. Because of this, a drug such as pentobarbital must often be used at dangerously high doses to achieve adequate anesthesia for major surgery. The ultrashort-acting barbiturates are most often used in larger species such as dogs and cats for anesthetic induction, followed by placement of an endotracheal tube and maintenance of anesthesia with an inhalant agent. They may also be used to provide anesthesia of short duration for diagnostic or minor surgical procedures. These agents are commonly used as 1 to 5% solutions which are alkaline and, therefore, specified for IV administration. Local tissue damage can result from inadvertent perivascular infiltration of these solutions. When inducing anesthesia with one of these agents, one third to two thirds of the calculated dose is initially administered IV, with the remainder titrated until the desired anesthetic effect is achieved. When premedications such as acetylpromazine or xylazine are used, the dose should be reduced accordingly. Anesthesia is generally produced within 30 to 90 s and lasts 5 to 15 min in dogs and cats.[3]

Pentobarbital sodium is still commonly used as an agent for anesthetic induction and maintenance in laboratory animals. Ideally it should be dosed IV to effect. However, it is often administered IP in the smaller species. Anesthetic effects are variable in rodents, especially when administration is via the IP route. Recovery from pentobarbital is generally prolonged in species other than small ruminants, making it suboptimal for use in survival procedures. Metabolism of the drug is affected by levels of liver microsomal enzymes which can be induced by certain types of cage bedding. Additionally, the barbiturates themselves are microsomal enzyme inducers.

Barbiturates are potent respiratory depressants. They have a narrow margin of safety in smaller animals such as rabbits and guinea pigs where anesthetic doses may be associated with respiratory arrest. Their duration of action is prolonged in the presence of hypothermia or if glucose is administered during the recovery period in some species.[5] Therefore, when using these agents it is important to titrate dosages carefully and to make provisions to support ventilation and body temperature.

c. Miscellaneous Agents

A variety of other injectable anesthetic agents have been used in various laboratory animal species. These include drugs such as chloral hydrate, alpha-chloralose, and urethane.

Chloral hydrate is a hypnotic agent that has been used for IP injection in rodents. In addition to being a poor analgesic with respiratory and cardiovascular depressant effects, it is irritating to tissues and has been associated with chemical peritonitis and adynamic ileus when administered intraperitoneally.[24] Its use as a general anesthetic for rodents has been largely replaced by safer, more effective techniques.

Alpha-chloralose is an agent with hypnotic properties that is believed to have minimal effects on the autonomic nervous system. It has been used in laboratory animals to produce basal narcosis of long duration for terminal physiological studies that do not involve significant surgical stimulation. Its use is generally limited to this type of study due to its questionable ability to produce a surgical plane of anesthesia at useful dosages and its prolonged recovery phase.[3, 6, 25]

Urethane or ethyl carbamate has been used similarly in laboratory animals. Because of mutagenic and carcinogenic properties, its use is considered hazardous to both laboratory personnel and animals and is no longer generally recommended.

2. Inhalant Anesthetic Induction Agents and Techniques

Anesthetic induction with inhalant agents can be achieved with the use of an induction chamber for the smaller species such as rodents, rabbits, and cats or with a mask for larger species such as dogs, pigs, and sheep. Administration of a volatile anesthetic via a mask can also be used to deepen anesthesia initiated by an agent such as ketamine, fentanyl-droperidol, or xylazine in preparation for endotracheal intubation. Inhalant anesthesia generally allows for better control of anesthetic depth and a relatively rapid recovery since most of the volatile agent is eliminated during the process of normal breathing.

Inhalant techniques require gas anesthesia equipment appropriate for the volatile agent and for the animal's size. This includes a source of carrier gas, which is usually oxygen, necessary pressure regulators, gas hoses, and flowmeters, a way of safely vaporizing the volatile anesthetic agent (vaporizer), an anesthetic breathing circuit appropriate for the animal's size which is capable of delivering the required amount of anesthetic gas and of removing waste gases including carbon dioxide (CO_2), and an endotracheal tube or mask connecting the breathing circuit to the animal.

A variety of anesthetic breathing circuits are available and are discussed in detail elsewhere.[4,12] In general, pediatric breathing circuits such as the Bain coaxial circuit or Ayre's T-tube which minimize equipment dead space and resistance to breathing should be used for animals weighing less than 10 kg. To minimize rebreathing of CO_2, it is recommended that gas flows of 2 to 2.5 times minute ventilation be used with these anesthetic circuits, where minute ventilation is the product of respiratory rate multiplied by the volume of gas inhaled with each breath (tidal volume). For larger animals, anesthetic "circle systems" are commonly used which allow the use of lower gas flows, reducing the amount of volatile agent used and the amount of waste gases produced. These systems have inhalation and exhalation hoses, unidirectional valves, a CO_2 absorber, a reservoir bag, and a release or "pop-off" valve for overflow waste gases which should be connected to a scavenging system.

Provisions should be made to scavenge waste anesthetic gases to minimize exposure of personnel to the inhalant agents. Potential for waste gases to escape into the laboratory is especially high when induction chambers and anesthetic masks are used. Therefore, it is preferable to limit their use as much as possible. Where masks are employed it is important to make sure that they fit properly and make adaptations to provide for scavenging if they are to be used for longterm anesthetic maintenance. Anesthetic chambers should be constructed of transparent material to allow observation of the animal during the induction period with sufficient structural strength to contain the animal safely. The chamber should be airtight when closed, with an inflow port for delivery of fresh anesthetic gases and an outflow port connected to a scavenging system. When the animal is removed from the induction chamber, care should be taken to minimize the release of anesthetic gases into the laboratory by such measures as turning off the anesthetic agent, utilizing a quick oxygen "flush" and opening the chamber under a ventilated hood.

Commonly used inhalant anesthetics include the volatile agents, halothane and methoxyflurane. Newer agents such as isoflurane and enflurane may offer advantages in certain cases. Halothane, isoflurane, and enflurane require the use of precision vaporizers for safe administration. These last three agents should not be used in closed chambers such as Bell jars for anesthetic induction as they can rapidly reach lethal concentrations under these conditions due to their high vapor pressures.

The potencies of the various inhalant anesthetics have been determined for a number of species. Potency in this case is defined as the minimum alveolar concentration (MAC) of the anesthetic required to prevent movement in response to a standardized noxious stimulus in 50% of the animals tested. Anesthetic concentrations of approximately 1.25 MAC are generally used for surgical anesthesia, but it is important to keep in mind that adjustments need to be made based on the requirements of individual animals and the amount and kind of premedication used.

a. Methoxyflurane

Methoxyflurane is a potent volatile anesthetic with analgesic properties which is nonflammable and nonexplosive at room temperature in air or oxygen. Because it is highly soluble in blood and fat, anesthetic induction and recovery is relatively slow when compared to other inhalants. Due to this quality, it is inconvenient to use for anesthetic induction in larger animals or in cases where rapid recovery is desired. Methoxyflurane anesthesia is frequently used in dogs and cats following anesthetic induction with another agent and in small laboratory rodents. The agent does not volatilize readily, reaching a maximal concentration of only about 3.5% at room temperature. Because of its low vaporization, it is commonly used in a closed chamber such as a Bell jar to induce anesthesia in rodents. For this purpose, a cotton pledget to which methoxyflurane has been applied is placed in the jar. The animal to be anesthetized is placed on a raised platform in the jar so that it does not contact the liquid methoxyflurane, and the lid is secured. The animal is removed when it is observed to be adequately anesthetized.

As with other volatile anesthetics, methoxyflurane produces a dose-related depression of the cardiovascular and respiratory systems. Metabolism of methoxyflurane produces flouride ions that may be damaging to the kidneys during prolonged exposure. Therefore, length of anesthesia should be limited as much as possible and, as for all volatile anesthetics, waste gases containing methoxyflurane should be properly scavenged.

b. Halothane

Halothane is a potent anesthetic which volatilizes readily and is neither flammable nor explosive. Since anesthetic induction and recovery are rapid with this agent, it can be easily used for induction with a mask or in a flow-through chamber. For safe anesthesia it should be administered at a carefully controlled concentration from a precision vaporizer manufactured for use with halothane. When used properly, halothane provides a safe, humane, manageable anesthetic for a wide variety of laboratory animal species.

Halothane produces dose-dependent depression of the cardiovascular system with decreased cardiac output, peripheral vasodilatation, and moderate hypotension observed at anesthetic levels. Administration of appropriate volumes of IV fluids helps support cardiovascular function during halothane anesthesia. Support of body temperature with a circulating warm water blanket or equivalent device is required to prevent the development of hypothermia secondary to vasodilatation.

Halothane also produces dose-dependent respiratory depression and sensitizes the myocardium to catecholamine-induced arrhythmias. Hepatic necrosis, which can occur in some humans following halothane anesthesia, has been reported with repeated halothane anesthesia in guinea pigs.[10] Halothane administration is associated with the development of malignant hyperthermia in susceptible people and pigs and should be avoided in these individuals.

When used properly, halothane provides a safe anesthetic for a wide variety of laboratory animal species.

c. Nitrous Oxide

Nitrous oxide (N_2O) gas may be used in concentrations of 50 to 70% to supplement other anesthetics in some species, but its limited potency in animals dictates that it should not be used alone to provide surgical anesthesia. Since it has analgesic properties and minimal cardiovascular and respiratory effects, N_2O can be used to advantage to reduce the amount of other agents required to produce surgical anesthesia, thus reducing their undesirable side effects. It can also be used to speed anesthetic induction with other volatile anesthetics via the "second gas" effect.[4] Nitrous oxide should be turned off 5 to 10 min prior to discontinuing oxygen administration. This is recommended to prevent hypoxia from developing as N_2O rapidly diffuses out of the blood, diluting adveolar gases. Nitrous oxide should not be used in cases where closed gas spaces are present in the body, as it will rapidly diffuse into these gas pockets, causing them to expand. Examples of gas pockets of this type include cases of pneumothorax, strangulated loops of bowel, and potentially, gas accumulated in the rumen and other organs. When N_2O is administered, provisions should be made to monitor inspired oxygen concentrations and arterial blood gas parameters so that hypoxia can be identified and corrected if it develops.

d. Ether

Diethyl ether has been used extensively as an anesthetic in the laboratory environment. It is flammable and explosive, posing a significant hazard to personnel. It is irritating to mucous membranes, causing coughing, increased salivary and respiratory secretions, laryngospasm, and exacerbation of preexisting respiratory disease. Due to its undesirable properties, ether is being replaced by the newer, safer inhalant anesthetics available for use in the laboratory.[3]

C. ANESTHETIC MAINTENANCE

Anesthesia, once induced, can be maintained with incremental doses or constant infusions of injectable agents or with inhalant agents. Inhalants may be administered via mask, endotracheal tube, or tracheostomy tube. The latter technique is reserved for use in special experimental protocols, in emergencies, or in terminal procedures using very small animals where it is difficult to place an endotracheal tube. In rabbits and larger animals it is recommended that an appropriately sized endotracheal tube be placed to facilitate support of ventilation and administration of inhalant anesthetics and oxygen, allow for emergency resuscitation, and protect the airway from aspiration of stomach contents and oral secretions. Placement of an endotracheal tube is recommended in certain cases for the smaller animals as well.

D. NEUROMUSCULAR BLOCKING AGENTS

These agents are occasionally used, especially in larger species, in conjunction with anesthetic agents to improve muscle relaxation. Examples of these drugs include pancuronium bromide and succinylcholine chloride. These drugs are paralytic agents, but are not considered to be anesthetics or analgesics. They are specifically not to be used without appropriate anesthesia and analgesia where painful stimuli are present.[2] Animals must be monitored during the use of these agents to assure adequate anesthesia and analgesia. Additionally, provisions must be made to support ventilation, usually with a mechanical ventilator, when neuromuscular blocking agents are used.

E. PREPARATION FOR ANESTHESIA

Careful preparation contributes to the success of any anesthetic. The animal should be evaluated and readied for the procedure, the anesthetic regimen should be selected based on the considerations discussed, and necessary equipment and supplies procured ahead of time.

Ideally, evaluation of the animal includes a physical exam and assessment of current weight, familiarity with the animal's medical history, and any diagnostics required for overall assessment of the animal's general health. In-depth evaluations may not be practical for smaller species and large numbers of animals. However, it is still important to be familiar with the overall health status of the animals to be sure that they are suitable for the procedure and to avert problems during anesthesia and postoperatively.

Preparation of the animal includes correction or stabilization of any underlying health problems where possible and appropriate fasting. Rabbits and rodents don't vomit, and preoperative fasting is not generally required in these animals. Guinea pigs commonly retain a bolus of food on the base of their tongues. An overnight fast prior to anesthesia is helpful in this species to clear the mouth of food. It also allows for a more accurate assessment of body weight by reducing the volume of gastrointestinal contents. Animals such as dogs, cats, small ruminants, pigs, and primates are generally fasted prior to anesthesia to reduce the danger of vomiting and aspiration of stomach contents. Food is commonly withheld overnight, but the period of fasting may vary from several hours for neonates and very small animals to 48 h for small ruminants or in preparation for certain kinds of bowel surgery.

The final step in preparation is to assemble the necessary drugs and equipment for administration of anesthesia, for monitoring, and for support of the animal during the procedure. A safety check should be run on the anesthetic equipment to assure that it is working properly.

F. MONITORING AND SUPPORTIVE CARE

Anesthetic depth should be continuously monitored in animals under anesthesia to be sure that they are adequately anesthetized, but not overdosed. Animals should not be aware of painful stimuli while anesthetized. Neither should they be so depressed physiologically that body systems are unable to support normal function.

The classical signs and stages of general anesthesia, progressing from induction through four planes of surgical anesthesia to anesthetic overdose and death, were originally described for diethyl ether anesthesia in humans by Guedel. However, it is generally recognized that signs of anesthetic depth vary with the animal species, the individual, the amount and kind of painful stimulus applied, and the pharmacological agents used. Signs of anesthesia specific to the various anesthetic agents are described in general anesthesia textbooks, and adequacy of anesthesia is further discussed in a recent review.[26]

General signs that may be observed to help assess the level and adequacy of anesthesia include responses to noxious stimuli presented in the form of surgical manipulation or a toe, tail, or ear pinch. Responses to such stimuli indicating light or inadequate anesthesia include an increase in muscle tone, swallowing, chewing, coughing, vocalizing, sweating, tearing, and shivering or spontaneous movement that may be random or purposeful. Changes in parameters such as heart rate, arterial blood pressure, and breathing pattern in response to stimuli may also be helpful in evaluating anesthetic level. Other useful indices include eye movement and position in the orbit, pupillary size and response to light, and the presence or absence of corneal and palpebral reflexes. Stable respiration, arterial blood pressure, and heart rate and rhythm with an absence of purposeful movement in response to surgical stimuli generally indicates an adequate level of anesthesia.[13] Respiratory depression with shallow or abdominal breathing patterns and signs of cardiovascular depression and decompensation such as significant hypotension, bradycardia, tachycardia, and other arrhythmias are associated with anesthesia that is dangerously deep. In general, a number of the above discussed parameters should be assessed at any given point in time to evaluate the level of anesthesia and analgesia and the overall physiological well-being of the animal.

Appropriate monitoring of physiological parameters in the anesthetized animal will be somewhat dependent upon the procedure, animal species, and anesthetic used. Parameters to monitor in addition to study parameters include body temperature and indicators of cardiovascular and respiratory status. Examples of measures of cardiovascular function include heart rate, EKG, arterial blood pressure, central venous pressure, and cardiac output. Measures of respiratory function include respiratory rate, rhythm, and depth, end-tidal CO_2, and arterial blood gases. Additionally, recording of the EEG as an assessment of neurological status may be helpful or germane to some procedures.

Supportive care for the anesthetized animal may include administration of IV fluids and therapeutic agents such as antibiotics and antiarrhythmic drugs. A bland ophthalmic ointment or artificial tears should be placed in the eyes to prevent drying of the corneas during the anesthetic period. In some cases, it may be helpful to lightly tape the eyelids closed to prevent inadvertent ocular trauma. Respiration and body temperature should be supported. Smaller animals with large surface to volume ratios are especially susceptible to rapid loss of body heat under anesthesia. A cool environment, exposure to cold surfaces, cool IV fluids and inspired gases, and open incisions during surgery accelerate this loss. Body temperature can be maintained with the aid of insulating drapes and warming blankets or heat lamps. At the same time, care must be taken to prevent overheating or burns from warming devices.

Administration of supplemental oxygen may be helpful to maintain tissue oxygenation when injectable anesthetic regimens are being used. Ventilation may be supported most effectively when an endotracheal tube is in place either by manual compression of the breathing circuit reservoir bag or with one of the numerous mechanical ventilators available.

G. RECORD KEEPING

Records of anesthetic and surgical events are considered to be part of the animal's medical history and may be helpful in the future selection of anesthetics and medical care of the animal. They are relevant to the individual research project and may be reviewed upon occasion by regulatory agencies. Anesthetic records may typically include the animal's identification number, species, weight, and age; a description of the procedure; the names of the anesthetist and surgeon; a description of the pharmacological agents used; and a chart of physiological parameters monitored during the procedure. The use of standardized forms is helpful for maintaining the operative record.

IV. ANESTHETIC TECHNIQUES FOR USE IN THE DOG

Anesthesia in the dog is relatively uncomplicated when compared to that in many of the other species found in the research laboratory. Many investigators are familiar with dogs and with principles of humane handling and restraint for these animals. Laboratory dogs are generally relatively easy to restrain for simple SC or IM injections of premedicating drugs and for placement of catheters for administration of intravenous anesthetics. The loose skin folds over the dorsum of the neck and shoulders and behind the axilla are used as SC injection sites. Intramuscular injections are given in the middle of large muscle masses such as those of the quadriceps, hamstring, and paralumbar muscle groups, away from vessels, nerves, and bone. The cephalic vein which runs along the front of the foreleg below the elbow is easily accessible and commonly used as a site for IV injections and percutaneous catheter placement. It is recommended that dogs be fasted for 12 h prior to induction of general anesthesia.

A variety of anesthetic regimens for the dog have been developed by research and veterinary personnel and presented in the literature. Some general concepts and techniques useful in canine anesthesia will be addressed in this section with representative drug dosages presented in Table 1.

Techniques commonly used for canine anesthesia utilize premedication with a tranquilizer or narcotic analgesic, anesthetic induction with an injectable agent, and placement of an endotracheal tube. Anesthesia is then maintained with an inhalant anesthetic or with incremental doses of an injectable anesthetic agent.

TABLE 1
Representative Drug Dosages for Dogs, Cats, and Macaques[a]

Drug	Dog (Ref.)	Cat (Ref.)	Macaque (Ref.)
Atropine	0.05 SC, IM (12)	As for dogs	As for dogs
Glycopyrrolate	0.011 SC, IM, IV (4)	0.011 SC, IM (3)	—
Acepromazine	0.03—0.1 SC, IM, IV (3)	0.03—0.1 SC, IM, IV (3)	0.2—0.55 IM (12)
Lenperone	0.22—0.88 IV (doubled IM) (3)	0.22—0.88 IM, IV (3)	—
Diazepam	—	—	0.5—1.0 IM (3,5)
Meperidine	2—10 SC, IM (3)	2—10 SC, IM (3,10)	—
Oxymorphone	0.2 SC, IM, IV (3)	0.2 SC, IM, IV (3)	—
Fentanyl-droperidol	1.0 ml/7—9 kg IM or 1.0 ml/10—25 kg IV (3)	1.0 ml/9 kg SC (20)	0.3 ml/kg IM (12)
Xylazine	0.5—1.0 IV (doubled IM) (3)	As for dogs	—
Thiamylal (2—4%)	10—20 IV (3)	As for dogs	—
Thiopental (1.5—2.5%)	10—30 IV (5)	10—20 IV (3)	15—20 IV (12)
Methohexital	5—9 IV (3)	As for dogs	10 IV; 20 IM (5,12)
Pentobarbital	20—30 IV (12)	25 IV (12)	25—35 IV (5)
Ketamine	—	2—5 IV; 20 IM (3)	5—25 IM (12)
Ketamine and xylazine	11 IV and 1.1 IV (32)	15—20 and 1.0 IM (12)	10 and 0.5 IM (12)
Ketamine and diazepam	10 and 0.5 IV (36)	—	15 and 1.0 IM (12)
Alpha-chloralose	80—110 IV (12)	80-90 IV (12)	—
Methoxyflurane	1—2% induction	As for dogs	—
	0.3—0.7% maintenance	As for dogs	0.3—1.0% maintenance (10)
Halothane	2—4% induction (4)	As for dogs	2—4% induction (4)
	0.5—1.5% maintenance (10)	As for dogs	0.5—1.2% maintenance (10)

[a] Doses given are in mg/kg unless otherwise specified.

A. PREMEDICATIONS

1. Anticholinergics

Anticholinergics such as atropine and glycopyrrolate are especially useful prior to the administration of xylazine, narcotics, and fentanyl-droperidol, and in other cases where bradycardia or excessive oral and airway secretions are anticipated or encountered. Atropine has a rapid onset of action and can be given shortly prior to anesthetic induction. Glycopyrrolate does not reach maximal effectiveness until 30 to 45 min after SC or IM injection.[4]

2. Tranquilizers

A variety of tranquilizing agents have been used successfully in the dog. These include phenothiazines such as acepromazine maleate, butyrophenones such as lenperone, and benzodiazepines such as diazepam. Acepromazine is given at a dosage of 0.03 to 0.1 mg/kg SC, IM, or IV.[3] At this dosage it may reduce the anesthetic maintenance requirements for halothane by 34 to 46% when compared to requirements for unpremedicated dogs.[27] In addition to sedation, acepromazine also produces antiemetic, antiarrhythmic, antispasmodic, hypotensive, and hypothermic effects.[4]

Lenperone was reported to be associated with a shorter recovery time when compared to acepromazine.[28] The recommended dosages for lenperone in the dog are 0.44 to 1.8 mg/kg IM or 0.22 to 0.88 mg/kg IV with onset of effect in 10 to 15 min or 3 to 5 min for the two routes, respectively.

Due to unpredictable behavioral effects, diazepam is not recommended for use alone as a tranquilizer in the dog.[3] However, it has been found useful when combined with ketamine for rapid induction of anesthesia in this species.

3. Narcotic Analgesics

Narcotic analgesics may be used to produce sedation and analgesia in the dog. Meperidine and oxymorphone are commonly used for this purpose. Meperidine is given at 2 to 10 mg/kg SC 30 min prior to induction of anesthesia. Oxymorphone at a dose of 0.2 mg/kg SC, IM, or IV produces good tranquilization and analgesia, allowing the reduction of barbiturates required for anesthesia by 30 to 60%.[4]

4. Neuroleptanalgesics

Fentanyl-droperidol can be used as an anesthetic premedication in the dog. The drug may provide sufficient analgesia, sedation, and immobilization for minor procedures, but is supplemented with other anesthetics for major surgery. The amount of barbiturate required for anesthesia subsequent to administration of fentanyl-droperidol may be reduced by 30 to 90%. Dogs under fentanyl-droperidol neuroleptanalgesia are hyperresponsive to auditory stimuli and are easily startled and stimulated to make spontaneous movements.

5. Sedative Analgesics

Xylazine is available as a 2% solution (20 mg/ml) for use as a sedative analgesic in dogs and cats. When used as a premedication, peak effects are seen in 3 to 5 min after IV injection and in 10 to 15 min after IM injection. A sleep-like state is produced, allowing a one-third to one-half reduction of subsequent anesthetic used. Dogs should be premedicated with an anticholinergic to minimize xylazine-associated bradyarrhythmias.

B. ANESTHETIC INDUCTION AGENTS AND TECHNIQUES

1. Injectable Techniques

Anesthetic induction following appropriate premedication is usually achieved with an intravenous agent administered via a catheter in the cephalic vein. An ultrashort-acting barbiturate such as thiamylal, thiopental, or methohexital sodium is commonly used for this purpose in the dog. An endotracheal tube is then placed, and anesthesia most commonly maintained with a volatile agent such as halothane or methoxyflurane. Alternatively, rapid induction may be achieved with combinations of ketamine and diazepam or ketamine and xylazine or with ketamine following premedication with acepromazine. These combinations are discussed in more detail in the following section on anesthetic maintenance.

Pentobarbital sodium may also be used for induction of anesthesia in the dog. Time to peak effect and duration of action are significantly longer for this agent compared to those of the ultrashort-acting barbiturates. Consequently, after the initial bolus of one half of the calculated induction dose is given, incremental doses should be given slowly to effect over at least a 5-min period. The induction dose of 20 to 30 mg/kg IV produces light surgical anesthesia for 45 to 60 min.[5]

Calculated dosages of induction agents should be adjusted according to the amount and type of premedicating drugs used. Since these induction agents have significant potential to depress respiration, provisions should be made to allow for rapid endotracheal intubation and support of ventilation until the animal is breathing well on its own.

2. Inhalant Techniques

Anesthesia in the dog can also be induced with a volatile anesthetic such as halothane administered in oxygen via a snug-fitting face-mask. A similar technique can be adapted for use in a flow-through induction chamber for small puppies. Since many dogs pass through a brief period of excitation during mask induction, careful restraint is required to prevent injury to the dog or animal handler, and induction agents with a rapid onset of action are preferred. Premedication with tranquilizing agents helps to attenuate the excitatory response. Due to its comparatively long onset of action, methoxyflurane is not commonly used for mask induction in the dog.

The dog is gently but firmly restrained and allowed to become accustomed to the mask while O_2 is delivered at a flow rate of 3 to 5 l/min. The level of volatile anesthetic is gradually increased at a rate of 0.5% per 10 s or as tolerated by the animal to levels of 2 to 4% required for anesthetic induction. Nitrous oxide at concentrations of 50 to 70% may be added to the gas mixture to speed the onset of anesthesia. Once the dog is sufficiently relaxed, a lubricated endotracheal tube is placed and anesthesia continued with the inhalant in oxygen administered at concentrations and flow rates suitable for maintenance of surgical anesthesia.

3. Endotracheal Intubation

Endotracheal intubation is easily performed with the anesthetized dog in sternal, dorsal, or lateral recumbency. Care should be taken to prevent trauma to the teeth, oral cavity, and airway during the intubation procedure. The animal is positioned with the neck extended, the mouth opened widely, and the tongue gently pulled forward to allow access to the larynx. A lighted laryngoscope and blade may be used to visualize the larynx and depress the epiglottis for placement of the tube. With practice, intubation may be performed without a laryngoscope by using the tip of the endotracheal tube to depress the epiglottis and guiding the tube into the larynx by palpation. An endotracheal tube of appropriate diameter and length for the individual animal should be selected. Generally, it is preferable to use a tube with the largest diameter that will comfortably glide into the airway to minimize resistance to breathing. Endotracheal tubes used in the dog range in size from 5 to 15 mm I.D. and are generally fitted with an inflatable cuff to protect the airway from aspiration of stomach contents and oral secretions. The tube should be placed just far enough into the trachea so that the cuff does not inflate in the larynx, but not so far as to impinge upon an opening to one of the bronchi. The cuff should be inflated gently to seal the airway around the tube. Excessive inflation can cause damage to the trachea and is to be avoided. Placement of the tube in the airway is best confirmed by direct visualization. Other techniques to check on proper placement of the tube include palpation[31] of the tube in the trachea as the tip passes over the tracheal rings and observation of condensation in the tube and gas expulsion at expiration.

C. ANESTHETIC MAINTENANCE

Anesthesia can be maintained with inhalant or injectable agents. Inhalant techniques employ anesthetics such as methoxyflurane, halothane, enflurane, and isoflurane administered in oxygen. Maintenance concentrations and carrier gas flow rates are significantly less than those required for initial induction. For example, halothane may be used at concentrations of 0.5 to 2.0% for anesthetic maintenance, depending upon the individual animal, the premedications used, and whether supplemental N_2O is added.

Thiopental and thiamylal can be used alone at the high end of the dosage range (Table 1) to provide surgical anesthesia for 10 to 20 min. Methohexital is not recommended for use in this manner unless it is accompanied by another anesthetic or tranquilizer to smooth recovery.

Pentobarbital has been replaced by inhalants in most cases, but is still used for anesthetic maintenance in selected situations in the laboratory. After the initial induction dose begins to wear off, anesthesia may be maintained with small incremental doses as needed to effect. Ventilation and body temperature should be supported. Combination regimens such as meperidine (5 mg/kg IM), diazepam (1.0 mg/kg IM), and pentobarbital (5 mg/kg IV to effect);[5] xylazine-pentobarbital; or fentanyl-droperidol-pentobarbital have been proposed to improve analgesia, reduce emergence delirium on recovery, and allow reduction of the amount of pentobarbital used. Because recovery from pentobarbital anesthesia is slow, it is not an agent of choice for survival procedures.

Ketamine is not recommended for use alone in the dog for surgical anesthesia, mostly due to the extreme muscle tone and occasional tonoclonic seizures observed when it is used in this manner.[29,30] A number of drug regimens combining ketamine with tranquilizers or xylazine have been proposed to alleviate the undesirable features of ketamine anesthesia in the dog.

Ketamine (5.5 mg/kg) and xylazine (2.0 mg/kg) were used to produce approximately 30 min of sedation and anesthesia for a dental polishing study in beagles.[31] Ketamine (10 mg/kg IV or 22 mg/kg IM) has been administered after acepromazine, xylazine, or diazepam to provide anesthesia of 15 to 40 min of duration. The shortest duration of effect was observed with ketamine-diazepam, and the longest with ketamine-acepromazine. The cardiopulmonary effects of the various combination regimens have been studied.[32-37] Administration of diazepam (0.5 mg/kg IV) was found to produce excitement in 36% of dogs studied. It was therefore recommended that diazepam and ketamine be administered simultaneously if this combination is to be used.[36] Diazepam (0.5 mg/kg) and ketamine (10 mg/kg) can be given together IV in the same syringe for smooth induction of anesthesia.[37]

The use of alpha-chloralose in the canine has been reviewed recently, and questions have been raised regarding its anesthetic and analgesic properties in this species.[25] Alpha-chloralose has been found most useful to provide basal narcosis or light anesthesia for terminal physiological and pharmacological experiments where surgical intervention is minimal. Under these conditions it produces light anesthesia with minimal disturbance of the autonomic nervous system for 6 to 10 h. Anesthetic induction with alpha-chloralose alone is a slow process, requiring the administration of large volumes of drug in an awake animal, often accompanied by excessive muscular activity and central nervous system stimulation. Because of this, the drug is most commonly used following induction with an ultrashort-acting barbiturate. Premedication with morphine (5 mg/kg) provided a quiet animal for anesthetic induction with alpha-chloralose (80 mg/kg IV). Stable anesthesia in ventilated dogs was maintained with supplemental doses of alpha-chloralose given at 28 mg/kg/h.[38]

V. ANESTHETIC TECHNIQUES FOR USE IN THE CAT

A number of the concepts and techniques discussed for the dog are relevant to the topic of feline anesthesia. However, cats present some special challenges to the anesthetist when compared to the dog. Their smaller size makes IV catheterization and endotracheal intubation more difficult. Because they may respond unpredictably to restraint in a strange environment, cats are best handled by experienced personnel for safe and humane induction of anesthesia. Cats are unable to tolerate a number of drugs or drug dosages used safely in other species. It is, therefore, necessary to be aware of their pharmacological sensitivities when planning an anesthetic regimen. Cats are generally fasted overnight prior to general anesthesia.

Commonly used techniques for feline anesthesia are designed to minimize the need for a long period of restraint in an unpremedicated cat. This reduces stress to the animal and to laboratory personnel. Intramuscular injection of a tranquilizer, xylazine, or ketamine may be used for sedation or immobilization, respectively. Xylazine alone provides adequate sedation to allow mask induction with an inhalant anesthetic. The combination of a tranquilizer or xylazine and ketamine provides adequate anesthesia and muscle relaxation for endotracheal intubation. Alternatively, tranquilization may be followed by placement of a catheter in the small cephalic vein and administration of an ultrashort-acting barbiturate for anesthetic induction. Finally, the cat may be placed in an anesthetic chamber and a volatile agent used for induction. The latter technique minimizes the amount of handling and restraint required but increases the risk of environmental contamination with anesthetic waste gases.

Anesthetic agents and techniques useful in the laboratory cat are described in greater detail below. Representative drug dosages for use in the cat are provided in Table 1.

A. PREMEDICATIONS

The anticholinergic drugs, atropine and glycopyrrolate, may be used for premedication in the cat as described for the dog.[39] An anticholinergic may be especially helpful prior to the use of xylazine or ketamine to prevent the bradycardia and excessive salivation associated with these agents, respectively. When atropine is used as a premedication, it is especially important to keep the eyes moist with a bland ophthalmic ointment or artificial tears, as its use has been associated with decreased tear production.[40]

Tranquilizers described for use in the dog are also useful to produce sedation and facilitate handling and restraint in the cat. Acepromazine and lenperone may be used to advantage for these purposes.[3,41]

Xylazine is used in the cat in a manner similar to that described for the dog. This drug usually causes vomiting within minutes of administration in the cat, followed by sedation and analgesia. Attention to support of body temperature following xylazine administration is important, as it has been shown to depress thermoregulation in the cat for up to 12 h.[42]

Potent narcotics or high doses of narcotics produce excitatory effects in cats and are not recommended for use in this species. Less potent narcotics such as meperidine or low doses of drugs such as oxymorphone or fentanyl-droperidol are occasionally used for premedication in the cat.[20] However, tranquilizers or xylazine are generally preferred for preanesthetic sedation.

B. ANESTHETIC INDUCTION AGENTS AND TECHNIQUES

1. Injectable Techniques

Anesthetic induction in the cat may be achieved with ketamine given IM or IV, with or without premedication. The IV dose is significantly lower and has a shorter duration of action, being suitable for such minor surgical procedures as the lancing of abscesses. When given IM, anesthetic onset and duration of action are 1 to 8 min and 30 to 45 min, respectively.[4] Ketamine provides good somatic analgesia in the cat, but animals retain high muscle tone with ketamine alone. It is usually supplemented with a tranquilizer, xylazine, or another anesthetic to improve muscle relaxation and analgesia for major surgical procedures.

Ultrashort- or short-acting barbiturates may be administered via a catheter placed in the cephalic vein after tranquilization. As in the dog, dosages should be reduced according to premedications used and provisions made for rapid endotracheal intubation and support of ventilation and body temperature postinduction.

2. Inhalant Techniques

Induction of anesthesia with an inhalant is easily achieved in the cat by use of an induction chamber with an inflow port for fresh anesthetic gases and an outflow port for the scavenging of waste gases. Anesthetic concentrations and gas flow rates are similar to those used for mask induction in the dog. The cat may be premedicated with an anticholinergic to reduce salivation. Tranquilization, while not required for success, is helpful. The cat is gently placed in the chamber, the lid secured, and the oxygen turned on. Volatile anesthetic is gradually added until induction levels are achieved. The cat is observed carefully and removed from the chamber as soon as anesthesia has been induced. Anesthesia may then be deepened as required to allow endotracheal intubation by using a mask to administer the volatile anesthetic. Mask induction may also be used in debilitated animals or in healthy animals following heavy tranquilization or premedication with xylazine. Anesthetic waste gases should be properly scavenged.

3. Endotracheal Intubation

Endotracheal intubation in the cat may be achieved with or without the assistance of a laryngoscope fitted with a lighted pediatric blade. A lubricated 3.5 to 6.0 mm Cole or 3.0 to 5.5 mm OD cuffed endotracheal tube will be suitable for use in most adult cats. The tube should be long enough to assure passage into the trachea, but not so long as to enter or occlude the entrance to a bronchus. The anesthetized cat is positioned with the neck extended and the mouth held open. The tongue is pulled forward and the tip of the laryngoscope blade used to depress the base of the tongue just in front of the epiglottis, bringing the laryngeal opening into view for placement of the tube. When the tube is in place, the cuff is inflated and the tube secured with gauze or tape tied behind the cat's ears. Cats are very susceptible to laryngospasm and the development of laryngeal edema and airway obstruction secondary to airway trauma. It is therefore necessary to avoid causing trauma to the airway during intubation. The tendency to develop laryngospasm can be reduced by having the cat adequately anesthetized before attempting intubation and by minimizing manipulation of the epiglottis and laryngeal structures during intubation. Topical application of a lidocaine aerosol (2 to 10%) has been shown to be useful for desensitization of the feline larynx, facilitating intubation.[43]

C. ANESTHETIC MAINTENANCE

Anesthesia in the cat can be maintained with inhalant or injectable agents. Inhalant techniques employing methoxyflurane, halothane, isoflurane, and enflurane are similar to those described for the dog. Maintenance concentrations of inhalants are considerably lower than those required initially. Anesthetic circuits which minimize dead space and resistance to breathing such as the Ayre's T-tube or Bain circuit are suitable for use in the cat due to its small size. Small animal ventilators are available for use where controlled or assisted ventilation is desired.

Prolongation of anesthesia with injectables can be achieved by careful sequential dosing. Ketamine can be repeated as needed to extend anesthesia beyond the initial 30 to 45 min period, beginning with one half of the original dose administered IM. Xylazine (1 mg/kg IM) followed by ketamine (20 mg/kg IM) produces good immobilization and analgesia for 30 to 40 min. Small (2 mg/kg) IV doses of ketamine can be given every 30 min for maintenance to a maximum total dose of 30 mg/kg. Recovery may take up to 12 h.[5]

An induction dose of sodium pentobarbital (25 to 30 mg/kg IV) provides approximately 2 h of surgical anesthesia in the cat. As for many other species, recovery from pentobarbital is prolonged in the cat, and attention to support of body temperature and ventilation is required during this period. Cats given pentobarbital (30 mg/kg IP) averaged 400 min to return of the righting reflex. This period was extended dramatically when chloramphenical was administered during anesthesia.[44] Due to excessively long recovery periods, the use of pentobarbital for survival procedures in the cat is being supplanted by inhalant anesthesia.

As described for the dog, alpha-chloralose may be used for 6 to 10 h of basal narcosis after anesthetic induction with an ultrashort-acting barbiturate. The addition of 50% N_2O to the inhaled O_2 may suppress excessive muscular activity. There is some indication that alpha-chloralose may be more acceptable as a surgical anesthetic in the cat than in the dog.[25]

VI. ANESTHETIC TECHNIQUES FOR USE IN LABORATORY MACAQUES

A number of non-human primate species are used in biomedical research. Because of significant hazards inherent in handling the macaques and other larger species while conscious, these animals are often worked with while immobilized under the effects of chemical restraint. Experimental protocols and preventative health programs may necessitate frequent chemical restraint in some animals. Other manipulations may require surgical levels of anesthesia. Humane considerations, coupled with the valuable and sometimes rare or endangered status of the non-human primate, dictate that chemical restraint and anesthetic techniques used for these animals be carefully planned and executed to insure safe recovery.

Personnel working with primates should be properly trained in techniques of primate handling and restraint and follow established protocols, including the use of gloves, masks, and other protective clothing. Procedures should be used that minimize stress in the animals, as it is undesirable for both humane and experimental reasons. The use of a squeeze cage with a back that can be moved forward to confine the animal to the front of the cage greatly facilitates IM injections. Animals can be trained to extend a limb for venipuncture and IV injections. A pole and collar technique can be used to move a trained monkey from its cage to a specially designed humane restraint device useful for a variety of noninvasive manipulations.[45]

Primates used for research should be in overall good health and properly acclimated to the research environment. An institutional veterinary care program should be in effect for these, as for other research animals.

The following discussion will address several techniques useful for chemical restraint and general anesthesia for laboratory macaques. Representative drug dosages for use in macaques are presented in Table 1. It is generally desirable to fast macaques for 12 to 16 h prior to general anesthesia. The fasting period should be considerably shorter for smaller species such as squirrel monkeys to avoid the development of a hypoglycemic crisis in these animals.

A. CHEMICAL RESTRAINT

Chemical restraint is easily, safely, and reliably achieved with an IM injection of ketamine at 5 to 25 mg/kg.[5,10] Injection sites include the large muscles of the anterior and posterior thigh and the upper arm. Onset of effect occurs in 2 to 6 min and duration of effect is 30 to 60 min. Animals recover in 1 to 6 h, allowing return to normal activity and feeding. Some animals salivate profusely after ketamine is administered. This effect can be reduced with atropine premedication (0.04 to 0.1 mg/kg).[10] Ketamine provides analgesia in primates, but, as in other species, individual animals may retain high muscle tone and move spontaneously. An occasional animal may exhibit seizure-like activity which can be controlled with diazepam. Swallowing and cough reflexes are generally preserved. The development of hypothermia may occur, and steps should be taken to monitor and support body temperature while animals are immobilized. Occasional vomiting may occur the day following ketamine administration, and tolerance may be observed with repeated dosing.

A number of studies have been done to evaluate the various effects of manual restraint, chemical restraint, and anesthesia on study parameters such as indices of cardiopulmonary function and levels of stress and reproductive hormones. The conditions under which these parameters are evaluated may effect experimental results and should be specified in the protocol for clarity. For example, when compared to manual restraint for blood collection, immobilization with ketamine was associated with a hemogram characterized by a lower total plasma protein, hematocrit, and white blood cell count.[46] Interestingly, when the effects of handling for injections were separated from effects due to ketamine administration, ketamine did not affect plasma insulin, glucose, cortisol, or mean arterial blood pressure.[47]

Tranquilizers are not commonly used alone in primates for several reasons. First, since they produce sedation without reliable chemical restraint, their administration may produce a tranquil animal that is still easily roused and potentially dangerous to personnel. Additionally, the stress induced by injection of the tranquilizer is no less than that induced by injection of an immobilizing agent such as ketamine. Tranquilizers such as acepromazine or diazepam may be used with ketamine to improve muscle relaxation.[48,49]

Different combinations of ketamine and xylazine have been evaluated for quality of anesthesia and muscle relaxation produced and duration of effect.[50] Ketamine (10 mg/kg) and xylazine (0.5 mg/kg) given IM produced surgical anesthesia and good relaxation for 30 to 40 min.[12] However, a similar ketamine-xylazine combination or xylazine (0.6 mg/kg IM) alone was shown to significantly depress mean arterial blood pressure, heart rate, and body temperature when compared to ketamine alone or saline controls. These effects were attributed mainly to xylazine.[51] Premedication with atropine may help alleviate the bradycardia associated with the use of xylazine. Body temperature should be supported when this drug is used in macaques.

Fentanyl-droperidol can be used at a range of doses (0.05 to 0.3 ml/kg) to produce varying degrees of sedation and analgesia.[5,10,12,52] The higher dose produces good sedation and analgesia lasting 30 to 60 min.[5] Bradycardia and salivation can be controlled with atropine.

B. ANESTHETIC INDUCTION

Premedication with atropine at 0.05 mg/kg SC or IM is generally recommended prior to general anesthesia in primates. As discussed, primates may be easily chemically restrained with ketamine. With practice, an endotracheal tube can be placed with ketamine immobilization and anesthesia continued with a volatile agent such as halothane or isoflurane in oxygen with or without the addition of nitrous oxide. If needed, anesthesia may be deepened with a volatile agent administered with a facemask or with 2.5% thiopental administered in small doses IV to effect prior to placement of the tube.

Intravenous induction agents are not commonly used except in small doses to deepen ketamine anesthesia. Intravenous induction techniques necessitate increased handling of unanesthetized animals and are, therefore, the exception rather than the rule. When these techniques are used animals should be properly restrained to prevent injuries to personnel and inadvertent perivascular injection of the anesthetic. Vessels most accessible for injection are the cephalic vein of the arm and the saphenous vein of the leg. Thiopental (15 to 20 mg/kg) or methohexital (10 mg/kg) may be administered IV to effect to produce 5 to 10 min of anesthesia. Pentobarbital (30 mg/kg IV) produces surgical anesthesia lasting 40 to 60 min with recovery in 2 to 6 h.[5] Signs of barbiturate anesthesia have been described for macaques.[53]

It is occasionally desirable to induce anesthesia with a volatile anesthetic in an anesthetic induction chamber. A technique using halothane in nitrous oxide and oxygen, similar to that described for the cat, has been adapted for use in primates. To minimize animal handling, the anesthetic chamber is designed with a sliding door that can be attached to the opening to the animal's home cage. The monkey can be trained to move into the chamber when both doors are open. Once inside, the sliding door is closed, confining the animal to the anesthetic chamber.[54]

C. ANESTHETIC MAINTENANCE

Chemical restraint in primates can be prolonged with repeated doses of ketamine up to a total dose of 40 mg/kg. Pentobarbital can be administered at 5 to 15 mg/kg IV slowly to effect to deepen or prolong anesthesia.[5]

Surgical levels of anesthesia are usually maintained with volatile anesthetics administered in oxygen, with or without the addition of nitrous oxide. As for other species, nitrous oxide has been shown to contribute only about one third of the MAC of required anesthetic when administered at a concentration of 75% in macaques.[55] It should not be used alone to provide total anesthesia in these animals. However, the addition of nitrous oxide allows some reduction in the amount of other agents required for anesthesia. Nitrous oxide can be used in a 1:1 ratio with oxygen and 0.5 to 1.0% halothane, isoflurane, enflurane, or methoxyflurane for maintenance anesthesia.

Volatile anesthetic agents should be administered carefully at the lowest levels that provide adequate anesthesia because they depress the cardiovascular system in a dose-related manner. It has been suggested that macaques may be more sensitive than some other animals to the depressant effects of halothane. In rhesus macaques anesthetized with halothane (0.5 to 0.7%) in nitrous oxide (40%) and oxygen, cardiac output and arterial blood pressure were reduced 20 to 22%.[56] A similar finding of progressive cardiovascular depression with increasing depth of halothane anesthesia was reported for stump-tailed macaques.[57]

Supportive care during general anesthesia is similar to that discussed for other species. An endotracheal tube should be placed whenever possible to maintain an open airway and facilitate administration of inhalants, oxygen, and ventilatory support as needed. Because many macaques weigh less than 10 kg, pediatric anesthesia circuits are usually used for these animals.

D. ENDOTRACHEAL INTUBATION

Placement of an endotracheal tube in a macaque is easily accomplished with the anesthetized animal in dorsal or sternal recumbency. Techniques described for use in the cat can be directly applied to the macaque. A laryngoscope fitted with a lighted pediatric blade is used to hold the fleshy tongue down to allow visualization of the laryngeal opening. A 4-mm cuffed endotracheal tube, trimmed to an appropriate length to minimize dead space, is usually satisfactory for medium sized rhesus monkeys. When placing the endotracheal tube, care should be taken not to pass it into the trachea any further than necessary to avoid impinging upon an opening to a bronchus.

VII. ANESTHETIC TECHNIQUES FOR USE IN THE RABBIT

The rabbit has a reputation for being difficult to safely anesthetize for a number of reasons. For example, its small size and long, narrow mouth make endotracheal intubation a difficult procedure, and its sensitivity to anesthetic doses of some of the commonly used agents predisposes it to the development of hypotension, hypoventilation, and hypothermia during anesthesia. Nevertheless, a number of techniques can be safely used to provide a spectrum from sedation and analgesia for noninvasive procedures to general anesthesia for major surgery in the rabbit.

Personnel working with rabbits should be instructed in techniques of proper handling and restraint for this species. When handling rabbits special care should be taken to provide support to and control of the hind legs and lower back to prevent injuries to the animal's spine. The loose soft tissues over the shoulders are used for SC injection sites, the paralumbar or large hind limb muscles for IM injections, and the veins running along the margins of the ears for IV injections.

Rabbit ear veins are extremely delicate, and injections should be made slowly using non irritating solutions whenever possible to prevent damage to the vessels. Small catheters can be easily placed into these veins for more reliable administration of IV fluids and drugs. Proper restraint of the animal is important when using these veins for injection as rabbits will often try to shake their heads vigorously when drugs such as ketamine or pentobarbital are administered.

Rabbits should be acclimated to the research environment before being placed on a project. They should exhibit healthy appetites and freedom from respiratory disease, which is common in rabbits and may be associated with increased anesthetic risk.

Preanesthetic fasting is optional in the rabbit, as vomiting is not a problem in this species. Commonly used anesthetic techniques for the rabbit include injectable techniques combining xylazine or a tranquilizer with ketamine and inhalant techniques. Drugs and techniques useful for rabbit anesthesia are discussed in the following section. Representative drug dosages for use in the rabbit are presented in Table 2.

A. PREMEDICATIONS

1. Anticholinergics

Premedications which may be useful in the rabbit include the anticholinergic, atropine sulfate, and tranquilizers such as acepromazine. Atropine may be useful to reduce oral and airway secretions during induction with an inhalant and to control bradycardia associated with the use of fentanyl-droperidol or xylazine. However, atropine is not routinely used as a premedication in rabbits, partly due to the fact that a significant proportion of these animals possess atropinesterase enzymes that rapidly inactivate atropine.[58] When atropine is used in rabbits, dosages and frequencies of administration may need to be increased for individuals exhibiting signs of rapid metabolism of the drug.

2. Tranquilizers and Sedative Analgesics

Tranquilizers such as acepromazine, chlorpromazine, and diazepam may be administered 15 to 30 min before anesthetic induction to alleviate anxiety and sedate rabbits. Acepromazine causes peripheral vasodilatation and may be used to facilitate blood collection from the marginal ear veins or the central ear artery.

Xylazine may be used as a premedication to provide sedation and analgesia. When given 15 min prior to induction of anesthesia with ketamine, analgesia and muscle relaxation are improved.[59]

3. Neuroleptanalgesics

Fentanyl-droperidol may be used in the rabbit for sedation and analgesia at lower doses and analgesia and immobilization at higher doses. At a dosage of 0.125 ml/kg SC, dilatation of ear vessels becomes apparent, peaking approximately 20 min after drug administration. Animals appear sedated and analgesic and blood collection from the central ear artery is greatly facilitated.[16] At a dose of 0.19 ml/kg IM, the righting reflex is lost.[60] A dose of 0.22 ml/kg IM was judged adequate for some surgical procedures with a time to onset and duration of effect of approximately 12 min and 45 to 60 min, respectively.[61] At doses of 0.22 to 0.44 ml/kg IM, surgical anesthesia is achieved, but duration of effect and risk of mortality increase with the higher doses.

Rabbits dosed with fentanyl-droperidol exhibit pupillary constriction, persistence of corneal reflexes, and variable responsiveness and ability to move spontaneously.[61] A dose of 0.2 ml/kg is associated with moderate reductions in heart rate, mean arterial blood pressure, and ventricular contractility, with a return to normal of all but the blood pressure within 45 min.[62] Fentanyl-droperidol may be associated with a decrease in the respiratory rate of 50% or more with minute volume decreased and arterial carbon dioxide increased significantly for at least 20 min following drug administration.[63]

TABLE 2
Representative Drug Dosages for Rabbits and Rodents[a]

Drug	Rabbit (Ref.)	Guinea Pig (Ref.)	Rat (Ref.)	Mouse (Ref.)	Hamster (Ref.)
Atropine	0.05—0.5 IM, SC (10)	0.05 SC, IM (12)	0.05 SC, IM (12)	0.04 SC, IM (12)	0.04 SC, IM (12)
Acepromazine	0.5—1.0 IM (69)	—	—	—	—
Diazepam	5—10 IM (69)	—	—	—	—
Fentanyl-droperidol	0.125—0.33 ml/kg IM (16,59,60)	0.44—0.66 ml/kg IM (78)	0.33 ml/kg IM (75)	0.002—0.005 ml/gm of 1:10 dilution IM (76)	—
Xylazine	2—5 IM (59)	—	—	—	—
Ketamine	50 IM (64)	100—200 IM (12)	100 IM (12)	200 IM (12)	200 IP (12)
Ketamine-xylazine	Ket. 22 + xyl. 3 IV (64) Ket. 35—50 + xyl. 5 IM (66)	40 + 5 IP, IM (12)	90 + 10 IM (12) 33 + 15.4 IM (83)	200 + 10 IP (12)	50—200 + 10 IP (89)
Ketamine-diazepam	Ket. 25 + diaz. 5 IM (12)	100 + 5 IM (12)	—	200 IM + 5 IP (12)	50 IM + 5 IP (12)
Thiopental	30 IV (12)	—	30 IV (12)	30—40 IV (12)	—
Methohexital	10 IV (12)	31 IP (12)	7—10 IV (12)	6 IV (12)	—
Pentobarbital	20—45 IV (10)	37 IP (12)	30—40 IP (12)	40 IP (12)	50—90 IP (12)
Alpha-chloralose	80—100 IV (12)	—	—	—	—
Tribromoethanol	—	—	—	125 IP (0.25%) (12)	—
Methoxyflurane	1—2% induction (10)	As for rabbits	As for rabbits	As for rabbits	As for rabbits
	0.3—1% maintenance (10)	As for rabbits	As for rabbits	As for rabbits	As for rabbits
Halothane	4—5% induction (10)	3—5% induction (10)	As for rabbits	As for rabbits	As for rabbits
	0.5—1.5% maintenance (10)	0.75—1.5% maintenance (10)	As for rabbits	As for rabbits	As for rabbits

[a] Doses are given in mg/kg unless otherwise specified.

B. ANESTHETIC INDUCTION AGENTS AND TECHNIQUES

1. Injectable Techniques

Ketamine (22 to 50 mg/kg IM or SC) may be used for immobilization of the rabbit. Dosed at 50 mg/kg it produces loss of the righting reflex for 45 min.[64] Administered IV at 15 to 20 mg/kg, it has been used to allow placement of an endotracheal tube.[65] However, ketamine alone is not considered satisfactory for surgical anesthesia in the rabbit due to inadequate muscle relaxation and questionable adequacy of analgesia for procedures associated with significant pain, especially when of visceral origin.[66]

Ketamine (22 mg/kg) and xylazine (3 mg/kg) given IV produced surgical anesthesia lasting 35 min.[64] The combination of ketamine (35 to 50 mg/kg IM or SC) and xylazine (5 mg/kg IM or SC) can be used to immobilize rabbits for easy endotracheal intubation. This combination used at the lower ketamine dosage of 35 mg/kg provided adequate anesthesia for abdominal incision and exposure of the femur with an almost 100% abolition of palpebral, corneal, pedal, and pinprick reflexes 10 to 20 min postadministration. Duration of surgical anesthesia was 20 to 75 min.[66] The ketamine-xylazine combination is capable of producing a 30% decrease in blood pressure, 77% decrease in respiratory rate, and 28% decrease in heart rate, similar to results found with xylazine alone.[67]

Ketamine can also be used at a dose of 60 to 80 mg/kg, depending upon the size of the rabbit, following premedication with a tranquilizer.[68,69] At our facility, an IM or SC combination of ketamine (50 mg/kg) and xylazine (5 mg/kg) following an acepromazine premedication (0.5 mg/kg) has been used to provide anesthesia for major surgery in the rabbit. It may be useful for anesthetizing a large number of animals at once. However, it is not recommended for use in debilitated animals which may not be able to tolerate its depressant effects.

Barbiturates are still used for anesthesia in the rabbit, although they have a narrow margin of safety in this species. It is recommended that barbiturates be administered as dilute solutions given slowly IV to effect via a catheter in an ear vein. As for other species, ventilation and body temperature should be supported.

The ultrashort-acting barbiturates may be used to provide anesthesia for endotracheal intubation, but are generally of limited use in rabbits due to their very short duration of action in this species. Intravenous pentobarbital may be especially useful to maintain a light level of anesthesia when minimal noxious stimulation is anticipated. An initial dose of 20 to 45 mg/kg may last for 20 to 45 min. Small additional doses may be used to maintain light anesthesia if mechanical ventilatory support is provided.

2. Inhalant Techniques

Anesthetic induction with inhalant agents such as halothane or isoflurane administered in an induction chamber can be used in rabbits in a manner similar to that described for cats. Premedication with atropine is helpful to reduce oral and airway secretions so that the larynx may be more easily visualized for endotracheal intubation. Also similar to the cat, administration of volatile anesthetic via a small facemask can be used to induce anesthesia in heavily sedated or debilitated rabbits or to deepen anesthesia to allow endotracheal intubation.

3. Endotracheal Intubation

In spite of difficulties presented by the rabbit's small size, oral anatomy, and tendency to develop laryngospasm, a number of techniques have been described for successful endotracheal intubation of the rabbit. The anesthetized rabbit may be positioned in sternal, dorsal, or lateral recumbency, with the neck extended and the endotracheal tube introduced using direct visualization, a stylet as a guide, or a "blind" introduction technique. Cole, straight, or small cuffed endotracheal tubes from 3.0 to 4.0 mm I.D. in size are generally suitable for use in laboratory rabbits.

It is possible to visualize the laryngeal opening in larger rabbits with a laryngoscope fitted with a lighted pediatric blade such as a #1 Miller blade. The laryngoscope blade is introduced at the diastema between the incisors and premolars on one side and gently advanced over the base of the tongue until the larynx is brought into view. If laryngospasm is a problem, the larynx may be swabbed with a cotton-tipped applicator that has been sprayed with a topical anesthetic. The endotracheal tube is then introduced into the larynx. Use of a semirigid, atraumatic stylet to stiffen the tube may aid in accurate placement. Alternatively, a small stylet such as a 3.5 to 5.0 Fr. canine urinary catheter may be placed into the larynx with the aid of the laryngoscope, the laryngoscope removed, and the endotracheal tube advanced over the stylet into the trachea. The stylet is then quickly removed so that the airway is not obstructed.[59]

For "blind" placement, the rabbit is positioned as described above with the neck extended. The endotracheal tube is advanced along the floor of the mouth toward the larynx while the anesthetist listens through the tube for breath sounds. The tube is slipped into the larynx as the glottis opens during inspiration. Loss of breath sounds generally indicates that the tube has passed into the esophagus. Often the rabbit will respond to correct placement of the tube by "bucking" or coughing. Correct placement can be verified with the laryngoscope, by auscultation of the chest during manual ventilation, by testing for air passage through the tube with a few rabbit hairs, or by observing condensation on a shiny surface such as the end of the laryngoscope handle.

C. ANESTHETIC MAINTENANCE

Anesthesia in the rabbit may be maintained with inhalant or injectable agents. Inhalants are preferred where a more precise control of anesthetic level is required. Agents such as methoxyflurane, halothane, and isoflurane, with or without 50 to 60% nitrous oxide, have been used successfully for maintenance of surgical anesthesia in the rabbit.[65,70-73] Administration via an endotracheal tube is preferred, but masks may also be used. Tracheostomy tubes are occasionally used for nonsurvival procedures. In general, equipment and techniques suitable for use in the cat are appropriate for use in the rabbit.

Anesthesia may be prolonged with small incremental doses of injectables such as ketamine, xylazine, or fentanyl-droperidol. For example, prolongation of the IM acepromazine-xylazine-ketamine anesthesia described previously can be achieved by redosing with one half of the original dose of ketamine, alternating with onehalf of the original dose of xylazine each time the rabbit begins to awaken from anesthesia. Light levels of anesthesia may be maintained with small incremental doses of pentobarbital if ventilatory support is provided. Additional injectable maintenance regimens have been developed for use in the rabbit, including regimens for nonsurvival procedures using agents such as alpha-chloralose.[5] Some useful signs of lightening anesthesia observed in the rabbit include spontaneous chewing movements and the return of responses to an ear or toe pinch.

VIII. ANESTHETIC TECHNIQUES FOR USE IN LABORATORY RODENTS

A large number of anesthetic techniques have been developed for use in laboratory rodents. These animals present a challenge to the anesthetist due to their small size, which makes peripheral vascular access and endotracheal intubation difficult and predisposes them to the development of hypothermia while anesthetized. Species, strain, and sex differences in response to various agents may be observed. For example, female rats have longer sleeping times than males for a given dose of ketamine or pentobarbital, while some male mice sleep longer than females with a given dose of pentobarbital.[5,74] Often it is helpful to test a proposed anesthetic regimen in a few animals to be sure that it works well before using it for a larger group of animals.

Laboratory rodents should be in overall good health, free of respiratory disease, in particular, and allowed time to acclimate to the research environment before being assigned to a project. They should be properly and humanely restrained by trained laboratory personnel for injections, induction of anesthesia, and other manipulations. It is often helpful to dilute drugs for injection 1:10 with physiological saline or sterile water to allow for more accurate dosing in the smaller animals and reduce local tissue damage at the injection site. Small 25- or 27-gauge needles are used for injections in the smaller rodents.

Injection sites are limited in these animals due to their small size. Lateral tail veins may be used for IV injections in rats and mice. Intramuscular injections in rats, guinea pigs, and hamsters may be made using small volumes given in the quadriceps muscles along the front of the thigh. Care should be taken to avoid injection of irritating solutions close to the sciatic nerve, which runs caudal to the femur, as damage to the nerve may ensue. Intramuscular injections in guinea pigs may also be given in the paralumbar muscles. Subcutaneous injections can be given in the loose tissues over the neck and shoulders.

The IP route is commonly used for administration of anesthetics in rodents. To properly administer an IP injection, the animal should be carefully restrained with the head tilted down and the tail up to allow the abdominal contents to fall toward the diaphragm. The injection is made in the caudal ventral abdomen away from the midline to avoid the bladder. The syringe should be checked for aspiration of bowel contents, urine, or blood before making the injection. If any of these are aspirated, the needle should be withdrawn and replaced with a clean one and another attempt at proper placement made. The use of the IP route may be associated with variable anesthetic uptake, onset, and duration of effect, especially in inexperienced hands. Peritonitis, adynamic ileus, and abdominal adhesions may result if bowel is punctured or irritating or contaminated solutions are used.

Rodents are not routinely fasted prior to anesthesia as vomiting is not a problem in these animals. It is helpful, however, to fast guinea pigs for 6 to 12 h before anesthesia to allow them time to clear their mouths of the food bolus commonly carried on the base of the tongue. This also allows for a more accurate assessment of weight by reducing the volume of gastrointestinal contents somewhat.

A variety of combination injectable techniques similar to those used for the rabbit are useful in rodents. Inhalant techniques may also be used in these animals. A number of approaches to rodent anesthesia are discussed in this section. Representative drug dosages for use in rodents are presented in Table 2.

A. PREMEDICATIONS

Atropine may be useful to reduce salivary and bronchial secretions. It is especially recommended as a premedication in guinea pigs prior to induction with fentanyl-droperidol, xylazine, and ether or other inhalants that stimulate secretions. Tranquilizers are not commonly used as premedications in these species due to the ease with which they may be restrained in the unpremedicated state for injection of anesthetics.

B. NEUROLEPTANALGESICS

Fentanyl-droperidol may be used in rodents in a manner similar to that described for the rabbit. It is not recommended for use in hamsters due to unpredictable effects in these animals.[75] In the rat, a dose of 0.13 ml/kg lasts 30 to 40 min, producing sedation and analgesia sufficient for orbital bleeding and experimental dentistry.[60,76] Doses of 0.2 to 0.33 ml/kg IM produce analgesia and immobilization adequate for some surgical procedures in the rat.[75,77]

In the guinea pig, fentanyl-droperidol given IM at a dose of 0.08 ml/kg produces 20 min of sedation for cardiac bleeding. An IM dose of 0.44 ml/kg produces neuroleptanalgesia sufficient for laparotomy with loss of righting reflexes for 60 min.[78] A dose of 0.88 ml/kg produces more profound neuroleptanalgesia. However, when this dose is administered in the caudal thigh muscles, lameness and self-mutilation of the injected leg have been reported. This is believed to be due to damage to the sciatic nerve resulting from the injection of a relatively large volume of an acidic solution into the relatively small muscle mass surrounding the nerve.[79,80] Consequently, this high dose of fentanyl-droperidol is not recommended for use in survival procedures in the guinea pig. Cranial thigh or paralumbar muscles are preferred sites for injections of even the lower doses of the drug.

Fentanyl-droperidol can be diluted 1:10 in physiological saline to a concentration of fentanyl (0.04 mg/ml) and droperidol (2.0 mg/ml) for use in mice. A dose of 0.002 ml/g was used to provide neuroleptanalgesia for orbital bleeding and skin grafting, while 0.005 ml/g was required for splenectomy.[76]

A neuroleptanalgesic combination of fentanyl-fluanisone is also recommended for use in laboratory rodents. When combined with diazepam or midazolam, surgical anesthesia of 20 to 45 min of duration results.[12]

C. INJECTABLE ANESTHETIC TECHNIQUES

1. Ketamine and Ketamine Combinations

Ketamine alone may be used for chemical restraint, but does not produce reliable analgesia or muscle relaxation in rodents.[64] Therefore, it is often combined with tranquilizers and xylazine to provide better anesthesia in these animals.

Combinations of ketamine (87 to 90 mg/kg IM) and xylazine (10 to 13 mg/kg IM) have been used in rats to produce surgical anesthesia lasting 15 to 30 min.[81,82] Another IM regimen of ketamine (33 to 45 mg/kg) and xylazine (15.4 to 21 mg/kg) produced surgical anesthesia lasting at least 30 min in the rat. Polyuria, bradycardia, and slowing of respiration were observed. Yohimbine (1.0 to 2.1 mg/kg IP) was used to shorten the duration of anesthesia and reverse the side effects.[83] Additional combinations of ketamine-xylazine and ketamine-diazepam have been compared to pentobarbital and fentanyl-droperidol for their anesthetic and antinociceptive effects in the rat.[84,85] A different combination of ketamine (60 mg/kg IM) followed in 5 min with pentobarbital (21 mg/kg IP) produced surgical anesthesia in 25 min lasting for 60 min and requiring 2 to 3 h for recovery.[86]

Ketamine (44 mg/kg) and diazepam (0.1 mg/kg) or ketamine (25 mg/kg) and xylazine (5 mg/kg) can be given IM to produce immobilization in the guinea pig.[87] Higher doses of these combinations are required for anesthesia (Table 2). Another combination of ketamine (27 mg/kg) and xylazine (0.6 mg/kg) was supplemented with local anesthesia for laparotomy and intestinal surgery.[88]

The combination of ketamine (22 to 44 mg/kg), xylazine (5 mg/kg), and acepromazine (0.75 mg/kg) has been used for anesthesia in the mouse.[59] Ketamine (50 to 200 mg/kg) plus xylazine (10 mg/kg) was administered IP to provide anesthesia of 27 to 72 min of duration for abdominal surgery in the hamster.[89] This combination has been shown to cause local tissue damage when administered IM in this species.[90]

2. Barbiturates

Barbiturates are still used extensively in rodent anesthesia. Pentobarbital sodium, usually administered IP, is the barbiturate most commonly used for anesthesia in rodents. It can also be titrated IV to effect in animals with accessible peripheral veins. Pentobarbital has a narrow margin of safety in these animals, with anesthetic doses of the agent associated with respiratory depression, hypothermia, and significant mortality in rats, mice, and guinea pigs.[91,92] As briefly alluded to earlier, the anesthetic effects of barbiturates may be very variable in rodents depending upon route of administration; age, sex, and strain of the animal; stress levels; nutritional status; levels of microsomal enzymes; and the time of day of drug administration.

Combination regimens such as the ketamine-pentobarbital anesthetic described above for rats have been developed to improve the quality of barbiturate anesthesia and minimize the side effects. A similar regimen utilizing premedication with fentanyl-droperidol and atropine followed by pentobarbital anesthesia has also been used for rats. Droperidol-fentanyl-atropine was given IP at a dose of 9 mg/kg, 0.18 mg/kg, and 0.018 mg/kg, respectively. A 1:10 dilution of fentanyl-droperidol was used with atropine at 0.004 mg/ml. Pentobarbital (15 mg/kg IP of a 1.25% solution) was administered 15 min later to produce deep anesthesia lasting 30 to 60 min. Anesthesia could be prolonged by repeating the dose of pentobarbital upon return of the corneal reflex.[93]

3. Miscellaneous Injectable Agents

A variety of other injectable agents have been used for anesthesia in rodents. Chloral hydrate, a hypnotic with poor analgesic properties, has been administered to rodents at high doses intraperitoneally to produce surgical anesthesia. When used at these doses, chloral hydrate causes marked respiratory and cardiovascular depression. When given IP at 400 mg/kg, solutions of greater than 12.5% are associated with the development of adynamic ileus and variable mortality up to 36 d postadministration.[24] It is not recommended for use in this manner as a general anesthetic, but may be useful at lower dosages or concentrations to produce hypnotic effects.

Tribromoethanol (TBE) solutions given IP have also been used for rodent anesthesia. A dose of 300 mg/kg of a 1.25% solution produced surgical anesthesia for 30 min in gerbils. However, use of more concentrated solutions or dosages ≥350 mg/kg was associated with the development of visceral adhesions and death.[94] Although used successfully in some laboratories, TBE cannot be recommended for general use as a rodent anesthetic because of problems that may be associated with the preparation and storage of this compound and the apparently narrow range of dosages and concentrations which can be safely used.[95]

D. INHALANT ANESTHETIC TECHNIQUES

Inhalant anesthesia using a variety of volatile agents can be readily adapted for use in rodents. While prior tranquilization is not required, premedication with atropine may be used to decrease oral and respiratory secretions during induction with volatile agents. As described for the cat, anesthetic induction can be achieved by delivering the volatile anesthetic in oxygen to a small anesthetic induction chamber containing the animal(s). For safe administration, an agent-specific precision vaporizer and a flow-through anesthetic chamber should be used with halothane, isoflurane, or enflurane. Methoxyflurane may be administered similarly. However, it is commonly used in small closed chambers such as Bell jars for induction of anesthesia in the smaller rodents due to its low vaporization. For this purpose, a cotton pledget to which methoxyflurane has been applied is placed in the jar. The animal to be anesthetized is placed on a raised platform in the jar so that it does not contact the liquid methoxyflurane and the lid is secured. The animal is carefully observed and removed as soon as it is adequately anesthetized. Alternatively, mask inductions with inhalants may be used for rodents that can be easily restrained.

Maintenance of anesthesia with inhalants may be preferred where precise control of anesthetic level is required. Inhalant anesthesia is commonly used with injectable agents such as ketamine or fentanyl-droperidol to provide supplemental anesthesia, improving muscle relaxation and analgesia. For smaller rodents, inhalants are commonly administered via masks. Commercially available feline anesthesia masks are suitable for guinea pigs. Masks suitable for rats and mice can be fashioned from the open end of a syringe case that will fit over the animal's nose. Other more sophisticated designs allow for efficient scavenging of anesthetic gases.[96] In some cases it may be desirable to place an endotracheal tube for administration of inhalants and oxygen. Tracheostomy tubes may be utilized in some nonsurvival procedures.

Gas anesthesia equipment for use in rodents should be appropriate for the small size of these animals. Anesthesia circuits such as the previously mentioned Bain circuit or Ayre's T-tube may be adapted for use in rodents. A recent review discusses breathing circuit considerations for small animals in detail.[97] Additionally, a number of special anesthesia machines and ventilators, some of which are available commercially, have been designed for use in small laboratory animals.[98-100]

1. Endotracheal Intubation

Endotracheal intubation is difficult to perform in small rodents, but may be desirable in certain situations, especially when working with larger rodents such as rats and guinea pigs. A number of innovative techniques have been described for the endotracheal intubation of rodents. Special equipment such as small laryngoscope blades or other light sources, oral speculums, and small 1.5 to 2.5 mm Cole or straight endotracheal tubes are required. The endotracheal tube selected for use should be cut so that the adapter is close to the tip of the animal's nose, and a special pediatric adapter should be used whenever possible to minimize dead space. The animal must be properly anesthetized and relaxed so that the mouth can be easily opened and the tongue pulled forward before placement of the endotracheal tube should be attempted. Premedication with atropine may be helpful, as it will reduce pharyngeal secretions making it easier to visualize the opening to the larynx.

Intubation of a large rat may be performed with the animal in sternal recumbency and the neck hyperextended. The larynx may be visualized with a size 0 Miller lighted laryngoscope blade. If laryngospasm presents a problem, the back of the pharynx and the epiglottis may be swabbed with a cotton-tipped applicator moistened with a few drops of topical anesthetic. A 1.5 to 2.0 mm Cole or straight endotracheal tube, 12 to 16 gauge intravenous catheter, or other specially designed endotracheal tube is then directed into the airway. Alternatively, the rat may be positioned in dorsal recumbency with the upper incisors and tongue used to hold the mouth open. A head-mounted light source or fiberoptic illuminator is used to visualize the larynx with or without the aid of an oral speculum.[101,102] Another technique, also adaptable for use in hamsters, utilizes a lighted otoscope with a 3- to 4-mm cone and a semirigid stylet such as a 3.5 Fr. canine urinary catheter. The otoscope cone is used to open the mouth and visualize the airway, facilitating placement of the stylet tip into the larynx. The otoscope cone is then removed and the endotracheal tube advanced over the stylet and into the larynx. During this maneuver, care should be taken to prevent inadvertent advancement of the stylet much beyond the larynx, as trauma to the airway may result. Proper placement of the tube into the airway will usually stimulate a "gag" reflex that can be detected by the anesthetist. After placement of the tube, the stylet is quickly removed to prevent obstruction of the airway and the tube is secured to the animal's head.[59]

Endotracheal intubation of the guinea pig presents special challenges due to the unusual anatomy of this species. The soft palate of the guinea pig extends down to the base of the tongue. The palate is perforated by a small opening, referred to as the palatal ostium, behind which is the opening to the larynx.[103] Care must be taken not to blindly introduce instruments into the back of the guinea pig pharynx, as trauma to the soft palate with resultant bleeding and aspiration of blood may result. Additional obstacles to intubation in the guinea pig are presented by the fleshy elevated bulge of the base of the tongue and the food bolus often carried in the animal's mouth. Fasting the animal for 6 to 12 h or longer prior to anesthesia will help to empty the mouth of food, and proper positioning of the animal with the neck in extension will minimize obstruction by the tongue. The stylet technique of intubation described above can be used for the guinea pig. The otoscope cone is used to visualize the palatal ostium and gently hold it open so that the stylet can be directed into the larynx. A 1.5 to 2.5 mm Cole or straight tube can then be advanced over the stylet, through the ostium, and into the larynx as described for the rat.

IX. ANESTHETIC TECHNIQUES FOR USE IN PIGS

Pigs have earned a reputation for being difficult to anesthetize due to a number of factors including their temperament, which is generally unaccepting of injections and restraint techniques used in other species; anatomical features which make peripheral venous access and endotracheal intubation difficult; and a susceptibility to malignant hyperthermia which can be triggered by stress and several anesthetic agents in certain breeds of swine. Many of the pigs used in the laboratory are small, young conventional pigs which are at risk of developing hypothermia, hypotension, and hypoventilation under general anesthesia if proper monitoring and supportive care is not provided. Adult conventional pigs present other problems to the anesthetist due to their extremely large body size, relatively small heart, and stiff chest wall. For example, they may require large volumes of injectable agents and special equipment for endotracheal intubation and ventilatory support due to their size. Finally, a number of breeds of swine are used in research. These different breeds and strains may have different requirements for anesthetics, and dosage regimens may need to be adjusted accordingly.

Anesthetic techniques for use in laboratory swine should be selected with the above considerations in mind. Whenever possible swine should be acclimated to the laboratory environment before being subjected to experimental manipulations. They should be handled by personnel familiar with principles of humane swine management and restraint. Well-designed facilities and special equipment such as transport cages, slings, and other minimum stress devices can be used to advantage to confine a pig for administration of tranquilizers and anesthetics.

Intramuscular injections can be given in the neck muscles behind the ear, the paralumbar muscles, or the leg muscles. Most peripheral veins are deep and difficult to access in the pig. The ear veins are most commonly used for IV injections and catheter placement. Dilute solutions should be used and administered slowly into these small delicate veins to prevent damage to the vessels and surrounding tissues.

Anesthetic protocols designed to minimize the need for long periods of restraint and painful injections are desirable. Commonly used techniques include mask induction in smaller pigs with or without premedication with a tranquilizer, followed by maintenance inhalant anesthesia; tranquilization followed by induction with an IV barbiturate or an immobilizing dose of ketamine and anesthesia maintained with injectables or inhalants; or immobilization with ketamine followed with supplemental injectables or inhalants as required.

A 12- to 24-h preoperative fast is recommended for pigs. During this period, bedding which could be ingested by the animals should also be removed. Up to 36 h of fasting may be required to prepare the pig for gastrointestinal surgery. An oral glucose and electrolyte preparation may be added to the animal's drinking water during a prolonged fast. Neonatal pigs should only be fasted for a short period of time to avoid hypoglycemic episodes.

Several reviews of anesthetic techniques for use in research swine which complement sections on swine anesthesia in veterinary anesthesia texts have been recently made available.[104-107] The following discussion presents a number of practical approaches to swine anesthesia in the laboratory. Representative drug dosages are presented in Table 3.

A. PREMEDICATIONS

1. Anticholinergics

Premedication with atropine is especially helpful to reduce salivation. This facilitates endotracheal intubation when mask induction with an inhalant or agents that may produce heavy salivation, such as azaperone and ketamine, are used. It is also helpful to prevent bradycardia associated with vagal stimulation and the administration of fentanyl-droperidol. Glycopyrrolate (0.01 mg/kg SC, IM, IV) may be alternatively used for these purposes.[3]

TABLE 3
Representative Drug Dosages for Swine and Small Ruminants[a]

Drug	Swine	Sheep/Goats
Atropine	0.04 SC, IM (125)	As for pigs
Glycopyrrolate	0.005—0.01 SC, IM (125)	As for pigs
Acepromazine	0.03—0.22 IM (3)	0.05—0.1 IM, IV (130)
Azaperone	2—4 IM (premed) (108)	—
	5—8 IM (immobilization) (108)	—
Lenperone	0.4—1.6 IM (125)	—
Diazepam	1—8 IM (3)	0.5-1.0 IM, IV (125)
Fentanyl-droperidol	1 ml/10—15 kg IM (10)	—
Xylazine	0.5—3.0 IM (125)	0.05 IV; 0.1 IM (125)
Ketamine	10—20 IM (133)	5—22 IV (sheep) (133)
Ketamine and xylazine	20 IM and 2 IM (133)	11.0 IM and 0.22 IM (133)
Ketamine and diazepam	10—18 IM and 1—2 IV (133)	4 and 1 IV (12) 22 and 0.88 IM (133)
Thiopental	10—20 IV (133)	10—16 IV (133)
Thiamylal	6—18 IV (133)	8—14 IV (133)
Methohexital	5 IV (133)	4 IV (12)
Pentobarbital	10—30 IV (133)	28—33 IV (133)
Halothane:		
Induction	2—4% (3)	2—4% (134)
Maintenance	1—2% (3)	1—2% (134)

[a] Doses given are in mg/kg unless otherwise specified.

2. Tranquilizers

A number of tranquilizers, including acepromazine, azaperone, and diazepam, are useful preanesthetics in the pig. Azaperone is a butyrophenone tranquilizer marketed for control of aggression in swine. When used at the label dose, it produces tranquility but pigs are readily roused and salivate heavily. Dosages of 2 to 4 mg/kg IM can be used for preanesthetic sedation.[108] Doses of 5 to 8 mg/kg IM are recommended for deep sedation and immobilization.[3] The use of diazepam in pigs is limited by the large volume required for sedation of larger animals. Tranquilizers normally require 20 to 30 min to attain peak sedative effects in swine. During this period, stimulation of the animals should be kept to a minimum to assure smooth onset of tranquilization.

3. Neuroleptanalgesics

Fentanyl-droperidol is useful in the pig to produce sedation and analgesia, facilitating handling, anesthetic induction with a mask, and IV catheter placement. An anticholinergic should be used with fentanyl-droperidol to prevent bradycardia.

B. ANESTHETIC INDUCTION AGENTS AND TECHNIQUES

1. Injectable Techniques

Intramuscular administration of ketamine (10 to 20 mg/kg) is often used for initial immobilization, especially in larger swine. The use of ketamine alone provides inadequate analgesia and muscle relaxation for major surgical procedures.[109] Excessive salivation and jaw muscle tone makes endotracheal intubation extremely difficult. Thus, ketamine is most useful for immobilizing animals for minor procedures, for mask induction or administration of an IV induction agent, or when used in combination injectable anesthetic regimens. Peak anesthetic effects and return of ability to stand occur at approximately 8 min and 90 to 100 min postadministration, respectively.

Premedication with an anticholinergic and tranquilizer improves muscle relaxation and controls excessive salivation. A regimen employing atropine and premedication with fentanyl-droperidol or a tranquilizer followed by ketamine (11 mg/kg IM) provides surgical anesthesia for 30 to 45 min. Pentobarbital (2.2 to 6.6 mg/kg IM) may be used to improve muscle relaxation and control excitement during recovery.[110] Small doses of IV pentobarbital have been similarly used.[111] However, since this regimen combines several agents with respiratory depressant effects, ventilation should be supported as needed.

While xylazine is not an effective sedative when used alone in swine, a combination of xylazine (2 mg/kg IM) and ketamine (20 mg/kg IM) can be used to allow endotracheal intubation.[112] Another combination of xylazine (2 mg/kg), ketamine (2 mg/kg), and oxymorphone (0.075 mg/kg) given IV produces anesthesia for minor surgery lasting 20 to 30 min. Doses of each drug are doubled for IM administration.[113]

A regimen utilizing premedication with meperidine and azaperone, each given at 2.2 mg/kg IM, followed by ketamine (22 mg/kg IM) and morphine sulfate (1.7 mg/kg IM), provides approximately 60 min of anesthesia. Repetition of the original ketamine and morphine doses upon return of the toe pinch reflex doubles the duration of anesthesia.[114]

Intravenous induction techniques can be successfully used in swine if animals are properly premedicated and restrained for placement of an IV butterfly or catheter in an ear vein. Ultrashort- or short-acting barbiturates can be used for IV induction, with the choice dependent upon the duration of effect desired. Dilute solutions should be used and drugs carefully titrated to effect with dosages adjusted according to premedications used.

2. Inhalant Techniques

Mask induction of anesthesia with inhalants such as halothane or isoflurane administered in oxygen is easily adapted for use in smaller swine. Premedication with a sedative or fentanyl-droperidol is helpful but not essential. This technique can also be used in larger animals that have been conditioned to a sling or to deepen anesthesia to allow placement of an endotracheal tube in larger animals that have been tranquilized or immobilized with ketamine.

A small untranquilized pig may be held in a transport cage or sling while the mask is firmly placed over the snout. The handler can help to calm the animal by scratching it on the head and shoulders and talking in quiet, soothing tones. Care should be taken to provide a quiet environment to smooth the induction period. High carrier gas flows are used, as for the dog, but the anesthetic concentration can be increased to the induction level quickly. Nitrous oxide (60%) can be added to speed induction. When the pig is adequately relaxed, an endotracheal tube can be placed for administration of maintenance levels of the inhalant anesthetic. As a rule, halothane and isoflurane may be administered at 3 to 4% in 50 to 70% nitrous oxide for anesthetic induction and 1 to 2% for maintenance. The addition of nitrous oxide may allow reduction of the amount of volatile anesthetic required.

3. Endotracheal Intubation

A number of factors contribute to the difficulties that may be encountered during endotracheal intubation of the pig. Porcine anatomy presents a number of obstacles, including a long narrow mouth; a long soft palate which often entraps the epiglottis, and a larynx with a middle ventricle at the base of the epiglottis, a lateral ventricle on either side, and several bends along the sagittal plane. Additionally, the pig salivates heavily if not premedicated with an anticholinergic and is susceptible to laryngospasm if too lightly anesthetized during attempts at intubation. The airway is easily traumatized if intubation is improperly performed, with potentially serious consequences.

The pig may be positioned in sternal, dorsal, or lateral recumbency for placement of the endotracheal tube. However, it may be difficult to pull the epiglottis down from its position overlying the soft palate with the animal in sternal recumbency. Therefore, endotracheal intubation is usually performed with small pigs in dorsal recumbency, while larger pigs are positioned on their sides for practical reasons related to their size and weight. Equipment required includes a straight lighted laryngoscope blade, a lubricated cuffed endotracheal tube, and a semirigid stylet for the tube, all of proper length and size. Topical anesthetic such as 4% lidocaine can be sprayed into the larynx through a catheter perforated with numerous tiny holes to help desensitize the larynx. Succinylcholine (1.1 to 2.2 mg/kg IV) has been suggested for use to improve muscle relaxation for intubation, if required.[104]

The pig is positioned with the neck extended and the mouth held open with an oral speculum or lengths of padded rope. The laryngoscope blade is used to visualize the larynx. If the epiglottis is positioned above the soft palate, it can be released by putting pressure on the back of the palate. The epiglottis is pulled ventrally with the blade and the tip of the endotracheal tube with the stylet in place is advanced into the larynx. The stylet is then removed and the tube rotated gently 180° as it is advanced through the laryngeal lumen to negotiate the bends in the airway. If resistance is encountered, the tube should be removed and the procedure repeated. An alternate technique employs a long stylet that can be made of canine urinary catheters that is placed into the larynx with the aid of the laryngoscope. The endotracheal tube is guided over this stylet and advanced into the larynx, as described above. Once the tube is in place, the cuff should be inflated just enough to secure the airway and the tube tied or taped securely to the snout.

C. ANESTHETIC MAINTENANCE

Anesthesia in the pig can be maintained with inhalant or injectable agents. Inhalant techniques commonly employ agents such as halothane or isoflurane at 1 to 2% in oxygen. Methoxyflurane is not commonly used in research swine due to the practical limitations presented by its slower onset of effect and recovery. While both halothane and methoxyflurane have been shown to decrease cardiac output, stroke volume and work, hematocrit, and mean arterial blood pressure in miniature swine, these parameters returned to control levels within 2 h after halothane anesthesia, while remaining depressed for 24 to 96 h following methoxyflurane.[115] Nitrous oxide may be used at 50 to 70% to supplement other anesthetic agents, but should be discontinued if bowel distension becomes evident. A series of studies, performed to observe the distribution of systemic blood flow in pigs anesthetized with various inhalant agents, provides information that may be useful in selection of an anesthetic for a particular procedure.[116-121]

Anesthesia may be maintained with supplemental doses of injectables, as described for the regimen combining meperidine, azaperone, ketamine, and morphine. Constant infusion techniques can also be used in swine. An IV infusion of guaifenesin (50 mg/ml), ketamine (1 mg/ml), and xylazine (1 mg/ml) in 5% dextrose in water was used to provide 120 min of anesthesia in swine. After an initial dose of 1 ml/kg, an infusion rate of 2.2 ml/kg/h was used for anesthetic maintenance.[122]

Techniques have been described using constant infusions or repeated boluses of barbiturates to maintain anesthesia after induction with ketamine-xylazine or ketamine-acepromazine combinations.[107,112] Thiopental sodium 2.5% administered in small boluses as needed has been recommended for anesthetic maintenance for procedures lasting 2 to 6 h in pigs. Ventilatory support should be provided. Pentobarbital anesthesia is not recommended for survival procedures lasting more than 2 h due to prolonged recovery.[106] Others have reported prolonged recovery with both thiamylal and pentobarbital as well as decreased cardiac output and stroke volume and increased peripheral vascular resistance, mean arterial pressure, and heart rate.[115]

X. ANESTHETIC TECHNIQUES FOR USE IN SMALL RUMINANTS

Sheep, goats, and calves present their own complement of challenges to the anesthetist.[123] Because general anesthesia is associated with a number of problems in these animals, a variety of local and regional anesthetic techniques combined with appropriate tranquilizers or xylazine have been developed for use in ruminants.[124] Local and regional techniques are occasionally used in the laboratory environment. For example, epidural or spinal anesthesia may be preferred for certain caudal abdominal procedures in pregnant sheep. However, most often the special environment of the research laboratory necessitates the use of general anesthesia in research subjects. With this in mind, the following discussion will present some of the considerations relevant to small ruminant anesthesia and a number of commonly used anesthetic approaches for use in these animals. The emphasis will be on general anesthesia for mature sheep and goats and for calves over 2 months of age with functional rumens. Representative drug dosages for use in these animals are presented in Table 3.

Problems presented to the anesthetist by the unique anatomy and physiology of the small ruminant include production of large volumes of saliva which can be easily aspirated or cause airway obstruction, a propensity for vomiting or passive regurgitation while under anesthesia, and susceptibility to the development of ruminal bloat and respiratory compromise from the pressure of abdominal contents on the diaphragm during recumbency. Endotracheal intubation can be difficult in a small ruminant due to the long, narrow mouth, excessive salivation, and active laryngeal reflexes which easily precipitate vomiting. Additionally, as in all species, care must be taken to consider species drug sensitivities when selecting anesthetic agents For example, ruminants are quite sensitive to the depressant effects of xylazine and should be dosed accordingly.

A number of strategies are used to minimize the adverse effects of the problems discussed above. First, it is important to limit the length of the procedure and time of recumbency as much as possible to minimize the development of bloat and consequent respiratory embarrassment. If bloating is a consistent problem, the diet should be reevaluated and green grass or highly fermentable feed stuffs omitted preoperatively. A preanesthetic fast of 12 to 48 h with water withheld for 6 to 8 h will reduce the volume of ruminal contents, reducing bloat, regurgitation, and ventilatory problems. The fasting period should be shortened or eliminated in young animals, and care should be taken in reducing food intake in near-term females, as this may precipitate pregnancy ketosis in these animals.

Since fasting will not empty the rumen, but only reduce the volume of ruminal contents, regurgitation and aspiration of ruminal contents is still a hazard. Therefore, placement of a properly sized cuffed endotracheal tube and careful positioning of the animal to reduce the chances of passive regurgitation are important aspects of safe ruminant general anesthesia.

Premedication with atropine may help to reduce salivary secretions enough to allow improved visualization of the larynx for intubation. Additionally, before attempting placement of the endotracheal tube, the animal should be deeply enough anesthetized with the vocal cords either topically anesthetized or briefly paralyzed so that active "gag" reflexes that can lead to vomiting are abolished. The animal's head should be held up to prevent passive regurgitation until the endotracheal tube is in place with the cuff inflated. Once the tube is in place, a large-bore stomach tube may be passed into the rumen if necessary to help relieve bloat and remove excess ruminal contents.

If the animal is to be recumbent for the procedure it is helpful to tilt the tip of the muzzle down relative to the oral pharynx to allow excess saliva to drip out of the mouth. Whenever possible, the hind end of the animal should be slightly lowered relative to the chest to reduce pressure from abdominal contents on the diaphragm and discourage passive regurgitation from the rumen. Because of the various factors that act to compromise spontaneous respiration in small ruminants, the use of a mechanical ventilator is helpful during general anesthesia in these animals.

A. PREMEDICATIONS

1. Anticholinergics

The use of anticholinergics for premedication in ruminants is controversial for a number of reasons. Atropine, while it may reduce the volume of salivary secretions, is shortlived in its effect in ruminants, increases the viscosity of secretions, and can cause sinus tachycardia. Prolonged adminstration of atropine is associated with ruminal atony and bloat. While some anesthetists may elect to use atropine to reduce salivation prior to endotracheal intubation, its major use in these species may be to combat bradycardia associated with vagal stimulation and the use of drugs such as xylazine. For this purpose, atropine may be administered at 0.04 mg/kg IM or SC.[125]

2. Tranquilizers and Sedative Analgesics

While small ruminants are generally docile and easy to restrain for induction of anesthesia, premedication with a tranquilizer or xylazine may be useful in some cases to reduce anxiety, facilitate restraint and transport, and reduce the amount of subsequent anesthetic required. Tranquilizers that may be used in small laboratory ruminants include acepromazine and diazepam.

Xylazine can be used to produce a spectrum of effects in small ruminants, from preanesthetic sedation to recumbency and light surgical anesthesia. Ruminants are quite sensitive to xylazine, requiring a dose that is approximately one tenth of that used in horses. Goats may be more sensitive to xylazine than sheep, with a dose of 0.05 mg/kg IV causing profound sedation and analgesia.[125] Consequently, careful dosage calculation and the small animal formulation containing xylazine 20 mg/ml should be used in small ruminants. Xylazine can produce a variety of side effects including cardiovascular and respiratory depression, increased salivation, decreased ruminal motility, impaired thermoregulation, and hyperglycemia and glucosuria. Administration of atropine will at least partially reverse the cardiovascular effects.[126] However, the combination of atropine (0.1 mg/kg IM) and xylazine (0.2 mg/kg IM) resulted in synergistic depression of ruminal motility.[127] A number of agents including yohimbine, tolazoline, 4-aminopyridine, and doxapram have been investigated for their abilities to reverse the effects of xylazine in small ruminants.[128-130]

B. ANESTHETIC INDUCTION

1. Injectable Techniques

Induction of anesthesia in laboratory small ruminants is most commonly achieved with injectable agents administered IV. The jugular, cephalic, or auricular veins may be used for injections in these species. Percutaneous placement of an IV catheter is helpful for the administration of IV fluids and anesthetic induction agents. After prepping the area selected for catheter placement, it may be necessary to make a small nick in the tough skin of the ventral neck to facilitate passage of the catheter through the skin for placement into the jugular vein.

Ultrashort-acting barbiturates such as thiopental or thiamylal (2 to 4%) can be titrated IV to effect (7 to 25 mg/kg) to produce approximately 10 min of general anesthesia. Required doses will be reduced if animals are premedicated with tranquilizers or xylazine.

Pentobarbital appears to be metabolized more quickly in sheep than in some other species. An IV dose of 25 to 30 mg/kg can be titrated to effect in sheep or goats to produce 15 to 30 min of anesthesia. In small ruminants, the use of solutions containing propylene glycol has been associated with hemolysis and is therefore not recommended. In general, barbiturates are not recommended for use in very young ruminants such as calves less than 6 to 8 weeks old due to prolonged recovery in these animals. If a barbiturate is to be used in such an animal, methohexital (1 to 4 mg/kg IV) is preferred.[5]

Ketamine administered IV or IM can be used for anesthetic induction in small ruminants. An IV dosage of 5 mg/kg provided 10 to 15 min of surgical anesthesia in pregnant ewes. It was associated with increased maternal arterial blood pressure and uterine blood flow and mild hypoventilation without a metabolic acidosis.[131]

Ketamine is usually combined with xylazine, diazepam, or acepromazine to provide a smoother anesthetic with improved muscle relaxation in small ruminants. Ketamine (2 to 10 mg/kg IV) can be combined with diazepam (1 mg/kg IV) or xylazine (0.05 to 0.1 mg/kg IV) to produce light surgical anesthesia in sheep and goats. The upper end of the ketamine-xylazine dose ranges may be administered IM to produce 30 to 40 min of anesthesia. Duration of anesthesia may be extended by administration of smaller doses of ketamine and/or xylazine as needed. The administration of atropine may be required to counteract bradycardia associated with the use of xylazine.

2. Inhalant Techniques

In certain situations it may be preferable to induce anesthesia with a volatile anesthetic administered with a facemask. Proper restraint is necessary for the success of this technique. Sheep are quite tolerant of the anesthesia mask, however, anxiety appears to be reduced if they are kept blindfolded in a quiet environment during anesthetic induction. Premedication with a tranquilizer is helpful in fractious animals.

High oxygen flows of 4 to 6 l/min are used for anesthetic induction. The concentration of volatile anesthetic is slowly increased to induction levels as tolerated since breath-holding may be a problem in these animals if initial anesthetic concentrations are too high. Addition of 50 to 70% nitrous oxide after 5 min of preoxygenation may be used to speed induction.

C. ENDOTRACHEAL INTUBATION

It is desirable to have a cuffed endotracheal tube in place while a ruminant is anesthetized to protect the airway from aspiration of saliva and rumen contents and to allow for administration of volatile anesthetics, oxygen, and ventilatory support. Placement of an endotracheal tube in a ruminant is facilitated by use of a lighted laryngoscope blade of appropriate length. A straight blade (180 to 200 mm in length) suitable for use in the largest dogs is often adequate for use in small- to medium-sized sheep and goats. Specially constructed extended laryngoscope blades or blades manufactured for use in large animal practice may occasionally be required. Endotracheal tubes of appropriate length and diameter (7.5 to 12 mm) should be selected. Large animal endotracheal tubes may be obtained commercially or constructed in the laboratory of polyvinylchloride tubing.[132]

The animal should be well anesthetized with the jaw relaxed and laryngeal reflexes minimal, to reduce the risk of inducing vomiting during intubation. The larynx can be desensitized with 2 to 4% lidocaine spray. Positioning the animal in sternal recumbency with the head held up helps to prevent passive regurgitation. With the neck extended and the mouth held open by an assistant, the laryngoscope blade is used to visualize the larynx. The lubricated endotracheal tube can then be advanced into the larynx. A semirigid stylet placed in the tube helps in accurate placement. Alternatively, a long stylet such as a canine urinary catheter can be placed in the larynx as a guide and the tube advanced over it into the airway. A "blind" technique of intubation can also be used in small ruminants. With the neck in extension, the larynx is palpated and stabilized through the soft tissues of the ventral neck with one hand, and the tube directed into the larynx with the other. Once the endotracheal tube is in place, the cuff should be inflated and the tube secured with tape or gauze to the animal's jaw or muzzle.

D. ANESTHETIC MAINTENANCE

Anesthesia can be maintained with injectables or inhalants in small ruminants as in other species. Redosing with ketamine and/or xylazine has been discussed. Recently, a technique has been described using an IV infusion of guaifenesin, ketamine, and xylazine prepared by adding ketamine (500 mg) and xylazine (50 mg) to 500 ml of guaifenesin solution (5% in 5% dextrose). Anesthesia is induced with 0.55 ml/kg of the preparation and maintained with an infusion of 2.2 ml/kg/h for adults and 1.65 ml/kg/h for young animals.[133] Endotracheal intubation, administration of oxygen, and availability of ventilatory support are desirable when using injectable techniques in ruminants.

Inhalant anesthesia for sheep and goats may be administered with anesthesia equipment suitable for use in adult humans or large dogs.[134] Halothane is the most frequently used volatile anesthetic in small ruminants. Nitrous oxide may be used to supplement halothane anesthesia, but its use is controversial as it can accelerate the development of ruminal bloat. If nitrous oxide is in use, it should be discontinued if bloating becomes apparent intraoperatively.

E. RECOVERY

In addition to the usual monitoring and supportive care required during recovery, the ruminant should be positioned in sternal recumbency with the head slightly elevated to allow eructation of accumulated ruminal gases and relieve bloat. The endotracheal tube should be left in place until the animal is actively swallowing. The nose and mouth should be cleared of debris prior to extubation, and the tube should be removed with the cuff still partially inflated to remove secretions and ruminal contents that may have accumulated in the airway above the cuff during the anesthetic period.

REFERENCES

1. Title 9: Animals and Animal Products, Subchapter A, Animal Welfare, in *Code of Federal Regulations (CFR)*, U.S. Department of Agriculture, Hyattsville, MD, 1985.
2. **Pakes, S. P.,** Chairman, *Guide for the Care and Use of Laboratory Animals,* Committee on Care and Use of Laboratory Animals, Institute of Laboratory Animal Resources, NIH Publication No. 85-23, U.S. Department of Health and Human Services, Public Health Service, National Institutes of Health, Bethesda, MD, 1985.
3. **Short, C. E., Ed.,** *Principles and Practice of Veterinary Anesthesia,* Williams & Wilkins, Baltimore, 1987.
4. **Lumb, W. V. and Jones, E. W.,** *Veterinary Anesthesia,* Lea and Febiger, Philadelphia, 1984.
5. **Green, C. J.,** *Animal Anaesthesia,* Laboratory Animals, London, 1979.
6. **Hall, L. W. and Clarke, K. W.,** *Veterinary Anaesthesia,* Bailliere Tindall, London, 1983.
7. **Sawyer, D. C.,** *The Practice of Small Animal Anesthesia,* W.B. Saunders Company, Philadelphia, 1982.
8. **Booth, N. H. and McDonald, L. E.,** *Veterinary Pharmacology and Therapeutics,* The Iowa State University Press, Ames,.1982.
9. **Gilman, A. G., Goodman, L. S., Rall, T. W., and Nurad, F., Eds.,** *The Pharmacological Basis of Therapeutics,* Macmillan, New York, 1985.
10. **Clifford, D. H.,** Preanesthesia, anesthesia, analgesia, and euthanasia, in *Laboratory Animal Medicine,* Fox, J.G., Cohen, B.J., and Loew, F.M., Eds., Academic Press, New York, 1984, chap. 18.
11. **Harkness, J. E. and Wagner, J. E.,** *The Biology and Medicine of Rabbits and Rodents,* Lea and Febiger, Philadelphia, 1983, chap. 3.
12. **Flecknell, P. A.,** *Laboratory Animal Anaesthesia,* Academic Press, New York, 1987.
13. **Heavner, J. E.,** Anesthesia, analgesia, and restraint, in *Methods of Animal Experimentation, Vol. 7,* Gay, W.I. and Heavner, J.E., Eds., Academic Press, New York, 1986, chap. 1.
14. **Kitchell, R. L.,** Problems in defining pain and peripheral mechanisms of pain, in *Colloquium on Recognition and Alleviation of Animal Pain and Distress, J. Am. Vet. Med. Assoc.,* 191 (10), 1195, 1987.
15. Panel report on the colloquium on recognition and alleviation of animal pain and distress, *J. Am. Vet. Med. Assoc.,* 191 (10), 1186, 1987.
16. **Tillman, P. and Norman, C.,** Droperidol-fentanyl as an aid to blood collection in rabbits, *Lab. Anim. Sci.,* 33 (2), 181, 1983.
17. **Knight, A. P.,** Xylazine, *J. Am. Vet. Med. Assoc.,* 176 (5), 454, 1980.
18. **Wallner, B. M., Hatch, R. C., Booth, N. H., Kitzman, J. V., Clark, J. D., and Brown, J.,** Complete immobility produced in dogs by xylazine-atropine: antagonism by 4-aminopyridine and yohimbine, *Am. J. Vet. Res.,* 43 (12), 2259, 1982.
19. **Hsu, W. H., Lu, Z.X., and Hembrough, F. B.,** Effect of xylazine on heart rate and arterial blood pressure in conscious dogs, as influenced by atropine, 4-aminopyridine, doxapram, and yohimbine, *J. Am. Vet. Med. Assoc.,* 186 (2), 153, 1985.
20. **Grandy, J. L. and Heath, R. B.,** Cardiopulmonary and behavioral effects of fentanyl-droperidol in cats, *J. Am. Vet. Med. Assoc.,* 191 (1), 59, 1987.
21. **Sawyer, D. C.,** Neuroleptanalgesia and anesthesia, in *Proc. Second Int. Congr. Veterinary Anesthesia,* Veterinary Practice Publishing, Santa Barbara, CA, 1985, 1.
22. **Muir, W. W.,** Cyclohexanone drug mixtures: the pharmacology of ketamine and ketamine drug combinations, in *Proc. Second Int. Congr. Veterinary Anesthesia,* Veterinary Practice Publishing Company, Santa Barbara, CA, 1985, 5.
23. **Wright, M.,** Pharmacologic effects of ketamine and its use in veterinary medicine, *J. Am. Vet. Med. Assoc.,* 180 (12), 1462, 1982.
24. **Fleischman, R. W., McCracken, D., and Forbes, W.,** Adynamic ileus in the rat induced by chloral hydrate, *Lab. Anim. Sci.,* 27 (2), 238, 1977.
25. **Holzgrefe, H. H., Everitt, J. M., and Wright, E. M.,** Alpha-chloralose as a canine anesthetic, *Lab. Anim. Sci.,* 37 (5), 587, 1987.
26. **Steffey, E. P.,** Concepts of general anesthesia and assessment of adequacy of anesthesia for animal surgery, in *Animal Pain: Perception and Alleviation,* Kitchell, R.L. and Erickson, H.H., Eds., American Physiological Society, Bethesda, MD, 1983, 133.
27. **Heard, D. J., Webb, A. I., and Daniels, R. T.,** Effect of acepromazine on the anesthetic requirement of halothane in the dog, *Am. J. Vet. Res.,* 47 (10), 2113, 1986.
28. **Muir, W. W. and Hubbell, J. A. E.,** Blood pressure response to acetylpromazine and lenperone in halothane anesthetized dogs, *J. Am. Anim. Hosp. Assoc.,* 21, 285, 1985.
29. **Haskins, S. C., Farver, T. B., and Patz, J. D.,** Ketamine in dogs, *Am. J. Vet. Res.,* 46 (9), 1855, 1985.
30. **Kirkpatrick, R. M.,** Use of xylazine and ketamine as a combination anesthetic, *Canine Pract.,* 5 (3), 53, 1978.
31. **Stephenson, J. C., Blevins, D. I., and Christie, G. J.,** Safety of Rompun®/Ketaset® in dogs: a two year study, *Vet. Med. Small Anim. Clin.,* 73 (3), 303, 1978.

32. **Kolata, R. J. and Rawlings, C. A.,** Cardiopulmonary effects of intravenous xylazine, ketamine, and atropine in the dog, *Am. J. Vet. Res.*, 43 (12), 2196, 1982.
33. **Farver, T. B., Haskins, S. C., and Patz, J. D.,** Cardiopulmonary effects of acepromazine and of the subsequent administration of ketamine in the dog, *Am. J. Vet. Res.*, 47 (3), 631, 1986.
34. **Haskins, S. C., Patz, J. D., and Farver, T. B.,** Xylazine and xylazine-ketamine in dogs, *Am. J. Vet. Res.*, 47 (3), 636, 1986.
35. **Clark, D. M., Martin, R. A., and Short, C. A.,** Cardiopulmonary responses to xylazine/ketamine anesthesia in the dog, *J. Am. Anim. Hosp. Assoc.*, 18, 815, 1982.
36. **Haskins, S. C., Farver, T. B., and Patz, J. D.,** Cardiovascular changes in dogs given diazepam and diazepam-ketamine, *Am. J. Vet. Res.*, 47 (4), 795, 1986.
37. **Kolata, R. J.,** Induction of anesthesia using intravenous diazepam/ketamine in dogs, *Canine Pract.*, 13, 8, 1986.
38. **Rubal, B. J. and Buchanan, C.,** Supplemental chloralose anesthesia in morphine premedicated dogs, *Lab. Anim. Sci.*, 36 (1), 59, 1986.
39. **Short, C. E., Martin, R., and Henry, C. W., Jr.,** Clinical comparison of glycopyrrolate and atropine as preanesthetic agents in cats, *Vet. Med. Small Anim. Clin.*, 78 (10), 1557, 1983.
40. **Arnett, B. D., Brightman, A. H., and Musselman, E. E.,** Effect of atropine sulfate on tear production in the cat when used with ketamine hydrochloride and acetylpromazine maleate, *J. Am. Vet. Med. Assoc.*, 185 (2), 214, 1984.
41. **Cloyd, G. D. and Gilbert, D. L.,** Dose calibration studies of lenperone, a new tranquilizer for dogs, cats, and swine, *Vet. Med. Small Anim. Clin.*, 68 (4), 344, 1973.
42. **Ponder, S. W. and Clark, W. G.,** Prolonged depression of thermoregulation after xylazine administration to cats, *J. Vet. Pharmacol. Ther.*, 3, 203, 1980.
43. **Dyson, D. H.,** Efficacy of lidocaine HCl for laryngeal desensitization: a clinical comparison of techniques in the cat, *J. Am. Vet. Med. Assoc.*, 192 (9), 1286, 1988.
44. **Adams, H. R. and Dixit, B. N.,** Prolongation of pentobarbital anesthesia by chloramphenicol in dogs and cats, *J. Am. Vet. Med. Assoc.*, 156 (7), 902, 1970.
45. **Anderson, J. H. and Houghton, P.,** The pole and collar system: a technique for handling and training nonhuman primates, *Lab Anim.*, 12, 49, 1983.
46. **Loomis, M. R., Henrickson, R. V., and Anderson, J. H.,** Effects of ketamine hydrochloride on the hemogram of rhesus monkeys *(Macaca mulatta)*, *Lab. Anim. Sci.*, 30(5), 851, 1980.
47. **Castro, M. I., Rose, J., Green, W., Lehner, N., Peterson, D., and Taub, D.,** Ketamine-HCl as a suitable anesthetic for endocrine, metabolic, and cardiovascular studies in *Macaca fascicularis* monkeys (41292), in *Proc. Soc. Exp. Biol. Med.*, 168, 389, 1981.
48. **Porter, W. P.,** Hematologic and other effects of ketamine and ketamine-acepromazine in rhesus monkeys *(Macaca mulatta)*, *Lab. Anim. Sci.*, 32 (4), 373, 1982.
49. **Connolly, R. and Quimby, F. W.,** Acepromazine-ketamine anesthesia in the rhesus monkey *(Macaca mulatta)*, *Lab. Anim. Sci.*, 28 (1), 72, 1978.
50. **Naccarato, E. F. and Hunter, W. S.,** Anaesthetic effects of various ratios of ketamine and xylazine in rhesus monkeys *(Macaca mulatta)*, *Lab. Anim.*, 13, 317, 1979.
51. **Reutlinger, R. A., Karl, A. A., Vinal, S. I., and Nieser, M. J.,** Effects of ketamine HCl-xylazine HCl combination on cardiovascular and pulmonary values of the rhesus macaque *(Macaca mulatta)*, *Am. J. Vet. Res.*, 41 (9), 1453, 1980.
52. **Marsboom, R. and Mortelmans, J.,** Some pharmacological aspects of analgesics and neuroleptics and their use for neuroleptanalgesia in primates and lower monkeys, in *Small Animal Anaesthesia*, Graham-Jones, O., Ed., Macmillan, New York, 1964, 31.
53. **Bywater, J. E. C. and Rutty, D. A.,** Simple techniques for simian anesthesia, in *Small Animal Anaesthesia*, Graham-Jones, O., Ed., Macmillan, New York, 1964, 9.
54. **Cohen, B.,** An anaesthetic method for oral surgery in monkeys, in *Small Animal Anaesthesia*, Graham-Jones, O., Ed., Macmillan, New York, 1964, 19.
55. **Steffey, E. P., Gillespie, J. R., Berry, J. D., Eger, E. I., Jr., and Munson, E. S.,** Anesthetic potency (MAC) of nitrous oxide in the dog, cat, and stump-tail monkey, *J. Appl. Physiol.*, 36 (5), 530, 1974.
56. **Lees, M. H., Hill, J., Ochsner, A.J ., and Thomas, C.,** Regional blood flows of the rhesus monkey during halothane anesthesia, *Anesth. Analg., (Cleveland)*, 50 (2), 270, 1971.
57. **Steffey, E. P., Gillespie, J. R., Berry, J. D., Eger, E. I., II, and Rhode, E. A.,** Cardiovascular effects of halothane in the stump-tailed macaque during spontaneous and controlled ventilation, *Am. J. Vet. Res.*, 35 (10), 1315, 1974.
58. **Liebenberg, S. P. and Linn, J. M.,** Seasonal and sexual influences on rabbit atropinesterase, *Lab. Anim.*, 14, 297, 1980.
59. **Sedgwick, C. J.,** Anesthesia for rabbits and rodents, in *Current Veterinary Therapy VII: Small Animal Practice*, Kirk, R.W., Ed., W.B. Saunders, Philadelphia, 1980, 706.

60. **Walden, N. B.,** Effective sedation of rabbits, guinea pigs, rats, and mice with a mixture of fentanyl and droperidol, *Aust. Vet. J.,* 54, 538, 1978.
61. **Strack, L.E . and Kaplan, H. M.,** Fentanyl and droperidol for surgical anesthesia of rabbits, *J. Am. Vet. Med. Assoc.,* 153, 822, 1968.
62. **Brill, R. W. and Jones, D. R.,** On the suitability of Innovar, a neuroleptic analgesic, for cardiovascular experiments, *Can. J. Physiol. Pharmacol.,* 59, 1184, 1981.
63. **Khanna, V. K. and Pleuvry, B. J.,** A study of naloxone and doxapram as agents for the reversal of neuroleptanalgesic respiratory depression in the conscious rabbit, *Br. J. Anaesth.,* 50, 905, 1978.
64. **Green, C. J., Knight, J., Precious, S., and Simpkin, S.,** Ketamine alone and combined with diazepam or xylazine in laboratory animals: a ten year experience, *Lab. Anim.,* 15, 163, 1981.
65. **Lindquist, P. A.,** Induction of methoxyflurane anesthesia in the rabbit after ketamine hydrochloride and endotracheal intubation, *Lab. Anim. Sci.,* 22 (6), 898, 1972.
66. **White, G. L. and Holmes, D. D.,** A comparison of ketamine and the combination ketamine-xylazine for effective surgical anesthesia in the rabbit, *Lab. Anim. Sci.,* 26 (5), 804, 1976.
67. **Sanford, T. D. and Colby, E. D.,** Effect of xylazine and ketamine on blood pressure, heart rate, and respiratory rate in rabbits, *Lab. Anim. Sci.,* 30 (3), 519, 1980.
68. **Mulder, J. B.,** Anesthesia in the rabbit using a combination of ketamine and promazine, *Lab. Anim. Sci.,* 28 (3), 321, 1978.
69. **Sedgwick, C.J .,** Anesthesia for rabbits, in *The Veterinary Clinics of North America: Food Animal Practice,* 2 (3), Thurman, J.C., Ed., W.B. Saunders, Philadelphia, 1986, 731.
70. **Sartick, M., Eldridge, M. L., Johnson, J. A., Kurz, K. D., Fowler, W. L., and Payne, C. G.,** Recovery rate of the cardiovascular system in rabbits following short-term halothane anesthesia, *Lab. Anim. Sci.,* 29 (2), 186, 1979.
71. **Davis, N. L., Nunnally, R. L., and Malinin, T. I.,** Determination of the minimum alveolar concentration (MAC) of halothane in the white New Zealand rabbit, *Br. J. Anaesth.,* 47, 341, 1975.
72. **Wass, J. A., Keene, J. R., and Kaplan, H. M.,** Ketamine-methoxyflurane anesthesia for rabbits, *Am. J. Vet. Res.,* 35 (2), 317, 1974.
73. **Freeman, M. J., Bailey, S. P., and Hodesson, S.,** Premedication, tracheal intubation, and methoxyflurane anesthesia in the rabbit, *Lab. Anim. Sci.,* 22 (4), 576, 1972.
74. **Waterman, A. E. and Livingston, A.,** Effects of age and sex on ketamine anaesthesia in the rat, *Br. J. Anaesth.,* 50, 885, 1978.
75. **Thayer, C. B., Lowe, S., and Rubright, W. C.,** Clinical evaluation of a combination of droperidol and fentanyl as an anesthetic for the rat and hamster, *J. Am. Vet. Med. Assoc.,* 161 (6), 665, 1972.
76. **Lewis, G. E. and Jennings, P.B .,** Effective sedation of laboratory animals using Innovar-Vet®, *Lab. Anim. Sci.,* 22 (3), 430, 1972.
77. **Jones, J. B. and Simmons, M. L.,** Innovar-Vet® as an intramuscular anesthetic for rats, *Lab. Anim. Care,* 18 (6), 642, 1968.,
78. **Rubright, W. C. and Thayer, C. B.,** The use of Innovar-Vet® as a surgical anesthetic for the guinea pig, *Lab. Anim. Care,* 20 (5), 989, 1970.
79. **Leash, A. M., Beyer, R. D., and Wilber, R. G.,** Self-mutilation following Innovar-Vet® injection in the guinea pig, *Lab. Anim. Sci.,* 23, 720, 1973.
80. **Newton, W. M., Cusick, P. K., and Raffe, M. R.,** Innovar-Vet®-induced pathologic changes in the guinea pig, *Lab. Anim. Sci.,* 25 (5), 597, 1975.
81. **Van Pelt, L. F.,** Ketamine and xylazine for surgical anesthesia in rats, *J. Am. Vet. Med. Assoc.,* 171 (9), 842, 1977.
82. **Stickrod, G.,** Ketamine/xylazine anesthesia in the pregnant rat, *J. Am. Vet. Med. Assoc.,* 175 (9), 952, 1979.
83. **Hsu, W. H., Bellin, S. I., Dellman, H. D., and Hanson, C. E.,** Xylazine-ketamine-induced anesthesia in rats and its antagonism by yohimbine, *J. Am. Vet. Med. Assoc.,* 189 (9), 1040, 1986.
84. **Wixson, S. K., White, W. J., Hughes, H. C., Lang, C. M., and Marshall, W. K.,** A comparison of pentobarbital, fentanyl-droperidol, ketamine-xylazine and ketamine-diazepam anesthesia in adult male rats, *Lab. Anim. Sci.,* 37 (6), 726, 1987.
85. **Wixson, S. K., White, W. J., Hughes, H. C., Marshall, W. K., and Lang, C. M.,** The effects of pentobarbital, fentanyl-droperidol, ketamine-xylazine, and ketamine-diazepam on noxious stimulus perception in adult male rats, *Lab. Anim. Sci.,* 37 (6), 731, 1987.
86. **Youth, R. A., Simmerman, S. J., Newell, R., and King, R. A.,** Ketamine anesthesia for rats, *Physiol. Behav.,* 10, 633, 1973.
87. **Gilroy, B. A. and Varga, J. S.,** Use of ketamine-diazepam and ketamine-xylazine combinations in guinea pigs, *Vet. Med. Small Anim. Clin.,* 75 (3), 508, 1980.
88. **Dawson, D. L. and Scott-Conner, C.,** Adjunctive use of local anesthetic infiltration during guinea pig laparotomy, *Lab. Anim.,* 17, 35, 1988.

89. **Curl, J. L. and Peters, L. L.,** Ketamine hydrochloride and xylazine hydrochloride anaesthesia in the golden hamster *(Mesocricetus auratus), Lab. Anim. Sci.,* 17, 290, 1983.
90. **Gaertner, D. J., Boschert, K. R., and Schoeb, T. R.,** Muscle necrosis in Syrian hamsters resulting from intramuscular injections of ketamine and xylazine, *Lab. Anim. Sci.,* 37 (1), 80, 1987.
91. **Wixson, S. K., White, W. J., Hughes, H. C., Lang, C. M., and Marshall, W. K.,** The effects of pentobarbital, fentanyl-droperidol, ketamine-xylazine and ketamine-diazepam on arterial blood pH, blood gases, mean arterial blood pressure and heart rate in adult male rats, *Lab. Anim. Sci.,* 37 (6), 736, 1987.
92. **Wixson, S. K., White, W. J., Hughes, H. C., Lang, C. M., and Marshall, W. K.,** The effects of pentobarbital, fentanyl-droperidol, ketamine-xylazine and ketamine-diazepam on core and surface body temperature regulation in adult male rats, *Lab. Anim. Sci.,* 37 (6), 743, 1987.
93. **Garcia, D. A., Wrenn, C. E., Jansons, D., and Maire, K. E.,** Deep anesthesia in the rat with the combined action of droperidol-fentanyl and pentobarbital, *Lab. Anim. Sci.,* 25 (5), 585, 1975.
94. **Norris, M. L.,** Anaesthesia of small laboratory mammals, with special reference to the Mongolian gerbil *(Meriones unguiculatus), J. Inst. Anim. Technol.,* 34 (1), 37, 1983.
95. **Norris, M. L. and Turner, W. D.,** An evaluation of tribromoethanol (TBE) as an anaesthetic agent in the Mongolian gerbil *(Meriones unguiculatus), Lab. Anim.,* 17, 324, 1983.
96. **Levy, D. E., Zwies, A., and Duffy, T. E.,** A mask for delivery of inhalation gases to small laboratory animals, *Lab. Anim. Sci.,* 30 (5), 868, 1980.
97. **Flecknell, P. A.,** Anaesthetic breathing circuits for small laboratory animals, *J. Inst. Anim. Technol.,* 38 (1), 1, 1987.
98. **Norris, M. L. and Miles, P.,** An improved, portable machine designed to induce and maintain surgical anaesthesia in small laboratory rodents, *Lab. Anim.,* 16, 227, 1982.
99. **Ventrone, R., Baan, E., and Coggins, C. R. E.,** Novel inhalation device for the simultaneous anaesthesia of several laboratory rodents, *Lab. Anim.,* 16, 231, 1982.
100. **Mulder, J. B. and Hauser, J. J.,** A closed anesthetic system for small laboratory animals, *Lab. Anim. Sci.,* 34 (1), 77, 1984.
101. **Thet, L. A.,** A simple method of intubating rats under direct vision, *Lab. Anim. Sci.,* 33, 368, 1983.
102. **Alpert, M., Goldstein, D., and Triner, L.,** Technique of endotracheal intubation in rats, *Lab. Anim. Sci.,* 32, 78, 1982.
103. **Timm, K. I., Jahn, S. E., and Sedgwick, C. J.,** The palatal ostium of the guinea pig, *Lab. Anim. Sci.,* 37 (6), 801, 1987.
104. **Riebold, T. W. and Thurmon, J. C.,** Anesthesia in swine, in *Swine in Biomedical Research,* Tumbleson, M.E., Ed., Plenum Press, New York, 1986, 243.
105. **Eisele, P. H.,** Inhalant anesthesia for research swine, in *Swine in Biomedical Research,* Tumbleson, M.E., Ed., Plenum Press, New York, 1986, 255.
106. **Swindle, M. M.,** Anesthesia in swine, Charles River Tech. Bull., Charles River Laboratories, Wilmington, MA, 1985, 3 (3).
107. **Swindle, M. M., Horneffer, P. J., Gardner, T. J., Gott, V. L., Hall, T .S., Stuart, R. S., Baumgartner, W. A., Borkon, A. M., Galloway, E., and Reitz, B. A.,** Anatomic and anesthetic considerations in experimental cardiopulmonary surgery in swine, *Lab. Anim. Sci.,* 36 (4), 357, 1986.
108. **Porter, D. B. and Slusser, C. A.,** Azaperone: a review of a new neuroleptic agent for swine, *Vet. Med.,* 80 (3), 88, 1985.
109. **Thurmon, J. C., Nelson, D. R., and Christie, G. J.,** Ketamine anesthesia in swine, *J. Am. Vet. Med. Assoc.,* 160 (9), 1325, 1972.
110. **Benson, G. J. and Thurmon, J. C.,** Anesthesia of swine under field conditions, *J. Am. Vet. Med. Assoc.,* 174 (6), 594, 1979.
111. **Bauck, S. W.,** An evaluation of a combination of injectable agents for use in pigs, *Can. Vet. J.,* 25, 162, 1984.
112. **Kyle, O. C., Novak, S., and Bolooki, H.,** General anesthesia in pigs, *Lab. Anim. Sci.,* 29 (1), 123, 1979.
113. **Breese, C. E. and Dodman, N. H.,** Xylazine-ketamine-oxymorphone: an injectable anesthetic combination in swine, *J. Am. Vet. Med. Assoc.,* 184 (2), 182, 1984.
114. **Hoyt, R. F., Hayre, M. D., Dodd, K. T., and Phillips, Y. Y.,** Long-acting intramuscular anesthetic regimen for swine, *Lab Anim. Sci.,* 36 (4), 413, 1986.
115. **Sawyer, D. C., Lumb, W. V., and Stone, H. L.,** Cardiovascular effects of halothane, methoxyflurane, pentobarbital, and thiamylal, *J. Appl. Physiol.,* 30 (1), 36, 1971.
116. **Manohar, M.,** Regional distribution of porcine brain blood flow during 50% nitrous oxide administration, *Am. J. Vet. Res.,* 46 (4), 831, 1985.
117. **Manohar, M. and Parks, C.,** Porcine regional brain and myocardial blood flows during halothane-O_2 and halothane-nitrous oxide anesthesia: comparisons with equipotent isoflurane anesthesia, *Am. J. Vet. Res.,* 45 (3), 465, 1984.

118. **Tranquilli, W. J., Manohar, M., Parks, C. M., Thurmon, J. C., Theodorakis, M. C., and Benson, G. J.,** Systemic and regional blood flow distribution in unanesthetized swine and swine anesthetized with halothane and nitrous oxide, halothane, or enflurane, *Anesthesiology,* 56, 369, 1982.
119. **Manohar, M. and Parks, C.,** Regional distribution of brain and myocardial perfusion in swine while awake and during 1.0 and 1.5 MAC isoflurane anaesthesia produced without or with 50% nitrous oxide, *Cardiovasc. Res.,* 18, 344, 1984.
120. **Lundeen, G., Manohar, M., and Parks, C.,** Systemic distribution of blood flow in swine while awake and during 1.0 and 1.5 MAC isoflurane anesthesia with or without 50% nitrous oxide, *Anesth. Analg. (Cleveland),* 62, 499, 1983.
121. **Manohar, M. and Parks, C.,** Porcine brain and myocardial perfusion during enflurane anesthesia without and with nitrous oxide, *J. Cardiovasc. Pharmacol.,* 6, 1092, 1984.
122. **Thurmon, J. C., Tranquilli, W. J., and Benson, G. J.,** Cardiopulmonary responses of swine to intravenous infusion of guaifenesin, ketamine, and xylazine, *Am. J. Vet. Res.,* 47 (10), 2138, 1986.
123. **Steffey, E. P.,** Some characteristics of ruminants and swine that complicate management of general anesthesia, in *The Veterinary Clinics of North America: Food Animal Practice,* 2(3), Thurmon, J.C., Ed., W.B. Saunders, Philadelphia, 1986, 507.
124. **Skarda, R. T.,** Techniques of local analgesia in ruminants and swine, in *The Veterinary Clinics of North America: Food Animal Practice,* 2(3), Thurmon, J.C., Ed., W.B. Saunders, Philadelphia, 1986, 621.
125. **Short, C. E.,** Preanesthetic medications in ruminants and swine, in *The Veterinary Clinics of North America: Food Animal Practice,* 2(3), Thurmon, J.C., Ed., W.B. Saunders Company, Philadelphia, 1986, 553.
126. **Kokkonen, U. M. and Eriksson, L.,** Cardiovascular and allied actions of xylazine and atropine in the unanesthetized goat, *J. Vet. Pharmacol. Ther.,* 10, 11, 1987.
127. **Seifelnasr, E., Saleh, M., and Soliman, F. A.,** In-vivo investigations on the effect of Rompun® on the rumen motility in sheep, *Vet. Med. Rev.,* 2, 158, 1974.
128. **Doherty, T.J ., Pascoe, P. J., McDonnell, W. N., and Monteith, G.,** Cardiopulmonary effects of xylazine and yohimbine in laterally recumbent sheep, *Can. J. Vet. Res.,* 50, 517, 1986.
129. **Hsu, W. H., Schaffer, D. D., and Hanson, C. E.,** Effects of tolazoline and yohimbine on xylazine-induced central nervous system depression, bradycardia, and tachypnea in sheep, *J. Am. Vet. Med. Assoc.,* 190 (4), 423, 1987.
130. **Trim, C. M.,** Special anesthesia considerations in the ruminant, in *Principles and Practice of Veterinary Anesthesia,* Short, C.E., Ed., Williams & Wilkins, Baltimore, 1987, 288.
131. **Levinson, G., Shnider, S. M., Gildea, J. E., and deLorimer, A. A.,** Maternal and Foetal cardiovascular and acid-base changes during ketamine anaesthesia in pregnant ewes, *Br. J. Anaesth.,* 45, 1111, 1973.
132. **Kissinger, J. T. and Hughes, H. C.,** Fabrication method for endotracheal tubes for sheep, goats, and calves, *Lab. Anim. Sci.,* 34 (1), 97, 1984.
133. **Thurmon, J. C.,** Injectable anesthetic agents and techniques in ruminants and swine, in *The Veterinary Clinics of North America: Food Animal Practice,* 2(3), Thurmon, J.C., Ed., W.B. Saunders, Philadelphia, 1986, 567.
134. **Tranquilli, W. J.,** Techniques of inhalation anesthesia in ruminants and swine, in *The Veterinary Clinics of North America: Food Animal Practice,* 2(3), Thurman, J.C., Ed., W.B. Saunders, Philadelphia, 1986, 593.

Chapter 18

LABORATORY ANIMAL ANALGESIA

G. J. Benson, J. C. Thurmon, and L. E. Davis

EDITOR'S PROEM

Perhaps the most significant, dramatic, and salubrious component of the new legislation regulating animal research is the requirement that pain not essential to a protocol be eliminated. This, in turn, has catalyzed an interest in laboratory animal analgesia, a subject essentially neglected throughout most of the 20th century. Proper use of analgesics is a moral, legal, and scientific necessity; yet few researchers are trained in this area. In this chapter, Drs. Benson, Thurman, and Davis present a detailed guide to analgesics in laboratory animals, discussing various aspects of their use and presenting current, practical information on a variety of analgesic regimens.

Pain is a perception or unpleasant sensation arising from activation of a discrete set of receptors and neural pathways by noxious stimuli that are actually or potentially damaging to tissues. It is a subjective experience accompanied by feelings of fear, anxiety, and panic. Pain elicits protective motor actions, results in learned avoidance, and may modify species-specific traits of behavior, including social behavior.[1]

While it is becoming increasingly accepted that pain invokes a similar experience in animals as in people, methods for relieving animal pain have not always been given adequate consideration. There are several reasons for this. Variable responses to pain among species and individuals make recognition of pain difficult. Animals cannot speak, thus the stoic individual or species, or the individual who lies quietly, is often assumed not to be experiencing pain. Because pain is a highly subjective and emotional experience, some are reluctant to equate the pain of animals with that of people. In addition, many people regard anthropomorphic judgments or assumptions as to whether or not an individual animal is in pain as unscientific or inappropriate. Some have even denied the presence of pain in animals, regarding their responses to be purely reflexive and mechanical. This latter attitude is rooted in the belief that animals have no souls and are merely machines, as first espoused by St. Thomas Aquinas and later by René Descartes. Thus, animals would have no moral rights or moral status, and man could do with them as he wished.[2] The theory of behaviorism and the rejection of consciousness in animals by the early ethologists led to the denial of thought and feeling in animals in the early 20th century.[3] While such philosophies have often prevailed in the scientific community and, until recently, suppressed the recognition of pain and suffering in animals, they are certainly no longer universally accepted.

Animals, while not viewed as being equal to human beings, have been considered to be worthy of moral concern by some thinkers because they were capable of suffering. In 1789, Jeremy Bentham stated, "The question is not, can they reason? nor, can they talk? but, can they suffer?" If an animal were a sentient being, then it could suffer and therefore would have moral status. Sentience refers to the capacity to suffer pain or distress and/or enjoy pleasure. While mental activities cannot be directly measured, animals probably have at least some of the same features of consciousness as people. Animals display distinct preferences when presented with choices in such things as diet, temperature, etc. Numerous examples exist suggesting that animals consciously interact with their environment and, if given the opportunity will alter it. For example, if allowed to manipulate a thermostat, pigs will control environmental temperature, such that early afternoon temperatures on average are 11°C higher than in the early morning.[4] Veal calves will control illumination to provide light 67% of the day while chickens will choose illumination 80% of the day.[5] Psychological stimuli such as frustration or conflict are as effective as physical stimuli in activating the pituitary-adrenal axis in animals and in people.[6] Comparative studies have revealed that the anatomic structures and neurophysiologic mechanisms of nociception leading to the perception of pain are remarkably similar in human beings and animals, as is the pain detection threshold.[1] Therefore, it is reasonable to assume that if a stimulus were painful to people, were damaging or potentially damaging to tissues, and induced escape and emotional responses in an animal, it must be considered to be painful to that animal.[2] That animals exhibit signs of distress, learn avoidance behavior, and vocalize in response to noxious (painful) stimuli is further evidence of their capacity to suffer pain. Further, it may be argued that any being with the capacity to suffer, a capacity to reason and display purposeful behavior, and the development of self-awareness has a life that matters to it and has a "right" to live a life free of unnecessary suffering. Thus, man has moral obligations regarding animals and their treatment.[2,3]

Society has not achieved a consensus view regarding our moral obligations to animals. There is no agreement on the value of animal life or our obligations to animals similar to the one which is shared on the value of human life and our obligations to our fellow man. The moral code of society recognizes the essentials of individual human nature, i.e., we are thinking, social beings who feel pleasure and pain, value freedom, etc. From this recognition comes legal and moral protection of the individual, called rights, which protect him as an object of moral concern. Just as humans beings have needs and interests which are essential to their nature, animals have certain basic needs that are genetically encoded, environmentally modified, and expressed which are essential to the survival of their species. Fulfillment of their needs is attended by joy and comfort, and thwarting these needs induces fear, pain, and anxiety. Their needs are as essential to them as ours are to us. What we do to animals matters to them. If there were no morally relevant difference justifying our withholding of our moral concern for animals, and we protect the essential nature of individual human beings, then we also are obligated to do so for animals. If we are obligated to relieve pain and human suffering, then we are morally obligated to do so whenever possible for animals.[7]

This obligation would appear to be especially true for animals with whom we establish a bond of companionship and/or stewardship, such as pets or those whose health and well-being are under our personal care. Being animals ourselves, we can understand the perception of pain in our animal charges. Thus, we sympathize with them when they are in pain. Does the existence of sympathy entail an obligation to relieve the pain? As human beings, we are capable of feeling and appreciating sympathy and framing an ideal of behavior. An ideal is a standard by which we judge how we should behave. To have ideals is essential to morality itself. Human beings are capable of possessing ideals; other animals are not. Our capability, not merely for sympathy, but for articulating it as an ideal, is sufficient reason for pursuing it. Once you conceive the ideal, you ought to attempt to attain it. The obligation springs from the ideal itself. Therefore, if we understand that animals suffer from pain and understand that our pain and theirs is the same, then it is incumbent on us to relieve their pain, especially if we have been the cause of it. Sympathy is not the only human ideal, but it is one which we neglect at the cost of some portion of our humanity.[8]

If one accepts that animals are sentient beings capable of feeling pain and having basic needs and interests peculiar to their species, then they do indeed have moral status that we are morally obligated to recognize. Further, if we subscribe to the ideal for human behavior of being caring and compassionate, we are obligated by that ideal to strive to provide relief of pain in animals. This is particularly true of those animals that have been entrusted to our care or for whose pain we are responsible. While pain may not always be overtly expressed and only may be evidenced by subtle changes in behavior or posture, one should be vigilant, especially in situations which are known to cause pain in people. Thus, a degree of anthropomorphism is appropriate and desirable.[9,10]

TABLE 1
Analgesic Drugs Used in Animals

Classification	Examples
Opioids	Morphine, meperidine, methadone, oxymorphone, pentazocine, butorphanol, nalbuphine, buprenorphine
Salicylates	Aspirin, salicylate
Para-aminophenol derivatives	Acetanilid, acetaminophen, phenacitin
Nonopioid, nonsalicylate	Phenylbutazone, dipyrone meclofenamic acid, flunixin
Local anesthetic agents	Procaine, lidocaine, mepivicaine, tetracaine
$Alpha_2$ adrenergic agonists	Xylazine, detomidine, medetomidine, azepexole

The administration of analgesic agents should not be reserved for only those animals displaying obvious signs of severe pain and suffering, but should be administered to any patient under circumstances in which the observer would desire an analgesic for himself. Just as antibiotics are administered prophylactically to prevent infection, it is appropriate to administer analgesics to prevent pain where it is likely to occur. The commonly stated reasons for withholding analgesics, e.g., to avoid opioid-induced respiratory depression or because pain relief would result in increased activity leading to self-injury, are seldom valid and should be carefully examined before a decision is made to withhold analgesic drugs. Accurate selection and dosing of analgesic drugs provides relief of pain without severe respiratory depression. Where pulmonary function is compromised, monitoring for signs of respiratory depression will provide all the information that is required to prevent hypoventilation or apnea. Appropriate splinting, bandaging, or confinement will prevent self-injury. Animals should not have to endure pain because of real or imagined sequelae to its relief.

Pain-induced alterations in metabolism, endocrine, and cardiopulmonary function are well recognized and of serious consequence to the animal. That these can have a major impact on research studies has been recognized by research workers who have stated, "…the wages of inhumanity" are "paid in ambiguous or otherwise unsatisfactory experimental results."[11] Thus, for both moral and scientific reasons, there is increasing interest in analgesic therapy for laboratory animals.

Analgesia in the strictest sense is an absence of pain, but clinically is the reduction in the intensity of pain perceived. Analgesics are those drugs whose primary effect is to suppress pain or induce analgesia (Table 1). Although actions and effects of most other drugs differ little among mammalian species,[12] there are marked differences in response to selected analgesics (e.g., opioids) that are independent of pharmacokinetics among species.[13,14] Some classes of analgesic drugs are relatively contraindicated in some species, e.g., opioids, because of their effects. The variable response among species to opioids appears to be related to the affinity of the drugs for opioid receptors or binding sites and their specific location and density within the CNS.

Opioids elicit their effects by acting at specific receptor sites in the CNS and other tissues. The most important are the mu (μ), sigma (σ), and kappa (κ) receptors. Activation of μ-receptors causes supraspinal analgesia, hypothermia, euphoria, miosis, bradycardia, respiratory depression, and physical dependency. The κ-receptor mediates spinal analgesia, miosis, sedation, and dysphoria, while σ-receptors mediate excitement, hyperkinesis, and mydriasis.[13,15] The effects of various opioid drugs have been related to their affinity for and specific activity (efficacy) at the various receptors. There is little difference among species as to anatomic location and density of opioid receptors, with one major exception. The concentration of opioid receptors in the amygdala and frontal cortex of species that are depressed by opioids, e.g., dogs and primates, is nearly twice as great as in those species that become excited in response to opioids, e.g., horses, cats.[16] Opioid-induced responses vary among species and are dose dependent. By decreasing the dose, excitement can be avoided in those species prone to bizarre reactions. The opioid response has been described as a pharmacologic amygdalectomy because bilateral amygdolectomy induces manic excitement and ferocity in cats and depression in monkeys, as do opioids.[16] Because analgesia and excitement are mediated by different receptors, i.e., μ analgesia and σ excitement, they can occur concurrently and are not mutually exclusive.

Opioid analgesics induce CNS depression, characterized by miosis, hypothermia, bradycardia, and respiratory depression in primates, dogs, rats, and rabbits. Stimulation occurs in horses, cats, ruminants, and swine, characterized by mydriasis, panting, tachycardia, hyperkinesis, and sweating in horses.[13,14] Systemic effects of opioids include release of ADH, prolactin, and somatotropin; inhibition of the release of luteinizing hormone; increased vagal tone; release of histamine and attendant hypotension; decreased motility and increased tone of the gastrointestinal tract; spasm of the biliary and pancreatic ducts; spasm of ureteral smooth muscle and increased bladder tone; and decreased uterine tone.

Opioid-induced excitement may result indirectly from increased release of norepinephrine and dopamine.[14] This may explain the mechanism whereby dopaminergic and noradrenergic blocking drugs such as phenothiazine and butyrophenone tranquilizers suppress clinical evidence of opioid-induced excitement in cats. Phenothiazine-derived tranquilizers and xylazine, an $alpha_2$ agonist, are effective in preventing opioid-induced excitement in horses and ruminants, but butyrophenone-derived tranquilizers appear to be more effective in swine than phenothiazines. In contrast to its efficacy in horses, xylazine is less effective in preventing opioid-induced excitement in pigs.[17]

Opioids raise the pain threshold or decrease the perception of pain by acting at receptors in the dorsal horn of the spinal cord and mesolimbic system, i.e., brainstem-nucleus raphe magnus and locus ceruleus, midbrain periaquaductal gray matter, and several thalamic and hypothalamic nuclei. Opioids interfere with cellular calcium influx and thus inhibit neurotransmitter release presynaptically. Acetylcholine, norepinephrine, dopamine, substance P, glutamic acid, and 5-hydroxytryptamine are involved.[18,19]

Opioid analgesics are classified as agonists or agonist-antagonists based on their relative efficacies at opioid receptors. The agonists are potent analgesics. The agonist-antagonists are also analgesic, but have some ability to antagonize full agonists. In addition, their analgesic and respiratory depressant effects are not dose dependent in that they exhibit the so-called "ceiling effect". That is, increasing the dose beyond a certain point does not further increase the level of analgesia or respiratory depression. In addition, they are not controlled substances because they have a low potential for human abuse and thus do not require the scrupulous records that must be kept for the agonists (e.g., morphine).

Successful use of opioids requires appropriate selection of the drug and dose for the given species to avoid undesirable side effects. They must be used with caution in animals having impaired pulmonary function because they depress the respiratory and cough centers, decrease secretions, and may induce bronchospasm secondary to histamine release. In species that can freely vomit, nausea and vomiting may occur. Repeated doses can result in constipation and ileus and urinary retention. Mice and rats rapidly develop tolerance and physical dependence to opioid agonists.[16-18] Morphine decreases the number and phagocytic function of macrophages and polymorphonuclear leukocytes in mice and may alter their immune function. Opioids are the analgesic drugs of choice for treatment of severe acute pain.

While opioids induce analgesia by interfering with nociceptive neural transmission centrally, the nonopioid, nonsteroidal, anti-inflammatory analgesics (NSAIA) act peripherally at the site of tissue response to impair generation and conduction of impulses which give rise to pain. When tissues are damaged, mediators are synthesized or released which activate nociceptors and primary afferent neurons leading to the sensation of pain. Trauma, inflammation, and hypoxia can cause release of these substances. Kinins (i.e., bradykinin and kallidin), substance P, histamine, 5-hydroxytryptamine, and eicosanoids (i.e., prostaglandins, leukotrienes, and eicosatetranoic acid derivatives of arachiodonic acid) are all involved in initiating and maintaining nociceptor activity.[23,24] The nonopioid analgesics induce analgesia by suppressing inflammation and the production and elaboration of kinins and prostaglandins. These drugs are only effective against pain of low to moderate intensity associated with inflammation and release of prostaglandins.[17] They are generally regarded as being useful for treating chronic pain of somatic or integumental origin, but of little use for visceral pain. An exception is flunixin, which appears to effectively blunt visceral pain in horses.[20]

Pharmacokinetics of these drugs vary widely among species. Following oral administration, wide species variations in plasma concentration results in part from size of the GI tract and gastric emptying time affecting rate of absorption, and also in part from rate of metabolism and elimination.[26] For example, because cats are relatively deficient in microsomal glucuronyl transferase, their rate of hepatic biotransformation of salicylate is slow, and salicylate half-life is four to five times longer in cats than in dogs. Thus, while the oral dose is the same for the dog and cat (10 mg/kg PO), the dose interval for the cat is 48 h vs. 12 h for the dog. The oral dose of salicylate for the cow is 100 mg/kg as a result of ruminal dilution. The dose interval is 12 h because of absorption rate rather than elimination half-life.[27] Salicylate clearance is 30 times greater in goats and ponies than in cats. The high clearance in ponies is due to rapid excretion of salicylate in the pony's alkaline urine and due to rapid biotransformation and excretion in the goat.[28]

The rate of elimination of phenylbutazone varies widely among species and illustrates the potential danger of extrapolating data from one species to another. The half-life of phenylbutazone ranges from 72 h in people to 42 to 55 h in cattle; 14 to 19 h in goats and cats; and 3 to 6 h in rabbits, horses, baboons, dogs, swine, and rats.[26] Because of the foregoing, toxicity of the NSAIAs also varies widely among species and drugs and deserves some consideration.[20] Most common toxic side effects include gastric and intestinal ulceration, with secondary anemia and hypoproteinemia. Impaired platelet function and delayed parturition have been reported. Nephropathy occurs in patients with hypovolemia, congestive heart failure, or other cardiovascular impairment due to inhibition of renal prostaglandin function in the face of increased norepinephrine and angiotensin II. Chronic or repeated use has been associated with chronic interstitial nephritis and renal papillary necrosis. Phenylbutazone and dipyrone have been associated with blood dyscrasias.

Salicylates can cause acute hepato- and nephrotoxicoses. Acute salicylism induces CNS stimulation followed by depression, nausea, vomiting, hyperpnea, and respiratory alkalosis followed by hypoventilation, carbon dioxide retention and metabolic acidosis, and finally cardiovascular collapse.[29-31]

Alpha$_2$ adrenergic agonists (e.g., xyalzine and detomidine) are generally regarded as sedative-hypnotic drugs and are most commonly administered to induce sedation.[32,33] They are, however, potent analgesics, xylazine having been shown to be the most potent analgesic agent in the horse for the relief of both visceral and somatic pain.[34,35] These drugs exert their effects through stimulation of alpha$_2$ adrenoceptors in the brain. The highest concentration of these receptors is in the locus coeruleus, which also has a high concentration of opioid receptors. Alpha$_2$ adrenergic stimulation results in decreased neural transmission by inhibiting norepinephrine release. Sedation results from decreased activity of ascending neural projections from the locus coeruleus to the cerebral cortex and limbic system.[36,37] Analgesia appears to be the result of cerebral and spinal effects, possibly in part mediated by 5-hydroxytryptamine and the descending endogenous analgesia system.[38] Alpha$_2$ adrenergic and opioid receptors appear to interact in ways that are not fully understood.[32,38,39] Administration of alpha$_2$ agonists (i.e., clonidine) has been shown to relieve symptoms of withdrawal in opioid- dependent humans.[40,41] Clinically, we have found the combination of an opioid and an alpha$_2$ agonist to induce profound analgesia in dogs and cats.[42] They have been used in combination for some years in horses.[43,44]

Although xylazine is the most commonly used sedative-analgesic in veterinary medicine, its comparative pharmacokinetics have not been studied extensively.[33] When administered intravenously, xylazine has a rapid onset of action of short duration. There is a wide variation in species sensitivity and response to xylazine. Plasma concentration required to induce a given response is six to seven times greater in horses than in cattle.[33] Only mild responses are observed in pigs receiving twice the dose administered to the horse.[45] Xylazine is the most effective analgesic for relief of colic pain in the horse currently available in the U.S. When so used, its duration of analgesia is approximately 90 min.[34,35] Detomidine is more potent and longer acting than xylazine and is currently available in Europe and the U.K. Medetomidine and azepexole are currently under investigation. These agents are the most potent $alpha_2$ agonists to date, being capable of reducing the minimal alveolar concentration of halothane in dogs by more than 85% to more than 90%, respectively. Thus, these agents are nearly complete anesthetics. Their role as analgesic agents for clinical use in relief of pain remains to be established. In addition to having profound sedative and analgesic activity, $alpha_2$ agonists induce cardiovascular and metabolic responses also related to their peripheral adrenergic effects. Following their administration, arterial blood pressure increases, then decreases;[46,47] cardiac output is decreased.[46] Insulin release is inhibited, resulting in hyperglycemia.[48,49] Urinary output is increased as a result of decreased ADH and vasopressin release.[48,51-53]

Analgesia can be induced by local infiltration or by regional nerve blocks with local anesthetic agents. Analgesia so induced is complete in the area blocked. Intercostal nerve blockade has been advocated to relieve pain following thoractomy and is said to result in better alveolar ventilation postoperatively than when opioid-induced analgesia is present.[54,55] Lastly, nonpharmacologic methods of pain relief can be utilized to good effect. These include immobilization and support with casts, splints, or bandages; appropriate use of hot or cold packs; and physical therapy such as massage and stretching.[56]

When pain is severe and acute, opioids are some of the most effective analgesics under most circumstances. These drugs do, however, have relatively short half-lives, requiring supplemental dosing. Most of the opioid agonists will provide analgesia within 30 min of administration. Duration of action varies, but is usually on the order of 2 to 4 h. The agonist-antagonists butorphenol and buprenorphine appear to induce longer periods of analgesia, i.e., 4 to 5 h for butorphenol and up to 12 h for buprenorphine.[21] Surgically induced pain may require the use of opioids for 24 to 48 h postoperatively. The nonsteroidal anti-inflammatory agents are not sufficient by themselves to relieve severe postoperative pain. Studies in our laboratory have shown that morphine (0.1 mg/kg i.v.) effectively reduces catecholamines in cats following declawing, while 40 mg/kg salicylate had no effect. The cats which received morphine purred, appeared euphoric, and moved freely in their cage, whereas cats receiving salicylate refused to move and assumed a guarding posture with their feet held under their body. Nevertheless, NSAIA can be used in combination with opioids postoperatively to good effect because they have differing mechanisms and sites of action. Further, they can be continued when opioids are no longer necessary. NSAIAs provide good analgesia where pain is chronic and of integumentary or musculoskeletal origin, or where pain is induced by inflammation. These drugs may be selected for a given species such that they may be administered at 12- to 24-h intervals. In addition, they may be administered orally in food or water, making their use convenient.

Recommended doses and dose intervals (where available) for selected analgesics and species are given in Table 2. As with any therapeutic regimen, the response should be monitored, and where an adequate response-relief of pain has not occurred, additional drug or drugs should be given. Extrapolation of data from one species to another should be avoided where specific information is available. Investigators should not be permitted to conduct studies unless they fully understand the potential for the procedure to induce pain and the appropriate selection and use of analgesics in the species being utilized.

TABLE 2
Doses of Analgesics That Have Been Used to Ameliorate Pain in Selected Species[a]

Dose in mg/kg Unless Otherwise Indicated

	Mouse	Rat	Guinea pig	Rabbit	Dog	Cat
Acetaminophen	300 IP	110—300 PO	NA[b]	NA	10—15 PO q 8 h	NR[c]
Anileridine	5 SC	5 SC	NA	NA	5 SC	1 IV
Aspirin	120—300 PO 200 IP	100 PO 20 PO	20 IP	10 PO	10 PO q 8—12 h	10 PO q 48—72 h
Buprenorphine	2 SC q 12 h 2.5 IP q 6—8 h	0.02—0.8 SC/IP q 6—8 h 0.1—0.5 SC q 8—12 h	0.05 SC q 8—12 h	0.02—0.05 SC/IV q 8—12 h	0.01—0.2 SC/IM q 8—12 h	0.005—0.01 IM q 12 h
Butorphanol	5—10 PO 0.05—0.5 SC q 1 h	0.05—2.0 SC q 6 h	NA	NA	0.05 SC 0.2—0.5 SC q 3—4 h	0.1—0.4 IM q 6 h
Codeine	40 IP 20 SC 60—90 PO	60 IP 60 SC q 4 h	10 PO	10 PO	5 PO 2 SC q 6 h	
Etorphine	NR	NR	NR	NR	NR	NR
Fentanyl	0.001 IM	0.006 IM	0.03 IM	0.09 IM	0.04—0.08 IM/IV q 1—2 h	NR
Flunixine	NA	NA	NA	NA	1 IM do not repeat	NR
Ibuprofen	7.5 PO	10—30 PO	NA	NA	10 PO q 24—48 h	NR
Meclofenamate	NA	NA	NA	NA	NR	NR
Meperidine	40 IP 20 SC/IM q 2—3 h	50 IP 20 SC/IM q 2 h	2 IM 20 SC/IM q 2—3 h	25 IM 10 IM q 2 h	2—6 IM q 1 h 10 IM 10 SC/IM q 2 h	5 IM 10 SC/IM q 2 h
Methadone	5 IP	1.3 SC	3—10 SC	5 SC	0.25 IM	0.1 SC
Morphine	2—8 SC q 2—4 h	2 IP 6—12 SC/IM	2 IP 6—12 SC/IM	5 SC 5 SC/IM	0.5—1 IM/SC q 4 h	0.05—0.1 IM/SC q 4 h

TABLE 2 (continued)
Doses of Analgesics That Have Been Used to Ameliorate Pain in Selected Species[a]

Dose in mg/kg Unless Otherwise Indicated

	2.3 IP 10 SC q 2—4 h	q 2—3 h	q 2—3 h	q 2—3 h	0.25 IM q 6 h 0.5—5 SC/IM q 2—4 h	0.1 SC q 4 h	
Nalbuphine	NA	NA	NA	NA	0.5—2 IM/IV q 3—8 h	NA	
Oxymorphone[d]	d d	d d	0.04 IM	NR	0.2 IV/IM/SC q 4—6 h	0.2 IV/IM/SC q 4—6 h	
Pentazocine	10 SC q 3—4 h	10 SC q 4 h	NA	10—20 SC/IM q 4 h	2 IM 1.5—3 IM q 4 h	2—3 IM/IV q 4 h	
Phenylbutazone	NA	10 PO	NA	15 PO	22 PO 9 PO q 8 h	NR	
Propoxyphene	NR	60 IP	NR	3.3 IV	2.5 PO	2.2 IM	
Xylazine	50 IM	13 IM	N	5 IM	1.1 IM	1.1 IM	
Detomidine	100 µg SC	NA	NA	NA	NA	NA	
Medetomidine	100 µg SC	NA	NA	NA	10—40 µg IM	NA	
Innovar-Vet (ml/kg)	0.02/g IM[e]	0.1—0.5 IM	NR	0.13—0.22 IM	0.05 IM	NA	
	Swine	**Sheep/goat**	**Cattle**	**Horses**	**Primates**	**Hamsters**	**Ferrets**
Acetaminophen	NA	NA	NA	NA	10 PO	NA	NA
Anileridine	NA	NA	NA	NA	1.25 IM	NA	NA
Aspirin	10 PO q 4 h	100 PO q 12 h	100 PO q 12 h	25 PO q 12 h then 10 PO q 24 h	10 PO 10—20 PO q 6 h	NA	NA
Buprenorphine	0.005—0.01 IM q 12 h	0.005 IM q 12 h	NA	0.004—0.006 IV	0.01 IM/IV q 8—12 h	NA	NA
Butorphanol	NA	NA	NA	0.022 IM q 2—4 h	0.03 IM	NA	NA

Codeine					2 PO		
Etorphine	0.025 IM	NA	0.002 IM	0.025 IM	0.001 IM	NA	NA
Fentanyl	NR		NR	0.002 IM	0.04 IM		
Flunixine	NA	1.1 IV/IM	1.1 IV/IM q 8 h	1.1 IV/IM	NA	NA	NA
Ibuprofen	NA	NA	NA	NA	NA	NA	NA
Meclofenamate	NA	NA	2 IV	2 IV PO q 12 h	NA	NA	NA
Meperidine	2 IM q 4 h	Do not exceed 200 mg total dose	2 IM	2—4 IM/IV	11 IM 2—4 IM q 3—4 h	2 IM q20 SC/IM q 2—3 h	
Methadone	0.3 IM	NA	NA	0.25 IM q 6 h	0.25 IM	3—6 SC	NA
Morphine	0.2 IM	Do not exceed 10 mg total dose IM	NR	0.22 IM	0.5 IM/IV 1—4 SC/IM q 4 h	6—12 SC/IM q 2—3 h	NA
Nalbuphine	NA	NA	NA		NA	NA	NA
Oxymorphone[d]	0.02 IM	NR	NR	0.022 IM	dNA	NA	
Pentazocine	2 IM q 4 h	NA	NR	0.33 IV	2—5 IM/SC q 4 h	NA	NA
Phenylbutazone	10 PO	5 IV 9 PO q 48 h	5 IV 10 PO q 48 h	5 PO q 24 h 3—6 IV q 12 h	10 PO	NA	NA
Propoxyphene	NR	NR	NR	0.5 IV	1 PO	NA	NA
Xylazine	10 IM	0.05—0.1 IM	0.1 IM	2.2 IM q 30—60 min	0.5 IM		1 IM
Detomidine	NA	NA	NA	10—40 µg IV	NA	NA	NA
Medetomidine	NA	NA	NA	NA	NA	NA	NA
Innovar-Vet (ml/kg)	0.07 IM	NA	NA	NA	NA	0.15/100 g IM	NA

[a] These doses have been compiled from many sources. The (References 20, 21, 26, 32, and 56 to 58) wide variability of recommended doses appears related to the variety and severity of stimuli used to establish the analgesic activity of the individual drug in a given species. Therefore, these doses are intended as guidelines only. Clinical judgement must be exercised to provide effective analgesia in a given situation.

[b] NA = not available.

[c] NR = not recommended.

[d] Calculated as 0.1 dose of morphine.

[e] 10% solution, i.e., 1 part Innovar-Vet in 9 parts saline.

REFERENCES

1. **Kitchell, R. L.,** Problems in defining pain and peripheral mechanisms of pain, *J. Am. Vet. Med. Assoc.,* 191, 1195, 1987.
2. **Rowan, A. and Tannenbaum, J.,** Animal rights, in *National Forum, 66,* Phi Kappa Phi, 1986, 30.
3. **Rollin, B.,** Animal pain, scientific ideology, and the reappropriation of common sense, *J. Am. Vet. Med. Assoc.,* 191, 1222, 1987.
4. **Curtis, S. E. and Morris, G. L.,** Operant supplemental heat in swine nurseries, in *Proc. 2nd Int. Livestock Environment Symp.,* Iowa State University Press, Ames, 1982, p 295.
5. **Mench, J. A. and van Tienhoven, A.,** Farm animal welfare, *Am. Sci.,* 74, 598, 1986.
6. **Levine, S.,** A definition of stress? in *Animal Stress,* Moberg, G. P., Ed., American Physiological Society, Bethesda, MD, 1985, chap. 4.
7. **Rollin, B. E.,** Veterinary and animal ethics, in *Law and Ethics of the Veterinary Profession,* Wilson, J. F., Ed., Priority Press, Yardly, PA, 1988, chap. 2.
8. **Warnock, M.,** The philosophical approach to pain and pain relief, in *Proc. First Int. Congr. of Veterinary Anesthesia,* Taylor, P. M., Ed., Association of Veterinary Anaesthetists of Great Britain and Ireland, Cambridge, England, 1982, chap. 6.
9. **Soma, L. R.,** Behavioral changes and the assessment of pain in animals, in *Proc. 2nd Int. Congr. of Veterinary Anesthesia,* Grandy, J., Hildebrand, S., and McDonnell, W., Eds., American Colleege of Veterinary Anesthesiologists, Sacramento, CA, 1985, chap. 34.
10. **Breazile, J. E., Kitchell, R. L., and Naitok, Y.,** Neural basis of pain in animals, in *Proc. 15th Research Conf. of the American Meat Inst. Found.,* American Meat Institute Foundation, Chicago, 1963, chap. 53.
11. **Russell, W. M. S. and Burch, R. L.,** *The Principles of Humane Experimental Techniques,* Methuene, London, 1959, chap. 3.
12. **Davis, L. E., Neff-Davis, C. A., and Wilke, J. R.,** Monitoring drug concentrations in animal patients, *J. Am. Vet. Med. Assoc.,* 176, 1156, 1980.
13. **Jaffe, J. H. and Martin, W. R.,** Opioid analgesics and antagonists, in *The Pharmacologic Basis of Therapeutics,* 7th ed., Gilman, A. G., Goodman, L. S., and Rall, T. W., Eds., Macmillan, New York, 1985, 491.
14. **Booth, N. H.,** Neuroleptanalgesics, narcotic analgesics, and analgesic antagonists, in *Veterinary Pharmacology and Therapeutics,* 5th ed., Booth, N. H. and McDonald, L. E., Eds., Iowa State University Press, Ames, IA, 1982, 267.
15. **Snyder, J. H.,** Drug and neurotransmitter receptors in the brain, *Science,* 224, 22, 1984.
16. **Simon, E. J.,** The opiate receptors, in *Receptors in Pharmacology,* Smythes, J. R. and Bradley, R. J., Eds., Marcel Dekker, New York, 1977, chap. 257.
17. **Benson, G. J. and Thurmon, J. C.,** Species differences as a consideration in alleviation of animal pain and distress, *J. Am. Vet. Med. Assoc.,* 191, 1227, 1987.
18. **Höllt, V.,** Opioid peptide processing and receptor selectivity, *Annu. Rev. Pharmacol. Toxicol.,* 26, 59, 1986.
19. **Frank, G. B.,** Stereospecific opioid drug receptors on excitable cell membranes, *Can. J. Physiol. Pharmacol.,* 63, 1023, 1985.
20. **Jenkins, W. L.,** Pharmacologic aspects of analgesic drugs in animals: an overview, *J. Am. Vet. Med. Assoc.,* 191, 1231, 1987.
21. **Flecknell, P. A.,** The relief of pain in laboratory animals, *Lab. Anim.,* 18, 147, 1984.
22. **Kumar, R. and Stolerman, I. P.,** Morphine dependent behavior in rats: some clinical implications, *Psychol.Med.,* 3, 225, 1973.
23. **Guyton, A. C.,** *Textbook of Medical Physiology,* 7th ed., W. B. Saunders, Philadelphia, 1986, 592.
24. **Douglas, W. W.,** Polypeptides-angiotensin, plasma kinins and others, in *The Pharmacologic Basis of Therapeutics,* 7th ed., Gilman, A. G., Goodman, L. S., and Rall, T. W., Eds., Macmillan, New York, 1985, 639.
25. **Dubinsky, B., Begre-Mariam, S., Capetola, R. J., et al.,** The antianalgesic drugs: human therapeutic correlates of their potency in laboratory animal models of hyperalgesia, *Agents Actions,* 20, 50, 1987.
26. **Davis, L.E.,** Species differences in drug distribution vs. factors in alleviation of animal pain, in *Animal Pain—Perception and Alleviation,* Kitchell, R. L. and Erickson, H. H., Eds., American Physiological Society, Bethesda, MD, 1983, 161.
27. **Baggot, J. D.,** Disposition and fate of drugs in the body, in *Veterinary Pharmacology and Therapeutics,* 5th ed., Booth, N. H. and McDonald, L. E., Eds., Iowa State University Press, Ames, 1982, 36.
28. **Davis, L. E. and Weatfall, B. A.,** Species differences in the biotransformation and excretion of salicylate, *Am. J. Vet. Res.,* 33, 1253, 1972.
29. **Sedor, J. R., Davidson, E. W., and Dunn, M. J.,** Effects of nonsteroidal anti-inflammatory drugs in healthy subjects, *Am. J. Med.,* 81, 58, 1986.
30. **Mazué, G., Rickey, P., and Berthe, J.,** Pharmacology and comparative toxicology of nonsteroidal anti-inflammatory agents, in *Veterinary Pharmacology and Toxicology,* Ruckenbrusch, Y., Toutain, P. L., and Koritz, G.D., Eds., MTP Press, Boston, 1983, 321.

31. **Goodwin, J. S.,** Immunologic effects of nonsteroidal anti-inflammatory agents, *Med. Clin. North Am.,* 69, 793, 1985.
32. **Short, C. E.,** Neuroleptanalgesia and alpha-adrenergic receptor analgesia, in *Principles and Practice of Veterinary Anesthesia,* Short, C. E., Ed., Williams & Wilkens, Baltimore, 1987, 47.
33. **Garcia-Villar, R., Toutain, P. L., Alvineric, M., et al.,** The pharmacokinetics of xylazine hydrochloride: an interspecific study, *J. Vet. Pharmacol. Ther.,* 4, 87, 1981.
34. **Pippi, N. L. and Lumb, W. V.,** Objective tests of analgesic drugs in ponies, *Am. J. Vet. Res.,* 40, 1082, 1979.
35. **Muir, W. W. and Robertson, J. T.,** Visceral analgesia: effects of xylazine, butorphanol, meperidine and pentazocine in horses, *Am. J. Vet. Res.,* 46, 2081, 1985.
36. **Martin, P. R., Ebert, M. H., Gordon, E. K., et al.,** Effects of clonidine on central and peripheral catecholamine metabolism, *Clin. Pharmacol. Ther.,* 35, 322, 1984.
37. **Stenberg, D.,** The role of alpha-adrenoceptors in the regulation of vigilance and pain, *Acta Vet. Scand.,* 82, 29, 1986.
38. **Lewis, J. W. and Liebeskind, J. C.,** Pain suppressive systems of the brain, *TIPS,* 73, 1983.
39. **Browning, S., Lawrence, D., Livingston, A., et al.,** Interactions of drugs active at opiate receptors and drugs active at alpha-2 receptors on various test systems, *Br. J. Pharmacol.,* 11, 487, 1982.
40. **Bakris, G. L., Cross, P. D., and Hammarstem, J. E.,** The use of clonidine for management of opiate abstinence in a chronic pain patient, *Mayo Clin. Proc.,* 57, 657, 1982.
41. **Lal, H. and Fielding, S.,** Clonidine in the treatment of narcotic addiction, *TIPS,* 70, 1983.
42. **Benson, G. J., Thurmon, J. C., and Tranquilli, W. J.,** Intravenous sedation-analgesia (neuroleptanalgesia?) induced by morphine or butorphanol and xylazine in pointer dogs, in *Proc. American Coll. Veterinary Anesthesia,* American College of Veterinary Anesthesia, Las Vegas, 1986.
43. **Klein, L. V. and Baetjer, C.,** Preliminary report: Xylazine and morphine sedation in horses, *Vet. Anesthesiol.,* 1, 2, 1974.
44. **Tranquilli, W. J., Thurmon, J. C., Turner, T. A., et al.,** A preliminary report: butorphanol tartrate as an adjunct to xylazine-ketamine in the horse, *Equine Pract.,* 5, 26, 1983.
45. **Benson, G. J. and Thurmon, J. C.,** Anesthesia for swine under field conditions, *J. Am. Vet. Med. Assoc.,* 174, 594, 1979.
46. **Klide, A. M., Calderwood, H. W., and Soma, L. R.,** Cardiopulmonary effects of xylazine in dogs, *Am. J. Vet. Res.,* 36, 931, 1975.
47. **Muir, W. W. and Piper, F. S.,** Effect of xylazine on indices of myocardial contractility in the dog, *Am. J. Vet. Res.,* 38, 931, 1977.
48. **Thurmon, J. C., Steffey, E. P., Zinkl, J. G., et al.,** Xylazine causes transient dose-related hyperglycemia and increased urine volumes in mares, *Am. J. Vet. Res.,* 45, 224, 1984.
49. **Thurmon, J. C., Neff-Davis, C. A., Davis, L. E., et al.,** Xylazine hydrochloride-induced hyperglycemia and hypoinsulinemia in Thoroughbred horses, *J. Vet. Pharmacol. Ther.,* 5, 241, 1982.
50. **Benson, G. J., Thurmon, J. C., Neff-Davis, C. A., et al.,** Effect of xylazine hydrochloride upon plasma glucose and serum insulin concentrations in adult pointer dogs, *J. Am. Anim. Hosp. Assoc.,* 20, 791, 1984.
51. **Thurmon, J. C., Nelson, D. R., Hartsfield, S. M., et al.,** Effects of xylazine hydrochloride on urine in cattle, *Aust. Vet. J.,* 54, 178, 1978.
52. **Greene, S. A., Thurmon, J. C., Tranquilli, W. J., et al.,** Effect of yohimbine on xylazine-induced hypoinsulinemia and hyperglycemia in mares, *Am. J. Vet. Res.,* 48, 676, 1987.
53. **Greene, S. A.,** The Mechanism of Increased Urine Output in Mares Receiving Xylazine, M.S. thesis, University of Illinois, Urbana, 1986.
54. **Haskins, S. C.,** Use of analgesics postoperatively and in a small animal intensive care unit, *J. Am. Vet. Med. Assoc.,* 191, 1266, 1987.
55. **Berg, R. J. and Orton, E. C.,** Pulmonary function in dogs after intercostal thoractomy: comparison of morphine, oxymorphone, and selective intercostal nerve block, *Am. J. Vet. Res.,* 47, 471, 1986.
56. **Crane, S. W.,** Perioperative analgesia: a surgeon's perspective, *J. Am. Vet. Med. Assoc.,* 191, 1254, 1987.
57. **Harvey, R. C. and Walberg, J.,** Special considerations for anesthesia and analgesia in research animals, in *Principles and Practice of Veterinary Anesthesia,* Short, C.E., Ed., Williams & Wilkins, Baltimore, 1987, 380.
58. **Lumb, W. V. and Jones, E. W.,** Anesthesia of laboratory and zoo animals, in *Veterinary Anesthesia,* 2nd ed., Lumb, W.V. and Jones, E.W., Eds., Lea and Febiger, Philadelphia, 1984, 413.
59. **Soma, L. R.,** Anesthetic and analgesic considerations in the experimental animal, in *The Role of Animals in Biomedical Research,* Sechzer, J. A., Ed., Annals of the New York Academy of Sciences, Albany, 406, 1983, 32.

Part VIII
Manipulative Techniques

Chapter 19

HANDLING, RESTRAINT, AND COMMON SAMPLING AND ADMINISTRATION TECHNIQUES IN LABORATORY SPECIES

M. Lynne Kesel

EDITOR'S PROEM

Few researchers think much about routine procedures such as restraint, handling, and bleeding of animals. Yet improper practice in these areas can result in pain, fear, distress, and injury to the animals, as well as in unwitting training of the animals to resist any such manipulation. In addition, it is now known that the stress induced by any manipulation, let alone improper manipulation, can have major effects on many variables directly relevant to an experiment. (See the discussions in Chapters 2 and 7.) Obviously, then, it is in the interest of both animals and researchers to develop smooth and minimally invasive techniques for routine procedures and to disturb the animals as little as possible when manipulating them. In this chapter, Dr. Kesel outlines some proven techniques in this area. All researchers and their staff ought to be proficient in such techniques and also be constantly searching for ways to refine these techniques in their own situations. Most particularly, the use of tender loving care should always be preferred to force; animals should be trained to procedures if possible; and anesthesia and/or sedation should always be considered if at all possible.

TABLE OF CONTENTS

I. INTRODUCTION

It would be imprudent to attempt to manipulate a complicated, delicate scientific instrument without sound knowledge of its use because of the likelihood of causing it damage. However, a broken machine suffers no discomfort when damaged, whereas the most sensitive of scientific "instruments", laboratory animals, do. Persons employing experimental animals should always bear in mind their moral obligation to minimize pain and discomfort during manipulations of animals, whether during simple handling or restraint, or while administering substances or taking biological samples. Happily, the use of less invasive techniques almost always corresponds to maximum efficiency in the research effort. The reasons for this are twofold: first, animals that do not experience pain or discomfort are easier to handle, reducing manipulation time; second, more invasive techniques increase the stress response of the animals, which can produce profound changes in physiologic parameters.

Many animals can be trained to bear uncomfortable procedures by stroking and talking to them and/or giving them food as rewards after a good performance. Dogs and primates, for example, are often successfully trained to stay still or present a limb for venipuncture for a reward. Even small rodents and rabbits will be easier to manipulate at a later date or even during a procedure if liberally stroked during and after the unpleasant experience. Frequent handling and stroking of experimental animals apart from uncomfortable manipulations will produce in them a positive attitude toward people, and, because of their familiarity with being handled by personnel, cause them less stress when experimentally manipulated. In this chapter we will explore proper manipulation methods with research animals species by species, with emphasis on the least invasive techniques.

II. GENERAL TERMINOLOGY AND INJECTION TECHNIQUES

Some discussion of terms which are familiar to veterinary or even human medical personnel, but which may be obscure to researchers in other fields should be discussed here. In a four-legged animal, the terms cranial and anterior refer to the direction of the head; similarly, caudal and posterior refer to the direction of the tail. Caudal in bipeds corresponds to dorsal in quadrupeds. Dorsal refers to the area of the back, and ventral to the area of the belly. Proximal is toward the center of the animal, and distal away from the center. Referring to an imaginary plane that passes through the backbone and sternum and extends from head to tail, medial is toward the center of the animal, lateral toward the side. Paws or feet have dorsal and plantar surfaces (plantar is the part that is toward the floor). Digits of the feet are numbered from the medial side as they correspond to those of human hands or feet (i.e., the thumbs and great toe are digits one and the little fingers and toes digits five).

Injection of substances by syringe and needle into the body of an animal may be by several routes, which are commonly abbreviated. Injection directly into a vein is called intravenous, or IV. Injection into the subcutanous tissues, the loose connective tissues between the skin and deeper structures like muscle and bone, is often abbreviated as SC. Intradermal, a route seldom used except for skin sensitivity or antigenicity studies, is administration into the layers of the skin; it makes a "bleb" or blister appear in the skin. Occasionally one administers directly into the heart, or intracardiac, which is sometimes abbreviated IC (IC is also sometimes used as an abbreviation for intracranial, but the intracranial route is seldom used except for the route of administration of infectious agents into the brains of suckling mice). Intraperitoneal injection (IP) is often used in animals which do not have accessible external veins. Intramuscular (IM) is the common method of administering many therapeutic and anesthetic drugs.

The different modes of administration are used because of the speed or retardation of absorption of substances. The fastest uptake, and therefore the fastest onset of effect of

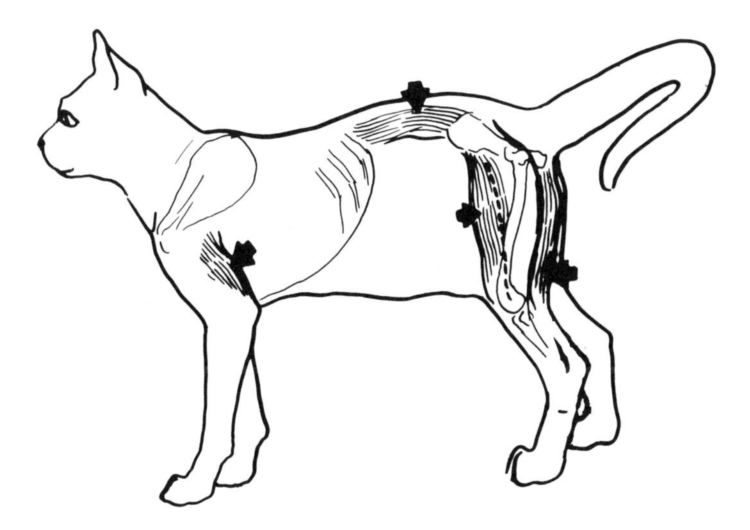

FIGURE 1. Diagram of muscle masses in a quadruped. The position of the femoral (dashed line) and sciatic (solid line) nerves is indicated. The femoral is shallow on the medial aspect of the hind limb. Arrows indicate the sites of intramuscular injection.

substances, is IV or IC, where the substance is placed directly into the blood. IP is not as quick as IV, but is more rapid than IM, as the absorptive surface of the gut and the lining of the abdominal cavity is much greater than that of a muscle. IM is swifter than SC, which is mainly used for vaccinations and sensitizing doses of antigens, where long-term release from the tissue is desired. Those tissues which take up a substance rapidly are also depleted of the substance rapidly, decreasing the duration of effect. It should be obvious that the uptake of a substance is dependant on the degree of circulation of blood in the target tissue; therefore, in an animal that is chilled and has poor peripheral circulation, SC injection will induce a pharmacologic effect slower than if the animal had warm or heated extremities where bloodflow was increased.

Methods of injection do not vary significantly among species, except for positioning of cardiac puncture. SC injection, probably the most common and usually the least painful, involves tenting the skin and inserting the needle under the tent. One always aspirates the syringe (that is, pulls back on the plunger) when injecting by any route to ascertain whether the needle is in the proper position, and not, for instance, within a hidden blood vessel. When injecting SC it is done mainly to make sure that the needle has not passed through the skin tent to outside air. In this case air would be sucked into the barrel of the syringe. If one does inadvertently pass through the skin tent, the syringe should be withdrawn and air removed, and then repositioned. Generally the best site for SC injection is over the neck region, where the skin is almost always the loosest.

IM injection is relatively simple, requiring only that one be certain that the needle tip is within the muscle mass. If animals are to survive a procedure, however, it is important that the injection, especially if it is of an irritating substance, not be given near a major nerve trunk. The most convenient IM site, away from major nerves, is low in the posterior thigh. Alternate sites include the heavy back muscles (Figure 1) over the loin and the muscle mass just above and behind the elbow. In small animals, the muscle is palpated and grasped, if possible, by the free hand while the needle is directed. On extremely small animals an appropriately short needle should be employed so that one does not pass through muscle and into bone or back out through the other side of a limb. As in other types of injection, one aspirates after the needle is positioned. If blood appears in the hub of the syringe, it indicates that one has inadvertently punctured a blood vessel, and the needle should be repositioned. The slower a substance is injected IM, the less pain to the

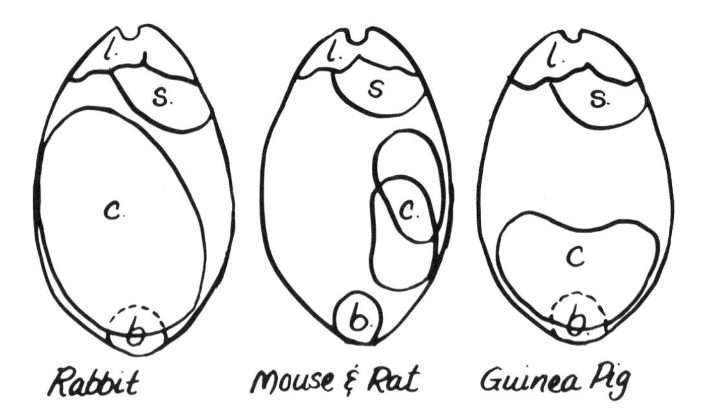

FIGURE 2. Diagram of large and fixed structures in the abdominal cavities of rabbits, rats and mice, and guinea pigs. The liver, stomach, cecum, and urinary bladder should be avoided when giving IP injections.

animal (fast injections tear the muscle fibers apart). One must balance the speed of the injection with the possibility of reaction from the animal; in those animals which routinely retaliate with claws or teeth when even slightly hurt, it makes more sense to inject more rapidly and get out of their way. However, if truly adequate restraint is employed, avoidance of a bite or scratches is achieved.

IP penetration (Figure 2), which is commonly used for administration of liquid anesthesia or withdrawal of peritoneal fluid, depends on the abdominal contents moving aside from the needle so that the injection is made into the peritoneal cavity rather than into an organ. Consequently, the anterior abdomen should be avoided, as the stomach (if full) and liver tend to stay in place rather than move aside from a needle. The urinary bladder, in the posterior abdomen, will not move readily if full, so this area should be avoided as well. When injecting IP, the skin and body wall is penetrated with the needle, while avoiding deep penetration which might impale the gut. Aspiration will indicate if the needle is within the gut, as ingesta will enter the syringe. Should this occur, the needle, syringe and contents should be discarded, as injection of gut contents would cause peritonitis. If the bladder has been entered, large quantities of amber or yellowish fluid will readily enter the syringe. A successful peritoneal stick would aspirate nothing at all or a slight amount of clear fluid from a normal abdomen. Animals with ascites will have abdominal distension, and fluid may be readily aspirated after entering the abdominal cavity. The fluid is usually clear to yellowish and watery to slightly viscous. The size of the syringe for aspiration of ascitic fluid should correspond to the expected volume available.

Cardiac puncture will be discussed for the individual species. In general, palpation of the chest wall will indicate the position of the strongest beat, which will be the area of the ventricles. In larger species, one enters from the right side of the chest and directs the needle toward this beat. In smaller species, the needle is inserted alongside the posterior sternum (the xiphoid cartilage) and directed cranially and slightly dorsally along the midline. Aspiration will indicate whether one has pierced the heart, when blood readily enters the syringe. Usually the needle will enter the right ventricle, which is the largest cavity in the heart, in which case the blood will have a bluish cast. Bright red blood usually means the needle is in the left ventricle. Air indicates lung or airway.

Venipuncture, for IV administration or withdrawing blood, requires precise positioning of the needle and complete lack of movement in the subject for success. The vein must be either seen or palpated before puncture or the effort is doomed to failure. Usually the vein is occluded proximally so that it is distended and therefore a better target. The needle is to be inserted "bevel up". When one is ready to inject, or is finished withdrawing blood, the vein is released. After the needle is removed, light pressure is applied to the skin over the puncture site. The pressure should not be strong enough to occlude the vessel, or rebound flow will cause the puncture site to open up upon release. Rolling the skin puncture away from the site of the hole in the vein assists in exposing the blood platelets to clot-forming substances in the subcutaneous tissues, therefore forming a plug on the vessel wall more readily. Sites of venipuncture differ with species, and each will be discussed separately.

III. THE MOUSE

The mouse, *Mus musculus,* the most common and smallest of the frequently used research animals, is relatively easy to manipulate if one is aware of its natural inclinations and defenses. The reactions of the mouse when frightened or anticipating pain are first to flee, and then, if flight is not an option, to bite. Many mice will bite readily, depending on the aggressiveness of their particular strain and their previous experiences with being handled. However, the natural propensity towards biting in mice can be modified by gentle handling and stroking.

The first consideration for manipulating any animal is how to pick up the creature. Mice are generally picked up by the tail, which is the most secure way to do it; the tail provides a convenient handle, easily grasped. Encircling the body with the hand may also be used, but the mouse can squirm out of the hand and escape fairly easily if not also restrained by the tail. The mouse's tail may be grasped anywhere along its length, as few mice are heavy enough to pull the skin off. However, if a mouse should grasp some object, it could potentially pull hard enough to deglove its tail if the tail is held near the tip. Therefore, the tail is best held near the body, at the base. Although it is always a good idea to wear at least disposable latex gloves for hygienic reasons, wearing heavy gloves to protect the hands from bites is not a good idea with mice. One needs to feel the tail to grasp it effectively. Also, heavy gloves may be so stiff that the force of the grip may be harmful to small animals like mice. Although many people use tongs to pick up mice, even rubber-covered tongs are more painful to the animals than fingertips and should not be used for picking up mice except for special reasons, such as containment or barrier conditions.

Like its larger cousin, the rat, the mouse can get uncomfortable hanging upside-down for any significant amount of time. It may, if youthful and athletic, try to climb back up its tail in order to bite. To avoid being bitten, one may spin the mouse gently, which will force it away from the tail and also tend to disorient it slightly due to vertigo. A better procedure is to put the mouse down on a surface as soon as possible, holding the tail so that it does not escape. The opposite forearm or hand serves well as a surface on which to rest the mouse while examining it.

In order to perform any action which is unpleasant to the mouse, one needs to hold the animal securely. The most frequently used hold for injection, gavaging, etc., is the scruff hold (Figure 3). Mice which have been "scruffed" previously generally have not found the procedure pleasant and will attempt to avoid having the dorsal skin pinched. However, if the operator places the mouse on a cage lid or other surface which the mouse may grasp, its natural inclination is to pull away from the tail, on which the operator maintains tension. The thumb and forefinger, in a closed "vee" shape, are pressed down on the pelvic region of the mouse, and moved forward until just behind the head, where the skin is pinched firmly. This hold will, if tight enough, prevent the mouse from biting a person. In addition, however, the body needs to be kept motionless, which may be effected by catching the tail or posterior torso with the little finger against the palm of the hand. Alternately, the remainder of the fingers can catch the loose skin along the mouse's back against the shaft and the ball of the thumb.

There are commercial restrainers available for mice. Some of these are variations of a tunnel, with a rumpboard for the tail to pass through and various ports for injections. Generally the hold on the tail should be maintained, or the animal will move freely about the tunnel, turning around, upside-down or otherwise attempting to avoid a needle. Often mice do not want to leave the

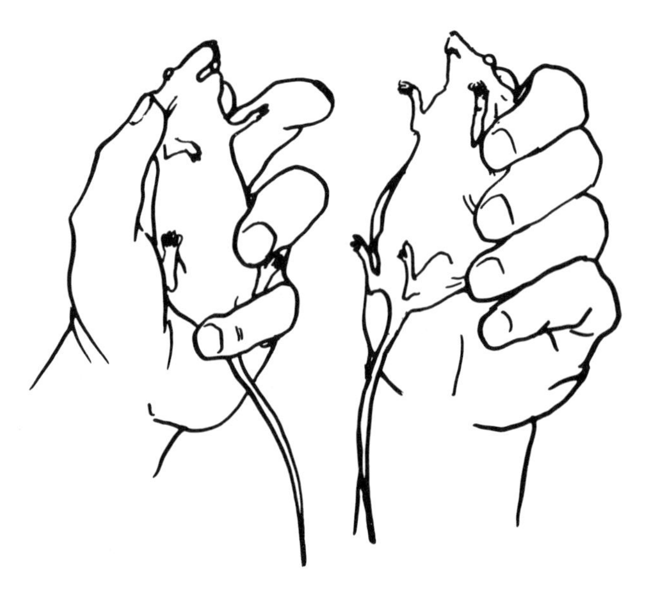

FIGURE 3. Two methods of scruffing a mouse. (Left) The neck scruff is caught between thumb and forefinger, and the little finger holds down the tail or pelvic region. (Right) The loose dorsal skin from the neck posterior is grasped against the thumb.

security of a tunnel restrainer and will grab onto the holes in the walls to prevent being pulled out. Rather than exerting a pull on the tail that might be damaging, one should try tipping the tunnel head-downward. Most mice will back out of the restrainer if it is in this position. Another possibility is to place the restrainer within the cage and leave it until the mouse decides to leave.

Tail vein administration in mice is a very common technique, which requires that the mouse be completely still. Any type of slot from about 0.5 to 0.75 cm wide may be used to restrain the mouse by pulling the tail through and continuing tension while performing venipuncture. Even wire cage tops may be employed if nothing else is available, but open slots are easier to use because one never needs to release the hold on the tail while positioning the mouse.

A 25-gauge or smaller needle about 1.5 to 2 cm long on a 1-ml syringe is used for most techniques with mice, including tail veining. Generally not more than 0.2 ml of fluid should be administered IV to a mouse at one time. The tail veins are the only readily available external veins in mice. There are two lateral tail veins; there are also arteries on the dorsal and ventral aspects of the tail. The arteries, although usually easy to visualize, are impossible to cannulate with a hypodermic needle because of their network configuration. Placing the cage on a heating pad or hanging a heat lamp over the cage several minutes before venipuncture will heat the animals and cause peripheral vasodilation, increasing the size of the vessels. Once the tail is pulled through a slot, the animal can be effectively restrained from movement by tension on the tail. The tail is held between thumb and forefinger, and the needle is inserted into the skin, bevel up, directly over the vessel where the tail bends over the forefinger. The vessel lies within a plane of muscle and connective tissue, which lends a sensation of gliding as the needle is advanced if it is in the proper position. The volume of blood in a mouse's tail is so miniscule that aspiration will not reveal blood in the syringe; one may withdraw the needle to see if a spot of blood will appear at the puncture site and then replace it or rely on visualization and sensation to decide whether the needle is in fact in the vein before injection. Test injection of the substance in the syringe may also be used to see if the needle is in the vein. The vessel will blanch of blood and the substance will flow readily from the syringe if it goes IV. Subcutaneous blanching will also occur if it is injected SC, the tail will be slightly swollen, and the resistance in the syringe will be greater. Mouse tail veining is a specialized skill that will improve with practice. Learning to hold the syringe barrel in the fingers and push the plunger with the little finger will avoid jarring the syringe as one changes the hand position to inject, which often tears the vessel.

Chemical restraint, or light anesthesia, may be used on mice. The simplest of these is the use of a gas anesthesia chamber. Methoxyflurane may be freely vaporized in a closed container, and the animal can be placed inside until it loses the righting reflex. The animal can then be removed and quick procedures, such as ear notching, performed. The animal is insensitive to mild pain and will remain still for several seconds to a minute. Methoxyflurane, a halogenated ether, is a volatile liquid which vaporizes to barely above a lethal level; exposure to the gas must be prolonged in order to cause death. This is very similar to the effect of ethyl ether, but there are several advantages to methoxyflurane. Unlike ether, methoxyflurane is not flammable or explosive. It provides much better pain reduction than ether, and is less irritating to respiratory mucosa. Mainly due to the nonreactivity of the gas, it has replaced ethyl ether as a small laboratory anesthesia. However, all gas anesthesias must be used in well-ventilated areas (such as fume hoods) with good scavenging systems for waste gas, as all are potentially toxic to people. Injectable restraint is hardly worthwhile in mice, as the degree of handling needed for giving the injection is as extreme as any needed for any other technique.

Oral administration of fluid substances may be accomplished with the use of a metal feeding needle ("gavage needle"). These needles fit onto ordinary syringes and come in sizes for all sizes of animals and birds. A size commonly used for mice is about 3 cm long, 18 gauge, with a 3-mm rounded tip. The round tip keeps the needle from readily piercing the esophagus. Three mm is larger than the trachea in mice, so there is no chance of placing the needle in the airway. One almost always performs gavage (the term literally means administration directly into the stomach) or esophageal administration (which is usually what is actually happening) with animals that are awake rather than tranquilized or anesthetized, so that they will not aspirate fluid which might be regurgitated or deposited in the oral cavity. Also, an alert animal will always swallow a tube rather than allow it to enter the trachea. The animal is restrained by a scruff hold, and the tip of the needle inserted into the side of the mouth behind the incisors, then directed gently down the esophagus. Keeping the head and neck as straight as possible and directing the needle somewhat dorsally in the back of the mouth will facilitate passage of the needle. Gentle manipulation of the needle may be needed to make the mouse swallow it; when it is swallowed, the slight amount of resistance previously met disappears, and the needle may be inserted to its hub, which will place the tip in or near the stomach. The needle should never be forced down the throat, as this invariably ruptures the esophagus. Substances may be injected if the needle is inserted beyond the level of the angle of the jaw, as this is beyond the opening to the trachea and peristalsis of the esophagus will carry the substance down the esophagus. If the needle is not forced down the mouse's throat, gavaging may be performed with little discomfort to the mouse and no permanent damage. The common sequelae to rough gavaging is subcutaneous emphysema; tearing of the esophagus and/or trachea allows air from the upper airways to be pumped under the skin as the animal respires. The taut air pocket often occurs in the neck, shoulder or back region. Animals with this condition should be euthanatized. Another negative outcome of gavaging is penetration of the esophagus and deposition of substance into the thoracic cavity, which usually leads to difficulty in breathing and death. Although this technique may appear dangerous and difficult, it really is quite simple and safe if the needle is not forced. One way to maintain a delicate touch is to hold the syringe barrel so that it rolls between fingers and thumb rather than in a normal injection position with the thumb on the plunger.

Very small blood samples may be acquired from mice by cutting off the end of a toenail or cutting across the tail vessels, either individually or by slicing off the end of the tail. It is important that the animal be warm and vasodilated when using these vessels for blood, or the bleeding will be negligible. These bleeding techniques may be used on a restrained animal, but investigators should consider anesthesia if the protocol allows it to reduce discomfort to the mouse. For larger blood samples or sterile samples, a jugular cutdown or a cardiac puncture may be used. Some studies may call for a sterile sample, in which case the mouse should receive a surgical prep, which includes shaving, cleansing and draping, if necessary (on a small area such as the jugular cutdown field, one can keep an undraped field uncontaminated for the short time it takes to perform the procedure). Both jugular cutdown and cardiac puncture may be done either aseptically or not (should the procedure be terminal). Both absolutely require general anesthesia.

For the jugular cutdown the animal is placed on its back. The forelegs are stretched to the side so that the pectoral muscles are taut across the anterior thorax (the stretched pectoral muscles will put pressure on the jugular veins and make them distend). The forelegs may be secured by tape. The hindquarters should also be secured to keep the animal in a symmetrical position, which will aid in puncture of the vessel. Incision is most easily made by picking up the skin with forceps and cutting across the skin tent with scissors, being careful not to damage deeper structures. The skin is incised over the jugular, if visible, or over the anterior edge of the pectoral muscle forward 1 or 2 cm along an imaginary line drawn from the center of the chin to the first teat.

Various subcutaneous tissues may be present under the skin incision: submandibular salivary gland, mammary tissue, and loose connective tissue. The salivary gland is dissected to a medial position and the other tissues laterally. Dissection should be done very carefully by picking up the tissue and cutting after being sure that there are no small vessels included. After the jugular has been exposed for 3 or more mm anterior to the pectoral muscle, the venipuncture may be performed. After making certain that the syringe plunger is sliding readily and then replacing it in the seated position, one supports the syringe barrel with one hand while advancing the syringe with the other. The needle is inserted bevel up into the front edge of the pectoral muscle and aimed for the center of the vessel anterior to it. If the vessel is not turgid, the forelegs may need to be stretched further to the side. When in place, the needle is visible within the lumen of the vessel. Withdrawing of blood should be done fairly slowly, or the vessel will collapse against the bevel of the needle. About 0.2 ml is all that should be taken from an adult mouse that is to be recovered, but up to 1 ml can be withdrawn in a terminal procedure with this technique, although after the first 0.2 ml the blood will come much more slowly because of vascular hypotension. If the procedure is to be a survival technique, the skin may be closed with suture or staples (staples may last longer, as mice are less able to groom them out). One should take care that staples or suture needle do not puncture the vessel when closing the skin, especially if the procedure is to be repeated later. Recovery of the mouse can be enhanced by IP administration of warm sterile physiologic saline or lactated Ringers solution to replace the volume of blood removed and avoid hypotensive shock. Up to about 0.5 ml of fluid can be given without causing a fluid volume overload after removing 0.2 ml of blood.

Cardiac puncture may be performed on an open or closed thorax, depending on whether the animal is to be recovered. Obviously, in such a potentially painful procedure, the mouse must be anesthetized. The skin should be prepped for a closed recovery procedure. After the mouse is positioned in dorsal recumbancy, the needle is placed along the right side of the xyphoid cartilage and advanced anteriorly and slightly dorsally toward the midline. When the tip is approximately mid-thorax the syringe should be aspirated. In this closed, "blind" technique more than one attempt may be necessary. If the thorax is to be opened, however, the heart can be visualized and the needle stuck directly into the right ventricle. The chest is opened from the posterior aspect. After incising the abdomen, the xyphoid cartilage is held, and the ribs cut 2 or 3 mm to either side of it (to avoid the veins that course about 1 mm lateral to the edges of the sternum), then the diaphragm cut if it has not torn while the chest wall is opened. Usually at least 1 ml of blood can be taken by this nonrecovery technique. It should be stressed that the operator should always be certain that the animal is well anesthetized during any surgical or invasive technique like cardiac puncture, not merely restrained. Cardiac puncture may result in blood leaking into the pericardial space, leading to cardiac tamponade, or tearing of the myocardium, either of which may lead to the death of the animal. These conditions are very painful, which is why the animal should be unconscious. Also, once the chest is opened, the animal will not be able to inflate the lungs and will experience suffocation if not fully anesthetized. Because of the seriousness of the sequellae, cardiac puncture is best chosen as a terminal procedure, performed under deep anesthesia.

Another blood collection technique, which should be used as a recovery technique only when other bleeding techniques are impossible, is orbital bleeding. Once frequently utilized in research with rodents, this technique has been recognized as potentially painful and damaging to the animals, and restrictions have been placed on its use in many institutions. It is based on the fact that there is a vascular plexus present behind the eye which can be ruptured with a small glass tube. The tube is inserted into the medial canthus (corner) of the eye socket and gently rotated until blood flows; holding the animal upside-down sometimes augments the flow. When the desired volume of blood is acquired, the tube is withdrawn and light pressure applied on the eye to aid hemostasis. Bleeding is controlled after orbital bleeding by pressure behind the eye, a condition that must be considered painful, like the puncture itself. For this reason, mice or other rodents must always be anesthetized for this technique and, preferably, euthanatized before regaining consciousness. Euthanasia must also be performed if faulty manipulation should cause the tube to enter the nasal area, which will result in copious nose bleeding. Another danger in this technique is the possibility of damaging orbital structures such as the ocular muscles or nerve, causing blindness. This technique should be avoided if at all possible.

Urine and fecal samples are simple to acquire from mice. Most mice will urinate and/or defecate when caught by their tails or when restrained by the scruff. If one is ready for the elimination with a test tube or petrie dish it can be caught as it is produced. Feces may also be collected from the cage; leaving cages without litter for a few hours will facilitate collection. There are also "metabolic" cages available for mice which have devices that separate urine and feces.

IV. THE RAT

The rat is not merely a large mouse. Most laboratory strains of rat are exceptionally docile and do not bite readily, although rats are capable of inflicting deep bite wounds that can be contaminated by harmful bacteria. However, unless their only contact with people is negative, rats are cooperative, responsive, and friendly animals. Most enjoy being stroked over the back or head.

All of the techniques used in the mouse can be used in the rat, with adjustments, if necessary. Because the rat is approximately ten times larger than the mouse, similar procedures are often easier. Volumes of substances to be administered are larger and sample volumes greater due to the tenfold difference in size. Because of its size, the rat should not be routinely picked up by the tail, but instead be grasped around its torso. However, if a rat indicates aggression by squawking, hissing, or opening its mouth as if to bite when a hand approaches it, it should be caught by the tail and picked up by the base of the tail rather than near the tip. The tails of large and heavy rats will readily deglove if the animals are lifted by the distal portion. An obstreperous animal can be moved by the tail base to a surface, placed in an anesthesia chamber, or given the opportunity to enter a tunnel restrainer while being held safely by the tail. Most rats, like mice, readily enter a tunnel restrainer, although animals that have had negative experiences in tunnel restrainers may refuse to enter them. One may use heavy leather gloves to restrain biting rats, remembering that most rats can pierce leather with their sharp incisors. Lack of adequate sensation for nonharmful restraint is always a problem with heavy gloves and small animals, however.

FIGURE 4. The rat is grasped around the thorax, with the thumb and first finger under the jaw. Holding the tail will keep the animal from scratching one's hand with the hind claws. In this position the animal can be injected in several sites, or gavaged.

To restrain a rat by hand (Figure 4), one holds the animal securely around the shoulder girdle, with thumb and/or forefinger under the jaw (snug around the neck) to hold the head up and prevent biting. Rats in this way are fairly secure as far as the anterior portion of the body is concerned, but may claw at the hand with the hind feet. If so, the tail may be held with the other hand, which prevents the rat from flexing the body enough to reach forward with the hind feet. Rats may be gavaged while held around the shoulders, with the hindquarters pressed against one's chest if it is necessary to restrain the hind feet. The feeding needle used for rats is longer than that for mice (about 10 cm) and may be of a similar or a larger diameter.

Rats are similar to mice in terms of entering tunnel restrainers and tail venipuncture. Tail veining a rat may require more than a slot, as they are stronger than mice and can squirm over a barrier to bite. Cage lids, however, are usually adequate. The wire cage lid is turned upside down over the rat and the tail pulled between the wires. Only the type of cage top that has an angled portion as a food holder (in which the rat is held captive) is appropriate, of course. Older rats develop considerable scale on their tails, and the skin is quite tough. The scale can be removed by scraping the skin against the grain, but this may cause enough irritation of the local skin that the underlying vessel is obscured. If scale is to be removed, it should be done a day or two before venipuncture. Since the skin is tougher and the vessels are larger on a rat as opposed to a mouse, a larger needle is used for venipuncture, 22 gauge. Length may vary from about 1.5 to 2.5 cm. The actual venipuncture is only a larger, tougher version of what is done with the mouse. Again, since the vessels are rather small, one is almost never able to withdraw blood on aspiration. Since rats are about ten times the size of mice, about ten times the volume of blood may be taken or substances administered by tail vein; that is, approximately 2 ml. Blood is most easily acquired, in amounts less than 1 ml, by restraining the rat with the tail dependent and obliquely cutting across the tail vein..Application of a thin layer of petrolatum to the skin before cutting reduces the amount of clotting that occurs on the skin surface. Of course, as in the mouse, the animal must be warm enough to be vasodilated in its extremities, and success usually depends on warming the animals to a panting state, where the ears and tail are obviously flushed. Cardiac puncture and jugular cutdown are essentially the same as in the mouse, but on a larger scale.

There are also gavage needles adjusted to the size of the rat. Administration is also similar to the mouse, except that the rat is held around the thorax rather than a scruff hold.

Rats readily defecate and urinate when picked up, or if not then, when restrained or experiencing noxious stimuli, such as being pricked by a needle during injection. They are even more suited to the use of a metabolism cage than mice, as the volumes of urine and feces are that much greater.

V. THE RABBIT

The laboratory rabbit is a research animal which offers the advantage of larger size than the rat or mouse, and in addition is remarkably docile. Biting is almost unheard of in rabbits, and striking with the front claws is a generally ineffectual aggressive act by territorial females. Rabbits may kick with the hind legs in an attempt to flee and inflict fairly severe scratches on handlers, but uncontrolled kicking is more of a danger to the rabbit and must be carefully avoided. Rabbits respond favorably to petting to reduce anxiety; the forehead seems to be a favorite spot to have scratched, followed by behind the ears. Stroking the fur of the back can also be calming to rabbits.

Catching a rabbit is usually a simple matter of opening the cage and grasping the usually crouching animal in a far corner. Since some flighty rabbits may suddenly decide to explode in movement away from the advancing hand, and, after rapidly circling the cage, exit out the front to the handler's startled amazement, it is important to guard the cage door opening when catching a young and athletic or unknown animal. Most flighty rabbits will indicate this tendency before the cage is even open by jumping around as soon as they perceive someone present.

Adult rabbits are first caught firmly by the scruff. The skin over the shoulders and neck is fairly loose, and skin and hair are easily grasped without evidence of distress. Should the rabbit attack the hand as it enters the cage (standing on the hind legs and striking with the forelegs while growling through the nose is the commonest way rabbits threaten), its attention can be diverted with one hand while darting in for the scruff hold with the other. Pushing down at the scruff toward the cage floor can be used to subdue a fractious or aggressive animal; any rabbit that threatens to jump or fight should be held down in this way until it relaxes.

The one most important thing to remember when handling rabbits is to control the hindquarters. Rabbits have fragile bones and a powerful posterior muscle mass. Kicking while being held may result in overextension of the lumbosacral region of the back, resulting in fracture or subluxation of the spinal column. This unfortunate situation, commonly called "broken back", may result in flaccid paralysis of the hind limbs, the urinary bladder, or other posterior structures. Paralysis may not be immediate, as pressure caused by hemorrhage and fluid infiltration of the spinal cord may occur some time after the event. Only severing of the spinal cord will cause immediate paralysis.

When picking up a rabbit (Figure 5A to C), therefore, the hindquarters must be supported. One way to do this is to slide one's hand under the rump from the rear as the other is lifting the scruff. This flexes the rabbit's spine into a position from which it cannot hurt itself. Another way, which depends upon the natural tendency of animals, especially neonates, to hang docilely while being suspended by the back, lifts the rabbit by scruff and rump skin, like a suitcase with two handles. Although esthetically less pleasing, this hold results in a motionless animal and has resulted in a zero incidence of "broken back" in our facility over the last 8 years that animal caretakers have been taught the technique.

A

FIGURE 5. Methods of picking up the rabbit. (A) A very young rabbit, weighing less than about 1.5 kg, can be picked up with one hand by the loose skin over the back. (B) Adult rabbits can be lifted by the scruff and a hand slid under the rump to support the hindquarters and to round the back. (C) Picking up the rabbit by the loose skin over the shoulders and neck, as well as that over the rump, will result in the animal hanging motionless.

Immature rabbits, those less than about 2 kg, can be lifted by a one-handed hold of the skin of the middle back. Rabbits, regardless of age, should never be lifted by the ears alone, although restraining the ears within the scruff hold calms some rabbits.

Rabbits should be held down when placed on any surface until they relax, to avoid kicking. A rubber mat is convenient for an exam table or counter, as traction will reassure the rabbit and probably avoid the mad scrambling which occurs on slick surfaces. The most likely time for a rabbit to scramble is when it is being replaced in its cage; there is a tendency to leap to security, thereby endangering the back. Placing the rabbit in the cage facing outward and holding it down until it relaxes will reduce the chance of this happening.

When carrying a rabbit more than a short distance (Figure 6), the rabbit may be gathered in the forearms in front of the abdomen; many rabbits find it more secure in this position to have their heads tucked into the crook of the handler's elbow. Rabbits will have to be gathered from a waist-level surface and replaced at the same level. Also, in this position, the rabbit can scratch the handler severely in the abdomen should it begin to struggle and kick.

Restraint of rabbits for several procedures may be effected by holding on a surface or by placing the animal in a restraint device. The rabbit may be simply held by scruff and over rump or can be held facing away from the handler with forearms pressing along its sides, thumbs behind the ears, and fingers lifting up slightly from under the jaws. Most rabbits are more secure being held by hand than in a restraint device. However, if one does not have the convenience of a handler, the device is necessary. Rabbit restrainers (Figure 7) are equipped with head stocks

FIGURE 5B.

FIGURE 5C.

and a rump board of some type. They should never be used without the rump board firmly in place, as a rabbit could harm itself fighting if only its head was held. When the rabbit is placed in the restrainer, the rump board is pushed forward until the rabbit's back is bowed and the rabbit is as far forward in the device as possible. Once the rump board is secured, the head stocks may

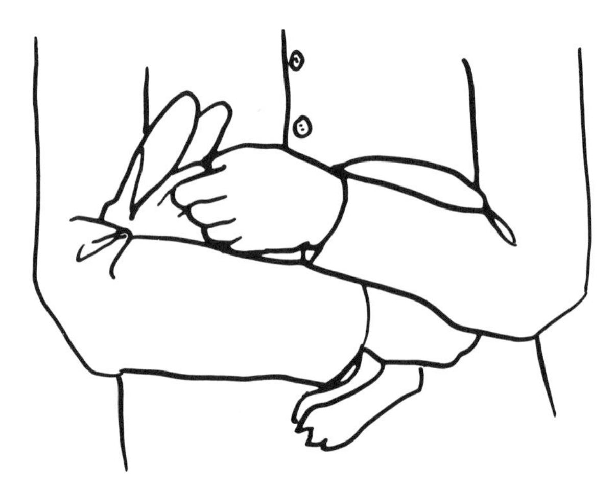

FIGURE 6. For longer distances, the rabbit should be carried in the forearms. The use of long-sleeved clothing is strongly recommended.

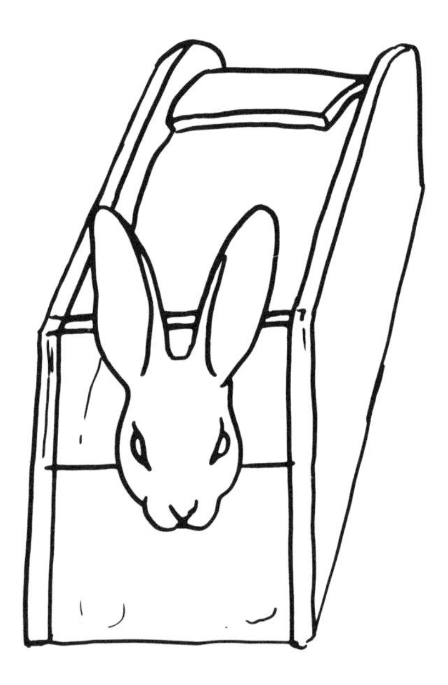

FIGURE 7. Rabbit restrainers come in several styles. They include a head stock and rump board. The neck should not be restrained without the rump board firmly in place.

be fastened. A rabbit restrained in this way is in perfect position for venipuncture or bleeding from the lateral marginal ear vein, assuming that it has been properly placed in the restrainer and cannot jerk or move away as its ear is manipulated.

Venipuncture is a simple procedure in rabbits (Figure 8), as the ear veins, although small, are easily visualized and just beneath the skin. If the rabbit is cold or is frightened, the ears will blanch and vessels will either be tiny or invisible due to peripheral vasoconstriction. Placing the rabbit under a heat lamp or in a warm room will cause the ear vessels to dilate (one should always be *careful* warming rabbits, as too much heat for too long a time can result in heat stroke). For local vasodilation of an ear vein, even in the presence of cold or fear, one can irritate it slightly. Some people advocate the use of xylene or alcohol for this purpose, but xylene can cause necrosis of the ear if not completely removed, and alcohol, which creates only minimal vasodilation, will

FIGURE 8. The marginal ear vein (arrow) as it is held for venipuncture. The central artery is also indicated.

cause a sharp stinging sensation when the needle enters the skin if not yet evaporated. A simple and noninvasive manner in which to cause vasodilation, which rabbits do not even seem to notice, is to pluck the hairs along the path of the vessel.

For the actual venipuncture, the vein must be occluded to make it turgid enough to enter easily with the needle. The vessel may be held off with thumb and forefinger, with the distal portion held between the third and last fingers. The needle is directed, bevel up, directly over the vein and at a very shallow angle. The translucent nature of the skin and vessel usually allows visualization of the needle in the vein. In very large veins, aspiration of the syringe may pull blood into the hub of the needle, but the volume of blood is often so slight that it will not show on aspiration. The hold on the vessel at the base of the ear is always released before injecting, or the vessel may rupture. If the needle appears in the vessel, a test injection of a tiny amount of substance will blanch the vessel and will inject with little or no resistance. For people with small hands, a handler may hold off the vessel, or a paper clip may be slid over it — but it is very important to remove the pressure before injecting.

Bleeding of rabbits is almost always performed via the lateral ear veins. Warmth and vasodilation are even more important for bleeding than for IV injection. The ear is prepared by plucking the hairs in the vicinity of the bleeding site and applying a thin layer of petrolatum to both front and back of the ear. The petrolatum will help keep the blood from clotting on the surface of the ear. The vessel is occluded close to the base of the ear and then incised lengthwise 2 to 4 mm with a small scalpel blade (#15 works well). Blood will flow freely and can be directed into a tube or other container. If clotting occurs before the desired quantity is collected, the clot should be wiped away with a bit of cotton or gauze. Up to about 50 ml can be collected this way, although at a drop at a time it can be quite time-consuming. Vacuum apparatuses have been employed to speed the process. Another way to collect blood faster from the ear is to use the central artery. The artery can be cannulated with a small bore needle and the flow directed into a container. There are two disadvantages to this technique, however. One is that it takes more skill, and the other is that hemostasis is more difficult when an artery is punctured. Getting even a well-dilated ear vein to stop bleeding can be challenging. The petrolatum should be removed and a bit of cotton pressed lightly over the site. Slightly rolling the skin incision away from the vessel also helps. Firm pressure over the incision site will only occlude the vessel, which will rebound in pressure and rebleed when the pressure is removed.

It is always a good idea to watch the rabbit at least 15 min after ear vein bleeding to make sure the blood clot does not come loose and rebleeding does not occur. Rabbits may scratch their ears with hind feet or shake their heads and dislodge a clot. Although they are unlikely to induce life-threatening bleeding in this way, their cages or environs will be splashed with blood, a situation which indicates inadequate postprocedural care.

Rabbits may be bled by cardiac puncture, but such a procedure always carries risk and should probably be reserved for terminal use. The rabbit must be in a surgical plane of anesthesia for cardiac bleeding. IM ketamine (40 mg/kg) and xylazine (8 mg/kg) will induce adequate anesthesia in about 10 min in most rabbits; they should be tested with a painful pinch for response before proceeding. The approach for a cardiac bleeding may be either from the xyphoid, similar to the rat, but with a 3- to 4-cm-long needle, or from the side. The needle should be attached to a syringe of the desired volume, or the largest one available. If the lateral approach is used, the rabbit is placed on its side and the heart palpated for the strongest beat. The needle is directed between the ribs toward the beat. It actually makes little difference which side of the chest one enters, as the right ventricle is the largest chamber and is almost always where the needle lands. Once the needle is in place, the syringe is aspirated, and if blood does not appear, the needle is advanced forward or backward. If that does not work, the needle is withdrawn to the skin or outside of it before redirecting it. If this is not done, the needle has a much larger chance of tearing a rent in the myocardium. If the quantity of blood needed is more than one full syringe, the syringe may be removed while the needle is left in place and another or the same reattached after emptying.

Occasionally the blind cardiac approach is not successful; in this case, if it is a terminal bleed, the thorax may be opened with scissors or scalpel. The thorax is opened from the diaphragm, and the heart can usually be visualized without cutting ribs. The needle is directed into the right ventricle. With the thorax open, the animal can no longer respire and will die within a few minutes from asphyxiation. Obviously the procedure must be performed rapidly and the animal must be well anesthetized prior to beginning.

On occasion, a study requires the catheterization of veins or arteries. Small animal (cat) catheters are appropriate for the rabbit. Unfortunately, except for the vessels of the ear, the rabbit does not have external vessels that can be catheterized through the skin. Jugular veins and carotid arteries or femoral arteries or veins must be catheterized via cutdown techniques.

IP administration in the rabbit is best performed with the rabbit's head downward so that the heavy and enormous cecum, which covers nearly the entire floor of the abdomen, is displaced anteriorly. The needle is inserted in the rear of the abdomen off of the midline and 4 to 5 cm anterior to the pubic bone so that the urinary bladder will not be pierced. Holding the rabbit head downward may be accomplished two ways. A handler can grasp the hind legs, a finger in between them so that they do not slip out, and slowly and smoothly drop the head while holding onto the neck scruff. Alternately, should there be no assistant handy, one grasps the hind legs and lets the rabbit's head down as before, but in addition places the rabbit's head and torso between one's knees and thighs, releases the scruff hold, and stretches the rabbit slightly by the legs. This leaves the other hand free to do the injection.

Oral administration to rabbits (Figure 9) may involve placing substances directly into the mouth or using a feeding tube. To administer directly into the mouth (or to examine the anterior chamber of the mouth), one holds the rabbit firmly with one hand, thumb hooked behind the ears and the fingers under the jaw, while the rabbit is pressed to the edge of a working surface between forearm and upper arm and chest. Any thick or sticky substance can be squeezed into the mouth by using a syringe or other dosing instrument. The instrument is inserted into the mouth from the side just behind the incisors in the toothless gap called the diastema.

FIGURE 9. Positioning for oral administration in a rabbit. With the rabbit's head firmly restrained, the substance is inserted into the mouth behind the incisor teeth.

Oral intubation of rabbits is performed on alert animals. A flexible tube of approximately 7 mm O.D. is used. It must be resilient enough that it does not collapse easily, or it can turn back on itself as it is being passed. It is useful to mark on the tube the approximate distance from the nose to the posterior border of the ribs, as the tube should never be forced further than this, the level of the stomach. To avoid the rabbit's chewing on the tube, a mouth gag should be used. The simplest of these is fashioned from a syringe case or plastic syringe container which is wide enough for the tube to be easily passed through it. This rigid tube can be cut off about 3 cm long. It should be wrapped with several layers of adhesive tape to strengthen it and give it a nonslippery surface for the incisors. Any sharp edges should be smoothed (the rabbit's mouth could get cut if the gag should slip). Before the tube is passed it should be lubricated; a water-soluble lubricant is best. The rabbit must be well-restrained for this procedure, as not surprisingly, all rabbits resist having a tube passed down their throats. A handler may restrain the rabbit with his forearms and hands, as previously described, except that in addition, he or she must place the gag. The gag is slipped between the incisors and held in place while the tube is inside it. The tube may need to be directed slightly dorsally to avoid being trapped at the base of the tongue, or may need gentle nudging when in the back of the mouth in order that the rabbit swallow it, but almost always the tube passes smoothly down the esophagus. When removing the tube it is important that the syringe stay on the tube or that the tube be crimped so that fluid contents will not flow out as the end passes over the larynx.

Rabbits do not defecate as readily as rats and mice when handled, and unless they have suffered enervation of the bladder they never urinate. Fresh feces may be collected from under the cage; rabbits will continue to defecate as long as they have food to nibble on, but will cease when food is withdrawn, as bulk in the nonmuscular stomach is required to push the ingesta into the rest of the intestinal tract. If cecotrophs are desired, the rabbit must be prevented from ingesting them by means of mechanical restraint. So-called Elizabethan collars, which are

FIGURE 10. The adult guinea pig must be picked up with both hands around the body. Immature pigs may be lifted with one hand.

usually rigid plastic cones fastened around the neck to prevent the mouth from access to the rest of the body, are available in cat sizes from veterinary suppliers.

If the bladder is full, it can be palpated in the posterior abdomen as a turgid, round structure. To express a rabbit's bladder it must be stabilized before squeezing by grasping the abdominal mass at the pelvis, and, retaining the grip, advancing the hand anteriorally. If the bladder has significant urine in it it will bulge into the hand as it is moved. Once the bladder is palpated and secured in this fashion it can be squeezed. The pressure exerted should be less than that required to break the average chicken egg by squeezing in the same fashion. Obviously, excess pressure can damage the bladder. Expressed urine will pick up debris and microorganisms from the urethra and genitalia. Sterile samples must be procured by cystocentesis — that is, by aspiration with needle and syringe. The skin over the posterior abdomen must be shaved and surgically prepped, or a bladder infection could result. The animal is preferably anesthetized, rather than restrained, and positioned on its back; the bladder is stabilized and the needle directed into its lumen. The smaller the needle, the better (22 gauge or smaller). Urine is extremely irritating to the peritoneum, and urine leakage from the bladder is more likely to occur with a larger needle.

VI. THE GUINEA PIG

The guinea pig, or cavy, is even less aggressive than the rabbit. It is unusual among rodents in that it readily vocalizes at a level audible to humans when apprehensive or in pain. Sometimes merely holding a guinea pig will elicit loud, shrieking whistles, and many pigs will sound off if rolled upside-down while being restrained. The guinea pig is a social animal that learns to enjoy being held snuggly on one's forearm, sometimes tucking its head under one's elbow. It is immediately evident when observing guinea pigs that they lack tails, unlike rats and mice, so lifting by the tail is not an option. However, their lack of aggression (biting by guinea pigs is virtually unheard of, even during painful procedures) allows for easy grasping around the thorax. With young animals this is easily accomplished with one hand, but with obese adults (Figure 10), two hands must be used, either both around the body or one cradling the hindquarters (support of the hindquarters is especially important with obviously pregnant females with extremely pendulous abdomens). Pigs which have not been extensively handled tend to be rather jumpy; they will either gallop about a cage immediately or wait until the hand is near, then explosively evade capture. Quiet, slow movements and trapping in the corner of the cage are the key to taming wild guinea pigs. When they discover that they are not harmed when picked up they soon learn not to run away.

Guinea pigs are easily restrained by hand for almost any procedure by grasping them around the pectoral and pelvic girdles. They are difficult to gavage with a needle (size should be similar to that used on rats), but will accept a gavage needle in the rear of the mouth (the mouth has two compartments, as in all rodents, that of the incisors, and that of the molars or cheek teeth). Most substances can be delivered in this manner. For oral administration, the guinea pig is either held in the same way as a rat for gavaging or is merely lifted by the anterior body with the forefeet off of the base surface and hind feet in contact with it while the gavage needle is put into the mouth from the side behind the incisors (diastema). Small volumes should be delivered with adequate time for swallowing.

One severe limitation to using guinea pigs is the fact that they have no veins that are accessible without cutdown. One exception to this generality is the dorsal penile vein, which can be easily visualized by extruding the penis. The vessel is about 1 mm in diameter and can easily be torn. To extrude the penis, one presses on the swelling anterior to the genital papilla. The penis forms a "u" in this area when not turgid, and can be palpated before extrusion. Since this area is likely to be very sensitive, anesthesia should be used.

Because of the limitation on external veins, IV administration is seldom performed on guinea pigs. IC is an alternative, but requires anesthesia to be humane. A convenient injectable anesthesia for cardiac puncture is ketamine (20 mg/kg) and xylazine (4 mg/kg IM or IP). (Some animals may require up to twice this dose and should be medicated to their own level of insensitivity to the procedure.) Cardiac puncture can be performed as for rats or rabbits.

VII. HAMSTERS AND GERBILS

Hamsters and gerbils as species have almost as little in common as rats and guinea pigs, but they are convenient to lump together as rodents of a similar size and similar environmental background (desert). However, their personalities are quite different. The female hamster, once mature, is a solitary beast which will severely batter the smaller male except when she is in estrus. Gerbils, on the other hand, mate for life and cohabit in the same cage all the time; although the male starts about the same size as the female, he surpasses her in weight due to obesity as time goes by since he is not nursing the frequent litters. Gerbils seem to adjust to the human cycle of activity, but hamsters are stubbornly nocturnal. Their reputation for biting is probably due to being constantly wakened during their sleep period. Gerbils have more of a frenetic personality than hamsters, which, along with their tendency to sleep during the day, are less busy than gerbils. A common gerbil activity is to spend hours digging in a cage corner. Hamsters do have a tendency to squeeze or chew their way out of cages that is unsurpassed by other small rodents. Hamsters have a short, almost naked tail, while gerbils have a long, fully haired one.

Generally, the best way to pick up hamsters or gerbils (Figure 11) is to trap them against the side or corner of a cage and grasp them with the hand, similar to the way the rat is held. In this grasp they may be examined, gavaged, or otherwise manipulated. Seldom do these little animals like to be held still, however. Hamsters may be held by the scruff, which is quite distensible due to the cheek pouches that extend from the mouth to the shoulders. It is apparently an unpleasant sensation for the animal, though, and will induce them to bite, or attempt to bite, the next time the scruff is about to be grabbed. The dislike of the scruff hold can be modified by keeping the duration of the hold to a minimum and by cupping the animal in the hands and stroking it afterwards. This is an example of the practice of doing something pleasant to an animal as its last experience with people as a way of altering avoidance behavior (Figure 12).

Obviously the tail does not provide an adequate handle in the hamster. The gerbil may be lifted by the tail *with caution,* as the thin skin is easily pulled off. The tail is grasped at the thickest part of the base, and the remainder is passed through the hand to provide more surface area to grasp.

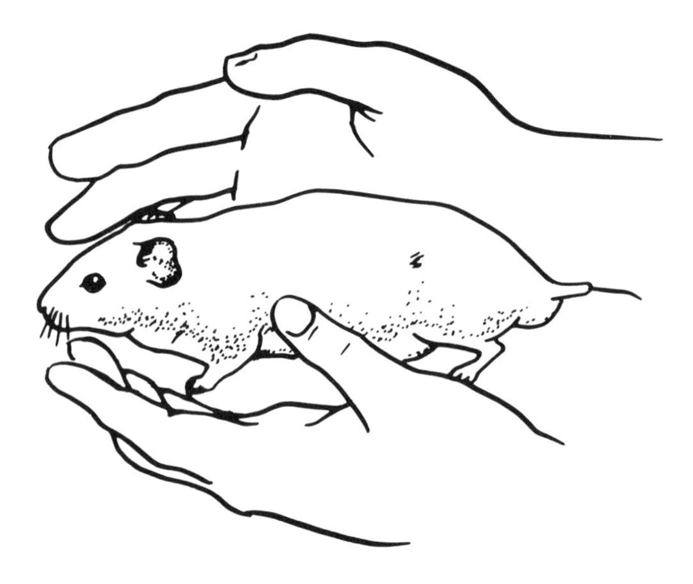

FIGURE 11. The hamster or gerbil is caught by surrounding with the hands or by catching it with one hand against the cage wall. Gerbils may also be caught (carefully, to avoid pulling off the tail skin) by the base of the tail.

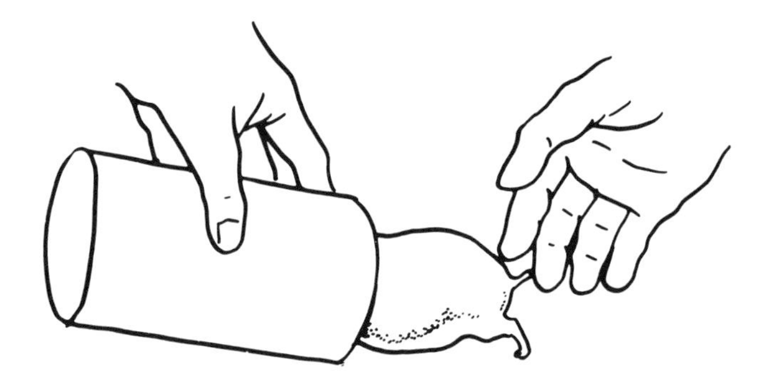

FIGURE 12. An alternate method for catching a hamster is to "haze" it into a can.

Both hamsters and gerbils may be restrained in tunnel-type restrainers made for small or immature rats. Tail veining is not an option with either species, and jugular cutdown or cardiac puncture are the available techniques for bleeding. Cephalic vein administration has been described for hamsters, however. They may be restrained by the scruff or anesthesthetized, and a rubber band can be placed around the upper arm near the elbow. Once the vessel is distended, it can be entered with a small-bore needle; the tourniquet is removed before injection, of course.

Hamsters and gerbils will sometimes defecate when handled, but never urinate. In fact, the urine of these animals is often quite concentrated and in very small quantities. Small metabolism cages may be used to gather feces and urine, or animals may be kept on a wire grate within a regular shoebox cage.

VIII. CATS

Cats, being typical household pets, are more familiar to many people than other research animals discussed above. Although often discussed along with dogs, with which they share some

FIGURE 13. Stretching the cat in lateral recumbancy. The cat is nearly helpless in this position.

characteristics (such as catching and eating prey), they are naturally of quite different personalities and needs. Cats, unlike dogs, are not needful of extensive exercise. Most cats are content to spend the major part of the time resting and, therefore, may be more amenable to cage housing than dogs. Cats, however, are curious creatures which seem to enjoy stimulation such as brief chase and pounce (mock hunting) interludes. Their attempts to escape from the cage are more likely to be for "intellectual" pursuits than for the pursuit of exercise. Less likely than dogs to need extensive human contact during formative weeks, cats are usually easy to tame. Certain personalities, most likely those that would be "aloof" in a household situation, however, may be intractable and vindictive in a research situation in which they experience even mild discomfort. Others are easily distracted from discomfort (as opposed to dogs, which are more likely to learn to bear discomfort for a reward of attention or treats). In terms of practicality, the cat has two major weapon systems for retaliation if it is hurt or held against its will: claws and teeth, usually employed in that order. Restraint of cats should be partially based on avoiding these potentially harmful reprisals and partially on the needs of the cats. Both bites and scratches are subject to severe infection.

Usually, if a procedure is done quickly, cats tend to act as if it hadn't happened. Liberally stroking the cat before or after a quick injection is one way to use this tendency. A cat should always be petted after it has been severely restrained, if possible. It is a type of apology for the affront to the cat's dignity; some cats, however, refuse to accept the apology.

Most cats are easy to catch in a cage; in fact, most will come to be petted as soon as the cage is opened. The most comfortable way to lift them is to cradle the torso in one or two hands. If a cat should hiss and flatten its ears when approached, it may or may not react by scratching and biting when grasped. Cats which will attack are candidates for heavy leather gauntlet gloves. The only difficulty with these heavy gloves is an inability to grasp the scruff well. The recalcitrant cat should be subdued by grasping the scruff, pushing the cat to the floor of the cage, and immediately grasping both hind legs. To get a firm hold on limbs of any animal it is important to keep a finger between the legs; the animal can pull one leg free if both are held together. As soon as the scruff and hind legs are secured, the handler "stretches" the cat (Figure 13). Most cats can do no harm when stretched, although some learn to reach up and claw the hand at their scruff. If any procedures are to be performed on a cat which must be caught in this manner (such as venipuncture), the cat should be tranquilized with ketamine at 11 mg/kg IM (is generally 0.2 to 0.3 ml for an adult cat) as soon as possible. A second person will be required to do the IM injection while the cat is held. The posterior thigh is the best site. Few cats will have anxiety attacks ("bad trips") when given light dosages of ketamine alone, but it does happen with some regularity in heavier dosages; the animal will show various fear or discomfort signs during recovery. These may be alleviated by administration of 0.25 mg/kg acepromazine IM. Animals that have bad reactions to ketamine should not receive it in the future.

Ketamine tranquilization, or immobilization, is commonly used in cats, as only the calmest will submit to venipuncture. As anyone who has had blood drawn can attest, the entry into the vessel stings, and the cat acknowledges it. One person can inject a cat which is not vicious IM with ketamine (or other substance, for that matter) by grasping a hind foot of the animal in the cage and sticking it in the thigh rear of the thigh as low as possible. Most cats, when they feel the needle and injection, think only of pulling away from the sensation. Two or three seconds later they turn around and claw at whatever is hurting them. Another method is to hold the scruff and then pull up one hind leg to tuck the foot under the thumb. Once the cat is in this position an IM injection into the low posterior thigh of the pinioned leg may be made. One should realize, however, that cats seldom will fall for this technique twice, and will writhe and try to get away as soon as the leg is pulled forward. Cats under the influence of ketamine lose their blink reflex, and it is important to put a slight amount of ophthalmic ointment in the eyes to avoid drying and damage to the cornea.

Cats can be carried short distances while being simply grasped around the torso as they were picked up. For longer distances, and for a possibly more secure carry, the cat is carried on one arm, the hand of which pinions both forefeet while the hindquarters are pressed to the body with the elbow. The opposite hand may rest on the scruff and be prepared to grasp it vigorously and hold the cat away from the body if the cat begins to gouge the handler with its hind legs or to bite the hand which restrains its forelegs. Grabbing the hindlegs and stretching the cat can also be efficacious if it must be held by the scruff.

Oral administration may be done on most alert, nontranqilized cats. If the head is lifted high enough, the lower jaw is forced to drop, or at least closes with decreased strength. The head may be forced back by pulling high on the neck scruff. A capsule may be dropped into the mouth and then pushed deep down the throat with a finger (the cat doesn't bite the finger because it can't close its mouth). Neither cats nor their throats are damaged by this treatment, unless the person has sharp or long fingernails; but if the material is not forced deep the cat will gag or push it up with its tongue and spit it out. There are also devices that will hold a capsule for insertion, which saves the finger from a potentially dangerous position; the eraser end of an ordinary pencil may also be used as a plunger. Liquid materials can be delivered into the posterior mouth. Caution must be taken in delivering more than about 1 ml of fluid that the cat has the opportunity to breath between doses. This is particularly important when delivering tasteless preparations, such as mineral oil, which might not signal the gag reflex and thereby be aspirated. Cats in general do not enjoy sweet tastes, but often crave the flavor of cod liver oil or other fish or meat extracts, and these preferences may be used to advantage when formulating a vehicle for administering a substance in solution. Large quantities of fluid must be administered by use of a feeding tube under light to moderate amounts of ketamine (above). If the cat still has its gag and swallowing reflex relatively intact it will readily swallow the tube. One should always make sure that the tube has not entered the trachea. The easiest way to do this is to open the cat's mouth, pull out the tongue, and visualize the empty laryngeal opening. The tube, of course, should never be advanced further than the level of the stomach, or it will curl back on itself or kink. As in other animals, the tube should not be allowed to leak contents as it is withdrawn (i.e., leave syringe attached and/or kink the tube while withdrawing.)

Holding a recalcitrant cat without chemical restraint can be accomplished by use of a cat bag. This is a canvas duck cylinder, usually with a zipper over the cat's back and hooks around its throat. Cats are generally not resistant to being bagged unless they have had negative experiences with them; unlike some species, they do not naturally fear or resist confinement. Cat bags are usually constructed so that one limb may be pulled out for manipulation, such as venipuncture, through a zippered slot. Jugular venipuncture, for administration or bleeding, is made relatively simple with a cat bag; the only difficulty is that the portion of the neck that is exposed may be rather short. The cat can be held on its back or in a relatively normal, upright position for venipuncture. Cats can also be rolled in a large towel to engage all four sets of claws, but most figure out a way to wriggle at least one limb free if they are determined.

FIGURE 14. The gauze muzzle on a cat. This same pattern, with a loop holding the nose piece up, can be used on pug-nosed dogs.

Although claws are usually more of a problem than teeth, cats may be muzzled (Figure 14) in the same way that short-muzzled dogs are (see below).

In extremely tractable cats, jugular venipuncture may be performed by restraining the head with one hand, the forefeet with the other, and the rear of the animal with the forearm and elbow as the cat crouches on a table. The cat is placed so that its sternum is in contact with the table by holding the forelegs straight downward over the edge of the table, a finger between the legs (Figure 15). The same arm restrains body and head. The cat is pressed to the holder's side and downward to the table with the arm while the hand holds the head at approximately 90° to the neck. The thumb and middle finger hold on either side of the head underneath the eyes, on the zygomatic bones. The index finger is then free to vigorously stroke the bridge of the nose. This is often enough distraction that the cat ignores the puncture. Blowing on the nose will also distract the cat. As mentioned above, the quicker the technique is performed, the less the cat seems to be offended. Several things will help to ensure success in this technique. Access to the jugular is difficult if the shoulders are too far forward; legs should be vertical or behind vertical. Lifting the head too high will stretch the skin, which may deflate the jugular vein. It is sometimes better to use a shorter rather than a longer needle (longer than 1 cm, shorter than 2 cm), as it tends to decrease the chances of threading a vessel and then passing through it again. Occasionally, in very small or short-necked cats, it may be advantageous to bend the needle at the hub for a cleaner entrance into the vessel, but use of a bent needle takes some practice to direct it straight. In all venipuncture it is important to see and/or feel the vessel. The person wielding the syringe must have the vessel occluded at the base of the neck in order to get it turgid enough to locate and puncture. Feline jugulars sometimes seem to take a long time to fill.

The jugular is the only vessel large enough for bleeding cats routinely, but cats, like dogs, have other accessible peripheral vessels for administration or catheterization, notably the cephalic and the saphenous.

IX. THE DOG

Very few dogs remain in a laboratory situation very long if they are vicious (i.e., actively aggressive). Larger dogs, particularly, can cause serious bite damage. Most biting by laboratory dogs results from fear, due to improper socialization as puppies. Unfortunately, fear biting has essentially the same effect on the person bitten, so even fear biters are culled in research projects.

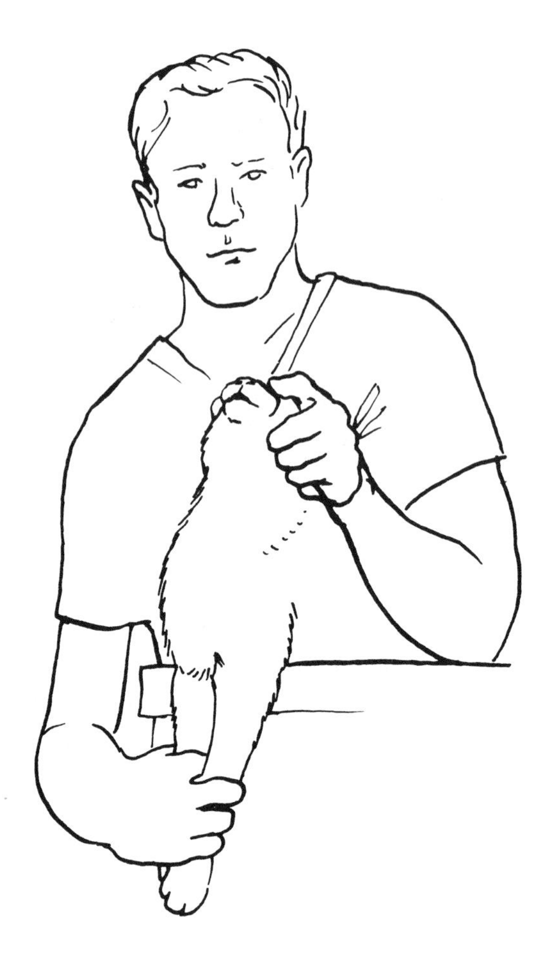

FIGURE 15. The cat positioned for jugular venipuncture. The handler is grasping the bones on the sides of the cat's face while stroking the nose with his index finger.

Restraint in research dogs, then, is a way to circumvent overexuberance or avoidance behavior. Dogs are by nature aerobic athletes, despite the impression of determined inactivity projected by many housepets. As such, they tend to be underexercised in the laboratory, leading to excitement when given attention or handled. Dogs can be trained not to struggle during mildly painful procedures such as venipuncture by forcing them physically to stay still and then rewarding with praise and petting after the procedure is completed. A romp around the room is also a form of reward for a caged or kenneled dog. Dogs which are properly trained are a joy to work with and enjoy their contact with people. A few extra moments of thoughtfully applied praise or petting will make dogs cooperative rather than obstructive of one's effort; both the dogs and the people benefit.

Dogs are such familiar companion animals that description of catching and carrying them seem unnecessary. However, there are things to remember. Dogs should never be allowed to jump out of cages or tables high above the floor. The surfaces necessary in animal facilities for hygiene are slippery, and a fall may result in a fracture or a broken tooth. Medium-sized dogs are usually used to periodic lifting and submit to being gathered fore and aft in the arms. Most dogs have redundant skin in the neck region and may be held, or even lifted (with the addition of a hand under the abdomen), by the scruff. The scruff hold is an important training device. It is a natural sign of dominance in dogs. When puppies misbehave, their elders grab and dangle

FIGURE 16. The gauze muzzle on a dog. The muzzle is tied in a bow at the back of the head for quick and easy removal.

them by the scruff. When a dog refuses to learn by being held in place, it may be dangled by the scruff, or even shaken to subdue it. The scruff hold is usually more effective than hitting a dog when it misbehaves, even in the case of a dog resorting to biting. Biting behavior, if it is nipped in the bud (so to speak) can be eliminated easily. If a dog is allowed to get away with biting it is being trained to that behavior. One should always retaliate immediately and forcefully if a dog tries to bite.

In some cases, when a dog is suspected of having a more aggressive personality, it is better to muzzle it when doing something uncomfortable than to take the chance on someone being bitten (Figure 16). There are commercial, cone-shaped muzzles that are very effective. However, if none is available, an effective muzzle may be fashioned from 2- to 4-in.-wide roll gauze. A half-hitch is tied in the middle of a 2- to 3-ft length of gauze and pulled tight around the dog's muzzle, the knot on the top. The ends are then brought together under the muzzle, and a square knot, or two half-hitches, are tied there. The muzzle is completed by tying the ends behind the head, under the ears, so that the muzzle will not drop off; the ends are usually tied in a bow for quick release. For short-nosed dogs or cats, the procedure is the same except that one of the ends in the last tie is brought forward over the forehead, looped around the nose noose, and tied back behind the head.

Most dogs can be easily orally dosed with a capsule or pill. The dog is made to sit and a hand brought over its back and top of head to press on either side of the upper lips to get it to open its mouth. Once the mouth is open, the other hand, holding the item in thumb and forefinger, presses the substance as deep into the mouth as possible, pushing it down the throat with the forefinger. The mouth should be held shut immediately afterward, and stroking the nose or neck sometimes encourages the dog to swallow. Some dogs are adept at bringing a capsule back from the depths and must be watched for several seconds to assure that they have swallowed. Although it may seem daunting to place the major portion of one's hand deep into a dog's mouth, the response of almost all dogs to this treatment is to open the mouth wider rather than to clamp down, and to try to move backwards to get away from it; the dog, however, cannot move backwards readily if it is sitting. Those dogs, rare as they are, which may bite down nearly always indicate their propensity by refusing to open their mouths to start with. Dogs may not be given drugs like ketamine to immobilize them (being stimulated to convulsions by this drug by itself), so in order to intubate them for oral administration they must be anesthetized. If they already have an endotracheal tube in place, a gavage tube will have no place to go except into the esophagus or stomach.

FIGURE 17. Cephalic venipuncture position. It is usually preferable for the dog to be lying down on its sternum, but tractable animals may be more comfortable sitting up. Note that the handler is pulling the dog's head to her shoulder and chest to steady it.

Subcutaneous administration in the dog is done in the dorsal neck area or high on the middle back where the skin is loosest. Preferable IM sites are are in the heavy muscle masses, especially the low posterior thigh. IV administration may be made in the saphenous, cephalic, or jugular veins (listed in increasing size and ease of administration). For saphenous venipuncture, the dog is held on its side on a table. The handler stands at the dog's back, and presses on its neck while restraining the two forelegs. The top hind leg is straightened by placing the palm of the hand against the anterior aspect of the stifle joint while pressing down on the dog's flank with the elbow and using the fingers to encircle the limb and hold off the vessel. Another person then makes the puncture of the vessel and signals the handler to release pressure on the vein when about to inject.

For cephalic venipuncture, the holder must restrain the dog's body, present the forelimb, and occlude the vessel (Figure 17). The animal is placed on a table near one end, facing the edge. The holder stands beside the table, facing the same direction as the dog, and nestles the animal on the table under the arm; the handler's forearm, upper arm, and elbow will exert pressure to pull the dog close and down to a sternal reclining position on the table. The hand of the same arm cradles the animal's elbow with the fingers while the thumb clamps down on the cephalic vein. In order to be sure that the vein is under the thumb and on the top (anterior portion) of the forearm, the holder grasps the limb just below the elbow with the thumb placed as far medially as possible. Then, with pressure applied, the skin is rotated laterally. The fingers should rest on the table, and the dog's elbow should be pushed slightly forward to stabilize the leg. The elbow should be at

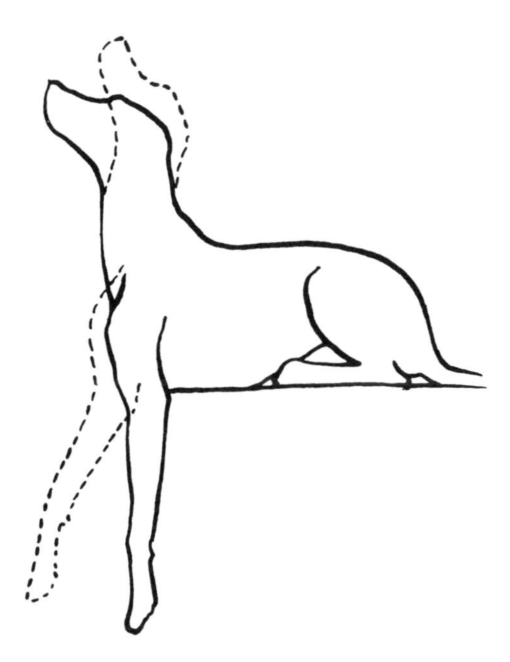

FIGURE 18. Position of the animal's body for jugular venipuncture; note the straight line from the angle of the jaw through the legs. The broken lines indicate improper position for the head and legs.

or near the edge of the table to allow good access to the vein. The free hand is used to restrain the animal's head, and for most dogs, this is best accomplished by pressing the head to one's chest by reaching under the neck and placing the hand behind the jaw. The same hand can also grasp the dog around the muzzle, if necessary. As in saphenous venipuncture, the holder must release the vessel before administration.

The cephalic vein in most dogs can be used for bleeding as well as administration, but the jugular is superior. In short-haired dogs clipping the site for venipuncture may not be necessary to visualize the engorged vessel, but if the vessel is not visible it is preferable to clip the hair. Jugular venipuncture (Figures 18 to 20) is almost the same as in the cat, although the head is held with the arm over the back and the muzzle nestled in the hand and the head pressed to the holder's chest. It is important not to close off the trachea or the nasal passages as the animal is held, which will cause understandable struggling in the dog. The vessel away from the handler is the one available for venipuncture.

Urine and feces may be collected by use of a mesh-bottomed cage. Clean urine samples may be collected by expressing the bladder. The dog is essentially a larger version of the rabbit in this regard, except for the position of the genitalia of male dogs which tend to get in the way. Cystocentesis can be performed on an anesthetized dog, with all the care for sterility as explained under the rabbit section. In some cases it can be performed on tranquilized or alert dogs, but if urine should leak around the needle it will cause immediate pain to the dog when it contacts the peritoneum, leading to movement and possible tearing of the bladder. Obviously, this should never be attempted unless the dog is securely restrained on its side.

Fecal samples are usually available within a few centimeters of the anus in dogs. A lubricated, gloved finger may be used to secure a digital sample. The dog, of course, must be securely held for such a procedure, as it is uncomfortable. A holder can usually hold a dog for anal exploration with arms under neck and abdomen as the dog stands on a table.

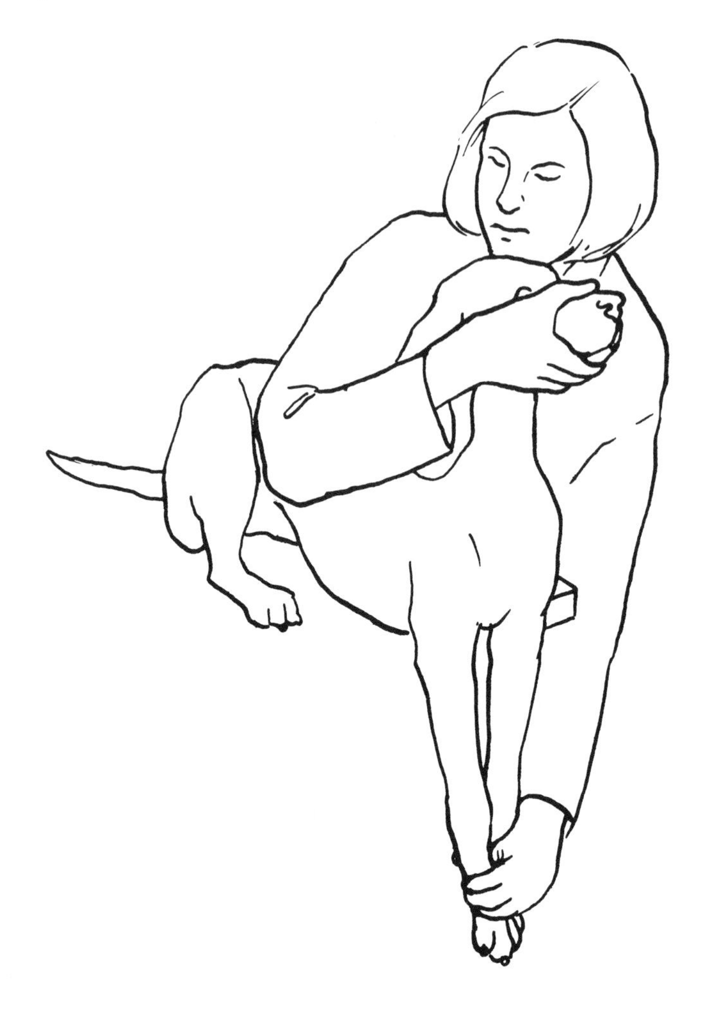

FIGURE 19. A dog positioned for jugular venipuncture.

X. CONCLUSION

The most appropriate restraint and handling technique is always the least invasive technique possible in a given situation. When we perform techniques such as taking blood, the animal, unlike a human, is unlikely to understand the need for the technique or that it will only last a short time. Reassurance, calm and gentle handling, and consistent training go far to improve the research environment for the animals, and, by extension, for investigators.

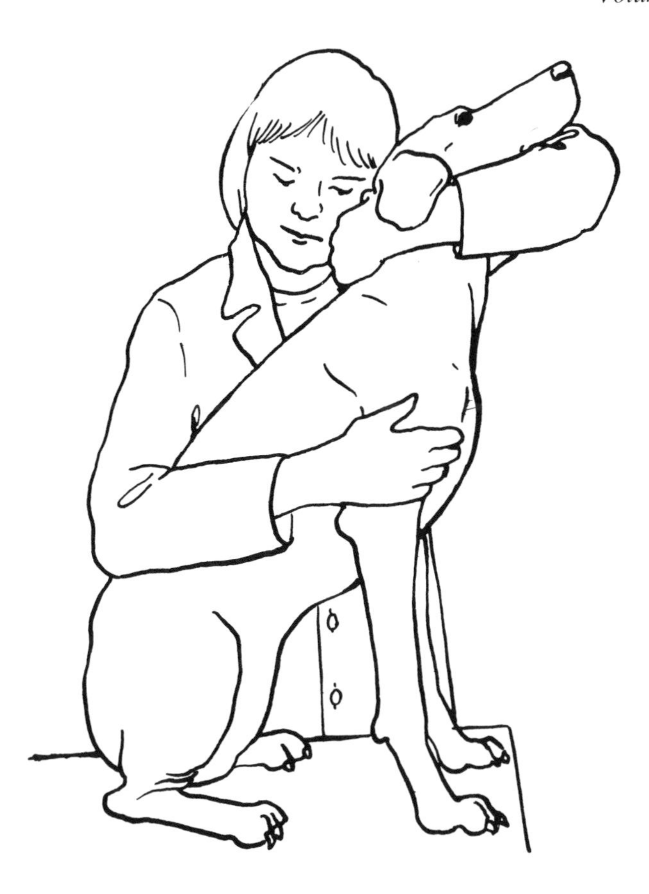

FIGURE 20. An alternate positioning of a large dog for jugular venipuncture. Large dogs can also be held on the floor for venipuncture.

Chapter 20

EXPERIMENTAL SURGERY

Anthony Schwartz

EDITOR'S PROEM

Probably no other type of experimental manipulation has the same potential for regularly occasioning pain and suffering as does experimental surgery. Historically, many researchers have not been trained in surgery and are consequently deficient in many surgical techniques and principles. This in turn may lead to problems such as pain, infection, improper healing, etc. which result in suffering and in wasted research. Recent legislation and regulations specifically target surgical procedures as an area demanding greater scrutiny and attention. In this chapter, Dr. Schwartz outlines some major concepts and considerations relevant to surgery undertaken in the course of experimentation.

TABLE OF CONTENTS

I. INTRODUCTION

Surgical research using animals began in the 19th century, when, for example, John Hunter (sometimes called the father of experimental surgery) successfully transplanted tissues in chickens.[1] It did not begin in earnest, however, until the late 19th and early 20th centuries, during which period Harvey Cushing helped found the Hunterian Surgical Laboratory at Johns Hopkins Hospital, in Baltimore, where the modern practice of experimental surgery (on dogs) was begun.[2] Experimental surgery continues to be fundamental to establishing principles and developing new techniques and instruments for clinical human and animal surgery.[3]

This discussion serves only as an introduction to the subject of experimental surgery. For further information, the reader may consult texts on human and veterinary surgery, laboratory animal medicine (the surgical sections), several of which are quoted below, and experimental surgery.[4-6]

II. PHILOSOPHICAL CONSIDERATIONS

The animal rights movement has led efforts to rescind pound seizure laws. This has had a negative impact on the ease of procuring dogs and cats, two of the most important species of animals used in surgical research. It also has resulted in an increase in the cost of surgical research employing these species. Although rescision efforts have, as their rationale, the elimination of (first) the use of former pets in research (and then, the cessation of animal research altogether), an unfortunate actual result has been that this group of dogs and cats still are killed by the millions annually, and now additional dogs and cats must be purpose-bred for research. This increased wastage of life is itself morally questionable. Some investigators now are using animals other than dogs and cats, e.g., swine,[7] in experimental surgery, which apparently is easier for the animal rights movement to accept. This raises yet other obvious philosophical questions.

In large part, however, from moral, ethical, and improved animal care standpoints, the animal rights movement clearly has had a positive effect. For example, the population's sensitivity to the message of the animal welfare movement has been raised, which has fostered the development of stronger federal, state, and local animal welfare laws and regulations. The sensitivity of investigators and administrators to our obligations to our experimental subjects also has been increased. Researchers now exercise greater care when considering whether the quality and likely outcome of the experiment warrants the animal use, pain, morbidity, and mortality involved.

The number of animals used in research should be reduced whenever possible by employing scientifically valid adjunctive methods or alternatives to animal experimentation. For example, the nonenzymatic hydrolysis of absorbable suture materials could be tested initially in an *in vitro* warm water bath rather than *in vivo*. Such approaches may or may not eliminate the eventual need for *in vivo* tests of materials, procedures, etc., prior to clinical studies, but the number of animals used certainly may be decreased.

Surgical models that involve substantial postoperative pain should be allowed to proceed only if no less painful model exists and if the likely importance of the outcome of the study is felt to justify the expected pain. In all protocols, evidence for postoperative pain must be searched for actively, and with a high index of suspicion of its presence (taking into consideration that certain relatively stoic strains, breeds, or individuals may exhibit minimal or no overt signs of pain), if it is to be recognized and treated properly. It is best to assess the likelihood of pain and treat it prophylactically (e.g., by an intercostal anesthetic nerve block given during chest wall closure after thoracic surgery). It is the obligation of the principal investigator, the laboratory animal medicine program, and the Institutional Animal Care and Use Committee (IACUC) to counsel all those associated with a project on the necessity to control postoperative pain. This is particularly important when novices and/or part-time investigators, such as residents, fellows, or students are involved.

III. LABORATORY ANIMAL MEDICINE AND EXPERIMENTAL SURGERY

Laboratory animal veterinarians can help assure the proper selection of surgical animal models because of their knowledge of the variations of anatomy and physiology across species, and their ready access to relevant literature and animal sources. They also know the capabilities and limitations of their facilities to support research projects. Both adequate technical support and centralized facilities for survival surgery should be under the control of the laboratory animal medicine program to assure humane care and compliance with laws, regulations, and guidelines pertaining to experimental survival surgery. In well-supported programs, personnel are available to provide optimal peri- and intraoperative care and sophisticated anesthesia. The staff can counsel investigators on the advisability, nature, and dosage of perioperative antibiotics and/or analgesic drugs for various species and procedures, and they can supply necessary veterinary drugs.

IV. WHO SHOULD PERFORM EXPERIMENTAL SURGERY?

Currently, there are no training programs specifically in experimental surgery in the U.S., so this discipline usually is learned "on the job". Because the training and background of those who perform experimental surgery vary widely, so do their abilities.

M.D. and dental surgical specialists and their residents and fellows have knowledge of the major clinical human disease entities warranting an experimental surgical approach and a general understanding of biology, and they are trained in surgical technique. Training in the surgical laboratory is their means of gaining proficiency in *animal* surgery. Especially at an early stage of their research careers, members of this group may lack an in-depth knowledge of anatomical and physiological differences among non-human species, so they may be handicapped when it comes to animal model selection in comparison with their veterinary surgical counterparts. Also, they often lack knowledge of the fine points of optimal anesthetic techniques in various species, which may lead to reliance on simple but inflexible anesthetic regimens, such as IV pentobarbital sodium, unless they have excellent laboratory support.

Ph.D.s and, non-surgeon M.D.s and veterinarians may be strapped by their lack of general surgical skills before becoming trained in the laboratory. This is less a problem for veterinarians because, unlike the others, they routinely receive hands on laboratory and clinical surgical training during their professional education. Further, some Ph.D.s may lack fundamental knowledge of anatomy and/or organ physiology.

Veterinary surgical specialists, unlike all of the above groups, have had special training and experience in surgery, anesthesia, analgesia, perioperative and intraoperative monitoring, postoperative normal and intensive care, restraint, infectious diseases, wound healing, anatomy, physiology and pathophysiology, clinical pathology, diagnostic imaging, etc. of a variety of animal species. Formal clinical veterinary surgical residency programs have existed for over 20 years, virtually all with an experimental component. A specialty board (The American College of Veterinary Surgeons [ACVS]) examines the trainees rigorously and certifies them. Many diplomates of the ACVS perform biomedical research at medical and veterinary medical schools and in industry. They have proven themselves as outstanding collaborators and principal investigators. Some also direct centralized biomedical surgical research facilities. Because of their interests and training, veterinary surgeons are comparative surgical scientists. Many have spent a component of their training in clinical human surgical programs, and all are educated in species (including human) comparisons. They have been brought up to be sensitive to and aggressive in the control of pain and suffering in clinical, research, and teaching animals. Thus, some would argue that veterinarians, especially ACVS diplomates, are the ideal individuals to direct experimental surgical research facilities and support programs and to collaborate in or direct surgical research programs.

Surgical technicians frequently become well trained and actively involved in surgical research. Some technicians, especially those with animal nursing education, develop to the point that they oversee the facilities, provide anesthetic and perioperative and intraoperative treatment and monitoring, and even produce home-made surgical implants, refine experimental procedures, etc. In short, they become essential to the program.

The Academy of Surgical Research was established in 1982 to help bring together all experimental surgeons, largely in response to the animal rights movement, which they saw as jeopardizing the ability to use animals in research.[8] The academy fosters the proper selection of animal models and surgical techniques, the treatment of laboratory animals with respect and humaneness, and recommends that all personnel performing surgery on laboratory animals be properly trained in surgery, anesthesia, and ethics. It also encourages individuals and institutions to follow laws and regulations pertaining to experimental animals and institutions to become accredited by The American Association for the Accreditation of Laboratory Animal Care (AAALAC), so the public may be assured of a commitment to the highest standards of research animal care.[8]

V. LAWS, REGULATIONS, AND GUIDELINES PERTAINING TO EXPERIMENTAL SURGERY

The Federal Animal Welfare Act of 1966, as amended in 1970, 1976, and 1985, and many state and local laws, regulations, and policies affect the performance of experimental surgery.

The *Guide for the Care and Use of Laboratory Animals*[9] (*Guide*) must be followed by all experimental programs funded by the U.S. Public Health Service (PHS) (e.g., the National Institutes of Health [NIH]) and institutions that are accredited by AAALAC. It also serves as a basis for new regulations being developed by the U.S. Department of Agriculture (USDA). Therefore, the portions of the *Guide* that pertain to experimental surgery are summarized in this section, with the exception of facilities requirements, which are presented below.

"Adequate veterinary care" includes monitoring the entire surgical program. There must be proper use of anesthetics, analgesics, and tranquilizers, with the choice and guidelines for use of the most appropriate drugs provided by the attending veterinarian. Deviation from the accepted drugs (because of purported adverse effects on the research) must be approved by the IACUC and be directly supervised by the principal investigator.

Aseptic surgery is to be accomplished only in facilities intended for that purpose, which are maintained, operated, and staffed by trained, experienced personnel. Surgery must be performed or directly supervised by trained, experienced personnel, and training should be provided as necessary. Aseptic technique is necessary for most animals undergoing major survival surgery, including lagomorphs, and rodents, although a restricted-use operating room is not required for survival surgery in rodents.

The animal should be observed postoperatively to ensure an uneventful recovery from anesthesia and surgery. Supportive fluids, analgesics, and other drugs should be given as required, the surgical wound should receive care, and appropriate medical records should be kept. For intensive care, heating pads, vaporizers, vacuum equipment, a respirator, a cardiac monitor, and oxygen are suggested, with proper monitoring during recovery provided by trained personnel. Minor surgical procedures, such as suturing a wound, cannulating a peripheral vessel, etc., may be performed under less stringent conditions.

VI. SURGICAL AND SUPPORT FACILITIES AND EQUIPMENT

The *Guide*[9] enunciates a set of minimum guidelines for aseptic surgical facilities and equipment.

The aseptic surgical suite should include a separate surgical support area, designed for storage of instruments and supplies and for washing and sterilizing instruments. Regularly used equipment and supplies, such as anesthetic machines and suture materials may be stored in the operating room. A preparation area should be present in which induction of anesthesia, removal of hair, and surgical preparation of the animal occur. There should be an area equipped with surgical scrub sinks (for the surgeon) which should be close to but not in the operating room. Also, a dressing area should be provided for personnel to change into surgical attire. At least one operating room must be present that is used exclusively for aseptic surgery. There must be an area in which to perform intensive and supportive care of animals. To minimize contamination, the operating room should be under positive pressure compared with surrounding spaces. The interior surfaces of the facility should be constructed of moisture-impervious and easily cleanable materials. If explosive anesthetic agents are used, electric outlets should be explosion proof and located at least 1.52 m (5 ft) off the floor. Anesthetic gas scavenging or waste gas exhaust should be provided in all rooms in which animals anesthetized with gas are located (including operating rooms, preparation, recovery, radiology, etc.) to decrease exposure of personnel to the gases. Survival rodent surgery does not require a separate surgery suite. Rather, the surgical area may be a room or portion of a room that is sanitized easily and is used for no other purpose during the time of surgery. Nonsurvival surgery requires neither special facilities nor aseptic technique. (From the standpoint of the sensitivity of other personnel, however, all surgical procedures should be performed with appropriate decorum and away from the view of personnel who have no need to see them.)

Beyond the above guidelines, some comments about aseptic surgical suite furnishings are in order.[10,11] It is helpful to have a "sink-table" in the operating room for pre-surgical scrubbing of the skin of animals up to about 100 kg and a sink for initial washing of surgical instruments and anesthetic tubing and bags. A laryngoscope is necessary for endotracheal intubation. Surgical vacuum should be available in the preparation, operating, and recovery rooms, not only for surgical suction, but also to enable aspiration of excessive secretions from the respiratory tract and to remove blood or air from the thoracic cavity after surgery. Suction may be built in or be generated by a portable pump. Oxygen, piped in or in tanks, and circulating warm water blankets and warm water bottles should be available in all three of the above rooms, as well. The preparation room also should have electric clippers to clean the surgical site of hair. A commercial-strength vacuum cleaner is necessary for picking up hair removed during preparation. There should be transportation tables on large casters for animals of approximately 100 kg or smaller and heavy duty carts for larger animals.

The surgeons' scrub sinks should be foot operated if possible, although faucets with large blades, operated with the elbow, may be adequate. The scrub sinks should be located in a separate room, but if space is short, they may be located in the preparation room. The door from the scrub room to the operating room should open into the operating room and not latch, so the surgeon can back in, using no hands. A microfilm dust pad applied to the floor outside the operating room helps remove dust from cart wheels, etc.

In addition to 100 to 150 fc of background room lighting, portable or small ceiling mounted surgical lamps are helpful in both the preparation and recovery rooms. High-quality surgical lamps are needed in the operating rooms. These should be capable of the following illumination ranges at 1 m above the table: flood, 1600 to 3800 fc; medium setting, 2000 to 5000 fc; and spot, 2500 to 6000 fc. Two lamps, that can be directed from different angles, are preferred to one.

Most of the usual equipment found in a normal human or clinical veterinary operating room is suitable for use with most of the larger laboratory animal species. The minimum size of the operating room for animals up to the size of sheep should be 180 to 220 ft^2. If large equipment, such as for cardiopulmonary bypass, is likely to be placed in the room, the room should be twice that size. For work on horses or other large animals, 400 ft^2 or more may be required.[12] For animals from the size of rabbits to sheep, a 6-ft surgery table of stainless steel, with hydraulic raise-lower and tilt capabilities is ideal. The table should either have a built-in V trough or a V trough-shaped thoracic positioner should be available, because of the long dorso-ventral diameter of the thorax of dogs and some other species. Larger animals, such as horses or cattle, require heavy duty tables that often have a hydraulic lift.[12] Instrument tables that can be raised and lowered and straddle the operating table are helpful for all but the most minor procedures, for which Mayo stands may be adequate. The furnishings of the operating room should be spartan, with most items on the floor removable to allow cleaning and disinfection. Shelves should be enclosed in cabinets with doors. If nitrogen-driven equipment is to be used, the gas may be piped in or a tank may be mounted on a portable cart which is brought into the operating room as necessary. A kick bucket, a clock (preferably a time-elapse type), and a radiographic viewer should be present. It is desirable to have ready access to postoperative radiography, and in an orthopedic laboratory, intraoperative radiographic capability is important. EKG equipment is necessary, and esophageal stethoscopes, central venous and arterial pressure, and continuous temperature monitors are highly desireable. Ready access to a direct current defibrillator with a discharge capability of 350 Ws and both external and internal paddles is important.

The surgical recovery area should have an intensive care capability. Cages should be readily cleaned and disinfected. There should be facilities for total environmental control of postoperative animals. This may range from a pediatric incubator, into which oxygen may be piped for small animals, to a glass door-fronted, sealed cage. In either enclosure, it should be possible to measure and control ambient oxygen and carbon dioxide concentrations, administer fluids, control humidity with the possibility of nebulization, and heat or cool the atmosphere. Equipment for monitoring the EKG, blood pressure, and body temperature should be available. A fluid infusion pump with warming capability should be present, as well. Access to a clinical pathology laboratory is necessary, and the recovery area itself probably should have a microhematocrit centrifuge, a microscope, a refractometer, urine and blood chemistry sticks, and, depending on the amount and type of surgery being done, a blood gas machine.

Other equipment useful in the surgical suite include an emergency drug cart, which also has the equipment needed to open the chest if emergency CPR in needed. A positive pressure ventilator is helpful for open-chest procedures. An electrosurgical unit is important for surgery and/or hemostasis.

Routine surgical instruments (and their cleaning and care) required for most procedures are the same as those used in clinical human and animal operating rooms.[14] The need for special instruments depends on the species and the procedure. Sterilization methods are presented below.

VII. SURGICAL PRINCIPLES[11,14]

Humane, successful survival surgery requires adherence to a set of basic principles. The list is the same for nonsurvival surgery, except for the requirement for strict asepsis and postoperative management:

1. *Strict asepsis.* Aseptic technique should never be broken. This is required to prevent a delay in wound healing and undue stress to the subject. Sterile technique involves attention to decontamination of the operating room, staff, equipment and the animal, plus some special surgical methods (see below). If a break in technique does occur, this should be remedied immediately. This may include changing gloves, replacing soiled or contaminated drapes, or discarding contaminated instruments.
2. *Hemostasis.* Hemorrhage must be controlled, to prevent shock, maintain visibility in the surgical field, and minimize postoperative adhesions caused by fibrin deposits.
3. *Exposure of the surgical area.* Proper positioning of the experimental subject and proper size and location of the incision facilitate presentation of the organ to be approached. The incision should not be made so large that the operation is prolonged unnecessarily.
4. *Knowledge of anatomy and physiology.* The surgeon must be aware of of normal tissue and organ function, structure, relationships, innervation, and blood supply. This allows quick and efficient surgery, and appropriate pre-, intra- , and postoperative supportive care.
5. *Gentleness of tissue handling.* The tissues should be handled gently and not be clamped unnecessarily. Tissue damage caused by pulling and tearing delays wound healing and may cause adhesions.
6. *Satisfactory anesthesia.* This is required for humane reasons and to provide immobilization and muscle relaxation for delicate procedures.
7. *Knowledge and understanding of surgical methodology.* It is necessary to know aseptic technique, suturing procedures, instrument use, etc.
8. *Incisions should be made with sharp blades, not with scissors* (this rule frequently is ignored in rodents), as crushed skin and delayed healing results.
9. *Incisions should preserve nerves and avoid unnecessary trauma to the muscles and blood supply.*
10. *Incisions should be decisive to allow rapid access to the desired area without damage to surrounding structures.*
11. *Tissue, including the skin, should not be undermined unnecessarily.*
12. *Fascial cleavage planes should be used (separated), whenever possible, to gain access to deeper structures.*
13. *Incisions should be closed one layer at a time,* although occasionally it is acceptable to suture multiple layers together.
14. *Dead space should be eliminated while closing the incision.*

VIII. PREOPERATIVE MANAGEMENT OF THE EXPERIMENTAL SUBJECT[15]

Experimental animals should be allowed to acclimate to their new surroundings for at least 1 week after arrival in the facilities prior to surgery. In some cases, several weeks or months may be required for preoperative evaluation, treatment of parasites, etc., and training of the subjects, and depending on the degree of their conditioning prior to arrival. As asepsis begins with cleanliness, whenever possible, larger animals should be bathed before surgery. This is easier with some species than others.

Presurgical evaluation should include observation for a good appetite and alertness and a complete physical examination. The experimental subject should be free from external and internal parasites, as determined by examining a fecal flotation and/or sediment. This could be of extreme importance, for example, in the case of heartworms in dogs to undergo cardiac or pulmonary surgery or intestinal parasites in animals to be subjected to an enteric protocol. Baseline hematology, a clinical chemistry profile (or selected chemistries), a urinalysis, and thoracic, abodominal, etc. radiographs also are important in many instances, depending on whether surgery will be survival or nonsurvival, the nature of the protocol, the species of animal, and the budget available.

If postoperative restraint devices are needed, animals should be trained to accept such devices prior to surgery. This makes restraint easier and places the subjects under less postoperative stress.

Food generally should be withheld from rabbits, dogs, and cats for 6 to 12 h prior to surgery to empty the stomach, minimizing the likelihood of vomition during anesthetic induction or during or after anesthesia. Otherwise, excessive stress to an abdominal incision could occur or aspiration of stomach contents may result in inhalation pneumonia. Water generally is withheld for 2 h prior to anesthetic induction.

In ruminants,[16-18] especially cattle, feed is withheld for 12 to 24 h to prevent regurgitation of rumen ingesta while under anesthesia, which may occur even before endotracheal intubation is possible. In cattle, pelleted rations are best replaced with hay for several days prior to surgery to further help prevent regurgitation. Withholding feed also decreases gaseous distension of the ruminoreticulum (bloat). Such distension may interfere with the ability to manipulate these organs during surgery and may cause respiratory compromise and a decrease in venous return to the heart due to pressure on the diaphragm and caudal vena cava, respectively. In sheep, if a 24 h fast is allowed before surgery, rumen gas production is negligible and volume of the rumen contents is smaller. Some surgeons prefer to withhold food for 48 h to allow greater emptying of the rumen.[17] Administration of surface active agents such as a silicone, prior to surgery, may further decrease the tendency toward gastric bloating. Care must be taken when starving ewes that are more than 100 d pregnant, as they may develop pregnancy toxemia, leading to brain damage in the ewe. Toxemia is unlikely to develop if starvation is limited to 24 h. Propionic acid or glucose given by stomach tube before surgery or glucose given IV during surgery helps prevent toxemia. Pregnancy toxemia in long-term fasted goats has been prevented by administration of sodium acetate, hydrolyzed casein, treacle, and milk by stomach tube on the evening and morning before and at the end of surgery.[17] Antibiotics have been advocated to decrease rumen bacterial fermentation and gas formation. This is to be discouraged, however, because normal flora and fauna are needed for postoperative nutrition.[16] Some surgeons cannulate the rumen of sheep in a prior surgery, to facilitate emptying the rumen on the day of surgery.[17] This necessitates either collecting and storing the rumen contents in an incubator at 37°C, for return to the rumen after surgery, or replacing the rumen contents with some from another cannulated sheep.

Prior to large intestinal surgery in dogs, cats, and swine, serial cleansing warm soapy water enemas may be given until up to 24 h before surgery. Later enemas result in liquid feces, which is undesirable during surgery. Food may be withheld for up to 48 h in healthy animals to decrease intestinal contents. Some surgeons prepare these animals with oral antibiotics, such as neomycin (20 mg/kg orally, every 6 h) and kanamycin (10 mg/kg orally, every 4 h). Antibiotics that affect gastrointestinal flora may be disastrous for some laboratory animal species.

If possible, the subject should have defecated and urinated just prior to anesthesia. If urination has not occurred, the bladder of small animals may be expressed manually while they are under anesthesia (if this is not possible and abdominal surgery is being performed, the bladder may be expressed or be aspirated via a needle and suction apparatus after celiotomy).

After anesthesia has been induced, the hair or wool must be removed from around the surgical site for a distance that allows an adequate margin to prevent contamination (approximately 20 cm in a 15-kg dog). (In some of the more tractable subjects, at least a preliminary removal of hair may be done the day prior to surgery, to decrease anesthesia time.) This *must* be done, along with aseptic preparation of the surgical site, *prior* to injection of local anesthetics, such as in diffuse field or nerve blocks. Sheep clippers are needed for members of this species. For other species, a number 15 Oster® clipper blade (Oster Mfg., Milwaukee, WI) normally is used for a first clip and a number 40 clipper blade for the finished clip.[11] Occasionally, finer clipper blades are needed to remove the thin hair of cats, rabbits, and some other species. Soap and a straight razor also may be used to effect a close shave after the initial clip. A depilatory cream, such as calcium thioglycolate, is needed for preparing locations that are difficult to clip or shave because of their topography (e.g., in interphalangeal regions). If an electric clipper is used, care should be taken not to allow the blade to get too hot, or a burn may result. Depilatories are the least traumatic to the skin, but may result in a mild inflammatory response. Birds' feathers should be plucked, not cut, as cutting may result in severe and sometimes fatal hemorrhage from the vascular shafts.

Prior to transporting anesthetized hooved animals into the operating room, it is necessary to clean their feet thoroughly and to cover them with a plastic bag to prevent spread of the contamination.[19]

If survival surgery is to be performed, the skin must be prepared with a germicidal scrub. An initial scrub is performed in the preparation room. Using germicide-soaked sponges, the surgical region is lathered well to remove all dirt and oils. In the process, the surrounding hair becomes wet, which decreases the likelihood that it will contaminate the field. Especially in smaller animals, the core body temperature starts to fall while it is being scrubbed, so warm water should be used and the animal should be surrounded with warm water bottles.

The animal is transported into the operating room and positioned with the aid of restraining ropes. The final "sterile" scrub employs sterilized gauze sponges and preparation bowls. Germicidal solutions are poured into the bowls and gloved hands or sponge forceps are used to handle the sponges. To prevent contaminating the center of the surgical area, the scrub occurs in a spiral pattern, beginning in the center of the incision site and moving outward, circumferentially. Upon reaching the periphery of the scrubbed area the sponge is discarded and a new one is used, again starting in the center. A minimum of a 5 min scrub is desirable.

A variety of scrub solutions have been used. These include hexachlorophene (not commonly used now due to its potential neurotoxicity), iodophors, chlorhexidine, alcohols, and quaternary ammonium salts (rarely used now as they are not as effective as iodophors or chlorhexidine). Alcohol usually is not used alone, although it kills bacteria fast and helps defat the skin. Iodophor scrubs (e.g., povidone iodine) are commonly used alone or in combination with ethyl or isopropyl alcohol (alternating alcohol-soaked sponges with iodophor-soaked sponges). After the final alcohol application, 10% povidone iodine solution is sprayed or painted onto the area, the excess is blotted off and drying is allowed. Chlorhexidine scrub (0.5% in 70% alcohol) effectively decreases bacterial counts after two 30-s applications, and it has a greater residual activity than iodophors.

After the surgeon is gowned and gloved, the site of the proposed incision is isolated by first surrounding it with towels and then a surgical drape. Four cloth towels usually are used, held in place by Backhaus towel clamps at the corners. This is then covered with a cloth surgical drape that has a hole pre-cut a few inches longer than the incision. Alternatively, paper drapes and towels are available. These tend to be less absorbent, which has the advantage of diminishing "strike through", i.e., leeching of contamination to the surface via soaked towels and drapes. The expense of using disposable paper drapes must be balanced against the labor involved in preparing and repairing cloth drapes. Adherent plastic drapes, some of which have antimicrobial activity built in (Ioban® 2 Antimicrobial Film, 3M Co., St. Paul, MN), may be used for certain procedures. The incision is made through the plastic, which isolates the skin incision from the surrounding skin better than any other technique, *if* the drape adheres well. Thorough drying of the skin is required before the drape is placed. A separate adhesive spray may be used to increase the degree of adhesion. A weeping wound, including lacerations due to clipping or shaving too close, or the continued seepage of bodily fluids, such as from the anus, diminish the degree of adhesion and, therefore, the usefulness of this method. Hair surrounding the clipped area should be covered with paper towels to avoid having the adhesive drapes stick to the hair.

IX. ASEPSIS AND WOUND INFECTION IN SURVIVAL SURGERY[20]

There is no such thing as sterile surgery, except when the surgeon is presented with a germ-free animal in a sterile environment, i.e., in an isolator. In the normal survival surgery situation, surgeons employ *aseptic* technique. Asepsis is the absence of pathogenic microbes in living tissue. This is effected by a series of actions which involve both antiseptic and aseptic components including: (1) *sterilization of anything to come in contact with the surgical wound or the surgeon's or assistant's hands* (instruments, surgical drapes and towels, the surgeon's gloves, etc.).[11] There must be access to a steam autoclave for sterilization of intruments, equipment, cloth towels and drapes, etc. As there is a limit to shelf time of sterile instruments, which varies with the method of their wrapping,[21] a schedule of routine resterilization is necessary. Ethylene oxide is another method of sterilization which is especially useful for plastics and sharp instruments. This gas is toxic, so there must be an excellent ventilation system in rooms in which ethylene oxide sterilizers and aerators are located. There should be at least ten air changes per hour, with air exhausted to the outside. Aeration of sterilized objects is necessary for varying periods, depending on the composition of the object,[21] but normally at least 2 d are required. "Cold sterilization" by immersion of instruments in one of a variety of disinfectant baths should be used routinely probably only for dental instruments, as the above methods are superior.[21] An ultrasonic instrument cleaner improves the cleaning of instruments and facilitates sterilization by removing organic material from their surfaces. Monitoring sterility is best done with biological indicators; minimal monitoring includes time and temperature recording; (2) *selection of the proper experimental subject* (see below) *and antiseptic preparation of its skin* (see above). The latter minimizes contamination with the subject's own skin flora (the most common source of bacterial contamination of surgical wounds); (3) *preparation of the surgeon and the surgical team*[10,22] by scrubbing their hands and arms with a disinfectant scrub solution such as an iodophore or chlorhexidine and properly donning clean (scrub suit, cap, mask, shoes, and shoe covers) and sterile (gown and gloves) clothing to shield the patient from the transmission of potential pathogens from the surgeon. Gowning and gloving techniques must be performed so as to assure asepsis (this can be a major source of contamination if a barrier fails, e.g., if a glove tears). Air in contact with the wound, carrying bacteria primarily from the animal and nonscrubbed personnel in the operating room, also is a common vehicle for bacteria contaminating the wound. All personnel should wear dry gowns and/or scrub suits; face masks are crucial as well. Decreasing traffic in the operating room helps lower the concentration of air-borne bacteria; (4) *cleanliness of the operating room equipment and facilities* (cleaning the operating room superficially between surgeries, thoroughly cleaning exposed surfaces, such as the surgical lamps, the walls, floors, tables, cabinets, etc., at least once daily and disinfecting the entire room at least once weekly decreases the hazard of environmental contamination). Isolating the operating room from all other traffic is important. If the procedure to be performed is particularly sensitive to environmental contamination, e.g., total hip replacement, a HEPA-filtered laminar flow system of air handling helps decrease the infection rate; (5) *careful attention to operative technique* (this includes preventing tissues from drying, avoiding external contamination, using fine instruments with sharp [less traumatic] dissection technique, minimal and skillful use of electrocoagulation, sutures, and ligatures, operating so as to maintain maximal vascularity of residual tissues [e.g., not placing sutures too tightly], using fewer sutures that are less reactive and preferably monofilament, closing down dead space, "packing off" contaminated viscera while making incisions into them to prevent spillage onto healthy tissues, removing any foreign materials and necrotic tissues prior to closure, etc.); (6) *proper aftercare of the experimental animal,* including monitoring and appropriately protecting the wound and implant by using bandages and restraint devices; and (7) *proper use of perioperative prophylactic antimicrobial agents.* For perioperative antibiotics to help prevent surgical wound infection, there must be a high tissue concentration prior to surgery. IV administration may be begun within 2 h of surgery, and most protocols terminate administration by 8 to 24 h after surgery, depending on the tissues invaded, unless there has been evidence of significant contamination of the tissues during surgery. In dogs and cats, a commonly used antibiotic is cephalothin (22 mg/kg, given every 6 h). For cattle, horses, and sheep, IV ampicillin is a common choice (22 mg/kg, given every 6 h).

The use of prophylactic antibiotics is not without danger. The generation of antibiotic-resistant strains may occur and, particularly when given orally and especially when the antibacterial agent selectively reduces anaerobic bacteria (e.g., by ampicillin, chloramphenicol, and cloxacillin) colonization with other, often multiply antibiotic resistant Gram-negative bacteria may result. This may thereby facilitate the occurrence and spread of nosocomial infections. Antiobiotic toxicity also must be considered in certain laboratory animal species.

In spite of all of these efforts, most surgical wounds become somewhat contaminated with bacteria. If contamination, trauma to the tissues, and the time the wound is open are minimal, foreign bodies have not been left, and little dead space has been created, the rate of infection is low. Tissue necrosis, often due to excessive trauma (due to clamps, ligatures, retractors, drying of the tissues, etc.), increases the likelihood of infection by serving as an excellent bacterial growth medium. Similarly, if foreign material is left in the tissues (inadvertently or intentionally in the form of a vascular or orthopedic implant, braided nonabsorbable suture material, etc.), infection can develop early or late after implantation. The likelihood of postoperative wound infection also increases if the procedure is prolonged, tissue dead space is not closed, or a hematoma develops. In each case bacteria may multiply without an adequate host response.

If the systemic resistance of the host to disease is decreased, the frequency of wound infection also increases. Very old or very young animals, those in poor physical condition, or those which are malnourished or have systemic diseases such as diabetes melitus, a malignancy, or an immunodeficiency disorder that is spontaneous or induced (e.g., treatment with cytotoxic drugs or corticosteroids) are predisposed to local and/or systemic infection. Anesthesia, hypothermia, and shock also have such an effect.

X. SURGICAL INSTRUMENTS, SUTURES, AND TECHNIQUES

Most of the instruments found in a human and/or veterinary clinical operating rooms are suitable. Special instruments for rodent surgery are discussed below. There are no major species-related differences in setting up instrument tables, handling instruments, surgical techniques, or sutures and ligature methods and materials. Information on these subjects may be found in many human,[23] veterinary,[24] and experimental[11] surgical texts.

XI. INTRAOPERATIVE MONITORING AND CARE[25]

Monitoring is the repeated assessment of the status of the experimental subject. Depending on the species of animal (e.g., it is difficult to monitor rodents), cardiac and respiratory function, fluid, electrolyte and acid-base balance, and body temperature should be monitored during anesthesia. In many instances, the quality and quantity of monitoring and intraoperative supportive therapy depend on whether the procedure is survival or nonsurvival. Even during nonsurvival surgery, monitoring and supportive therapy still may be required, at least to assure survival through the experimental period. The fact that monitoring helps direct the investigator to maintain the experimental animal in a physiologically and biochemically stable state is important both in survival and nonsurvival protocols.

Monitoring of core body temperature usually is done via a deep rectal (if the abdomen is closed) or esophageal (if the thorax is closed) probe. This may be compared to skin (i.e., toe web) temperature to help determine the degree of peripheral vasoconstriction. A drop in body temperature occurs often as the surgical site is being scrubbed and when body cavities are open for extended periods of time. This can be minimized by decreasing the time of preparation and surgery and by using circulating warm water blankets (placed between the animal and the table) and bottles (to surround the animal). In some species (e.g., swine and horses), an elevation of body temperature may raise the suspicion of malignant hyperthermia. In addition, hyperthermia may occur in small rodents if incandescent bulbs and/or heavy insulating drapes are used to maintain body temperature. In some instances, it may be desirable to have such animals placed directly on a steel surface, rather than on an insulating or heating pad. This must be balanced, however, with the concern that the latter may actually cause hypothermia. Young animals (e.g., pups under 2 weeks of age) are particularly likely to become hypothermic as they have poorly developed temperature regulating ability.

Measuring the rate and depth of breathing helps in evaluating the depth of anesthesia. The color of the mucous membranes (blue *vs.* pink) provides an estimate of the efficiency and adequacy of gas exchange, although not as accurately as an assessment of arterial and venous blood gases. The latter also allows assessment of acid-base status. Measuring the packed cell volume or estimating it by examining the color of the mucous membranes (pink *vs.* "white") helps evaluate the oxygen carrying capacity of the blood.

The heart's rate, electrical activity, and rhythm should be monitored, whenever possible, with an EKG oscilloscope and recorder. Auscultation for murmurs, rate, rhythm, and strength of cardiac contraction may be accomplished in an anesthetized animal by means of an esophageal stethoscope. Depending on the species, the rate, rhythm, and character of the arterial pulse may be monitored by palpating the femoral, lingual, maxillary, or other artery.

Arterial blood pressure may be recorded directly via an intra-arterial catheter, but indirect measurement via a Doppler ultrasonic unit now is being performed more frequently in animals. A central venous pressure (CVP) line may be placed as well. Loss of approximately 40% of the circulating blood volume results in hemorrhagic shock. Weighing all sponges after use and measuring the amount of blood aspirated helps in estimating blood loss.

Also monitored are capillary refill time (to assess tissue perfusion), serum protein or total solids, serum electrolytes, urine output (as a measure of tissue and, in particular, kidney perfusion), and, less commonly, pulmonary arterial and wedge pressures, cardiac output, and tissue pH.

A. FLUID THERAPY IN SURGICAL SUBJECTS

Larger laboratory animals undergoing experimental surgery should receive IV fluids, via a catheter, throughout the duration of surgery. Besides helping to support intravascular fluid volume and pressure, this provides an access port for giving emergency drugs, and if the catheter is inserted into a jugular rather than a peripheral vein, the line also can be used to monitor the CVP. In most healthy animals, an isotonic balanced electrolyte solution, such as lactated Ringers, usually is given to dogs and cats, etc. at a rate of 10 ml/kg/h, and to larger animals, such as horses and cattle, at 4 to 5 ml/kg/h. Signs of fluid overload should be searched for during fluid administration (e.g, elevation of the CVP to above 10 cm H_20 in dogs and cats and over 30 cm H_20 in horses and cattle, and evidence for pulmonary edema [e.g., auscultable moist rales and poor pulmonary compliance noted during ventilation]). Inadequate volume replacement may be detected by decreasing or absent urinary output, arterial hypotension, and/or peripheral vasoconstriction.

For some surgical protocols, it is advantageous to either have drawn and stored the animal's own anticoagulated blood some time prior to surgery, to have banked cross-matched blood, or to keep blood typed "universal donor" animals (e.g., dogs of the A^- blood group) which are free of infectious diseases (such as feline leukemia virus) or blood parasites (such as *Babesia* or *Hemobartonella* in dogs). Newer, less expensive methods of autotransfusion of blood lost into the abdominal or thoracic cavity now are available as well. In the absence of available blood, frozen plasma or dextran solution may be administered.

XII. POSTOPERATIVE CARE AND COMPLICATIONS[26]

Bandages may be needed to protect the surgical wound from contamination or trauma by the animal. They should be changed immediately if they become soiled or moist. The need for a postoperative bandage varies with the species, increases with the degree of contamination of the surroundings, e.g., in farm animals, and decreases if surgical technique is gentle, as less irritation leads to less self-trauma. Surgical drainage tubes always should be covered to minimize the chance of ascending infection. Surgical stockinettes often may be used to protect the wound or hold dressings in place. Plastic spray dressings may be used to protect noninfected wounds.

The recovery period begins when the surgical procedure ends and consciousness is regained. The endotracheal tube may be removed when the animal has regained the gag reflex, can swallow, has increased jaw tension, and responds to manipulation of the tube.

Animals should be placed in postoperative enclosures that are large enough for them to be comfortable in a completely outstretched position.[27] Large domestic animals should recover from anesthesia in a padded room. This is especially important for horses, which may thrash about and injure themselves or people around them.

Close observation is critical during the recovery period, because many drugs will not yet have been completely eliminated from the body by the time the animal is conscious. Some drugs have important side-effects, e.g., phenothiazines cause release of epinephrine, which decreases antidiuretic hormone (ADH), increases ACTH and serum glucose, and may cause cardiac arrhythmias, especially in the presence of the barbiturate thiamylal or the inhalant anesthetic halothane. Further, afferent stimuli, due to the injury of surgery, cause the release of epinephrine, corticosteroids, renin, aldosterone, and ADH. Pain, or bandages on the chest, may inhibit coughing or adequate ventilation. Tracheobronchial secretions may accumulate due to a depressed cough reflex and weak ciliary activity.

During the recovery period there should be continuous observation to include assessing body temperature, pulse, and respiration every 5 to 10 min in larger species. The airway should be kept free of secretions by postural drainage, suction, and wiping with gauze sponges. The ambient temperature should be maintained at 20° to 24°C. To prevent loss of body heat, any stainless steel surface on which the animal reclines should be covered with insulated pads or circulating warm water blankets, and the animal should be surrounded by warm water bottles, blankets, or heating pads. Electric heating pads or electrically heated cage bottoms should not be used, as they may result in severe burns, especially when the animal reclines directly on them, due to the insulating effect of its body. Electrically heated cage bottoms also tend to develop dangerous "hot spots". The animal should be turned frequently if it is unable or unwilling to change its own position. This helps prevent hypostatic pulmonary congestion, atelectasis, and decubital ulcers.

Tranquilizers or analgesics should be used if the animal is distressed or in pain, respectively. This must be done with great caution as these drugs may have side effects, including depression of respiration. Coughing should be induced, as necessary, by palpating the larynx and trachea in some species (e.g., dogs and cats). It may be necessary to assist respiration in some animals, so the presence of a mechanical ventilator is helpful. In some recovering animals the administration of oxygen is necessary. This may be done by face mask, nasal endotracheal intubation, or in an environmentally controlled chamber.

Surgical complications may include shock, usually due to hypovolemia; respiratory insufficiency, due to a variety of causes; cardiac abnormalities, including bradycardia, often due to drug interactions or hypothermia; tachycardia, due to pain, apprehension, hypoventilation, or hypothermia; arrhythmias due to electrolyte disturbances, hypoxia, acidosis, increased catecholamine levels, drug interactions, or prolonged anesthesia; electrolyte imbalance; oliguria and/or renal failure, due to hypotension with hypoperfusion of the kidneys; jaundice, due to hemolysis, hypoperfusion of the liver, drug or anesthetic toxicity, septicemia, etc.; acid-base abnormalities, most commonly hypoxemia and hypercarbia; alimentary tract dysfunction, e.g., vomiting, aspiration of vomitus or excessive salivation, ileus, fecal impaction, inadequate nutrition; disorders of consciousness, including lethargy, aggression, coma, and seizures, due to prolonged anesthesia, cerebral hypoxia, intracranial hemorrhage, thrombosis, or alterations of cerebral metabolism; hypothermia or hyperthermia; and infection.

Increased body temperature often is seen within 24 h of surgery. This is due to the response to trauma and tissue and protein absorption. There also may be a low-grade infection, but often temperature elevations due to infection do not start until approximately 48 h after surgery. A progressively increasing body temperature, in conjunction with signs of inflammation of the surgical wound and the finding of neutrophilic leukocytosis on a complete blood count, suggests infection. This may require surgical drainage and bacterial culture and antibiotic sensitivity testing. Treatment first may be with broad spectrum antibiotics and then with specific antibiotics selected on the basis of culture. Gram staining of the infectious exudate may suggest an appropriate first antimicrobial treatment. Fly strike also may occur during the postoperative period, especially in farm animals housed in a barn environment.

Restraint sometimes is required to protect implanted electrodes or cannulae, but this is not common.[27] There are commercially available Elizabethan collars which can prevent dogs and cats from chewing their body or scratching their head. Tranquilization also may diminish self-inflicted trauma. Simple hobbles (partially taping the rear legs together) can help prevent scratching. An aluminum collar may be fashioned to prevent turning of the head to traumatize wounds.[28] Jackets of leather or cloth have been used to protect wounds. Harnesses on metal frames may be used to support the body if there is partial limb paralysis or in some species, such as sheep, to protect the animals from injuring themselves during the anesthetic recovery period. Plaster or fiberglass casts also may be used to protect limb incisions.

Postoperative nutrition is important to the healing process. Fluid therapy should continue at 5 to 10 ml/kg/d after surgery, but 5% dextrose at this dosage rate will supply only 10 to 20% of the caloric needs of the experimental subject. Oral feeding of small quantities of food and water usually may be started the day following surgery, even of the digestive tract (except the esophagus), unless there is evidence of intestinal ileus. If the appetite does not return rapidly there should be a concern about infection. Enteral alimentation is preferable to parenteral hyperalimentation because of the difficulty of managing the latter, the possibility of catheter infection, etc. Enteral alimentation may be accomplished via nasogastric tube, pharyngostomy tube, or intermittent gavage. In some cases, gastrostomy or enterostomy feeding tubes may be indicated.

XIII. SOME SPECIAL SPECIES CONSIDERATIONS

A. RODENTS[29]

1. Surgical Technique

Many of the surgical procedures developed for dogs and cats have been adapted for use in rodents. By far, more sophisticated procedures have been performed in rats than in mice, due to the smaller size of the latter. Home-made cork or soft wood operating boards may be fashioned, but it is better to use a commercially available operating board with an impervious surface that may be cleaned and disinfected more readily.[30] Positioning is important; this is usually accomplished by means of clips or nooses placed around the feet and incisor teeth. The commercially available board also has sterilizable, built-in tissue retractors.

A list of basic instruments has been suggested for rat surgery,[30] but it is appropriate for rodent surgery, in general. This includes Mayo's scissors, "sharp-sharp" dressing scissors, two pairs of medium-width, curved, serrated (Pakes) forceps, medium width, curved toothed forceps, Michel skin clip application forceps, toothed forceps with Michel clip holder, Michel skin clips, an intestinal needle and suture, plus other instruments, including Michel clip extraction forceps, scalpel handles and blades, neurosurgical clips, a drill and assorted burrs, gauze sponges, cotton-tipped applicators, etc. Microsurgical technique, using ophthalmic or watchmaker's instruments[30] and ophthalmic sutures with swaged on needle, may be necessary for certain procedures, especially in mice. A binocular dissecting microscope or an operating microscope may be needed, depending on the procedure.

Little or no preoperative care is needed for rodents, unless surgery involves the alimentary canal, which should be emptied by fasting for from 12 to 24 h. Any food must be removed from the mouth after anesthesia. The methods of preparation of the animal and the surgeon vary with the experiment and the reason for surgery. As with other species, nonsurvival procedures pose few restrictions. Survival surgery does not require a cap, mask, gown, or operating suite, but sterile instruments, decontamination of the skin, and not handling the viscera bare handed are necessary. The hair should be removed from the surgical site. In mice, plucking rather than clipping the hair leaves some hair shafts, but there is less tendency for fine bits of hair to migrate into abdominal incisions. Depilatories are useful, as well. Following skin decontamination, the surgical site may be draped. Drapes may be fashioned from autoclavable plastic bags, disposable drapes designed for human or veterinary surgery, or cloth. Drapes are optional during some short procedures, but are recommended for procedures that require exteriorization of the viscera or on immunosuppressed animals.

Tissues should be handled with the tips of sterile instruments, and the surgeon should pick up the instruments by the handles and not touch the tips. If finger contact with the viscera is needed, sterile surgical gloves must be worn. Separate instruments should be used to perform different tasks, e.g., skin *vs.* deeper tissues. The instruments should be arranged on a sterile surface, such as a surgical towel, with all of the handles facing toward the surgeon and the tips covered with a sterile towel or drape. Between procedures on different animals, blood should be removed from the tips of the instruments with sterile saline-soaked sponges and then with alcohol-soaked sponges. It is acceptable to flame the alcohol-soaked tips of instruments (except for sharp-edged or other fine instruments, which may be damaged by this procedure) when there is the likelihood that contamination has occurred, and/or when a new group of animals is to undergo surgery. Fine instruments should be handled more gently, and, whenever possible, instrument packs should be changed between cages/groups.

In many cases, experimental surgeons prefer to incise the tissues with scissors (e.g., blunt-ended Mayo's) rather than with a scalpel, as is done for most larger species. Mouse skin and abdominal muscles are thin and fragile, and the cut edges tend to curl under during suturing. Everting, horizontal mattress, or simple interrupted sutures minimize tissue inversion. The abdominal muscle/peritoneal layer and the skin should be closed separately. Small (1 to 3 mm long) incisions of the dorsal abdominal musculature near the lumbar muscles, which are not directly under the skin incision, need not be closed. Michel wound clips may be used in the skin, as long as they evert the skin edges.

Usually, postoperative treatment of rodents is complicated by their size. Treatment with antibiotics such as tetracycline is often accomplished by water administration, but this does not assure appropriate dosage. It can be done on an individual basis by intraperitoneal, intravenous, subcutaneous, or oral routes, if necessary. Antibiotics must be given with care (see below).

Postoperatively, the animals should be placed back in their cages with clean bedding and fresh food and water available. Often, it is necessary to keep the environment warmer than usual during the immediate postoperative period. This may be accomplished by means of a heat lamp (e.g., a 60 W bulb placed about 60 cm above the cage for 30 to 60 min), with careful observation to avoid overheating or burning. Following recovery, the animals should be observed regularly and carefully, especially if Michel wound clips are used. Clips causing apparent discomfort should be replaced. Fluid therapy is usually done by intraperitoneal injection using a syringe and a 22- to 26-gauge needle. For some operations, restraint by means of a special cage or harness may be required.[30]

Prior to returning neonatal rats or mice to their cage, the dam might be tranquilized, the pups should be warmed to 37°C, and the application of pheromones to the young should be considered in an attempt to avoid cannibalism.[31]

2. Prophylactic Antibiotics

Fatal reactions are a major concern in association with antibiotic therapy of rodents,[32,33] which decreases the ability to use perioperative antibiotics or to treat postoperative infections. This occurs mainly due to direct alteration of the intestinal flora. Penicillins, bacitracin, and some other antibiotics should not be given to guinea pigs or hamsters. Such antibiotics suppress Gram-positive organisms, such as lactobacilli and streptococci, leading to clostridial or coliform enterotoxemia, with hemorrhagic cecitis, colitis, diarrhea, or deaths, beginning within 3 and 5 d and continuing for a week or more. Rats, mice, and gerbils are relatively resistant to this effect. A combination of 5 mg of neomycin and 3 mg of polymixin B may be given orally to guinea pigs twice daily for 5 d to attempt to counteract lethal antibiotic effects. Guinea pigs and hamsters can be treated with chloramphenicol (50 mg/kg of the palmitate per day, orally, or 30 mg/kg of the succinate IM twice daily), cephalosporins, neomycin, or other broad-spectrum antibiotics not selective for Gram-positive bacteria.[32] Tetracyclines, however, may cause fatal enterotoxemia in hamsters, if given SQ at 50 mg/kg, unless sulfaguanidine is given simultaneously.[32] Some penicillin preparations contain procaine, which may be fatal at 0.4 mg/kg in mice and guinea pigs. Streptomycin is toxic in mice and dihydrostreptomycin in gerbils, resulting in ascending paralysis, respiratory arrest, coma, and death.

B. RABBITS[32,33]

Similar to rodents, rabbits are also susceptible to fatal reactions to antibiotics due to effects on their intestinal microflora and of procaine. Lincomycin is contraindicated in rabbits, in which doses as low as 5 to 30 mg may cause enterotoxemia and death. Gentamycin may counteract this effect. Outbreaks of diarrhea after antibiotic treatment have been attributed to *Clostridium dificile* and *C. sordelli.*

C. NON-HUMAN PRIMATES

As monkeys and other non-human primates have hands, they may manipulate surgical appliances, tubes, etc., and remove their own sutures prematurely. Formerly, monkeys commonly were placed in restraining chairs for long periods of time after surgery to prevent such events. Due to the inhumaneness of such restraint techniques, this is now a rare event, requiring thorough prior review by the IACUC. Now, if chairs are used, this is usually only for one day or so, until complete recovery from the effects of anesthesia has occurred. If surgical technique is excellent, the tendency for self-mutilation is minimal and tranquilizers may be beneficial. The use of subcuticular instead of skin sutures also decreases the likelihood of self-trauma to the incision line.

REFERENCES

1. **Brieger, G. H.,** The development of surgery. Historical aspects important in the origin and development of modern surgical science, in *Davis-Christopher Textbook of Surgery,* 12th ed., Sabiston, D. C., Ed., W. B. Saunders, Philadelphia, 1981, 1.
2. **Cushing, H.,** Instruction in operative medicine with the description of a course given in the Hunterian laboratory of experimental medicine, *Bull. Johns Hopkins Hosp.,* 17, 123, 1906.
3. **Schwartz, A.,** Contributions of biomedical research to veterinary surgery, *J. Am. Vet. Med. Assoc.,* 193, 1145, 1988.
4. **Markowitz, B. B. E., Archibald, J., and Downie, H. G.,** *Experimental Surgery,* 5th ed., Williams and Wilkins, Baltimore, 1964.
5. **De Boer, J., Archibald, J., and Downie, H. G.,** Eds., *An Introduction to Experimental Surgery,* Elsevier, New York, 1975.
6. **Lupukhin, Y. M.,** *Experimental Surgery,* MIR publishers, Moscow, 1976.
7. **Swindle, M. M., Smith, A. C., and Hepburn, B. J. S.,** Swine as models in experimental surgery, *J. Invest. Surg.,* 1, 65, 1988.
8. **Swindle, M. M.,** The use of animals in surgical research, *J. Invest. Surg.,* 1, 3, 1988.
9. Institute of Laboratory Animal Resources, National Research Council, National Academy of Sciences, *Guide for the Care and Use of Laboratory Animals,* DHEW NIH Publ. No. 85-23, National Institutes of Health, Bethesda, MD, 1985.
10. **Hobson, H. P.,** Surgical facilities and equipment, in *Textbook of Small Animal Surgery,* Slatter, D. H., Ed., W.B. Saunders, Philadelphia, 1985, 285.
11. **Sumner-Smith, G.,** Surgical preparations, in *An Introduction to Experimental Surgery,* Elsevier, New York, 1975, chap. 6.
12. **Milne, F. J.,** General surgical considerations, in *Textbook of Large Animal Surgery,* 2nd ed., Oehme, F. W., Ed., Williams & Wilkins, Baltimore, 1988, chap. 1.
13. **Merkley, D. F. and Grier, R. L.,** Surgical instruments, in *Textbook of Small Animal Surgery,* Slatter, D. H., Ed., W.B. Saunders, Philadelphia, 1985, 301.
14. **Kaplan, H. M. and Timmons, E. H.,** *The Rabbit. A Model for the Principles of Mammalian Physiology and Surgery,* Academic Press, New York, 1979, 13.
15. **Powers, D. L.,** Preparation of the surgical patient, in *Textbook of Small Animal Surgery,* Slatter, D. H., Ed., W.B. Saunders, Philadelphia, 1985, 279.
16. **Doherty, R. W.,** *Experimental Surgery in Farm Animals,* Iowa State University Press, Ames, 1981.
17. **Hecker, J. F.,** Pre-operative and post-operative management, in *Experimental Surgery on Small Ruminants,* Butterworths, London, 1974, chap. 4.
18. **Hecker, J. F.,** Management of experimental sheep, in *The Sheep as an Experimental Animal,* Academic Press, New York, 1983, chap. 6.
19. **Mount, L. E. and Ingram, D. L.,** *The Pig as a Laboratory Animal,* Academic Press, New York, 1971, 91.
20. **McCurnin, D. M. and Jones, R. L.,** The principles of surgical asepsis, in *Textbook of Small Animal Surgery,* Slatter, D. H., Ed., W.B. Saunders, Philadelphia, 1985, 250.
21. **Berg, R. J. and Blass, C. E.,** Sterilization, in *Textbook of Small Animal Surgery,* Slatter, D. H., Ed., W.B. Saunders, Philadelphia, 1985, 261.
22. **Wagner, S. D.,** Preparation of the surgical team, in *Textbook of Small Animal Surgery,* Slatter, D. H., Ed., W.B. Saunders, Philadelphia, 1985, 269.
23. **Sabiston, D. C.,** Ed., *Davis-Christopher Textbook of Surgery,* 12th ed., W.B. Saunders, Philadelphia, 1981.
24. **Slatter, D. H.,** Ed., *Textbook of Small Animal Surgery,* W. B. Saunders, Philadelphia, 1985.
25. **Kolata, R. J.,** Monitoring the surgical patient, in *Textbook of Small Animal Surgery,* Slatter, D. H., Ed., W.B. Saunders, Philadelphia, 1985, 351.
26. **Brown, N. O.,** Patient aftercare, in *Textbook of Small Animal Surgery,* Slatter, D. H., Ed., W.B. Saunders, Philadelphia, 1985, 373.
27. **Bleicher, N.,** Care of animals during surgical experiments, in *Methods of Animal Experimentation,* Vol. 1., Gay, W. I., Ed., Academic Press, New York, 1965, 103.
28. **Foss, M. L. and Barnard, R. J.,** A vest to protect exposed chronic implants in dogs, *Lab. Anim. Care,* 19, 113, 1969.
29. **Cunliffe-Beamer, T.,** Biomethodology and surgical techniques, in *The Mouse in Biomedical Research,* Foster, H. L., Small, J. D., and Fox, J. G., Eds., Academic Press, New York, 1983, 402.
30. **Waynforth, H. B.,** Surgical technique, in *Experimental and Surgical Technique in the Rat,* Academic Press, New York, 1980, chap. 3.
31. **Baker, H. J., Lindsey, J. R., and Weisbroth, S. H.,** Research methodology, in *The Laboratory Rat,* Vol. 2, Academic Press, New York, 1980, 1.

32. **Harkness, E. J. and Wagner, J. E.,** Clinical procedures, in *The Biology and Medicine of Rabbits and Rodents,* 2nd ed., Lea and Febiger, Philadelphia, 1983, chap. 3.
33. **Hoar, R. M.,** Biomethodology, in *Biology of the Guinea Pig,* Wagner, J. E. and Manning, P. J., Eds., Academic Press, New York, 1976, 13.

Chapter 21

EUTHANASIA: ACCEPTABLE AND UNACCEPTABLE METHODS OF KILLING

H. C. Rowsell

EDITOR'S PROEM

The majority of research animals are eventually euthanized. In some cases, euthanasia provides an end point to an experiment which would otherwise engender considerable pain and suffering for the animal. In other cases, the animals are euthanized to obtain post-mortem samples for study. In still others, the animals are euthanized because the research is over and they are of no further use. In all of the above instances, it is surely morally incumbent upon researchers to assure that death is as close to painless as possible, and that stress, fear, and anxiety be avoided. In addition, it is now recognized that long-term involvement with euthanasia causes problems for laboratory personnel and may result in long-term stress which is harmful to the person's physical and mental health. In his paper, Dr. Rowsell discusses a number of aspects of euthanasia relevant to the welfare of animals, the stress on personnel, and the validity of research.

TABLE OF CONTENTS

I. INTRODUCTION

Euthanasia — the killing of an animal — is one of the most important, albeit probably the most distasteful, tasks faced by veterinarians, scientists, animal health technicians, animal control officers, and others involved with animal use.[1,2] The primary criteria for euthanasia (which means "easy or humane death", literally "good death") are that the method be "painless, must minimize fear in the animal, be reliable, reproducible, irreversible, simple and safe, rapid, and, if possible, aesthetically acceptable for the observer or operator."[3,4]

Euthanasia methods are commonly categorized as physical methods (e.g., cervical dislocation, decapitation, guillotining, stunning, electrocution); methods utilizing noninhalant pharmacologic agents (e.g., barbiturate derivatives, T-61); methods using inhalant anesthetics (e.g., halothane, methoxyflurane, chloroform, ether); and methods involving nonanesthetic gases (e.g., carbon monoxide, carbon dioxide, nitrogen flushing, and argon).

The scientific community is often criticized for the euphemistic terminology it uses for destroying an animal: the animal is said to be "sacrificed", "put down", or "put to sleep". However, it behooves us all to avoid euphemisms and "call a spade a spade" or, in this case, to call killing "killing". But if it is to be done, the most important consideration is that the killing be done humanely, i.e., so as to produce rapid loss of consciousness without fear or anxiety, and to be irreversible. In this regard, it is important to stress that the primary criterion to be met for the humane killing of animals is an initial depressive action on the central nervous system, to ensure unconsciousness and, thus, immediate insensitivity to pain. The AVMA Panel on Euthanasia Report (1986)* calls for rapidly occurring unconsciousness caused by cardiac or respiratory arrest. It notes that the presence of reflex action in an unconscious animal is not a reliable indication of whether the animal is feeling pain. Conversely, "an animal can experience pain, even though no body movements occur in response to noxious stimuli."

When assessing a euthanasia method, one considers possible signs of distress, including vocalization, attempts to escape, shivering, tremors, and muscle spasms, some of which may occur in animals which clinically are unconscious due to reflex nervous system reactions.

Guidelines for euthanasia methods are primarily found in the following sources: in the U.S., the 1986 American Veterinary Medical Association (AVMA) Report on Euthanasia,[5] and the Federal Animal Welfare Act, Revised 1985 (Public Law 99-198); in Canada, the Canadian Council on Animal Care's (CCAC) Guide to the Care and Use of Experimental Animals (1980),[6] (Federal) Agriculture Canada's Meat Inspection Act S.C., 1985 c.17 (May 1985) Part II, Item 39, SOR/85 1078, S14, and the Province of Ontario's Animals for Research Act as amended 1980, (Regulation 17, Sections 20, 21, 22). In Great Britain, the major guides are the Universities Federation for Animal Welfare (UFAW) report on Euthanasia of Unwanted, Injured or Diseased Animals for Education or Scientific Purposes (1986)[7] and the Animals (Scientific Procedures) Act 1986.

A. NUMBERS OF ANIMALS KILLED ANNUALLY

In the U.S., an Office of Technology Assessment estimate (1983) puts the numbers of animals used in research, education, and testing at 50 to 70 million. According to the U.S. Census Bureau, a further 4,285,400,000 animals were slaughtered for food in 1981. In 1983, the latest year for which we have figures, 20 million unwanted cats and dogs were destroyed in pounds and shelters in the U.S.

In Canada in 1987, 2,015,222 animals (including rodents) were used in research, teaching, and testing, down 3% from the previous year. These figures in this regard are compiled by the Canadian Council on Animal Care (CCAC) which has recently prepared and launched a

* Reproduced in Appendix II of this book.

computer program called Animal Research Protocol Management System, which will be used across Canada. In 1986, 369,918,775 domestic livestock and fowl were slaughtered; 193,833 unwanted pet animals were killed in pounds and animal shelters operated by affiliates of the Canadian Federation of Humane Societies (CFHS). However, this is a small proportion of the total numbers killed across Canada by municipal pounds and shelters operating independently; thus, it is difficult to give an accurate figure, although it is more likely well beyond 1 million. Although the numbers are important, the method by which they were killed is of even more significance.

In Great Britain, numbers of animals euthanized per se are not known; however, the number of experiments on animals in 1986 was 3.1 million. Michael Balls, chairman of the Board of Trustees, Fund for the Replacement of Animals in Medical Experiments (FRAME), has recently criticized this lack of information.[8] Although the term "euthanasia" is usually reserved for those animals killed in pounds, shelters, and laboratories, principles and practices used to produce a humane, acceptable death for this group of animals should also apply to all animals deliberately killed in any context. Thus, my studies in euthanasia methods[9] have involved use of various types of harpoons in killing whales,[10,11] use of the hakapik and bat to kill adult and young seals,[12,13] how vertebrate pesticides kill,[14] trapping methods,[15-17] and preslaughter stunning in abattoirs.[18]

B. HUMAN ASPECTS OF EUTHANASIA

Although the needs of the animal must always take precedence over all other considerations, so psychologically traumatic is the process of euthanasia to those administering it that the 1986 AVMA Report considered it an important personnel problem that must be addressed. Ironically, constant exposure to such procedures can result, among other things, in "careless and callous handling of the animal." "This is one of the principal reasons for turnover of personnel directly involved in euthanasia," states the report, which has been praised by Canadian veterinarians who commended the AVMA panel for "producing such a complete document."[19] (The Canadian Veterinary Medical Association (CVMA) has itself produced a thorough report on euthanasia of dogs and cats.[20] However, rather than duplicate the AVMA 1986 report, the CVMA has recommended it be regarded as the Canadian authority for acceptable euthanasia procedures.)

Owens et al. have noted that those who perform euthanasia have developed strategies for dealing with their emotions of isolation and sorrow; for example, they avoid unnecessary contact with the animals and believe that the animal is being spared additional suffering.[20] Rollin has discussed the moral stress engendered in humane society workers, laboratory animal veterinarians and caretakers, and others whose job requires extensive involvement with euthanasia.[21]

The aesthetics of the euthanasia method are important. For example, Canada has lost its $12 million sealing industry[22] primarily because the method of hitting the photogenic "whitecoat" seal pup over the head with a bat or hakapik appeared cruel. Little or no attention was paid to the fact that scientists, myself among them, found the method humane and effective after extensive clinical and post-mortem examinations.[12,13] The final blow to the industry came in December 1987, when Canada's Minister of Transport and Minister of Fisheries announced a complete halt to the commercial fishery.[23] Animal Rights activists, led by Brian Davies and the International Fund for Animal Welfare (IFAW) had, beginning in 1963, convinced Europe (and most Canadians as well) that the method of killing the seal pups was inhumane — because it looked brutal. Large-scale commercial hunting of seals effectively ended in 1983 when the European Economic Community, the largest market for the pelts, banned imports of whitecoat pelts "in wake of intense international pressure."[24] The U.S. had banned imports in 1971.

Issues regarding euthanasia affect all aspects of veterinary medicine. Recently, controversy erupted in the pages of the AVMA Journal regarding the veterinarian's duty (or lack of it) to euthanize pets on request. "Technically and legally, a pet is a personal possession and the owner does have the right to take its life," wrote Dr. William Dorsey. "But we, as animal doctors, do not have any obligation to be a party to it. To the contrary, we have an obligation to safeguard its life."[25] Others have argued that those refusing to kill a healthy pet at the request of its owner may be dooming the animal to greater pain and suffering due to painful trauma at the owner's hands in an inexpert attempt to kill it, or due to accidental injury, death, or starvation after being released in the countryside to fend for itself. Still others have written advice on how to help an owner cope with the death of a pet.[26,27] It is becoming increasingly clear that laboratory personnel may also experience grief and depression following euthanasia of animals.

Among the scientific community, the disposal of animals following a study is commonplace. Often animals are killed so that tissue can be examined. Most scientists such as myself have been involved in such studies. In the study of atherosclerosis, with pigs as the experimental animal, we examined the platelet and the encrustations considered platelet material that were appearing on the wall of the blood vessel. Thus, the animal was killed in order to study the pathogenesis of a disease, a common occurrence in scientific investigation. At other times, the animals used in research are killed as a therapeutic intervention, when pain and suffering cannot be controlled. As practicing scientists or laboratory animal veterinarians, it is our ethical responsibility to maintain meaningful life without pain or suffering. We must accept death when there is no other recourse, believing that when an animal is killed it is a final act which one cannot undo; thus, there have to be strong scientific reasons and moral justification for doing so.

C. THE "3 RS"

Today, ever-increasing effort is being made to abide by the Russell-Burch "3 R" tenet of reduction, replacement, and refinement.[28] To this end, attempts are being made to conduct studies using methods which are as nonconsumptive of animals as possible. For example, Dr. Jean Dodds, of the State of New York Health Department, is conducting research using clinical material supplied by veterinarians. Dr. J. Moore-Jankowski, director of New York's world-renowned Laboratory for Experimental Medicine and Surgery In Primates (LEMSIP), is also known for this approach in his work with chimpanzees. The past chairperson of the CCAC, Dr. Gail Michener, a biologist with the University of Lethbridge, Alberta, works primarily with Richardson's ground squirrels (incorrectly labeled "prairie dogs"), which she returns to the wild at the completion of the studies. This latter process must, however, be very carefully researched, for if animals are returned to the wild insufficiently prepared, and their habitat is not ensured, they will die in a far less humane fashion than if they were euthanized in the laboratory.

D. ANIMAL PAIN

Issues of animal awareness and behavior[29-32] are becoming of increasing importance as moral concern for animals increases in society and as our knowledge of animal pain escalates[33-37] (although much remains to be discovered). In this regard, the Canadian Council on Animal Care requires that "any animal observed to be experiencing severe, unrelievable pain should immediately be humanely killed, using a method providing initial rapid unconsciousness."[6]

The AVMA Report notes that pain must be defined before criteria for a painless death can be established. "Pain is that sensation (perception) that results from nervous impulses reaching the cerebral cortex via specific neural pathways called nociceptive pathways.... Endogenous chemical substances such as hydrogen ions, serotonin, histamine, bradykinin, and prostaglandin, as well as electric currents, are capable of generating nerve impulses in nociceptors."[5] The report notes that for pain to be experienced, the cerebral cortex and subcortical structures must be functional. If the impulses are interrupted by such means as hypoxia, depression by drugs, electric shock, or concussion, pain is not experienced.

Absence of the "blinking" reflex is often taken as indicating unconsciousness, and flattening of the electroencephalogram (EEG) as indicating brain death. The latter criterion has recently been adopted by the American Academy of Neurology as acceptable for establishing brain death in young children after other clinical criteria, such as deep coma, failure to breathe spontaneously, and absence of reflexes occur.[38]

II. METHODS OF EUTHANASIA

Another method of refinement is improvements in steps leading up to euthanasia. For example, gentler handling can reduce animal anxiety and distress, thereby leading to a smooth anesthetic induction. Gärtner notes that rough handling by a technician of a cage can cause the animal's stress levels to rise.[41] Landi describes the effect of shipping on the immune function of mice.[42] Therefore, animals that require transportation to the location where they are to be killed may suffer much pain, anxiety, and distress before their death, even though an acceptable, humane method is used. This applies to slaughter of domestic animals and birds as well as animals used in the laboratories. The tranquilizers used before anesthesia can chemically calm animals which resist the reassurance of gentle handling. Not only is the peaceful euthanasia more emotionally acceptable to personnel, more important, the animals are relieved of predeath anxiety.

A. NONANESTHETIC GASES

Nonanesthetic gases include carbon monoxide, carbon dioxide, and nitrogen. Although exhaust from gasoline combustion engines may be an efficient source of carbon monoxide, impurities in the fumes can produce irritation and discomfort, so delivery of an irritant-free product must be provided if the method is used. Also carbon monoxide is a potent, cumulative poison, and therefore, exposure to it by humans and other animals must be avoided. In addition, vocalization occurs,[6] which increases the stress on personnel. Because of these difficulties, carbon monoxide is not recommended for euthanasia.

Nitrogen is used as a flushing agent to reduce oxygen to a level of 1.5%, at which time the animal will collapse. However, dogs, cats, and rabbits have been observed to vocalize at this level as well as to show muscular activity and struggling. Moreover, nitrogen offers no advantages over carbon dioxide as a killing method.[6]

As regards the 3 Rs, refinement in euthanasia can be accomplished by improving methods of euthanasia or replacing those found to be inhumane. For example, in 1976, the American Humane Association (AHA) accepted high-altitude decompression chambers, carbon monoxide chambers, and nitrogen flushing chambers as humane euthanasia systems.[39] However, it later found that much of the equipment used to administer the gases did not provide a "safe and humane delivery system." It now recommends injection of an overdose of barbiturate as most acceptable.[40]

Carbon dioxide is frequently used to kill research rodents and birds, as induction of unconciousness occurs in a few seconds if the gas is used in high concentration, the carcasses are residue free, and escaped gases are nontoxic. However, the Ontario Ministry of Agriculture and Food (OMAF), with the Ontario Humane Society (OHS), has recently prepared a (draft) report which recommends that, under the province's Animals for Research Act, the killing or unwanted dogs and cats using carbon monoxide or carbon dioxide be phased out within 6 months. Carbon monoxide was criticized because of significant hazard to the operator, and the problem of producing gas free of contaminants if improperly tuned automobile engines are used. Carbon dioxide was criticized because it both stimulates the respiratory center and, in low dosages of up to 10% of inspired gas, is "a potent respiratory stimulant causing a ten-fold increase in ventilation rate and a feeling of profound respiratory distress. The stimulation can produce hyperventilation, thus limiting the onset of narcosis, and may thus cause distress. At approximately 40%, the gas induces anesthesia which is slow in onset and accompanied by prolonged involuntary excitement. Eventually, there is apnea, a fall in blood pressure and death."[43] In addition, the act requires that the competence of the administrators of any method must be proven to the satisfaction of inspectors under the act, who are employed by the OMAF.

It should be noted, however, that while the AVMA, UFAW, and CCAC accept use of carbon dioxide as a killing method, there are several cautions involved in its use; e.g., it should not be used with investigative animals, such as dogs, which often extend their heads above the zone of effective gas concentration. Britt asserts that slow induction is preferable, as undue haste can upset the animal, but that even though the period to unconsciousness is brief, "there are signs of distress.... Neither (slow nor swift) method was found to be stress-free, so no recommendations can be a counsel of excellence."[44]

B. INHALANT AGENTS

Inhalant agents include ether, chloroform, halothane, methoxyflurane, and nitrous oxide. Chloroform is universally out of favor because of its suspected danger as a carcinogen and hepatotoxic agent to human handlers;[45] it also is less than humane to the animal if not vaporized, because it irritates the mucous membranes. In addition, low levels of chloroform vapor in the air can kill certain inbred strains of male mice.[46]

The aforementioned Ontario government report recommends that the use of chloroform and ether be phased out immediately. Ether is considered unacceptable because of its explosive properties and the possible discomfort to the animal because of irritation of the mucous membranes.

Generally, since nitrous oxide is a weak anesthetic agent, it should not be employed alone for euthanasia. Both halothane and methoxyflurane can be freely vaporized in a cool container for euthanasia. Both will induce unconsciousness rapidly, particularly in small laboratory animals, although halothane is generally regarded as the quicker of the two. Halothane, freely vaporized, also exceeds by many times the lethal level for the gas, and animals are likely to die more rapidly; methoxyflurane is vaporized to just higher than the lethal concentration. Disadvantages of these two gases include mild irritation of the respiratory tract (as evidenced by aversive behavior displayed when animals detect the gas, as well as by pawing at the nose) and potential toxicity for humans. All gas anesthesias are controlled poisons, and humans must avoid exposure to them by use of adequate scavenging systems, for example, chemical hoods. Pregnant women can endanger their unborn children by exposure to anesthetic gases, and some people get violent headaches. Since these agents are neuroactive, they can induce chemical or physiological changes in the animals' brains which may be contraindicated in some studies. However, if residues are not a consideration, use of halothane or methoxyflurane is less stressful to the animals and should supplant the use of CO_2 for humane reasons if proper scavenging systems are available.

A final caution should be observed when using any gas anesthesia: animals must never be exposed to the liquid agent, as the rapid vaporization will produce uncomfortable cooling.

C. NONINHALANT PHARMACOLOGIC AGENTS

The noninhalant pharmacologic agents used for euthanasia include: barbituric acid derivatives, barbiturate mixtures, and T-61. All of these must be introduced intravenously as they are highly irritant to tissues when injected.

Pentobarbital sodium and other barbiturate derivatives are most often used and constitute the agents of choice for most euthanasia, on both esthetic and scientific grounds. Administration by the intravenous route is preferred because of the speed with which it induces unconciousness and produces death.

Chloral hydrate and ketamine hydrochloride are dissociative anesthetics, and there is no loss of the eye reflex or "blinking" reflex in the anesthetic state. Chloral hydrate is not recommended because of its slow onset of action and restraint difficulties, such as muscle spasm and vocalization (although some recommend its use if preceded by a tranquilizer); intramuscular is the only route recommended. Ketamine is not recommended as it is difficult to dose to a lethal level.

T-61* contains a local anesthetic, a strong agent which depresses the CNS causing unconsciousness (brain death), as well as a drug which has a paralytic effect on the respiratory center and a relaxing effect on skeletal muscle. It should be administered intravenously or intracardially. Rapid intravenous or intracardiac injection of T-61 may produce excitement and vocalization; therefore, it should be given at the recommended rate. In Canada and the U.S., this is not a "restricted" drug under the Bureau of Dangerous Drugs, Health and Welfare Canada, or the U.S. Food and Drug Administration and may therefore be obtained and used by nonmedical personnel. However, a prescription by a licensed veterinarian is required in Canada.[6]

Use of paralyzing curariform agents is universally condemned as unacceptable as a euthanasia method, as they have no depressing effect on the CNS; the animal dies by asphyxiation caused by paralysis of the respiratory muscles and can be in excruciating pain. Ulysses Seal, Department of Fisheries and Wildlife, University of Minnesota, claims that a rapidly administered, large overdose produces rapid unconsciousness and death (personal note). However, supporting scientific documentation is lacking. In addition, there are too many possibilities of underdosing, causing death by slow asphyxiation. Therefore, the scientific community in general justly disapproves of the use of such agents.

D. PHYSICAL METHODS

Physical methods of euthanasia include stunning, cervical dislocation, electrocution, pithing, decapitation, and shooting.

Stunning is used primarily in small laboratory rodents. A blow must be delivered to the central skull bones with sufficient force to produce massive cerebral hemorrhage and thus immediate depression of the CNS. This technique should not be undertaken in the presence of casual observers, as it is esthetically unpleasant. However, when properly applied, there is no question but that the animal is immediately rendered unconcious and insensitive to pain. Subsequent to stunning, the animals's major blood vessels should immediately be incised, the chest opened, and the heart muscle cut.

Cervical dislocation is practiced on mice, rats, or similar small species. The technique consists of a separation of the skull and brain from the spinal cord by pressure applied posterior to the base of the skull. When separation of cord from brain occurs, CNS stimulation of respiration and heartbeat is interrupted, leading to death The supply of blood to the brain continues to nourish the brain because the carotid arteries and jugular veins are intact, although blood will rapidly be depleted of O_2 and increase in CO_2 after respiration ceases, leading to brain dysfunction. Studies have demonstrated that the blinking reflex disappears immediately as the spinal cord separates, indicating that the animal is not sensitive to pain. In addition, the severed spinal cord does not deliver painful stimuli from areas posterior to the separation. Thus with the separation of the spinal cord from the brain, painful stimuli are not perceived. Significant muscular movement may take place; deprived of the control of the brain, muscles jump in spinal reflex response to stimuli. Devices are commercially available for cervical dislocation of mice, rats, similar small species, and rabbits. Light anesthesia or sedation should be used routinely before cervical dislocation, unless clear evidence shows that these drugs would skew relevant variables.

Pithing, which requires considerable dexterity and skill, is used to euthanize frogs by destroying the brain. After the animal has been cooled to 4°C (40°F) for anesthesia, a sharp, pointed probe is inserted through the skin and in between the skull and the atlas; it is then pushed forward through the foramen magnum into the cranial cavity, using a twisting motion. The technique should be attempted only after a period of training in frog anatomy using skeletons and after practice on dead animals, as it can cause pain and suffering if the proper region is not destroyed.

* Not available in the U.S. or the U.K.

Shooting is an effective means of humanely destroying animals in the field. Only experts should carry out this procedure. The subject must be shot at close range, and the bullet must strike the brain, so as to render the animal immediately insensitive to pain; 12- or 20-bore shotguns or 22-caliber rifles or revolvers may be employed depending upon the species and size of animal to be killed.

A captive bolt pistol may be used for pre-stunning domestic animals in agricultural or veterinary research where the meat may be salvaged after exsanguination. The use of the captive bolt pistol should be undertaken only by trained experts. Following shooting with this instrument, the major blood vessels should be severed and the animal exsanguinated.

The electrical method of killing is still in use and should involve two phases: the first shock passing through the brain (stunning the animal), the second fibrillating the heart (killing the animal).[6] In the Huelec cabinet, designed for killing dogs, for example, stunning is by means of a 500-v shock through the brain via a pair of electrodes attached to the ears, followed by a lethal shock at 1 kV passing from ear to hind leg electrode.[47] The AVMA, however, considers electrocution as a euthanasia method hazardous to personnel and difficult to administer. It is interesting to note that the World Society for the Protection of Animals has criticized the electrical stunning tanks being used in India to kill frogs.[48] The criticism relates to the concentration of frogs in the tank which allows the majority of frogs to recover from the stun before being killed. An electrical unit developed for killing lobsters has not been widely accepted.[49]

A fairly new euthanasia method is the use of microwaves, albeit not using a commercial home machine. This method is still under investigation; however, units are available for humane killing of rats and mice (Stoelting Co., 1350 South Kostner Ave., Chicago, IL 60623-1196, U.S.A.) It is hoped that it can be improved sufficiently to serve as an alternative to stunning, as the rapid increase of temperature in the brain to 45°C produces instant unconsciousness. This must be achieved without burning the skin, the eyes, or bone.

One of the most controversial methods of euthanasia, and one which is being debated at the present time, is the use of the guillotine for decapitating rats or rabbits.[50] The AVMA Report demands prior use of an anesthetic, based primarily on a 1975 study by Mikesa and Klemm.[51] The latter recently defended this paper.[52] The CCAC at the present time believes decapitation in small rodents is humane, stating that the method causes immediate cessation of the "blinking" reflex and a flattening of the EEG. It notes that studies have indicated that the effect is almost as rapid as intravenous injection of a barbiturate, citing Carney and Walker.[53] However, as in the AVMA report, it recently recommended that, unless contrary to the demands of the study, light anesthesia or sedation be given prior to administration of either guillotining or cervical dislocation.

Another recent action of interest was that of the World Society for the Protection of Animals (WSPA), which came out against decapitation as a means of killing amphibians and reptiles. Although considered humane in some mammals, WSPA says the differences in physiologies make the method questionable in amphibians and reptiles. It adds that it "cannot even recommend humane killing methods (for these species) with any degree of confidence."[54] It is worth noting that many experts in the study of reptiles and amphibians approve the use of cold for anesthesia of these animals. Subsequent freezing or decaptiation is not considered painful.

The CCAC has come out strongly against the use of high-altitude decompression as a method of euthanasia because animals with an upper respiratory problem or gastrointestinal disease may have difficulty venting the gases. Such criticisms have led to the demise of the once popular decompression chambers which were widely used by pounds and humane societies.

One method that has also come under recent criticism is the "punctilla" method used by matadors to dispatch the bull and by some Australian farmers in slaughtering sheep and cows. Australian researchers comparing this method with decapitation and captive bolt demonstrated that 80% severance of the spinal cord as a killing method was inhumane, as the animals (lambs) were "sensible and distressed until the major blood vessels of the neck were severed."[55] The authors found the method so inhumane that they recommended that no further similar experiments be conducted. In decapitation, the authors noted reflex actions in the severed head, gasping movements, eyelid movements, and body movement even after the carotid artery was severed. In contrast, death was considered instantaneous with the captive bolt pistol.

With regard to slaughter of livestock, an interesting recent development has been the scientific ratification of the fact that when cattle were hung upside down they "struggled less and the procedure became less distressing." According to Schoental, an explanation of the phenomenon became possible only after the discovery of endorphins, the endogenous polypeptides having an opioid morphine-like action. He notes that the fact that animals placed in an upside-down position appeared catatonic has been used for some time "even by scientists when injecting experimental animals."[56]

The use of rubbing boards as a means of calming chickens during the slaughtering procedure was observed by the author in 1984. In this method, the chicken passes through a "rubbing-board", a 6-ft piece of plastic which firmly touches the breast and wings. In passing through this device, only 1 in a group of 20 birds raised its head; 1 in a group of 14 squawked. However, the remainder lay quietly with their heads down and passed through the device, which cut vessels on one side of the head-neck region. It is not recommended for younger birds. Schoental notes again that these endogenous opioids appear responsible for the stress-induced analgesia which occurs in a number of species.[56]

Animals should never be allowed to witness the death of other animals; even exposure to a room recently used to kill animals may produce anxiety. It has been stated that "some animals being euthanized may emit distress vocalizations or stress pheromones even when unconscious and feeling no pain, that triggers stress reactions in other animals."[2] Clifford has stated that witnessing the death of a fellow creature may cause a fivefold increase in corticosterone in rats.[4] I have been in animal pounds where thousands of animals have been killed using an electrothanator, and I have seen dogs come in and lie down, and show no apparent awareness that other dogs have died there. However, we cannot get into the brain of the animal to see what it is thinking. Therefore, until we know more about awareness and thinking in animals and have evidence that animals witnessing euthanasia are not distressed, we should give them the benefit of the doubt and try to reduce the possibility of anxiety and painful stress to a minimum.

III. CONCLUSIONS

It behooves us all to ensure our "house is in order", and that everything possible is being done to encourage the humane treatment of the animals who do so much to improve our lives. It is hypocritical to provide and demand humane care and treatment of animals, and then cause the animal anxiety, distress, and pain by the method used to kill it.

We have hidden behind the euphemism, the tender term, "euthanasia", rather than the harsher and more realistic term "killing". When we take an animal's life, we must ensure that we have applied at least some research into one of the 3 Rs: refinement. This may be achieved by such simple practices as gentling, proper handling, a comfortable environment, and tender loving care. We cannot deny the need for more research into the production of an acceptable, comfortable death. When an animal is euthanized, it is the human animal who has made the decision, and that decision must be justified.

REFERENCES

1. **Urquhart, R.,** Euthanasia: A necessary evil, in *Proc. First Canadian Symp. Pets & Society,* Loew, F. M., McCutcheon, E., McWilliam, A., and Rowsell, H.C., Eds., Canadian Federation of Humane Societies, Ottawa, Canada, 1976, 81.
2. **Anon.,** Euthanasia: Providing a good death, in *The Biomedical Investigator's Handbook for Researchers Using Animal Models,* Loew, F. M. et al, Eds., Foundation for Biomedical Research, Washington, DC, 1987, chap. 4.
3. **Clifford, D. H.,** Preanesthesia, anesthesia, analgesia, and euthanasia, in *Laboratory Animal Medicine,* Fox, J. G., Cohen, B. J., and Loew, F. M., Eds., Academic Press, New York, 1984, chap. 18.
4. **Owens, C. E., Davis, R., and Smith, B. H.,** The psychology of euthanising animals: the emotional components, *Int. J. Stud. Anim. Prob.,* 2(1), 19, 1981.
5. **American Veterinary Medical Association,** 1986 Report of the AVMA Panel on Euthanasia, *JAVMA,* 188(3), 252, 1986.
6. *Guide to the Care and Use of Experimental Animals,* Vol. 1, Canadian Council on Animal Care, Ottawa, 1980.
7. *Euthanasia of Unwanted, Injured or Diseased Animals or for Educational or Scientific Purposes,* Universities Federation for Animal Welfare, Hertfordshire, U.K., 1986.
8. **Balls, M.,** The OTA report — a critical appraisal, *ATLA,* 14(4), 289, 1987.
9. **Rowsell, H. C.,** Euthanasia: the final chapter, in *Proc. Second Canadian Symp. on Pets & Society,* Standard Brands Food, Toronto, 1979, chap. 18.
10. **Rowsell, H. C.,** Assessment of harpooning as a humane killing method in whales, Report to the International Whaling Commission, 1979.
11. **Rowsell, H. C.,** Harpoon trauma, *Can. Rev. Lab. Med.,* 2(Abstr.), 11, 1980.
12. **Rowsell, H. C.,** Sealing operation — Gulf and Front 1979, Report to Committee on Seals and Sealing, 1979.
13. **Rowsell, H. C.,** Observations on the harp seal hunt — The Front 1980, Report to Committee on Seals and Sealing, 1980.
14. **Rowsell, H. C., Ritcey, J., and Cox, F.,** Assessment of pain and distress caused by vertebrate pesticides, Research Symp. Ontario Pesticide Advisory Committee, Toronto, 1980.
15. **Rowsell, H. C., Ritcey, H., Cox, F.,** Assessment of effectiveness of trapping methods in the production of a humane death, in *Proc. Worldwide Furbearer Conf.,* Vol. 3, Worldwide Furbearer Conference, 1980, 1647.
16. **Rowsell, H. C.,** Research for development of comprehensive trapping systems (Snare study, part 1), a report to the Federal-Provincial Committee for Humane Trapping, February, 1981.
17. **Rowsell, H. C.,** Killing systems: husbandry guidelines, Canadian Federation of Humane Societies Fur Farm Seminar, Toronto, 1981.
18. **Rowsell, H. C.,** Pre-slaughter stunning methods in Toronto abbatoirs, a report to the Humane Slaughter Review Committee of Agriculture Canada, 1979.
19. **Anon.,** New euthanasia guidelines adopted, *Can. Vet. J.,* 28(8) A2, 1987.
20. **Canadian Veterinary Medical Association Humane Practices Committee,** Statement on the euthanasia of dogs and cats, *Can. Vet. J.,* 19, 164, 1978.
21. **Rollin, B. E.,** Euthanasia and Moral Stress, in *Suffering: Psychological and Social Aspects in Loss, Grief, and Care,* **DeBellis, R. et al.,** Eds., Haworth Press, New York, 1986.
22. **Burns, H. F.,** Canada allows resumption of seal hunting from ships, *New York Times,* p. 1, March 25, 1987.
23. **Anon.,** Government finally kills seal hunt, *Ottawa Citizen,* p. A1, December 31, 1987.
24. **Loh, J.,** Replacing clubs with cameras at sealing grounds, *Sunday Telegram (Worcester, Leeds),* p. 12, October 18, 1987.
25. **Dorsey, W.,** (Letters) Thoughts on euthanasia, *JAVMA,* 191, 1252, 1987.
26. **Rowsell, H. C. and McWilliam, A. A.,** Anguish and grief: helping the aged to cope with the loss of a pet, in *Proc. Third Symp. Pets in Society,* Pet Food Manufacturers Association of Canada, Toronto, 1982, chap. 8.
27. **Thomas, C. J.,** Client relations: dealing with grief, *Vet. Tech.,* 8(8), 406, 1987 (Continuing Education. Article #3).
28. **Russell, W. M. S. and Burch, R. L.,** *The Principles of Humane Experimental Technique,* Charles C Thomas, Springfield, IL, 1959.
29. **Griffin, D. R.,** *The Question of Animal Awareness. Evolutionary Continuity of Mental Experience,* Rockefeller University Press, New York, 1981.
30. **Dawkins, M. S.,** *Animal Suffering. The Science of Animal Welfare,* Chapman & Hall, London, 1980.
31. **Dawkins, M. S.,** *Unravelling Animal Behaviour,* Longman Group, Harlow, U.K., 1986.
32. **Fraser, A. F.,** *Farm Animal Behaviour,* Balliere Tindall, London, 1974.
33. **Walzack, P. D. and Melzack, R. Eds.,** *Textbook of Pain,* Churchill Livingstone, Edinburgh, 1984.
34. **Morton, D. B. and Griffiths, P. H. M.,** Guidelines on the recognition of pain, distress and discomfort in experimental animals and an hypothesis for assessment, *Vet. Rec.,* 116(16), 431, 1985.
35. **Flecknell, P. A.,** The relief of pain in laboratory animals, *Lab. Anim.,* 18, 147, 1984.

36. **Kitchell, R. L. and Erickson, H. H. Eds.,** *Animal Pain: Perception and Alleviation,* American Physiological Society, Baltimore, MD, 1983.
37. **Hampson, J.,** *Pain and Suffering in Experimental Animals in the United Kingdom,* Royal Society for the Prevention of Cruelty to Animals, Causeway Horsham, West Sussex, 1983.
38. **Anon.,** EEGs used to establish brain death in children, *Med. Post,* September 15, 1987; as cited in *Neurology,* June 1987.
39. **Anon,** Euthanasia symposium highlights, *Am. Humane Mag.,* 64(12), December, 1976.
40. **Anon,** High altitude euthanasia not recommended, *Am. Humane Shoptalk.,* 2(2), 1, March/April, 1984.
41. **Gärtner, K., Buttner, D., Dohler, K,. Friedel, R., Lindena, J., and Trautschold, I.,** Stress response of rats to handling and experimental procedures, *Lab. Anim.,* 14, 267, 1980.
42. **Landi, M., Krieder, J. W., Lang, C. M., and Bullock, L. P.,** Effect of shipping on the immune functions of mice, in *Proc. ICLAS/CALAS Symp. The Contribition of Laboratory Animal Science to the Welfare of Man and Animals: Past, Present and Future,* Archibald, J., Ditchfield, J., and Rowsell, H. C., Eds., Gustav Fischer Verlag, Stuttgart, 1985, pg. 11.
43. Ontario Ministry of Agriculture and Food, Memo to pound operators and veterinarians in Ontario, August 12, 1987.
44. **Britt, D. P.,** The humaneness of carbon dioxide as an agent of euthanasia for laboratory rodents, in *Euthanasia of Unwanted, Injured or Diseased Animals or for Educational or Scientific Purposes,* Universities Federation for Animal Welfare, Hertfordshire, U.K., 1987, Sect. I, p. 19.
45. *Registry of Toxic Effects of Chemical Substances,* Publ. No. 83-107, National Institute of Occupational Health and Human Services, Washington, DC, 1981-82.
46. *Merck Veterinary Manual,* 6th ed., Merck, Rahway, N.J., 1986, 942.
47. **Bousfield, W. E. D.,** (Letters) Electrocution cabinets, *Vet. Rec.,* 112(25), 593, 1983.
48. **Anon,** Frog's legs in Calcutta, *Agscene,* 9, December, 1986.
49. **Baker, J. R. and Dolan, M. B.,** Experiments on the humane killing of lobsters (*Humarus vulgaris*) and crabs (*Cancer pagurus*). II. The exposure of lobsters to electric shock before boiling, *Sci. Pap. Humane Educ. Cent.,* 2, 1, 1975.
50. **Allred, J. B. and Berntson, G. C.,** Is euthanasia of rats by decapitation inhumane?, *J. Nutr.,* 116, 1859, 1986.
51. **Mikeska, J. A. and Klemm, W. R.,** EEG evaluation of humaneness of asphyxia and decapitation euthanasia of the laboratory rat, *Lab. Anim. Sci.,* 25(2), 175, 1975.
52. **Klemm, W. R.,** Correspondence, *Lab. Anim. Sci.,* 37(2), 148, 1987.
53. **Carney, J. A. and Walker, B. L.,** Mode of killing and plasma corticostorone concentration in the rat, *Lab. Anim. Sci.,* 23, 675, 1973.
54. **Anon,** Euthanasia of reptiles and amphibians, *Anim. Int.,* 19, 7, 1986.
55. **Tidswell, S. J., Blackmore, D.K., and Newhook, J. C.,** Slaughter methods: electroencephalographic (EEG) studies on spinal cord section, decapitation and gross trauma of the brain in lambs, *N.Z. Vet. J.,* 35, 46, 1987.
56. **Schoental, R.,** Endogenous opioids and slaughter-induced stress, *Vet. Rec.,* 119(9), 223, 1986.

Appendices

APPENDIX I

PANEL REPORT OF THE AVMA COLLOQUIUM ON RECOGNITION AND ALLEVIATION OF ANIMAL PAIN AND DISTRESS

Reprinted from JAVMA, Vol. 191, No. 10, November 15, 1987, with kind permission of the American Veterinary Medical Association.

Few premises are more obvious than that animals can feel pain, fear, loneliness, boredom, and other forms of distress. However, systematic scientific study of those phenomena has been undertaken only relatively recently. Divergent viewpoints reflect the inherent difficulty of developing objective methods for studying the experience of pain and distress by animals and a reluctance to accept anthropomorphic (ie, having human attributes) extrapolations as a legitimate source of scientific data. Recently, societal concerns about animal welfare have highlighted the need to develop consensus standards that take into account information from all available sources including anthropomorphic observations and scientific data available from biological and behavioral disciplines. In the final analysis, pain and distress must be understood on the basis of laboratory and behavioral data as well as from an anthropomorphic consideration.

The increased ethical concern on the part of society about the treatment of animals and the identification of society's responsibility to animals have given considerable impetus to the initiation of a program of study dealing with recognition and alleviation of animal pain and distress.

Animals can be used to good effect to study human pain and its management, and feelings that human beings experience as hurtful and noxious can logically be ascribed to animals. Such an attitude corresponds with the ever growing societal concern that pain and distress arising from human use of animals can be eliminated, alleviated, ameliorated, or controlled.

APPROPRIATENESS

The willingness of the veterinary medical profession to address the question of animal pain and distress is clearly illustrated in the convening of the *Colloquium on Recognition and Alleviation of Animal Pain and Distress*. The veterinarian's role requires ongoing evaluation on these important subjects starting with priority recommendations given by this panel. With the privilege of working with animals comes a responsibility—as is illustrated in the Veterinarian's Oath:

> Being admitted to the profession of veterinary medicine, I solemnly swear to use my scientific knowledge and skills for the benefit of society through the protection of animal health, the relief of animal suffering, the conservation of livestock resources, the promotion of public health, and the advancement of medical knowledge.
>
> I will practice my profession conscientiously, with dignity, and in keeping with the principles of veterinary medical ethics.
>
> I accept as a lifelong obligation the continual improvement of my professional knowledge and competence.

Society accepts the use of animals for food production, biomedical research, assurance of public health, companionship, and education. Therefore, the veterinary profession has the obligation of offering leadership in thoughtful and humane animal use. Society has assigned to the veterinary profession the moral stewardship of animals. Therefore, the veterinary profession is compelled to use scientific advances for the benefit of animals and human beings.

Much of our present biomedical knowledge has been extrapolated from one species to another—including data from human beings to other animals. Pharmacologic, physiologic, and behavioral principles are formed from quantitative data and anthropomorphic considerations. Inferences on pain and distress made on the basis of experience, clinical assessment, and intuitive judgements also can be sound, if not objectively verifiable.

WORKING DEFINITIONS

Avoidance, escape, or control of pain and distress are among the most important responses an animal, including a human being, will make to survive, to adapt to its environment, or to readapt to a changed environment. These homeostatic mechanisms may be covert, ie, internal physiologic or psychologic changes, or they may be overt, eg, expressed behaviorally. Such responses to achieve and maintain homeostasis may be short-term, or in cases of chronic pain or distress, they may become an integral part of the animal's physiologic and behavioral state.

Clinical signs of pain and distress appear to be different in different animal species. Therefore, recognition and differentiation of pain and distress are complicated. Determining precisely whether an animal is in a state of discomfort, pain, or distress or is suffering is difficult. Despite this difficulty, many clinical signs of pain and distress are shared by many animal species, and the tendency to highlight differences rather than similarities may make the task of recognizing these syndromes more complex than necessary.

Many terms involve internal mental states that are impossible to distinguish with precision, even in human beings. However, this lack of precision does not mean that certain domains of experience and perception that are generally recognized between individuals and across species lines, cannot be determined. Animal models of pain or anxiety have proven useful in investigating neurologic and neurochemical elements that seem to mediate pain or anxiety in human beings. Therefore, it seems appropriate that on the basis of analogy of human pain or anxiety, animals are subject to similar emotional phenomena and that some degree of anthropomorphism is essential to a discussion of these terms.

A variety of terms are in common colloquial use such as suffering, pain, anxiety, fear, distress, discomfort, and injury. These terms refer to phenomena that are different, yet may overlap in varying degrees.

Discussion of pain, distress, anxiety, and suffering is hampered by the lack of clear understanding of what these terms mean. To address such problems, the International Association for the Study of Pain drew up a definition of terms relevant to pain experience and research.[1] It is not possible to define terms and to develop a classification of syndromes that are acceptable to everyone; however, even if the present definitions and classifications are not perfect, they represent a starting point.

Adaptation of a simple classification scheme does not mean that these definitions are fixed. These terms will be evaluated and modified as new knowledge is acquired and further understanding of these processes is gained. These working definitions are guidelines to help recognize pain, distress, anxiety, fear, and suffering in animals and to facilitate a reasoned discussion of these phenomena. These definitions should form the basis for a selection of appropriate pharmacologic and nonpharmacologic approaches for alleviation of pain, distress, anxiety, fear, and suffering.

Pain—Pain occurs in human beings and in other animals, and it is clearly understood to have evolutionary survival value. Sharp acute pain signals the need to take rapid, evasive action to escape injury, whereas dull or chronic pain leads to rest and other recuperative behaviors.

Pain was defined by the International Association for the Study of Pain[1] as an unpleasant sensory and emotional experience associated with actual or potential tissue damage. Pain is a perception that depends on activation of a discrete set of receptors (nociceptors) by noxious stimuli, eg, thermal, chemical, or mechanical. Further processing in neural pathways, eg, spinal cord, brain stem, thalamus, and cerebral cortex, enables noxious stimuli to be perceived as pain.

Pain perception varies according to site, duration, and intensity of the stimulus and can be modified by previous experience, emotional states, and perhaps innate individual differences.

A pain-detection threshold can be established as the lowest intensity of a stimulus that is perceived as painful or that induces a response. A pain-tolerance threshold is the highest intensity of a stimulus that will be tolerated voluntarily. An animal may perceive a hypodermic puncture as painful, but may tolerate the injection if it is coupled with a food reward. Seemingly, the pain-detection threshold is uniform across species lines, whereas the pain-tolerance threshold may be more species-specific and subject to modification.

Anxiety and fear—Anxiety and fear may have the same evolutionary benefit as pain. Anxiety is activated by novel stimuli to increase the state of the animal's awareness. When placed in an unfamiliar environment, a rat initially will become motionless before cautiously beginning to explore its new surroundings. Anxiety can be defined as an emotional state involving increased arousal and alertness prompted by an unknown danger that may be present in the immediate environment. Fear can be defined similarly, except that fear would refer to an experienced or known danger in the immediate environment. Thus, anxiety appears to be a generalized, unfocused response to the unknown, and fear is a focused response to a known object or previous experience. A dog may tremble in a veterinarian's examination room during the first visit because of anxiety about what will happen. On the second visit, the dog may whine or try to escape from fear of a remembered event.

Stress—Stress can be defined as the effect of physical, physiologic, or emotional factors (stressors) that induce an alteration in the animal's homeostasis or adaptive state. The covert or overt response of an animal to such stressors may be seen as adaptive. This adaptive response acts to return the animal to a base-line behavioral and physiologic state. The response to stress often involves changes in the neuroendocrinologic function, in the autonomic nervous system, and in the mental state of the animal, as well as in its behavior. The response of the animal may vary according to its experience, sex, age, genetic profile, and physiologic and psychologic state.

Stress and subsequent responses may be categorized in 3 ways. Neutral stress is not in itself harmful to an animal and evokes responses that neither improve nor threaten the animal's well-being. Eustress involves environmental alterations that in themselves are not harmful to the animal but initiate responses that may, in turn, have potentially beneficial effects. Distress is a state in which the animal is unable to adapt to an altered environment or to altered internal stimuli. If such stressors are short-term, responses an animal will make to adapt to these changes do not usually, but may, result in long-term harmful effects. Prolonged or excessive distress may result in harmful responses, eg, abnormal feeding and social interaction behavior, inefficient reproduction, and can result in pathologic conditions, eg, gastric and intestinal lesions, hypertension, immunosuppression. Distress also may be induced through changes in internal states such as disease, nausea, excessive anxiety, and fear. Such responses may become a permanent part of the animal's repertoire and seriously threaten the animal's well-being.

Suffering—Suffering is a much used and abused colloquial term that is not defined in most medical dictionaries. Neither medical nor veterinary curricula explicitly address suffering or its relief. Therefore, there are many problems in attempting a definition. Nevertheless, suffering may be defined as a highly unpleasant emotional response usually associated with pain and distress. Suffering is not a modality such as pain or temperature. Thus, suffering can occur without pain, and although it might seem counterintuitive, pain can occur without suffering.

Suffering as well as other phenomena (eg, pain, fear, anxiety) range over a spectrum of effects, from mild to severe. Modulation of suffering may involve removing or changing some aspect of the internal or external environment or giving the animal the opportunity to avoid, escape, or control some aspect of that environment.

Comfort—Comfort is a state of physiologic and behavioral homeostasis in which the animal has adapted to its environment and has normal feeding, drinking, activity, sleep/wake cycles, reproduction, and social behavior. The behavior of such an animal remains relatively constant without remarkable fluctuation.

Discomfort—Discomfort represents a minimal change in the animal's adaptive level or homeostasis as a result of changes in its environment because of biological, physical, social, or psychologic alterations. Physiologic or behavioral changes may be observed in the animal, but these changes do not deviate markedly from the animal's previous behavior to indicate that the animal is experiencing pain or distress.

Injury—Injury is biological, chemical, or physical damage to tissue, which may be permanent or partly or completely reversible. The extent of injury can vary from having little to no effect on the normal function of an animal to marked impairment of the animal's function, thereby altering the state of the animal.

PHYSIOLOGIC BASIS OF ANIMAL PAIN AND DISTRESS

Human beings and animals interact—in a veterinary practice, in a laboratory, in a food production operation, and in the home. Therefore, anthropomorphic judgments are inevitable and often necessary to evaluate the animal's state. Nevertheless, these judgments must be made with appropriate caution. Judgments about human-animal interactions also depend on physiologic and behavioral observations that are independently verifiable. On the basis of human perceptions and language, the terms pain and distress can be used to describe different categories of aversive perceptions in animals.

In animals and human beings, anatomic physiologic substrates, underlying the perception of noxious stimuli, are basically the same. Neurotransmitter systems and physiologic mechanisms such as automatic responses, neuroendocrinologic changes, and alerted EEG responses also are similar. Behavioral responses are similar; animals avoid noxious stimuli that are painful to human beings. Threshold measures (eg, temperature, skin pressure) are comparable for aversive behaviors.

In animals, behavioral and physiologic responses to noxious stimuli are modified by drugs in ways that are qualitatively and often quantitatively similar to modifications seen in human beings. Pain varies on a continuous scale ranging from perception threshold to intensities that human beings described as intolerable. The intensity of pain perceived by animals should be judged by the same criteria that apply to its recognition and to its physiologic and behavioral observations in human beings. Seemingly, if a given condition causes pain in human beings it probably causes pain in other animals.

Qualitative and quantitative similarities in perception of distress are difficult to define in human beings as well as in animals. Nevertheless, animals do exhibit clinical signs of physiologic and behavioral distress, revealing that animals experience distress that manifests itself on a spectrum similar to that for pain.

ALLEVIATION OF ANIMAL PAIN AND DISTRESS

Obviously, the best mechanism to alleviate pain and distress is prevention. When prevention is not possible, alleviation is the next best alternative. However, alleviation of pain or distress does not necessarily mean elimination. The optimum method of alleviation depends on the source of the problem. Changes in the environment or treatments of disease states are remedies that can be applied when states of pain or distress are recognized. Pharmacologic measures can be applied in intraoperative and postoperative periods.

Many pharmacologic agents currently are available for the alteration of animal pain and distress. However, comparative data about kinetics, effectiveness, and deleterious effects of these drugs, when used in domestic, exotic, and laboratory animals, are limited.

Mechanisms of action of the major analgesics, tranquilizers, and anesthetics are fairly well elucidated. However, considerable differences in the deposition of these drugs in different species and in individual animals are known. A potential danger exists when pharmacologic data are extrapolated from one species to another, eg, the duration of a therapeutic or clinical dose of phenylbutazone in human beings is markedly different from that in horses. Veterinarians must be familiar with pharmacologic actions of these drugs and must use the most effective drug available. Unfortunately, most drugs available for the alleviation of pain are not approved by the Food and Drug Administration for use in all species.

Pain and distress associated with surgery can be alleviated. Unfortunately, when balanced anesthesia and/or muscle relaxant drugs are used, many indicators of pain and distress are lost during and after surgery.

The recognition of pain in animals is sometimes difficult. Questions as to whether the animal is experiencing pain should be resolved in favor of the animal with due recognition that any drug treatment can have deleterious effects. The important issue of pain and distress in human and animal neonates still needs to be evaluated.

EXPERIMENTAL SITUATIONS

Observations from animal experiments have contributed to recognition and subsequent alleviation of pain and distress in human beings. In animals, pain and distress may be unintended side effects of experimental design unrelated to the study's purpose. Pain and distress resulting from procedures such as chronic restraint, antibody production methods, toxicologic studies, infectious disease research, and deprivation studies should be minimized.

In toxicologic and carcinogenic research, animals may have to be euthanatized for humane and scientific reasons before they become moribund or develop unacceptably large tumors. Data obtained from moribund animals may be confounded by the existence of secondary conditions (eg, shock or malnutrition.) Relatively few investigations deal directly with elucidating mechanisms of pain or in which pain is an end point. Rarely do experiments have to be conducted without providing the animal an opportunity to avoid or to control aversive stimuli.

Considerable thought has been given to alleviation of animal pain and distress in guidelines issued by scientific societies and by the federal government, but more can be done. More specific guidelines seem to be required to define acceptable end points for toxicologic, carcinogenic, and infectious disease research. Guidelines that take specific account of the scientific as well as the humane considerations of laboratory animal research are of enormous benefit to investigators, animal care committees, and laboratory animal veterinarians.

FUTURE DIRECTIONS AND IMPLICATIONS

The Veterinarian's Oath charges graduates to use their skills and knowledge for the relief of animal suffering. This charge is no longer an option, but is a clearly defined obligation.

Society regards the veterinary profession as the guardian of animal health and well being. This stewardship is manifested in new federal laws and regulations that place specific responsibility on laboratory animal veterinarians for the health and welfare of animals used in research, testing, and education. The veterinary profession recognizes that public concern is not confined solely to the welfare of laboratory animals. Thus, new standards of practice must be applicable to all areas of veterinary medicine.

Adaptation to new standards that are evolving will require new scientific information and additional training, with more intensive emphasis on the recognition of animal pain and distress. The time seems to be ripe to develop a data base for analgesic, tranquilizer, and anesthetic drugs, including doses, side effects, and contraindications in all species. This data base should be implemented in a form (possibly a computerized data base) that can be easily accessed and updated. In the laboratory, there may be some reluctance to use anesthetic or analgesic drugs for fear that these drugs will interfere with the purpose of a given experiment. Wider dissemination of the relevant information would be useful in confirming or negating such fears.

RECOMMENDATIONS

For the veterinary profession—In addition to being committed to the relief of animal suffering, veterinarians must continue to be sensitive to situations in which animals may be in pain or distress. Private practitioners also must continually reassess preoperative, intraoperative, and postoperative procedures to ensure that pain is prevented, lessened, or abolished. Anesthesia always should be adequate to prevent animal awareness.

The veterinary profession should establish standards of accepted practices against which new methods and techniques can be measured. Standards must be established for the use of neuromuscular blocking agents and criteria for the assessment of pain and the recognition of the need for anesthetic agents and analgesia. Also, the best way to achieve these standards must be explored. Publication of such standards could serve as a guide to private practitioners, research workers, and the academic community.

Procedures commonly performed without anesthetic or analgesic agents, such as docking of tails in lambs and pups, castration, dehorning, and hot-iron branding must be reconsidered. New guidelines must be developed that take into consideration the age of the animal, the animal's threshold of pain, and the animal's tolerance of pain. Although objective data and methods may not be available to assess quantitatively the extent of pain or distress inherent in some of these practices in veterinary medicine, every effort should be made to increase the practitioner's awareness that procedures painful to human beings are probably painful to animals. Alternative procedures should be developed, and pharmaceutic agents should be administered when appropriate. Although all pain cannot, and perhaps even should not, be eliminated, the veterinary profession should provide leadership and guidance to ensure that animal pain and distress are minimized.

For the academic community—Schools and colleges of veterinary medicine should give appropriate attention to curricula that emphasize the recognition and alleviation of animal pain and distress. The field of laboratory animal medicine clearly needs additional emphasis. Laboratory animals must be given the status of the target species rather than simply be considered animal model species. Ethical understanding of, and concern with, issues relevant to pain should be incorporated into the veterinary curricula.

The veterinary profession is in the best position to educate animal technicians to evaluate objectively pain and distress in animals and to recognize conditions that generate unwanted stress. Often the animal technician is the person who has most daily contact with individual animals and therefore is most likely to recognize behavioral abnormalities that signal pain or distress.

For the teaching and research community—The use of animals in teaching and research is clearly under intense public review. The training of undergraduates and professionals alike must include animal welfare considerations. Students who intend to pursue research careers must be made aware of their responsibilities to promote animal welfare in their care and use of laboratory animals to guarantee the continued privilege to use animals in research settings.

Teaching and research communities should include attention to the concepts of refinement, reduction, and even replacement of animals where appropriate.[2] Alternative teaching methods, such as the use of videotaped demonstrations to replace living animals should be considered. The need for animals, the degree of invasiveness, and the intended audience for teaching demonstrations should be reassessed continuously.

Research investigators should reexamine their animal research designs to determine whether they will induce unintended pain or distress, and to minimize those aspects of the design when they cannot be eliminated. These investigators also should foster a spirit of respect for animals among their students, by precept and by example. Only in this way, will it be possible for the biological community to retain the high degree of public respect that is necessary to preserve continued support for the use of animals.

For laboratory animal medicine—The special responsibility of laboratory animal veterinarians has been underscored through the enactment of federal legislation regarding animal welfare. In response, laboratory animal veterinarians must make additional efforts to develop and disseminate the scientific information that is required. These veterinarians must take a leadership role to assist investigators to refine experimental designs when possible and to encourage sensitivity to animal pain and distress in research.

The laboratory animal veterinarian is in the best position to examine questions about the nature, magnitude, and necessity for pain and distress in the laboratory setting and to minimize unwanted sources of problems in animal husbandry. Laboratory animal veterinarians should maintain a collaborative and cooperative relationship with their research colleagues to foster optimum conditions of animal care and use. The laboratory animal veterinarian is an ideal source of information about drugs and research techniques that should be disseminated widely within and between institutions.

To pharmaceutic industry and regulatory agencies—Cooperation between the pharmaceutic industry and the FDA is essential to facilitate the development and approval of agents that can alleviate animal pain and distress. Manufacturers can be encouraged to submit applications for existing compounds that could be approved and to develop new compounds with better therapeutic advantages. The market for such products is clearly growing.

The FDA should be encouraged to give special status to abbreviated new animal drug applications for compounds targeted for relief of pain and distress in laboratory animals. This consideration eventually may need to be extended to other classes of animals as the need is perceived and the standard of veterinary practice in the private sector changes.

Regulatory agencies that inspect the use of animals in laboratory settings should understand the limitations of drug use in these situations, given the limited availability of data on drug use in various species. Until sufficiently reliable data are available, guidelines should be established by the research community, and regulatory supervision should be exercised with reliance on the professional judgment of laboratory animal veterinarians who provide immediate oversight of the projects in question. Regulatory agencies should not demand extra-label use of compounds in violation of existing laws and federal regulations.

For the Council on Biologic and Therapeutic Agents (COBTA) of the AVMA—The COBTA should work closely with the FDA to pursue the aforementioned recommendations, ie, to encourage FDA to give special (fast track approval) status to abbreviated new animal drug applications for those compounds that could have application in alleviating pain or distress in animals, especially on a long-term basis.

REFERENCES

1. Merskey H. Pain terms: a list with definitions and notes on usage. Pain 1979;6:249-252.
2. Rullell WMS, Burch RL. *The principles of humane experimental technique*. London: Methuen & Co, Ltd, 1959.

PANEL MEMBERS

Hyram Kitchen, *Chairman of the Panel*
College of Veterinary Medicine, University of Tennessee, Knoxville
Arthur L. Aronson, *Chairman of the Colloquium*
School of Veterinary Medicine, North Carolina State University, Raleigh
James L. Bittle
Scripps Clinic and Research Institute, LaJolla, Calif, *COBTA*
Charles W. McPherson
School of Veterinary Medicine, North Carolina State University, Raleigh
David B. Morton
University of Leicester, United Kingdom
Stephen P. Pakes
Southwestern Medical School, University of Texas, Dallas
Bernard Rollin
Colorado State University, Fort Collins
Andrew N. Rowan
School of Veterinary Medicine, Tufts University, Boston, Mass
Jeri A. Sechzer
Cornell University Medical College, White Plains, NY
Jack E. Vanderlip
University of California, LaJolla
James A. Will
University of Wisconsin, Madison
Ann Schola Clark
Publications Division, AVMA, Schaumburg, Ill
Joe S. Gloyd
Scientific Activities Division, AVMA, Schaumburg, Ill

APPENDIX II

1986 REPORT OF THE AVMA PANEL ON EUTHANASIA

Reprinted from JAVMA, Vol. 188, No. 3, February 1, with kind permission of the American Veterinary Medical Association.

MEMBERS OF THE PANEL

A. W. Smith, DVM, PhD (Chairman), College of Veterinary Medicine, Oregon State University, Corvallis, OR 97331
K. A. Houpt, VMD, PhD, New York State College of Veterinary Medicine, Cornell University, Ithaca, NY 14853
R. L. Kitchell, DVM, PhD, School of Veterinary Medicine, University of California, Davis, CA 95616
D. F. Kohn, DVM, PhD, University of Texas Medical School at Houston, PO Box 20708, Houston, TX 77225
L. E. McDonald, DVM, PhD, College of Veterinary Medicine, University of Georgia, Athens, GA 30601
Martin Passaglia, Jr., MS, Executive Director, The American Humane Association, 1979-1985
J. C. Thurmon, DVM, MS, College of Veterinary Medicine, University of Illinois, Urbana, IL 61801
E. R. Ames, DVM, PhD, Staff Coordinator, American Veterinary Medical Association, Schaumburg, IL 60196

PREFACE

In 1984, at the request of the AVMA Council on Research, the Executive Board of the AVMA appointed a Panel on Euthanasia consisting of six veterinarians and one public representative. The purpose of the panel was to review and update the third Panel Report published in 1978.[1] Since 1978 the panel has become aware of a need for additional information on some aspects of euthanasia. In this report the panel has expanded the information on research uses of animals wherein euthanasia is required, on public disposal of surplus animals, and on slaughter of food animals. The panel is aware that there are euthanatizing methods and agents not discussed, but has limited this report to those methods and agents supported by reliable information. The report will be targeted primarily to veterinarians, but will be understandable to a broad segment of the general population. Although the interpretation and use of this panel report cannot be limited, the panel would remind all who refer to it that our overriding commitment is to give professional guidance for relieving the pain and suffering of animals.

INTRODUCTION

Euthanasia is the act of inducing painless death. Criteria to be considered for a painless death are: rapidly occurring unconsciousness and unconsciousness followed by cardiac or respiratory arrest. The distress experienced by people when observing euthanasia or death in any form is an emotional response dependent on the background of the observer. Kinship of people with higher animals, however distant, serves to transfer the unpleasant reaction to human death to death of animals. Such distress occurs even though the observer experiences no physical pain. This distress may be minimized by perfection of the technique of euthanasia. Although not an adequate criterion, observers may mistakenly relate any movement with consciousness and lack of movement with unconsciousness. Techniques in which animals being euthanatized exhibit little or no movement are the most aesethetically acceptable to most people.

Pain must be defined before criteria for a painless death can be established. Pain is that sensation (perception) that results from nerve impulses reaching the cerebral cortex via specific neural pathways called nociceptive pathways. The term nociceptive is derived from noxious stimuli. Noxious stimuli threaten to, or actually do, destroy tissue. They initiate nerve impulses by acting upon a specific set of receptors, called nociceptors. Nociceptors respond to excessive mechanical, thermal, or chemical energies. Endogenous chemical substances such as hydrogen ions, serotonin, histamine, bradykinin, and prostaglandins as well as electrical currents are capable of generating nerve impulses in nociceptors.

Nerve impulse activity generated by nociceptors is conducted to the spinal cord or the brainstem via nociceptor primary afferent fibers. Within the spinal cord or the brainstem the nerve impulses are transmitted to two sets of neural networks. One set is related to nociceptive reflexes and the second set consists of ascending pathways to the reticular formation, thalamus, and cerebral cortex for sensory processing. The transmission of nociceptive neural activity is highly variable. Under certain conditions, both the nociceptive reflexes and the ascending pathways may be suppressed, as, for example, in deep surgical anesthesia. In another set of conditions, nociceptive reflexes may occur, but the activity in the ascending pathways is suppressed; thus the noxious stimuli are not perceived as pain, as, for example in light surgical anesthesia. It is incorrect to use the term pain for stimuli, receptors, reflexes, or pathways because the term implies perception whereas all of the above may be active without consequential pain perception.[2-6] Pain is divided into two broad categories: (1) sensory-discriminative, which indicates the site of origin and the energy source giving rise to the pain; and (2) motivational-affective in which the severity of the stimulus is perceived and the animal's response is determined. Sensory discriminative processing of nociceptive impulses is most likely to be accomplished by brain mechanisms similar to those utilized for processing of other sensory discriminative input that provides the individual with information about the intensity, duration, location, and quality of the stimulus.

Motivational-affective processing involves the ascending reticular formation for behavioral and cortical arousal. It also involves thalamic input into both the forebrain and the limbic system for perceptions such as suffering, fear, anxiety, and depression. The motivational-affective neural networks also have strong inputs into the hypothalamus and the autonomic nervous system for reflex activities of the cardiovascular, pulmonary, and hypophyseal adrenal systems. Responses activated by these systems feed back into the forebrain and enhance the perceptions derived via motivational-effective inputs. Based upon neurosurgical experience in people, it is possible to separate the sensory-discriminative components from motivational-affective components of pain.[3] On an anatomic basis, it would appear that in animals the sensory-discriminative pathways are smaller, compared with that of people, whereas motivational-affective pathways are more numerous and more diverse in animals.[3,7] Some have speculated that animals need, more than human beings, the motivational-affective input to warn them of impending danger.

For pain to be experienced, the cerebral cortex and subcortical structures must be functional. An unconscious animal cannot experience pain because the cerebral cortex is not functioning. If the cerebral cortex is rendered nonfunctional by any means such as hypoxia, depression by drugs, electric shock, or concussion, pain is not experienced.

Stimuli that might evoke pain in a conscious animal may elicit only reflex responses manifested by movement in an unconscious animal. For this reason, nonpurposeful movements of an animal are not reliable indicators of pain perception. Conversely, an animal can experience pain, even though no body movements occur in response to noxious stimuli, if the animal is given muscle-paralyzing agents such as curare, succinylcholine, gallamine, pancuronium, nicotine, or decamethonium. These muscle paralyzing agents do not induce unconsciousness or depress the cerebral cortex or any neural mechanism involved in pain perception.

As with other procedures applied to animals, euthanasia requires some physical control over the animal. The degree of control and kind of restraint needed will be determined by the animal species, breed, size, state of domestication, presence of painful injury or disease, degree of excitement, and method of euthanasia. Suitable control is vital to minimize pain in animals, to assure safety of the person performing euthanasia, and, frequently, to protect other animals and people.

Selection of the most appropriate method of euthanasia in any given situation is dependent on species of the animal involved, available means of animal control, skill of personnel, numbers of animals, economic factors, and other considerations. This report deals primarily with domestic animals, but the same humane considerations should be applied to all species.

BEHAVIORAL CONSIDERATIONS

The facial expressions and body postures that indicate various emotional states have been described.[8-10] Behavioral and physiologic responses to noxious stimulation include distress vocalization, struggling, attempts to escape, defensive or redirected aggression, salivation, urination, defecation, evacuation of anal sacs, pupillary dilation, tachycardia, sweating, and reflex skeletal muscle contractions causing shivering, tremors, or other muscular spasms. Some of these responses can occur in unconscious as well as conscious animals. Fear can cause immobility or freezing in certain species, particularly rabbits and chickens. This immobility response should not be interpreted as unconsciousness when the animal is, in fact, conscious.[10]

In very young animals, automatic and reflexive reactions are evident, although overt behavioral reactions may differ from those of adults.

The need to minimize fear and apprehension must be considered in determining the method of euthanasia. Distress vocalizations, fearful behavior, and release of certain odors or pheromones by a frightened animal may cause anxiety and apprehension in others. Therefore, whenever possible, animals should not observe euthanasia of others, especially of their own species. This is particularly important when vocalization or release of pheromones may occur during induction of unconsciousness. Gentle restraint, preferably in a familiar environment, careful handling, and talking during euthanasia often have a calming effect on companion animals. However, some of these methods may not be operative with wild animals or animals that are injured or diseased. Where capture or restraint may cause pain, injury, or anxiety to the animal or danger to the operator, the use of tranquilizing or immobilizing drugs may be necessary.

Animals for food or fur should be euthanatized as specified by the United States Department of Agriculture.[11] Painless death can be achieved by stunning the animal before exsanguination. Animals must not be restrained in painful positions before slaughter. Preslaughter handling should be as stress-free to the animals as possible. The use of electric prods or other devices to encourage movement of animals should be eliminated. Proper design of chutes and ramps enables animals to be moved into restraining chutes without undue stress.[12,13]

The ethical and psychologic issues involved with euthanasia of diseased, injured, or unwanted animals is within the purview of this discussion. Moral and ethical imperatives associated with individual or mass animal euthanasia should be consistent with acceptable humane practice. In all circumstances, choice of method should be selected and employed with the highest ethical standards and social conscience. Many issues involving diseased, injured, or unwanted animals and the need for euthanasia have been addressed in conferences cosponsored by AVMA.[14,15]

Distress may occur among personnel directly involved in performing repetitive euthanasia of diseased, injured, or unwanted animals. At the point of terminating the life of an animal, we should be prepared not only to treat the animal but also to consider the people attached to the animal.[16]Constant exposure to or participation in euthanasia procedures can cause a psychologic state characterized by a strong sense of work dissatisfaction or alienation, which may be expressed by absenteeism, belligerence, or careless and callous handling of the animals.[17] This is one of the principal reasons for turnover of employees directly involved with repeated performance of animal euthanasia. This should be recognized as a bona fide personnel problem related to animal euthanasia, and management measures should be instituted to decrease or eliminate the potential for this condition.

MODES OF ACTION OF EUTHANATIZING AGENTS

Euthanatizing agents terminate life by three basic mechanisms: (1) hypoxemia, direct or indirect; (2) direct depression of neurons vital for life function; and (3) physical damage to brain tissue (Table 1).

Agents that produce death by direct or indirect hypoxia can act at various sites and can cause different times of onset of unconsciousness. With some agents, unconsciousness may occur prior to cessation of motor activity (muscle movement). Thus, even though animals demonstrate muscular contractions, they are not perceiving pain. The uninformed observer may find this difficult to accept.

Conversely, muscle relaxants induce a flaccid muscular paralysis and the animal remains conscious until death eventually occurs as a result of hypoxemia and hypercapnea. While the animal appears relaxed, it is actually in a state of panic and can feel pain. Outwardly this would appear to be an ideal form of euthanasia, but it is not. Agents that do not induce rapid loss of consciousness prior to death include: curare, succinylcholine, gallamine, nicotine, magnesium or potassium salts, pancuronium, decamethonium, and strychnine. *Use of any of these agents alone for euthanasia is absolutely condemned.*

The second group of euthanatizing agents depresses nerve cells of the brain first, blocking apprehension and pain perception, followed by unconsciousness and death.

Some of these agents "release" muscle control during the first stage of anesthesia, resulting in a so-called "excitement or delirium phase," during which there may be vocalization and some muscle contraction.[18] Although these responses appear to be purposeful, they are not. Death is due to hypoxemia and direct depression of respiratory centers.

With the third group, physical damage to the brain, concussion, or direct electrical flow[19] through the brain produces instant unconsciousness. Exaggerated muscular activity may follow unconsciousness. When electrocution is properly performed, loss of motor function occurs concomitantly with loss of consciousness; with the other methods in this group, muscular contraction may occur. Although this may be unpleasant to the observer, the animal is not suffering.

Electroencephalograms and other physiologic measurements can be employed to measure responses to euthanatizing agents.[1]

CRITERIA FOR JUDGING METHODS OF EUTHANASIA

Several criteria were used in evaluating methods of euthanasia: (1) ability to produce death without causing pain; (2) time required to produce loss of consciousness; (3) time required to produce death; (4) reliability; (5) safety of personnel; (6) potential for minimizing undesirable psychologic stress on the animal; (7) nonreversibility; (8) compatibility with requirement and purpose; (9) emotional effect upon observers or operators; (10) economic feasibility; (11) compatibility with histopathologic evaluation; and (12) drug availability and abuse potential.

INHALANT AGENTS

Occupational exposure to inhalation anesthetics constitutes a human health hazard. An increased incidence of spontaneous abortion and congenital abnormalities results from exposure to trace amounts of inhalation anesthetic agents.[20] Human exposure levels for volatile liquid anesthetics (ether, halothane, methoxyflurane, ethrane, and isoflurane) should be less than 2 ppm, and less than 25 ppm for nitrous oxide.[21] While there are no controlled studies proving that such levels of anesthetics are "safe," these concentrations were established because they were shown to be attainable under hospital conditions. Effective procedures must be employed to protect personnel from anesthetic vapors.

TABLE 1
Mode of Action of Euthanatizing Agents

	Site of action	Classification	Comments
	Hypoxic Agents		
Carbon moxide	Carbon monoxide combines with hemoglobin of RBC, preventing combination with oxygen	Histotoxic hypoxia	Unconsciousness occurs rapid; motor activity persists after unconsciousness
Hydrogen cyanide	Depression of oxygen transport at tissue level; no O_2 available in tissues	Histotoxic hypoxia	Unconsciousness occurs rapidly; motor activity persists after unconsciousness
Curariform drugs; curare, succinyl-choline, gallamine, decamethonium	Paralysis of respiratory muscles; oxygen not available to blood	Hypoxic hypoxemia and hypercarbia	Unconsciousness develops slowly, preceded by anxiety and fear, no motor activity
Rapid decompression	Reduced partial pressure of oxygen available to blood	Hypoxic hypoxemia	Unconsciousness occurs rapidly; motor activity persists after unconsciousness
Nitrogen inhalation	Reduced partial pressure of oxygen available to blood	Hypoxic hypoxemia	Unconsciousness occurs rapidly; motor activity persists after unconsciousness
Electrocution when current does not pass through brain	Spastic paralysis of respiratory muscles and ventricular fibrillation; oxygen not available to blood in lungs	Hypoxic hypoxemia	Unconsciousness develops slowly, occuring after violent muscle spasms
	Direct Neuron Depressing Agents		
Anesthetic gases; ether, chloroform, methoxyflurane, halothane, nitrous oxide, and enflurane	Direct depression of cerebral cortex and subcortical structures and vital centers	Hypoxemia due to depression of vital centers	Unconsciousness occurs first; no anxiety or pain; possible involuntary motor activity after unconsciousness; no motor activity after brief period
Carbon dioxide	Direct depression of cerebral cortex, subcortical structures and vital centers; direct depression of heart muscle	Hypoxemia due to depression of vital centers	Unconsciousness occurs first; no anxiety or pain; possible involuntary motor activity after unconsciousness, no motor activity after brief period
Barbituric acid	Direct depression of cerebral cortex, subcortical structures and vital centers; direct depression of heart muscle	Ultimate cause of death is hypoxemia due to depression of vital centers	Unconsciousness reached rapidly; no anxiety; no excitement period, no motor activity; best to administer by

TABLE 1 (continued)
Mode of Action of Euthanatizing Agents

	Site of action	Classification	Comments
	Hypoxic Agents		
			intravenous or intracardiac administration
Chloral hydrate and chloral hydrate combinations	Direct depression of cerebral cortex, subcortical structures and vital centers; direct depression of heart muscle	Ultimate cause of death is hypoxemia due to depression of vital centers	Transient anxiety; unconsciousness occurs rapidly; no motor activity
T-61	Direct depression of cerebral cortex, subcortical structures and vital centers; direct depression of heart muscle	Hypoxemia due to depression of vital centers	Transient anxiety and struggling may occur before unconsciousness when given too rapidly; tissue damage may occur. Must be given intravenously at recommended dosage and rates
	Physical Agents		
Gunshot or captive bolt into brain or stunning	Direct concussion of brain tissue	Hypoxemia due to depression of vital centers	Instant unconsciousness; motor activity may occur after unconsciousness
Cervical dislocation	Direct depression of brain	Hypoxemia due to disruption of vital centers	Violent muscle contractions can occur after cervical dislocation
Decapitation	Direct depression of brain	Hypoxemia due to disruption of vital centers	Violent muscle contractions occur subsequent to decapitation
Exsanguination	Direct depression of brain	Hypoxemia	If this method is preceded by and occurs during unconsciousness, there should be no struggling or muscle contraction
Decompression	Direct depression of brain	Hypoxemia	There are numerous disadvantages associated with mechanical problems as well as painful physiologic problems associated with expanding gases
Electrocution through brain	Direct depression of brain	Hypoxemia	Violent muscle contractions occur at same time as unconsciousness

TABLE 1 (continued)
Mode of Action of Euthanatizing Agents

	Site of action	Classification	Comments
		Hypoxic Agents	
Microwave irradiation	Direct inactivation of brain enzymes by rapid temperature increase of brain	Brain enzyme inactivation	Currently used only in rodents
Rapid freezing	Direct depression of brain	Rapid or near instantaneous cessation of metabolism	Approved only in specialized, well-controlled cases, in small laboratory animals
Air embolism	Direct depression of brain	Hypoxemia due to circulatory collapse	No motor activity occurs when preceded by anesthetics

INHALANT ANESTHETICS: ETHER, HALOTHANE, METHOXYFLURANE, ENFLURANE, ISOFLURANE, CYCLOPROPANE, AND NITROUS OXIDE

The inhalant anesthetics, primarily ether, halothane, and methoxyflurane, have been used to euthanatize many species.[22] With these agents, the animal is placed in a closed receptacle containing cotton or gauze soaked with the anesthetic.[23-25] Vapors are inhaled until respiration ceases and death ensues (Table 2). Because the liquid state of most inhalant anesthetics is a topical irritant, animals should be exposed to vapors only. Also, air or oxygen must be provided during the induction period.[24]

Other inhalation anesthetics are seldom used for euthanasia, due to low potency (nitrous oxide), high cost (isoflurane, enflurane), or danger (cyclopropane). For example, cyclopropane is highly flammable and explosive, and requires special equipment for administration. Nitrous oxide may be used alone to produce mild analgesia, anesthesia, and death by hypoxemia. Nitrous oxide is nonflammable and nonexplosive, but will support combustion.

Advantages—(1) The inhalant anesthetics are particularly valuable for euthanasia of birds, rodents, cats, and young dogs, ie, animals in which venipuncture may be difficult, and (2) halothane, enflurane, isoflurane, methoxyflurane, and nitrous oxide are nonflammable and nonexplosive under ordinary environmental conditions.

Disadvantages—(1) Struggling and anxiety may occur during induction of anesthesia because anesthetic vapors are irritating and induce excitement; (2) ether is flammable and explosive and should not be used near an open flame or other ignition sources; (3) personnel and other animals can be injured by exposure to these agents; and (4) halothane, methoxyflurane, nitrous oxide, enflurane, and isoflurane are relatively expensive.

Recommendations—Under certain circumstances, ether, halothane, isoflurane, enflurane, methoxyflurane, and nitrous oxide administered by inhalation are acceptable for euthanasia of small animals (ie, birds, rodents, cats, and young dogs). Although acceptable, these agents generally are not used in larger animals because of their cost and difficulty of administration. In emergency situations, halothane, enflurane, or isoflurane may be administered to large animals.

TABLE 2
Characteristics of Agents and Methods of Euthanasia

Agent	Safety for personnel	Ease of performance	Rapidity	Economic considerations	Tissue changes	Efficacy	Species suitability	Remarks
				Inhalants				
Ether	Flammable and explosive	Easily performed with closed chamber or container	Slow onset of anesthesia	Relatively inexpensive	Slight changes may occur in parenchymatous organs	Highly effective provided that subject is sufficiently exposed	Suitable for cats, young dogs, birds, rodents, and other small species	Acceptable but dangerous
Halothane*	Nonflammable and nonexplosive; chronic exposure of animals or personnel to vapor may be harmful	Easily performed with closed container; can be administered to large animals by means of a mask	Rapid onset of anesthesia	Expensive	May occur in parenchymatous organs	Highly effective provided that subject is sufficiently exposed	Suitable for cats and young dogs, birds, rodents, and other small animals	Acceptable
Methoxyflurane[†]	Nonflammable and nonexplosive under normal environmental conditions; chronic exposure of animals or personnel to vapor may be harmful	Slow with closed container; may be difficult to obtain necessary vapor concentration; can be administered to large animals by means of a mask	Slow onset of anesthesia	Expensive	May occur in parenchyma-tous organs	Highly effective provided that subject is sufficiently exposed	Suitable for cats, young dogs, birds, rodents, and other small animals	Acceptable
Enflurane[‡]	Nonflammable and nonexplosive; chronic	Easily performed with closed container or	Rapid onset of anesthesia	Expensive	May occur in parenchymatous organs, particularly	Highly effective provided that subject is	Suitable for cats, young dogs, birds, and other	Acceptable, but not recommended because of

TABLE 2 (continued)
Characteristics of Agents and Methods of Euthanasia

Agent	Safety for personnel	Ease of performance	Rapidity	Economic considerations	Tissue changes	Efficacy	Species suitability	Remarks
				Inhalants				
Enflurane[‡] (continued)	exposure of animals or personnel to vapors may be harmful	chamber, can be administered to large animals by means of a mask			kidneys	exposed; deep anesthesia may be accompanied by motor activity (twitching)	small species	motor activity in deep plane of anesthesia
							Useful in large animals in emergency situations	Not acceptable except in emergencies
Isoflurane[‡]	Nonflammable andnonexplosive; chronic exposure of animals or personnel to vapors may be harmful	Easily performed with closed container or chamber; can be administered to large animals by face mask	High volatility and potency; rapid onset of anesthesia	Very expensive	May occur in parenchymatous organs	Highly effective provided that subject is sufficiently exposed; induction does not appear to be stressful	Suitable for all small animals including birds and rodents	Acceptable
Nitrous oxide	Nonflammable and non-explosive, but will support combustion; chronic exposure of animals or personnel to gas may be harmful	Easily performed with closed container or chamber	Rapid onset in 100% concentration	Relatively expensive	Hypoxic lesions may occur	Highly effective provided that subject is sufficiently exposed	Suitable for cats, small dogs, birds, rodents, and other small species	Acceptable
							Not recommended	Use in larger animals

							alone for larger animals	requires supplementation with other agents
Chloroform	Nonflammable and nonex-plosive; chronic exposure of animals or personnel is dangerous because of potential liver or kidney damage and carcinogenicity	Easily performed with closed container; can be administered to large animals by means of a mask	Rapid onset of anesthesia	Inexpensive	Extensive changes may occur in parenchymatous organs	Highly effective provided that subject is sufficiently exposed	Suitable for cats, young dogs, birds, rodents, and other small species	Acceptable only in controlled conditions (see text)
							Useful in large animals in emergency situations	Acceptable only in emergency situations
N_2	Safe if used with ventilation	Use closed chamber with rapid filling	Rapid	Inexpensive	Changes associated with hypoxemia may occur	Effective except in young and neonates	Suitable for most small species, including mink	An effective agent, but other methods preferable; not acceptable in most animals less than 4 months old
Hydrogen cyanide	Extremely hazardous to personnel	Airtight chamber required unless used under field conditions	Rapid	Inexpensive	Changes consistent with tissue hypoxia may occur	Effective and irreversible	Dens of foxes, badgers, rabbits, and rodents	Because of extreme danger to operators, other methods are preferred.
Carbon monoxide	Extremely hazardous; toxic and difficult to detect	Requires appropriately operated equipment for gas production	Moderate onset time, but insidious, so animal is unaware of onset	Inexpensive when proper equipment is in place	Changes associated with hypoxemia may occur	Effective	Most small species including dogs, cats, rodents, mink, and chinchillas	Acceptable only when properly designed and operated equipment is used

TABLE 2 (continued)
Characteristics of Agents and Methods of Euthanasia

Agent	Safety for personnel	Ease of performance	Rapidity	Economic considerations	Tissue changes	Efficacy	Species suitability	Remarks
				Inhalants				
Carbon dioxide	Minimal hazard	Used in closed container	Moderately rapid	Inexpensive	Changes associated with hypoxemia may occur	Effective	Small laboratory animals, birds, cats and small dogs	Acceptable, but time required may be prolonged in immature and neonate animals
				Noninhalants				
Barbiturates	Safe except human abuse potential; DEA-controlled substance	Animal must be restrained; personnel must be skilled in IV injection	Rapid onset of anesthesia	Relatively inexpensive	Drug residues	Highly effective when appropriately administered	All species	Acceptable IV (see text)
Secobarbital/ dibucaine	Safe	Animal must be restrained; personnel must be skilled in IV injection	Rapid onset of anesthesia	Expensive	Drug residues	Highly effective when appropriately administered	Dogs and cats and some laboratory animals	Acceptable IV
T-61	Safe	Animal must be restrained; personnel must be skilled in IV injection	Rapid onset of anesthesia	Expensive	Drug residues	Highly effective when appropriately administered	Dogs and cats and some laboratory animals	Acceptable IV (see text)
Mixture of chloral hydrate, $MgSO_4$, pentobarbital	Safe except human abuse potential; DEA-controlled substance	Animal must be restrained; personnel must be skilled in IV injection	Rapid onset of anesthesia	Relatively inexpensive	Drug residues	Highly effective when appropriately administered	Large animals, ie, horses and cattle	Acceptable IV
Strychnine/ nicotine/	...	...	...	...	...	...	...	Absolutely unacceptable

curariform drugs/ $MgSO_4$/KCl								when used alone (see text)
Physical Methods								
Captive bolt, gunshot and stunning	Safe, but some concern for mechanical injury; gunshot can be especially dangerous	Requires skilled individuals; however, skills easily developed	Rapid	Inexpensive	Trauma of brain tissue; others unchanged	Highly effective	Usually applied in larger agricultural animals, but can be used for rabbits and guinea pigs	Acceptable (see text for limitations)
Cervical dislocation	Safe	Requires training and skill	Moderately rapid	Inexpensive	Primarily useful if chemical residue-free tissues are needed	Irreversible	Suitable only in chickens, laboratory mice, and rats less than 200g or rabbits less than 1 kg	Acceptable with prior sedation or light anesthesia (see text)
Decapitation	Some concern for mechanical injury	Easily performed with minimal training	Moderately rapid; can have consciousness for 13 to 14 seconds	Inexpensive	Primarily useful if chemical residue-free tissues are needed	Irreversible	Suitable for rodents and small rabbits	Acceptable with prior sedation or light anesthesia (see text)
Exsanguination	Safe	Easily performed with minimal training	Moderately rapid	Inexpensive	Minimal	Irreversible	Recommended to ensure death following stunning by captive bolt, or gunshot in large domestic species or rabbits	Acceptable when preceded by other methods that relieve anxiety, consciousness

TABLE 2 (continued)
Characteristics of Agents and Methods of Euthanasia

Agent	Safety for personnel	Ease of performance	Rapidity	Economic considerations	Tissue changes	Efficacy	Species suitability	Remarks
				Physical Methods				
Decompression	Safe	Requires special equipment that must be operated and maintained by skilled personnel	Relatively slow	Inexpensive	Trauma can occur to tissues when air is trapped in middle ear and gut	Effective if equipment is properly operated and maintained	Primarily for dogs and smaller animals	Not recommended because of numerous disadvantages (see text)
Electrocution	Hazardous to personnel	Not easily performed in all instances	Can be rapid	Inexpensive	Petechial hemorrhages can occur	Highly effective if properly performed	Used primarily in farm animals	Acceptable; however, disadvantages far outweigh advantages in most appli-cations (see text)
Microwave	Safe	Requires training and highly specialized equipment	Very rapid	Equipment is expensive	Primarily used for brain tissue studies	Highly effective	Used in mice, rats, or animals of similar size	Acceptable only in small rodents (see text)
Rapid freezing	Safe	Easily performed	Rapid	Inexpensive	Arrest enzymatic reactions in tissue, usually brain	Effective	Used in small rodents	Acceptable (see text)
Air embolism	Safe	Easily performed	Moderately rapid	Inexpensive	Hypoxemic lesions may occur	Effective if closely monitored	Used primarily in rabbits and other small species	Acceptable only in fully anesthetized animals

* Flurothane, Ayerst Laboratories, New York, NY: Halocarbon Laboratories, Hackensack, NJ. †Metofane, Pitman-Moore, Inc, Washington Crossing, NJ; Penthrane, Abbott Laboratories, North Chicago, Ill. ‡Ethrane and Isoflurane, Anaquest, Division of BOC, Inc, Madison, Wis.

CHLOROFORM

Chloroform is a known potent hepatotoxin and is a suspected carcinogen.[26] Although chloroform is nonexplosive, its use in the presence of a flame may result in the production of phosgene gas. Because of its significant hazards to human beings, chloroform can be recommended for euthanasia only under conditions that prevent human exposure.

NITROGEN

Nitrogen (N_2), a colorless, odorless gas, constitutes 78% of normal atmospheric air. It is inert, nonflammable, and nonexplosive.

Euthanasia is induced by placing the animal in a closed container into which pure N_2 is introduced rapidly at atmospheric pressure. Nitrogen displaces oxygen in the container, thus inducing death by hypoxemia. Commercial equipment for N_2 euthanasia consists of a closed chamber, an oxygen monitor to ensure that atmospheric oxygen is 1.5% or less in 45 to 50 seconds, and a timer to assure adequate exposure to the N_2. Because N_2 mixes with room air when the chamber is opened, there is minimal danger to personnel when ventilation is adequate.

In studies by Herin et al,[27] dogs became unconscious within 76 seconds when an N_2 concentration of 98.5% was achieved in 45 to 60 seconds. The electroencephalogram became isoelectric (flat) in a mean of 80 seconds and arterial blood pressure was undetectable in an average of 204 seconds. Although all dogs hyperventilated prior to unconsciousness, the authors concluded that this method induced death without pain. Following loss of consciousness, yelping, gasping, convulsions, and muscular tremors occurred in some dogs. At the end of a 5-minute exposure period, all dogs were dead.[27] These findings were similar to those for rabbits[28] and for a single dog.[1]

Glass et al[29] reported that newborn dogs, rabbits, and guinea pigs can survive an N_2 atmosphere much longer than can adults. Newborn dogs and rabbits survived 31 minutes, and guinea pigs 6 minutes, whereas adult dogs and guinea pigs survived 3 minutes, and adult rabbits $1^1/_2$ minutes. Nitrogen has been used for euthanasia of mink.[30]

With N_2 flowing at a rate of 39% of chamber volume per minute, rats collapsed in approximately 3 minutes and respiratory arrest occurred in 5 to 6 minutes. Regardless of flow rate, rats exhibited signs of panic and distress before collapsing and dying.[31] Insensitivity to pain under such circumstances is questionable.[32]

Advantages—(1) N_2 gas is readily available, (2) can be rapid and reliable; and (3) hazards to personnel are minimal.

Disadvantages—(1) Very young animals are not euthanatized rapidly; (2) rapid N_2 flow may produce a noise that may frighten animals; (3) the responses by the unconscious animal may be aesthetically objectionable; (4) reestablishing a low concentration of O_2 (ie, 6% or greater) in the chamber before death will allow immediate recovery; (5) in some areas the use of N_2 for euthanasia has been prohibited by law; and (6) when flow rates are low, time to death can be excessively long.

Recommendations—The effect of N_2 can be rapid and is reliable when properly used. However, the manner of death may be aesthetically objectionable. Although N_2 is an effective agent, other methods of euthanasia are preferable. Nitrogen is not an acceptable euthanatizing agent in animals less than four months of age. The use of N_2 requires that the equipment be (1) properly constructed to rapidly attain high concentration, (2) correctly maintained, and (3) proficiently operated. In addition, an effective exhaust or ventilation system must be present to preclude exposure of personnel to high concentrations of N_2.

HYDROGEN CYANIDE GAS

Hydrogen cyanide gas is one of the most rapidly acting poisons.[33] Cyanide reacts readily with the ferric ion of mitochondrial cytochrome oxidase to form cytochrome oxidase-cyanide complex, inducing cytotoxic hypoxia. The maximum allowable concentration for 8 hours' occupational exposure in industry is 10 ppm. Hydrogen cyanide gas for euthanasia is produced by placing pellets of sodium cyanide in sulfuric acid.

There is no evidence to indicate that the effects of cyanide are painful.[34] Cyanide induces intense respiratory stimulation and may cause excitement resulting in sounds of distress before death. Cyanide possesses a pungent and rather unpleasant smell and, because of its high toxicity, an airtight chamber is required to confine the gas. In the United Kingdom, cyanide is used for euthanasia of rabbits, foxes, and badgers in dens.[35]

Advantages—(1) Hydrogen cyanide gas induces rapid death, and (2) it can be used under field conditions where other agents are impractical.

Disadvantages—(1) The animal manifests violent convulsive seizures and opisthotonos prior to death; (2) the responses are disagreeable to most observers; (3) the gas is very irritating to the respiratory mucosa; and (4) there is extreme danger to personnel.

Recommendations—Although the effect of hydrogen cyanide gas is rapid, reliable, and irreversible, it endangers the operator, and the manner of death is aesthetically objectionable.[14] Other methods of euthanasia are preferable.

CARBON MONOXIDE

Carbon monoxide (CO) combines with hemoglobin to form carboxyhemoglobin. This blocks the uptake of oxygen by erythrocytes, and leads to fatal hypoxemia.

Clinical signs of CO toxicosis, as originally described, are due to its action upon the circulatory system.[36] These result in generalized vascular dysfunction, characterized by extensive vasodilation and hemorrhage. In people, initial symptoms are headache sometimes combined with nausea, followed by depression progressing to unconsciousness. Because CO stimulates motor centers in the brain, unconsciousness may be accompanied by convulsions and muscular spasms.

Carbon monoxide is a cumulative poison.[37] Distinct signs of CO toxicosis are not induced until the concentration is 0.05% in air, and acute signs do not occur until the concentration is approximately 0.2%. In human beings, exposure to 0.32% and 0.45% of CO for one hour will induce unconsciousness and death.[38]

There are 3 practical methods of generating CO for mass euthanasia:

1) Chemical interaction of sodium formate and sulfuric acid,
2) Exhaust fumes from gasoline internal combustion engines, and
3) Commercially compressed CO gas in cylinders.

Carding[39] used sodium formate and sulfuric acid to generate CO to euthanatize dogs. He found a marked decrease in the average time to collapse and death by increasing CO concentration from 2% to 3% in air.

When CO is produced by combustion, oxides of nitrogen and hydrocarbons, oxygenates of hydrocarbons, and heat must be controlled to prevent discomfort to the animal. This may be done by passing exhaust gases through a water chamber and a metal gauze filter with a cloth screen. The water chamber cools the gas, removes some carbon particles, and entraps the oxides of nitrogen, hydrocarbons, and oxygenates of hydrocarbons, The cloth filter removes carbon particles, allowing relatively clean, nonirritating CO gas to enter the chamber.

An idling engine running on a rich fuel mixture will produce the highest percentage of CO in exhaust gas. Carbon monoxide produced by an internal combustion engine is just as effective as cylinder CO and considerably less expensive.[40] Chamber concentration of CO from piped exhaust gas can quickly reach 8%, resulting in 70% saturation of hemoglobin. Carbon monoxide must be considered extremely hazardous for personnel because it is highly toxic and difficult to detect. An efficient exhaust or ventilatory system is essential to prevent accidental exposure of human beings.

In a study by Ramsey and Eilmann,[41] 8% CO caused guinea pigs to collapse in 40 seconds to 2 minutes, and death occurred within less than six minutes.

Carbon monoxide has been used to euthanatize mink[40] and chinchillas. The animals collapsed in one minute, breathing ceased in 2 minutes, and they were considered dead when the heart stopped beating in approximately 5 to 7 minutes.

Blood et al[42] reported excellent results with CO for euthanasia of cats and dogs. Several chamber concentrations were tested: a 6% CO concentration gave fastest results. No advantage was observed by further increasing CO concentration.

In a study designed to evaluate the physiologic and behavioral characteristics of dogs exposed to 6% CO in air, Chalifoux and Dallaire[43] could not determine the precise time of unconsciousness. Electroencephalographic recordings revealed a 20- to 25-second period of abnormal cortical function prior to unconsciousness. It is during this period that agitation and vocalization occur. These reactions are not necessarily due to pain, but are probably caused by cortical hypoxemia. The authors concluded that CO meets most accepted criteria for mass euthanasia of adult dogs, but that pretreatment with a tranquilizer (phase I) followed by CO inhalation (phase II) might decrease or eliminate objectionable behavioral and physiologic responses associated with CO inhalation alone.[43] Subsequent studies have shown that premedication with acepromazine significantly decreased behavioral and physiologic responses of dogs euthanatized with CO.[44]

In a comparative study, CO (gasoline engine exhaust) and 70% CO_2 + 30% O_2 were used to euthanatize cats. Euthanasia was divided into 3 phases. Phase I was time from initial contact with gas until clinical signs of effect (eg, yawning, staggering, or trembling). Phase II extended from end of phase I until recumbency and phase III was from the end of phase II until total immobilization.[45] The study revealed that phase I responses were greatest with CO_2 + O_2. Convulsions occurred during phases II and III with both methods. However, when the euthanatizing chamber was pre-filled with CO (ie, "exhaust fumes"), convulsions did not occur in phase III. Time to complete immobilization was greater with CO_2 + O_2 (approximately 90 seconds) than with CO alone (approximately 56 seconds).[45] In piglets, excitation was less likely to occur when CO was combined with nitrous oxide (N_2O). If excitement did occur, it followed the onset of unconsciousness.[36]

Gasoline engine exhaust piped into an enclosed chamber quickly reaches CO levels in excess of 8%. Saturation of hemoglobin to the 70% level occurs more rapidly when concentrations are greater than 6%, which has been described as optimal. Use of compressed CO allows rapid attainment of an effective chamber concentration. Carbon monoxide produced by a gasoline internal combustion engine and cylinder CO are equally effective,[40] but the gasoline engine costs less to use.

Advantages—(1) Carbon monoxide induces rapid and painless death; (2) hypoxemia induced by CO is insidious so that the animal is completely unaware of it; and (3) unconsciousness occurs without pain or discernible discomfort, when properly administered.

Disadvantages—(1) Safeguards must be taken to prevent exposure of personnel; (2) during chemical generation by sodium formate and sulfuric acid, irritating vapors of sulfuric acid must be removed by passing the CO through a solution of 10% sodium hydroxide; and (3) exhaust gases must be filtered and cooled to prevent discomfort to animals.

Recommendations—Carbon monoxide used for individual or mass euthanasia is acceptable for small animals, including dogs and cats, provided that the following precautions are taken: Personnel using CO must be instructed thoroughly in its use and must understand its hazards and limitations; the CO generator and chamber must be located in a well-ventilated environment, preferably out-of-doors; the chamber must be equipped with internal lighting and viewports that allow personnel direct observation of animals; the gas generation process should be adequate to achieve a CO concentration throughout the chamber of at least 6% within no more than 20 minutes after animals are placed in the chamber; sodium formate- and sulfuric acid-generated CO must have the irritating acid vapors filtered out by passing it through a 10% solution of sodium hydroxide; if CO generation is by combustion of gasoline in an engine, (1) the engine must be operating at idling speed with a rich fuel-air mixture; (2) prior to entry into the chamber, the exhaust gas must be cooled to less than 125 F (51.7 C); (3) the chamber must be equipped with accurate temperature guages monitored by attendants to assure that internal temperature of the chamber does not exceed 110 F (41.3 C), and (4) the exhaust gas must be passed through water and cloth filtration processes to remove irritants and carbon particles before entering the chamber. Exhaust gas piped into a chamber from a cruising vehicle is not acceptable.

CARBON DIOXIDE

Room air contains 0.04% carbon dioxide (CO_2). Pure CO_2 is heavier than air and nearly odorless. Inhalation of CO_2 in concentrations of 7.5% increases the pain threshold, and higher concentrations of CO_2 have a rapid anesthetic effect.[45-49]

Inhalation of 60% CO_2 results in loss of consciousness within 45 seconds, and respiratory arrest within 5 minutes.[50] Carbon dioxide has been used to euthanatize groups of small laboratory animals, including mice, rats, guinea pigs, chickens, and rabbits,[2,51-55] and for humane slaughter of swine for human consumption.[11,56] According to Croft,[57] animals do not detect the CO_2 immediately, and its depressant action takes place almost unnoticed.

Leake and Waters[47] reported the experimental use of CO_2 as an anesthetic agent in the dog. Thirty percent to 40% CO_2 in oxygen induced anesthesia within 1 to 2 minutes, usually without struggling, retching, or vomiting. The combination of 40% CO_2 and approximately 3% CO has been used experimentally for euthanasia of dogs by Carding.[39] Carbon dioxide has been used in specially designed chambers to euthanatize cats[58-60] and other small laboratory animals.[51,60]

Studies in day-old chickens have shown that CO_2 is an effective euthanatizing agent. Inhalation of CO_2 caused little distress to the birds, suppressing nervous activity and inducing death rather quickly.[52] Because respiration begins during embryonic development, the unhatched chickens' environment may normally have a CO_2 concentration as high as 14%. Thus, CO_2 concentration for euthanasia for baby chickens and other neonates should be especially high. A CO_2 concentration of 60% to 70% with a 5-minute exposure time appears to be optimal.[52]

Carbon dioxide is used for preslaughter anesthesia of pigs. The undesirable side effect of CO_2, as used in commercial slaughter houses, is that the pigs experience a stage of excitement with vocalization for about 40 seconds before they lose consciousness.[56,61] For that reason, CO_2 preslaughter anesthesia may appear less humane than other techniques. The signs of effective CO_2 anesthesia are those associated with deep surgical anesthesia, such as loss of withdrawal and palpebral reflexes.[62]

Advantages—(1) The rapid depressant and anesthetic effects of CO_2 are well established. (2) Carbon dioxide may be purchased in cylinders or in a solid state as "dry ice"; (3) CO_2 is inexpensive, nonflammable and nonexplosive, and presents minimal hazard to personnel when used with properly designed equipment; (4) CO_2 does not result in accumulation of tissue residues in food producing animals; and (5) CO_2 euthanasia does not distort cellular architecture.[63]

Disadvantages—(1) Because CO_2 is heavier than air, incomplete filling of a chamber may permit a tall or climbing animal to avoid exposure and survive, and (2) in immature animals, the time required for euthanasia may be substantially prolonged.

Recommendations—Carbon dioxide is recommended in small laboratory animals such as birds, cats, and small dogs. Chamber design should allow for precharging with CO_2 and should enable cleaning and removal of dead animals with minimal loss of CO_2. Compressed CO_2 gas in cylinders is preferable to dry ice. Inflow to a euthanatizing chamber can be regulated precisely with compressed CO_2. Optimal flow rate appears to be one that will displace approximately 20% of chamber volume per minute.[31] When using compressed CO_2, O_2 can be added (for example, 30% O_2, 70% CO_2), thus decreasing the discomfort of hypoxia prior to onset of narcosis and anesthesia. If dry ice is used, animal contact must be avoided to prevent freezing or chilling.

NONINHALANT PHARMACOLOGIC AGENTS

Noninhalant agents that can be used for euthanasia are widely diverse in chemical composition.[64,65] Although death can be induced by administering these drugs via many routes (intravenous, intracardiac, intraperitoneal, intrathecal, intramuscular, intrathoracic, subcutaneous, and oral), intravenous administration is preferred because the effect is most rapid and reliable. Intrapulmonic injection should be avoided.

Oral, rectal, and intraperitoneal routes of administration of drugs for euthanasia are inadvisable because of prolonged onset of action, wide range in lethal doses, and potential irritation of tissues. An hour or more may elapse from time of administration to death. Some drugs, such as chloral hydrate and T-61, when given intraperitoneally are irritating and cause abdominal pain. Others may produce tissue changes, depending on dose and route of administration.[25,63]

Because crying and struggling may follow improper intracardiac injection, this route of administration is objectionable. Skill is required to penetrate the heart of an animal with one thrust of a hypodermic needle, especially if the animal is not easily restrained. Intracardiac injection of drugs is not recommended for euthanasia, except in depressed, anesthetized, or comatose animals. Intrathecal use of drugs in unanesthetized animals is not recommended because puncture of the cisterna magna without causing pain and struggling is not possible.

If the animal to be euthanatized is excitable or vicious, use of analgesics, tranquilizers,[44] narcotics, ketamine, xylazine, or other depressants is recommended before administration of the euthanatizing agent.

BARBITURIC ACID DERIVATIVES

Barbiturates depress the central nervous system in descending order, beginning with the cerebral cortex. Within seconds of intravenous administration, unconsciousness is induced and it progresses to deep anesthesia.[1] Apnea occurs due to depression of the respiratory center, and cardiac arrest quickly follows. Several barbiturates are acceptable, but pentobarbital sodium most commonly is used for euthanasia.

Advantages—(1) A primary advantage of barbiturates is speed of action. This effect depends on the dose, concentration, and rate of injection; (2) the barbiturates induce euthanasia smoothly, with minimal discomfort to the animal, and favorably impress the observer because the animal dies quietly[1]; a cost comparison by Lumb and Moreland[66] indicates that barbiturates are less expensive than most other injectable agents.

Disadvantages—(1) Intravenous injection is necessary for best results, necessitating trained personnel; (2) each animal must be restrained; (3) current federal drug regulations require strict accounting for the barbiturates and, by necessity, these must be used under the supervision of personnel registered with the US Drug Enforcement Agency; and (4) an aesthetically objectionable terminal gasp may occur in the unconscious animal.

Recommendations—The advantages of using barbiturates for euthanasia in small animals far outweigh the disadvantages. The intravenous injection of a barbituric acid derivative is the preferred method for euthanasia of dogs, cats, and other small animals; however, intraperitoneal administration is an acceptable alternative for laboratory rodents. Nervous or vicious animals may require tranquilization[44] or sedation prior to injection of a barbiturate.

SECOBARBITAL-DIBUCAINE COMBINATION[a]

This mixture contains a short-acting barbiturate to induce anesthesia. The cardiotoxic effects of dibucaine cause cardiac arrest. Limited trials suggest this combination is an acceptable barbiturate euthanatizing agent when given intravenously to dogs and cats.[67-68]

T-61[b]

T-61 is an injectable nonbarbiturate, non-narcotic mixture of three drugs used to induce euthanasia. These drugs provide a combination of general anesthetic, curariform, and local anesthetic actions. Death results from severe central nervous system depression, hypoxia, and circulatory collapse. T-61 has been withdrawn from the United Kingdom market since 1976, but because it is not currently listed as a controlled substance under federal drug regulations, it has gained some popularity in the United States in recent years.

A comparative study of T-61 and pentobarbital (at 57.1 mg/kg of body weight) indicated that either agent induced euthanasia smoothly.[71] Dogs given T-61 received two-thirds of the total dose (0.3 ml/kg of body weight) at a rate of 0.2 ml/second, with the last one-third given at the rate of 1.2 ml/sec. With both pentobarbital and T-61, the electroencephalogram changed from a normal awake pattern to one of low frequency and increased amplitude for approximately 5 seconds, followed quickly by electrical silence. The pentobarbital-treated dogs required 12 seconds longer for the occurrence of electrical silence. With both agents, electrocardiographic alterations developed immediately and arterial pressure dropped to zero. Results of this study indicate that painless death is induced by pentobarbital or T-61.[71]

A recent survey[73] and review[74] indicate a need for further controlled studies on the efficacy and humaneness of T-61 as a euthanatizing agent in several species.

Advantages—(1) T-61 may be used intravenously in dogs, cats, horses, laboratory animals, and birds; (2) the terminal gasp[75] that may accompany pentobarbital euthanasia in unconscious animals is not evident with T-61; and (3) it is not regulated or controlled by the US Drug Enforcement Agency.

Disadvantages—(1) T-61 must be administered intravenously because it is painful when administered extravascularly; (2) if T-61 is injected at too rapid a rate, the animal may appear to experience pain or discomfort immediately prior to becoming unconscious; (3) doses larger than recommended may cause pulmonary edema and other tissue lesions; and (4) T-61 is not approved for use in animals intended for human consumption.

[a] Each milliliter contains 400 mg of secobarbital and 25 mg of dibucaine. Dosage for dogs and cats, IV, 0.22 ml/kg body weight. Repose Diamond Laboratories, Des Moines, Iowa.

[b] Each milliliter contains 200 mg of N-(2-(n)-methoxy-phenyl)-2-ethylbutyl (1) gamma hydroxybutyramide, 50 mg of 4,4'-methylene-bis-(cyclohexyl-trimethyl-ammonium iodide) and 5 mg of tetracaine hydrochlorides with 0.6 ml dimethylformamide in distilled water. Produced by American Hoechst Corp., Somerville, NJ.

Recommendations—If barbiturates cannot be used, T-61 is a substitute, but only when it is administered intravenously by a highly skilled person at recommended dosages and at proper injection rates.[74] Intracardiac or intrathoracic administration is not recommended.

CHLORAL HYDRATE

Because chloral hydrate depresses the cerebrum slowly, restraint during induction may become a problem in some animals. Death is due to hypoxemia caused by progressive depression of the respiratory center. Death may be preceded by gasping, muscle spasms, and vocalization.

Recommendation—Chloral hydrate is not recommended for euthanasia of dogs, cats, and other small animals because the associated signs may be severe and are aesthetically objectionable.

COMBINATION OF CHLORAL HYDRATE, MAGNESIUM SULFATE, AND SODIUM PENTOBARBITAL

This combination has been used for anesthesia of large animals and may be administered in overdosage for euthanasia.[18]

Recommendations—This mixture is an acceptable large animal euthanatizing agent when administered intravenously.

STRYCHNINE

Strychnine or any of its salts, such as strychnine sulfate, increases the excitability of the central nervous system. The animal remains conscious and excessively responsive to stimuli. Strychnine causes violent convulsions and associated painful contraction of skeletal muscles.[76]

Recommendation—Strychnine is absolutely condemned for euthanasia.

MAGNESIUM SULFATE ($MgSO_4$) OR POTASSIUM CHLORIDE (KCl)

Magnesium or potassium ions exert little if any direct depressant effect on the central nervous system. Dosages of $MgSO_4$ previously recommended for euthanasia have been shown to cause complete neuromuscular block and death due to hypoxemia.[77]

Recommendation—Magnesium or potassium salts must not be used alone for euthanasia because of the lack of analgesic or anesthetic effect. Potassium chloride may be administered to an anesthetized animal as an efficient and inexpensive way to cause cardiac arrest or death.

NICOTINE

Nicotine is an alkaloid obtained from the tobacco plant. Nicotine as a sulfate compound is an odorless, clear, water-soluble liquid. Lethal doses can be absorbed through mucous membranes and intact skin, making it dangerous for personnel.

Nicotine sulfate has been used extensively in the past in capture equipment for immobilizing wild game and feral and domestic dogs. Other drugs have replaced nicotine sulfate because of the high mortality associated with its use.

Nicotine sulfate induces a short period of stimulation, followed by blockade of autonomic ganglia. When injected in sufficient quantity, it induces a muscle relaxant effect, with paralysis of the respiratory muscles and, subsequently, hypoxemic death. Salivation, vomiting, defecation, and convulsions commonly occur prior to death.

Recommendations—Because nicotine sulfate is an extremely dangerous drug for personnel, is inhumane for animals, and induces serious side effects prior to death, *it is absolutely condemned for euthanasia.*

CURARIFORM DRUGS

Drugs such as curare, succinylcholine, pancuronium, guaifenesin, and other neuromuscular blocking agents induce death by immobilizing the respiratory muscles, causing fatal suffocation. There is no depressant action on the brain.[1] Human patients given these drugs have described periods of full consciousness accompanied by complete muscular immobility and intense anxiety.[18]

Hicks and Bailey[78] administered massive doses of succinylcholine to dogs and observed changes in the electrocardiogram, electroencephalogram, and respiratory responses. They found bradycardia initially and elevated arterial blood pressure. As the myocardium failed, tachycardia was followed by a decrease in arterial pressure. The electroencephalogram indicated no loss of consciousness, as the normal awake wave form progressed to an activated electroencephalogram. The electroencephalographic activity continued long after cessation of respiration (7 minutes) before becoming isoelectric. Urination, salivation, and defecation commonly occurred along with muscular fasciculations. Asphyxial struggling was sometimes accompanied by violent fasciculations and head movements. These authors concluded that dogs given succinylcholine are conscious for prolonged intervals before dying of respiratory and cardiac arrest, and for this reason it should not be given as the sole agent for euthanasia.

Recommendations—Where standard methods of restraint are impractical, impossible, or dangerous, or where manual capture and restraint may cause pain and injury through struggling and anxiety, the use of immobilizing drugs is justified. Because the immobilized animal is fully conscious and subject to death by suffocation, euthanasia by chemical or physical methods must be accomplished immediately. *Curariform drugs alone are absolutely condemned for euthanasia.*

OTHER PARENTERAL PREPARATIONS

A number of injectable agents are currently available for immobilizing domestic and feral animals. Drugs such as ketamine HCl and xylazine are used for immobilization, analgesia, and anesthesia. The effect occurs within minutes following intramuscular injection, making these drugs useful for restraint (ie, phase I of euthanasia). Currently these agents are not included as controlled substances by federal drug regulations. Narcotics also may be used for sedation and analgesia, but detailed records must be kept, and their use properly supervised by a licensed professional registered with the US Drug Enforcement Agency. Although it is possible to cause death with high doses of these drugs, such use would be impractical and may produce convulsions before death.

PHYSICAL METHODS

Physical methods include captive bolt pistol, gunshot, stunning, cervical dislocation, decapitation, exsanguination, decompression, electrocution, microwave irradiation, rapid freezing, and air embolism. It is a requirement that any method of euthanasia be performed by knowledgeable, well-trained individuals. Because trauma often is associated with physical methods for inducing unconsciousness, this requirement is particularly vital. Physical methods are often aesthetically displeasing; however, when properly performed, unconsciousness is rapid, without distress or pain to the animal. In some applications the amount of apprehension of the animals will be less than if chemical means had been used.

CAPTIVE BOLT PISTOL

This method is used in ruminants, horses, and swine and is being developed for laboratory rabbits. The purpose of both the captive bolt and nonpenetrative percussion methods is to damage the cerebral hemisphere and brainstem so that the animal is unconscious.[79] Penetrating or nonpenetrating captive bolts are powered by gunpowder or compressed air. Nonpenetrating instruments usually are less effective in inducing unconsciousness. Animals must be adequately restrained to ensure proper placement of the captive bolt. The muzzle of the pistol should be placed at a right angle to the skull and directed toward the center of the brain in ruminant species. A multiple projectile has been suggested as a more effective technique, especially on large cattle. In swine, the captive bolt should be directed to the brain from a central point slightly above a line between the eyes.[80]

Evaluation of unconsciousness can be difficult. Some of the methods used are electroencephalography, loss of visually evoked responses, loss of the menace or blink response to a hand brought rapidly toward the eye, and loss of coordinated movements and pupillary dilation.[62,79,81,82] The signs of effective stunning are immediate collapse and a 15-second period of tetanic spasm, followed by slow hindlimb movements of increasing frequency.

Advantages—A humane method for use in abattoirs and in research facilities when the use of drugs is contraindicated.

Disadvantages—(1) Aesthetically displeasing; (2) nonpenetrating captive bolt pistols cannot be used effectively in mature swine; and (3) death may not occur.

Recommendations—Use of the penetrating captive bolt pistol, when followed by exsanguination or pithing, is an acceptable method of euthanasia for horses, ruminants, and swine when chemical agents cannot be used. Nonpenetrating captive bolt pistols are not recommended.

GUNSHOT

Under some circumstances, gunshot may be the only practical method of euthanasia, It should be performed by highly skilled and trained personnel utilizing a rifle or pistol appropriate for the situation. The projectile should be accurately placed to enter the brain, causing instant unconsciousness.[83,84]

Advantages—(1) Euthanasia is instantaneous; and (2) under field conditions, gunshot may be the only effective method available.

Disadvantages—(1) It is dangerous to personnel; (2) it is aesthetically unpleasant; and (3) under field conditions it may be difficult to hit the brain.

Recommendations—When other methods cannot be used, competently performed gunshot is an acceptable method of euthanasia. When the animal is appropriately restrained, the captive bolt pistol is preferred to gunshot.

STUNNING

Stunning can render an animal unconscious; however, unconsciousness will occur only if a blow to the head is properly executed. If not performed correctly, various degrees of consciousness with concomitant pain will ensue.

Advantages—(1) Stunning is humane when properly performed; and (2) enables collection of blood and other tissues without chemical contamination.

Disadvantages—(1) Stunning is inhumane if improperly performed; (2) it is impossible to ensure constancy of performance by personnel; and (3) stunning may be aesthetically displeasing for personnel performing or observing the procedure; and (4) it must be followed by other means to ensure the death of the unconscious animal.

Recommendations—Stunning of laboratory rodents and rabbits by a sharp blow to the head is strongly discouraged as a method of euthanasia because of the inherent risk of not rendering animals immediately unconscious. Stunning must be followed by other means to ensure death of the unconscious animal. Any use must be predicated on a case-by-case review by the institutional animal welfare committee or other responsible bodies.

CERVICAL DISLOCATION

Cervical dislocation is used to euthanatize poultry, mice, and immature rats and rabbits. For mice and rats, the thumb and index finger are placed on either side of the neck at the base of the skull or, alternatively, a rod is pressed at the base of the skull. With the other hand, the base of the tail or hindlimbs are quickly pulled, causing separation of the cervical vertebrae from the skull. For immature rabbits, the head is held in one hand and the hindlimbs in the other. The animal is stretched and the neck is hyperextended and dorsally twisted to separate the cervical vertebrae from the skull.[51,80]

Advantages—(1) Cervical dislocation is a technique that may induce immediate unconsciousness; (2) does not chemically contaminate tissues; and (3) it is rapidly accomplished.

Disadvantages—(1) May be aesthetically displeasing to personnel; and (2) its use is limited to poultry, mice, and immature rats and rabbits.

Recommendations—When properly executed, cervical dislocation may be a humane technique to euthanatize poultry, mice, and rats weighing less than 200 g, and rabbits weighing less than 1 kg. In heavier rats and rabbits, the greater muscle mass in the cervical region makes cervical dislocation physically more difficult and, accordingly, it should not be performed. Because unconsciousness may not occur immediately, it is preferable to lightly anesthetize or sedate the animal prior to cervical dislocation.

Institutional animal welfare committees or other responsible bodies must determine that personnel who perform cervical dislocation techniques have been properly trained.

DECAPITATION WITH GUILLOTINE

Decapitation is most often used to euthanatize rodents and small rabbits. It provides a means to recover tissues and body fluids that are chemically uncontaminated. It also provides neurobiologists with a means to obtain anatomically undamaged brain tissue for study.[63] In the latter case, the head is immediately placed in liquid nitrogen to halt metabolic processes.

Advantage—Guillotines that are well-designed to accomplish decapitations in a uniformly instantaneous manner are commercially available.

Disadvantages—(1) Decapitation may be aesthetically displeasing to personnel performing or observing the technique; and (2) data suggest that animals may not lose consciousness for an average of 13 to 14 seconds following decapitation.[85]

Recommendation—Until additional information is available to better ascertain whether guillotined animals perceive pain, the technique should be used only after the animal has been sedated or lightly anesthetized, unless the head will be immediately frozen in liquid nitrogen subsequent to severing.

EXSANGUINATION

Exsanguination is recommended to ensure death subsequent to stunning or electrocution. Rabbits and other laboratory animals may be exsanguinated to obtain hyperimmune antisera, but because of the anxiety associated with extreme hypovolemia, exsanguination should be done only in sedated, stunned, or anesthetized animals.[86]

DECOMPRESSION (HYPOXIA)

Decompression is a means to induce unconsciousness and death due to cerebral hypoxia. Decompression chambers simulate an ascent to an altitude high above sea level. The higher the altitude, the lower the ambient pressure and the greater the hypoxia. Regardless of altitude, the percentage composition of the atmospheric gases remains the same as at sea level. At sea level, the ambient or barometric pressure is 760 mm of Hg, whereas at 55,000 feet above sea level, the pressure is 68.8 mm of Hg. At sea level, the partial pressure of O_2 is 159 mm of Hg, and at 55,000 feet, it is only 14 mm of Hg. The mean oxygen tension in arterial blood (PaO_2) of dogs is normally about 95 mm of Hg; accordingly, at a simulated altitude of 55,000 feet, the rapid decrease in PaO_2 results in unconsciousness and death.[87]

Data from human studies indicate that decompression at 1,000 feet/minute results in excitement and euphoria, followed by sensory dullness, weakness, dyspnea, and unconsciousness. An optimal rate of decompression in the adult dog has been recommended to be 4,000 feet/minute and for other selected species, approximately 1,100 feet/minute.[88,89] The rate of decompression for equipment most commonly used by animal shelters is 1,000 feet/second for 45 to 60 seconds.[90]

There are adverse physical effects due to rapid decompression, since trapped gases in body cavities (eg, sinuses, eustachian tubes, middle ears, and intestines) follow Boyle's Law and expand proportionally to the level of decompression. The rate of decompression is important, since the slower the rate, the less the discomfort and pain due to expanding gases.

Decompression chambers have been widely used in municipal and humane society shelters as a method for euthanasia of dogs and cats. They are rarely used in laboratory animal facilities.

Advantages—(1) Decompression is safe for personnel; (2) it minimizes operator stress; and (3) it is cost-effective.

Disadvantages—(1) Commonly used chambers are designed to produce decompression at a rate 15 to 60 times faster than that recommended as optimum for animals.[88-90] Although animals are rendered unconscious approximately 10 seconds after exposure to a simulated altitude of 50,000 ft, pain and distress may occur due to expanding gases trapped in body cavities during a portion of the 50- to 60-second period of decompression; (2) immature animals are tolerant to hypoxia, and longer periods of decompression are required before respiration ceases; (3) if decompression rates that are used reflect recommended values, the time for onset of unconsciousness would be 14 minutes (4,000 feet/minute) to 60 minutes (1,100 feet/minute) and decompression then would be a time-inefficient method for shelters and pounds that euthanatize large numbers of animals; (4) accidental recompression, with recovery of injured animals, can occur due to malfunctioning of equipment or to personnel error; (5) bloating, bleeding, vomiting, convulsions, urination, and defecation, which are aesthetically unpleasant, may occur in the unconscious animal; (6) respiratory or middle ear infections, or both, may cause pain from unequalized pressure; and (7) there may be failure to understand the mechanisms of action of hypoxia and its effects on animals.

Recommendation—Decompression is not a recommended method for euthanasia of animals because of the numerous disadvantages.

ELECTROCUTION

Electrocution as a form of euthanasia or stunning has been employed for years in species such as dogs, cattle, sheep, and hogs.[19,64,84,91-95] Experiments in dogs have shown the necessity of directing the electrical current through the brain in order to produce instant stunning with loss of consciousness. In the dog, when electricity passes between fore- and hindlimbs or neck and feet, it causes the heart to fibrillate promptly, but does not produce unconsciousness. The dog does not lose consciousness for at least 12 seconds after cardiac fibrillation, which causes cerebral hypoxemia. These data dictate against use of the methods that direct current through the heart and not directly through the brain. An apparatus that applies electrodes to opposite sides of the head, or in another way directs electrical current immediately through the brain, is necessary to induce immediate unconsciousness. The signs of an effective electrical stun are: extension of the limbs, opisthotonos, downward rotation of the eyeballs, and a tonic spasm changing into a clonic spasm, with eventual muscle flaccidity. This effect should be followed promptly by electrically induced fibrillation of the heart, exsanguination, or other appropriate methods to ensure death.

Advantages—(1) Humane if current is directed through the brain; (2) does not chemically contaminate tissues; and (3) is economical.

Disadvantages—(1) Electrocution is hazardous to personnel; (2) it is not a useful method for mass euthanasia because so much time is required per animal; (3) it is not a useful method for a vicious, intractable animal; (4) because violent extension and stiffening of the limbs, head, and neck occur, electrocution is aesthetically objectionable; and (5) in small animals, electrocution may not result in death because ventricular fibrillation and circulatory collapse do not always persist after cessation of current flow.

Recommendations—Electrocution for euthanasia and preslaughter stunning requires special skills and equipment that will assure passage of sufficient current through the brain to produce unconsciousness followed by electrically induced fibrillation of the heart. Although the method is acceptable if the above requirements are met, the disadvantages far outweigh its advantages in most applications.

MICROWAVE IRRADIATION

Microwave irradiation is used by neurobiologists as a means to fix brain metabolites without the loss of anatomic integrity of the brain. There are microwave instruments that have been specifically designed or modified for use in the euthanasia of laboratory mice and rats. These instruments, which direct most of their microwave energy to the head of the animal, vary in maximal power output from 2 kw to 20 kw. The kilowattage required to halt brain chemical activity within a specific time depends on the size of rodent within the chamber. A 2,450-MHz instrument (6.5 kw) will elevate the brain temperature of a 30-g mouse to 90 C in 325 msec, whereas a 915-MHz instrument (25 kw) is required to achieve the same temperature within one second in a 300-g rat.[96]

Advantages—(1) Unconsciousness and death occur in less than one second, and (2) this is the most effective method to fix brain tissue chemical activity in small animals.

Disadvantages—(1) Instruments are expensive; and (2) only animals the size of mice or rats can be euthanatized with currently available instruments.

Recommendations—Microwave irradiation is a very humane method to euthanatize small laboratory rodents if instruments that induce immediate unconsciousness are used. Only instruments that provide appropriate kilowattage and directed microwaves can be used. *Microwave ovens designed for domestic and institutional kitchens are absolutely condemned for euthanasia use.*

RAPID FREEZING

Neurobiologists require means to accurately measure labile brain tissue metabolites. Rapid freezing of the brain has been used to euthanatize animals and to achieve inactivation of enzymatic activity within the brain. Several techniques are used in rodents: (1) immersion of intact animal into liquid nitrogen; (2) decapitation and immediate immersion of head into liquid nitrogen; (3) freeze-blowing; (4) in situ freezing; and (5) funnel freezing.[97]

Advantages—(1) Rapid freezing provides means to humanely euthanatize rodents if correctly executed and if animals are adequately anesthetized prior to freezing (in situ freezing and funnel freezing).

Disadvantages—(1) Immersion into liquid nitrogen may be used only in animals weighing less than 40 g because heavier animals are not rapidly rendered unconscious; (2) rapid freezing requires very well-trained personnel and appropriate equipment; and (3) rapid freezing may be aesthetically displeasing to personnel.

Recommendations—Animal welfare committees and investigators must ensure that animals are made immediately unconscious either from rapid freezing or prior anesthetic. If a paralyzing drug is given in conjunction with an anesthetic, the amount of anesthetic administered must reflect known dose-effect data for a surgical level of anesthesia. Immersion of unanesthetized animals into liquid nitrogen can be used only in animals weighing less than 40 g. Anesthesia must be employed in animals weighing more than 40 g.

AIR EMBOLISM

Intravenous injection of 5 to 50 ml/kg of air induces rapid death in rabbits. However, it may be accompanied by convulsions, opisthotonos, and vocalization.[80] It is an acceptable method only in anesthetized animals.

PRECAUTIONS CONCERNING USE OF EUTHANATIZING AGENTS IN ANIMALS INTENDED FOR HUMAN FOOD

In euthanasia of animals intended for human food, agents cannot be used that lead to tissue residues, unless approved by the US Food and Drug Administration.[98] Carbon dioxide is the only chemical currently used in euthanasia of animals (primarily swine) for human food that does not lead to tissue residue.[11]

PRECAUTIONS CONCERNING LESS COMMON SPECIES

When euthanasia of poikilothermic (cold blooded) animals and aquatic animals is performed, the differences in their metabolism, respiration, and tolerance to cerebral hypoxemia may preclude methods that would be acceptable in terrestrial mammals.

POSTFACE

This panel report summarizes contemporary scientific knowledge on euthanasia in animals and calls attention to the lack of scientific reports assessing pain and discomfort in animals undergoing euthanasia. Many reports on various methods of euthanasia are either anecdotal or testimonial narratives and are therefore not cited in this report. The panel unanimously endorses the need for well-designed experiments to more fully determine the extent to which each procedure meets the criteria used for judging the methods of euthanasia.

The Panel on Euthanasia is fully committed to the concept that whenever it becomes necessary to kill any animal for any reason whatsoever, death should be induced as painlessly as possible. It has been our charge to develop workable guidelines for addressing this need, and it is our sincere desire that these guidelines be conscientiously used by all who exercise stewardship over animals on earth.

REFERENCES

1. McDonald LE, Booth NH, Lumb WV, et al. Report of the AVMA panel on euthanasia. *J Am Vet Med Assoc* 1978:173:59-72.
2. Breazile JE, Kitchell RL. Euthanasia for laboratory animals. *Fed Proc* 1969;28:1577-1579.
3. Kitchell RL *Animal pain: perception and alleviation.* Carsten E, et al, eds. Bethesda, Md, Am Physiol Soc. 1983.
4. Kitchell, RL, Johnson RD. Assessment of pain in animals. In: Moberg GP, ed. *Animal stress.* Am Physiol Soc, 1983;113-140.
5. Willis WD. *The pain system. The neural basis of nociceptive transmission in the mammalian nervous system.* Basel, Switzerland: S Karger, 1985;346.
6. Zimmerman M. Neurobiological concepts of pain, its assessment and therapy. In: Bromm B, ed. *Pain measurement in man. Neurophysiological correlates of pain.* Amsterdam: Elsevier Publishing Co, 1984;15-35.
7. Dennis SG, Melzack R. Perspectives on phylogenetic evolution of pain expression. In: Kitchell RL, Erickson HH, Cartens E, et al, eds. *Animal pain: perception and alleviation.* Bethesda, Md, Am Physiol Soc, 1983.
8. Beaver B. *Veterinary aspects of feline behavior.* St Louis: CV Mosby Co, 1980.
9. Hafez ESE. *The behavior of domestic animals* 3rd ed. Baltimore: The Williams & Wilkins Co, 1975.
10. Houpt KA, Wolski TR. *Domestic animal behavior for veterinarians and animal scientists.* Ames, Iowa: Iowa State University Press, 1982;356.
11. Humane slaughter regulations. *Fed Register* 1979;44(232):68809-68817.
12. Grandin T. Observations of cattle behavior applied to the design of cattle-handling facilities. *Appl Anim Ethol* 1980;6:19-31.
13. Grandin T. Pig behavior studies applied to slaughter-plant design. *Appl Anim Ethol* 1982;9:141-151.
14. *Proceedings of the national conference on dog and cat control.* Denver: Am Vet Med Assoc, 1976.
15. *Proceedings of the national conference on the ecology of the surplus dog and cat problem.* Chicago, Ill: Am Vet Med Assoc, 1974.
16. Bustad LK. An educator's approach to euthanasia. *Lab Anim* 1982;11:37-41.
17. Wolfle TL. Laboratory animal technicians: their role in stress reduction and human-companion animal bonding. *Vet Clin North Am (Small Anim Pract)* 1985;15:441-454.
18. Lumb WV, Jones E. *Veterinary anesthesia.* Philadelphia: Lea & Febiger, 1985.
19. Warrington R. Electrical stunning, a review of literature. *Vet Bull* 1974;44:617-628.
20. *Occupational exposure to waste anesthetic gases and vapors.* Washington, DC: Department of Health, Education, and Welfare (National Institute for Occupational Safety and Health) No. 77-140, 1977.
21. Lecky JH, ed: Waste anesthetic gases in operating room air: a suggested program to reduce personnel exposure. *Special report.* Park Ridge, Ill: The American Society of Anesthesiologists, 1983.
22. Booth, NH Inhalant anesthetics. In: Booth NH, McDonald LE,eds. *Veterinary pharmacology and therapeutics.* 5th ed. Ames, Iowa: Iowa State University Press, 1982;175-202.
23. *Chloroform box construction and use.* Denver: American Humane Association, 1969.
24. *Humane killing of unwanted animals.* 2nd ed. Herts, England: Potters Bar, The Universities Federation for Animal Welfare,1968;19-20.
25. Sawyer DC: Comparative effects of halothane. *Gaines Dog Res Prog* 1976,2-3.

26. *Registry of toxic effects of chemical substances.* Washington,DC: National Institute of Occupational Health. Department of Health and Human Services Publication No. 83-107,1981-1982.
26a. Chloroform: IARC monographs on the evaluation of the carcinogenic risk of chemicals to humans: Vol 20.1979.
27. Herin RA, Hall P, Fitch JW: Nitrogen inhalation as a method of euthanasia in dogs. *Am J Vet Res* 1978;39:989-991.
28. Noell WK, Chinn HI. Time course of failure of the visual pathway in rabbits during anoxia. *Fed Proc* 1949;8:119.
29. Glass HG, Snyder FF, Webster E. The rate of decline in resistance to anoxia of rabbits, dogs, and guinea pigs from the onset of viability to adult life. *Am J Physiol* 1944;140:609-615.
30. Vinter FJ. *The humane killing of mink* London: The Universities Federation for Animal Welfare, 1957.
31. Hernett TD, Haynes AP. Comparison of carbon dioxide/air mixture and nitrogen air mixture for the euthanasia of rodents. Design of a system for inhalation euthanasia. *Anim Technol* 1984;35:93-99.
32. Stonehouse RW, Loew FM, Quinn JA, et al. The euthanasia of dogs and cats: a statement of the humane practices committee of the Canadian Veterinary Medical Association. *Can Vet J* 1978;19:164-168.
33. Klassen CD. Nonmetalic environmental toxicants: air polutants, solvents, and solvents and pesticides In: Gilman AC, Goodman LS, Gilman A, eds. *The pharmacological basis of therapeutics* 6th ed. New York: MacMillan Publishing Co, Inc, 1980;1638-1659.
34. Sanford J Euthanasia of domesticated animals by injection of drugs. In: *Humane destruction of unwanted animals.* Herts, England: Potters Bar, The Universities Federation for Animal Welfare, 1976,18-21.
35. Scott WN. The use of poisons in animal destruction. In: *Humane destruction of unwanted animals.* Herts, England: Potters Bar, The Universities Federation for Animal Welfare, 1976;33-42.
36. Lambooy E, Spanjaard W. Euthanasia of young pigs with carbon monoxide. *Vet Rec* 1980;107:59-61.
37. Haldane J. The action of carbonic oxide in man. *J Physiol* 1985;18:430-462.
38. Bloom JD. Some considerations in establishing diverse breathing gas purity standards for carbon monoxide. *Aerosp. Med* 1972;43:633-636.
39. Carding AH: Mass euthanasia of dogs with carbon monoxide and/or carbon dioxide: preliminary trials. *J Small Anim Pract* 1968;9:245-259.
40. Moreland AF. Carbon monoxide euthanasia of dogs: chamber concentrations and comparative effects of automobile engine exhaust and carbon monoxide from a cylinder. *J Am Vet Med Assoc* 1974;165:853-855.
41. Ramsey TL, Eilmann HJ. Carbon monoxide acute and chronic poisoning and experimental studies. *J Lab Clin Med* 1932;17:415-427.
42. Blood DC, Johnston DE, Blackwood JD. Carbon monoxide euthanasia. A report to the Committee of the Victorian Division of the Royal Society for Prevention of Cruelty to Animals. (Information provided for the 1972 Report of the AVMA Panel on Euthanasia.)
43. Chalifoux A, Dallaire A. Physiologic and behavioral evaluation of CO euthanasia of adult dogs. *Am J Vet Res* 1983;44:2412-2417.
44. Dallaire A, Chalifoux A. Premedication of dogs with acepromazine or pentazocine before euthanasia with carbon monoxide. *Can J Comp Med* 1985;49:171-178.
45. Simonsen HB, Thordal-Christensend AA, Ockens N. Carbon Monoxide and carbon dioxide euthanasia of cats: duration and animal behavior. *Br Vet J* 1981;137:274-278.
46. Klemm WR. Carbon dioxide anesthesia in cats. *Am J Vet Res* 1964;25:1202-1205.
47. Leake CD, Waters RM. The anesthetic properties of carbon dioxide. In: *Current researches in anesthesiology and analgesia.* 1929;8:17-19.
48. Mattsson JL, Stinson JM, Clark CS. Electroencephalographic power-spectral changes coincident with onset of carbon dioxide narcosis in rhesus monkey. *Am J Vet Res* 1972;33:2043-2049.
49. Woodbury DM, Rollins LT, Gardner MD, et al. Effects of carbon dioxide on brain excitability and electrolytes. *Am J Physiol* 1958;192:79-90.
50. Glen JB, Scott WN. Carbon dioxide euthanasia of cats. *Br Vet J* 1973;129:471-479.
51. Hughes HC. Euthanasia of laboratory animals. In: *Handbook of laboratory animal science.* Vol III. Melby, Altman, eds. Cleveland: CRC Press, 1976;553-559.
52. Jaksch W. Euthanasia of day-old male chicks in the poultry industry. *Int J Stud Anim Prob* 1981;2:203-213.
53. Kline BE, Peckham V, Hesit HE. Some aids in handling large numbers of mice. *Lab Anim Care* 1963;13:84-90.
54. Kotula AW, Drewniak EE, Davis LE. Experimentation with in-line carbon dioxide immobilization of chickens prior to slaughter. *Poult Sci* 1961;40:213-216.
55. Stone WS, Amiraian K, Duell C, et al. Carbon dioxide anesthetization of guinea pigs to increase yields of blood and serum. *Proc Care Panel* 1961;11:299-303.
56. Hoenderken R. Electrical and carbon dioxide stunning of pigs for slaughter. In: *Stunning of animals for slaughter.* Eikelenboom G, ed. Boston: Martinus Nijhoff Publishers, 1982;59-63.
57. Croft PG. Anaesthesia and euthanasia. In: UFAW *handbook on the care and management of laboratory animals.* 3rd ed. Baltimore: The Williams & Wilkins Co, 1967;160-172.

58. Euthanasia (carbon dioxide). In: *Report and accounts*. Herts, England: Potters Bar. The Universities Federation for Animal Welfare 1976-1977;13-14.
59. Hall HW. The Anaesthesia and euthanasia of neonatal and juvenile dogs and cats. *Vet Rec* 1972;90:303-306.
60. McArthur JA: Carbon dioxide euthanasia of small animals (including cats). In: *Humane destruction of unwanted animals*. Herts, England: Potters Bar. The Universities Federation for Animal Welfare, 1976;9-17.
61. Laursen AM. Choosing between CO_2 and electrical stunning of pigs. A preliminary examination of stress and ethics. In: *Stunning of animals for slaughter*. Eikelenboom G, ed. Boston: Martinus Nijhoff Publishers, 1983;64-72.
62. Blackmore DK, Newhook JC. The assessment of insensibility in sheep, calves, and pigs during slaughter. In: *Stunning of animals for slaughter*. Eikelenboom G, ed. Boston: Martinus Nijhoff Publishers, 1983.
63. Feldman DB, Gupta BN. Histopathologic changes in laboratory animals resulting from various methods of euthanasia. *Lab Anim Sci* 1976;26:218-221.
64. Hatch RC. Euthanatizing agents. In: Booth NH, McDonald LE, eds. *Veterinary pharmacology and therapeutics*. 5th ed. Ames, Iowa: Iowa State University Press, 1982;1059-1064.
65. Lumb WV. Euthanasia by noninhalant pharmacologic agents. *J Am Vet Med Assoc* 1974;165:851-852.
66. Lumb WV, Moreland AF. Chemical methods of euthanasia. *Lab Anim* 1982;11:29-35.
67. Herschler RC, Lawrence JR, Schiltz RA. Secobarbital-dibucaine combination as a euthanasia agent for dogs and cats. *Vet Med/Sm Anim Clin* 1981;1009-1012.
68. Wallach MB, Peterson KE, Richards RK. Electrophysiologic studies of a combination of secobarbital and dibucaine for euthanasia of dogs. *Am J Vet Res* 1981;42:850-853.
69. Eikmeier H. Experience with a new preparation for painless destruction of small animals. (T-61). *Die Blauen Hefte Tieraerztl* 1962;5:22-23.
70. Kuepper G: T-61 used in large animals. *Die Blauen Hefte Tieraerztl* 1964;8:32-33.
71. Lumb WV, Doshi K, Scott RJ. A comparative study of T-61 and pentobarbital for euthanasia of dogs. *J Am Vet Med Assoc* 1978;172:149-152.
72. Quin AH. Observations on a new euthanasia agent for small animals. *Vet Med* 1963;58:494-495.
73. Rowan AN. T-61 use in the euthanasia of domestic animals: a survey. *Adv Anim Welfare Sci*, in press.
74. Barocio LD. Review of literature on use of T-61 as an euthanasic agent. Inst *J Stud Anim Prob* 1983;4:336-342.
75. Fowler NG, Foster SJ. The last gasp. *Vet Rec* 1970;86:145.
76. Franz DN. Central nervous system stimulants. In: Goodman LS, Gilman A, eds. *The pharmacological basis of therapeutics*. 5th ed. New York: Macmillan Publishing Co, Inc, 1975;359-366.
77. Bowen JM, Blackman DM, Heavner JE. Effect of magnesium ions on neuromuscular transmission in the horse, steer, and dog. *J Am Vet Med Assoc* 1970;157:164-173.
78. Hicks T Bailey EM Jr. Succinylcholine chloride as a euthanatizing agent in dogs. *Am J Vet Res* 1978;39:1195-1197.
79. Blackmore DK. Energy requirements for the penetration of heads of domestic stock and the development of a multiple projectile. *Vet Rec* 1985;116:36-40.
80. Clifford DH. Preanesthesia, anesthesia, analgesia, and euthanasia. In: *Laboratory animal medicine*. Fox, Cohen, Loew, eds. New York: Academic Press, 1984;528-563.
81. Blackmore DK. Differences in behaviour between sheep and cattle during slaughter. *Res Vet Sci* 1984;37:223-226.
82. Blackmore DK. Non-penetrative percussion stunning of sheep and calves. *Vet Rec* 1979;105:372-375.
83. Anis GW. Euthanasia of domesticated animals by shooting. In: *Humane destruction of unwanted animals*. Herts, England: Potters Bar. The Universities Federation for Animal Welfare, 1975.
84. Carding T. Euthanasia of dogs and cats. *Anim Reg Stud* 1977;1:5-21.
85. Mikeska JA, Klemm WR. EEG evaluation of humaneness of asphyxia and decapitation euthanasia of the laboratory rat. *Lab Anim Sci* 1975;25:175-179.
86. Gregory NG, Wotton SB. Time to loss of brain responsiveness following exsanguination in calves. *Res Vet Sci* 1984;37:141-143.
87. Booth NH. Effect of rapid decompression and associated hypoxic phenomena in euthanasia of animals: a review. *J Am Vet Med Assoc* 1978;173:308.
88. Barber BR. Use of a standard autoclave for decompression euthanasia. *J Inst Anim Technol* 1972;23:106-110.
89. Smith DC. Methods of euthanasia and disposal of laboratory animals. In: *Methods of animal experimentation*, Vol 1. New York: Academic Press, 1965;167-195.
90. High altitude (low pressure) euthanasia. Denver, Colo, *Operational Guide for Euthanasia*, 1969.
91. Croft PG, Hume CW. Electric stunning of sheep. *Vet Rec* 1956;68:318-321.
92. Loftsgard G, Braathen S, Helgobostd A. Electrical stunning of mink. *Vet Rec* 1972;91:132-134.
93. Roberts TDM. Correspondence: electrocution cabinets. *Vet Rec* 1974;95:241-242.
94. Roberts TDM. Cortical activity in electrocuted dogs. *Vet Rec* 1954;66:561-567.
95. WHO Joint FAO/WHO Expert Committee on Meat Hygiene. Second Report. *WHO Tech Rep Ser 241* 1962.

96. Medina MA, Diam AP, Stavinoha WB. Inactivation of brain tissue by microwave irradiation in cerebral metabolism and neural function. Chapter 8. Passoneau RA, et al eds. Baltimore: The Williams & Wilkins Co, 1980.
97. Passoneau RA, Hawkins RA, Lust WD, et al, eds *Cerebral metabolism and normal function.* Chapters 2-8. Baltimore: The Williams & Wilkins Co, 1980.
98. Booth NH. Drug and chemical residues in the edible tissues of animals. In: Booth NH, McDonald LE, eds. *Veterinary pharmacology and therapeutics.* 5th ed. Ames, Iowa: Iowa State University Press, 1982;1065-1113.

Index

INDEX

A

D

F

O

P

S

T

U

V

W

X

Y

Z